Multidimensional hypergeometric functions and the representation theory of Lie algebras and quantum groups

ADVANCED SERIES IN MATHEMATICAL PHYSICS

Editors-in-Charge

H Araki (*RIMS, Kyoto*)
V G Kac (*MIT*)
D H Phong (*Columbia University*)
S-T Yau (*Harvard University*)

Associate Editors

L Alvarez-Gaumé (*CERN*)
J P Bourguignon (*Ecole Polytechnique, Palaiseau*)
T Eguchi (*University of Tokyo*)
B Julia (*CNRS, Paris*)
F Wilczek (*Institute for Advanced Study, Princeton*)

Published

Vol. 10: Yang-Baxter Equations in Integrable Systems
  edited by M Jimbo

Vol. 11: New Developments in the Theory of Knots
  edited by T Kohno

Vol. 12: Soliton Equations and Hamiltonian Systems
  by L A Dickey

Vol. 14: Form Factors in Completely Integrable Models of Quantum Field Theory
  by F A Smirnov

Vol. 15: Non-Perturbative Quantum Field Theory – Mathematical Aspects and Applications
  Selected Papers of Jürg Fröhlich

Vol. 16: Infinite Analysis – Proceedings of the RIMS Research Project 1991
  edited by A Tsuchiya, T Eguchi and M Jimbo

Vol. 17: Braid Group, Knot Theory and Statistical Mechanics (II)
  edited by C N Yang and M L Ge

Vol. 18: Exactly Solvable Models and Strongly Correlated Electrons
  by V Korepin and F H L Ebler

Vol. 19: Under the Spell of the Gauge Principle
  by G. 't Hooft

Vol. 20: The State of Matter
  edited by M Aizenman and H Araki

Advanced Series in Mathematical Physics
Vol. 21

MULTIDIMENSIONAL HYPERGEOMETRIC FUNCTIONS AND REPRESENTATION THEORY OF LIE ALGEBRAS AND QUANTUM GROUPS

A. Varchenko
Department of Mathematics
University of North Carolina

World Scientific
Singapore • New Jersey • London • Hong Kong

Published by

World Scientific Publishing Co. Pte. Ltd.
P O Box 128, Farrer Road, Singapore 9128
USA office: Suite 1B, 1060 Main Street, River Edge, NJ 07661
UK office: 57 Shelton Street, Covent Garden, London WC2H 9HE

MULTIDIMENSIONAL HYPERGEOMETRIC FUNCTIONS AND THE REPRESENTATION THEORY OF LIE ALGEBRAS AND QUANTUM GROUPS

Copyright © 1995 by World Scientific Publishing Co. Pte. Ltd.

All rights reserved. This book, or parts thereof, may not be reproduced in any form or by any means, electronic or mechanical, including photocopying, recording or any information storage and retrieval system now known or to be invented, without written permission from the Publisher.

For photocopying of material in this volume, please pay a copying fee through the Copyright Clearance Center, Inc., 27 Congress Street, Salem, MA 01970, USA.

ISBN 981-02-1880-X

Printed in Singapore by Uto-Print

Contents

1. Introduction — 1
    1.1 Example. Three points on the line — 2
    1.2 Brief description of the contents — 9
    1.3 Acknowledgements — 14

2. Construction of complexes calculating homology of the complement of a configuration — 15
    2.1 Configuration in a real space — 15
    2.2 Configuration in a complex space — 18
    2.3 Homology — 19
    2.4 Compatible orientation of cells of complexes Q, Q'/Q'', X — 22
    2.5 Local system $\mathcal{S}^*(a)$ and distinguished sections over cells — 22
    2.6 Quantum bilinear form of a configuration — 25
    2.7 Homomorphism of the projection on the real part — 29
    2.8 Action of a group preserving configuration — 33
    2.9 Abstract complexes of a real configuration — 35

3. Construction of homology complexes for discriminantal configuration — 39
    3.1 Discriminantal configuration — 39
    3.2 Facets of a discriminantal configuration — 40
    3.3 Centers of top-dimensional facets — 41
    3.4 Basic polytopes — 42
    3.5 Centers of facets of arbitrary codimension — 42
    3.6 Cells of the construction of Sec. 2 are convex polytopes — 43
    3.7 Description of basic polytopes by inequalities — 45
    3.8 Admissible monomials — 46
    3.9 Equalities and inequalities defining cells — 48
    3.10 Distinguished coorientation of a discriminantal configuration — 50
    3.11 Weights of $\mathcal{C}_{n,k}(z_1,\ldots,z_n)$ and the reduced quantum bilinear form for $n \leq 1$ — 53
    3.12 Complexes $C.(Q'/Q'', \mathcal{S}^*(a))$, $C.(X, \mathcal{S}^*(a))$ for a discriminantal configuration $\mathcal{C}_{n,k}$ — 54

	3.13	Action of the permutation group Σ_k on a discriminantal configuration $\mathcal{C}_{n,k}(z)$	58
	3.14	Complexes $\mathcal{X}.(\mathcal{C}_{n,k})$ and $Q.(\mathcal{C}_{n,k})$	61
	3.15	Homology of $Q.(\mathcal{C}_{n,k})$	62
	3.16	Action of the permutation group on $\mathcal{X}.(\mathcal{C}_{n,k})$, $Q.(\mathcal{C}_{n,k})$	62
4.	Algebraic interpretation of chain complexes of a discriminantal configuration		67
	4.1	Quantum groups	67
	4.2	Contravariant form	69
	4.3	Coalgebra structure on $U_q\mathfrak{n}_-$, algebra structure on $(U_q\mathfrak{n}_-)^*$	75
	4.4	Hochschild homology	79
	4.5	Two-sided Hochschild complexes connected with a discriminantal configuration	80
	4.6	Algebraic interpretation of the abstract complexes $\mathcal{X}.(\mathcal{C}_{n,k})$ and $Q.(\mathcal{C}_{n,k})$ of a discriminantal configuration	82
	4.7	Geometric interpretation of complexes $C.(^+U_q\bar{\mathfrak{n}}'_-;\overline{M};2;{}^+\Delta)_\lambda$ and $C.^*(^+U_q\bar{\mathfrak{n}}'_-;\overline{M};2;\mu)_\lambda$ defined in (4.5)	85
	4.8	Symmetrization	88
	4.9	Proof of Theorem (4.7.5)	91
5.	Quasiisomorphism of two-sided Hochschild complexes to suitable one-sided Hochschild complexes		93
	5.1	One-sided Hochschild complexes connected with a discriminantal configuration	93
	5.2	Complexes (5.1.1) and (5.1.24) as subcomplexes of complexes (4.5.5)	100
	5.3	Construction of a monomorphism $\varphi: C.(^+U_q\mathfrak{n}_-;M;{}^+\Delta)_\lambda \to C.(^+U_q\bar{\mathfrak{n}}'_-;\overline{M};2;{}^+\Delta)_{\bar{\lambda}}$	106
	5.4	**Theorem.** *The monomorphism* $\varphi: C.^*(^+U_q\mathfrak{n}_-;M;\mu)_\lambda \to C.^*(^+U_q\bar{\mathfrak{n}}'_-;\overline{M};2;\mu)_{\bar{\lambda}}$ *is a quasiisomorphism*	115
	5.5	**Theorem.** *The monomorphism* $\varphi: C.(^+U_q\mathfrak{n}_-;M;{}^+\Delta)_\lambda \to C.(^+U_q\bar{\mathfrak{n}}'_-;\overline{M};2;{}^+\Delta)_{\bar{\lambda}}$ *is a quasiisomorphism*	118
	5.6	Filtration in $C.(^+U_q\bar{\mathfrak{n}}'_-;\overline{M};{}^+\Delta)$	119
	5.7	Degree	120
	5.8	Proof of Theorem 5.6.12	121
	5.9	Proof of Theorem 5.6.11	123
	5.10	Remark	126

	5.11	Geometric interpretation of Theorems 5.3.43, 5.4, and 5.5	126
6.		Bundle properties of a discriminantal configuration	145
	6.1	Subordinated monomials	145
	6.2	Leaves	145
	6.3	Properties of leaves	147
	6.4	Proof of Theorem 6.2.4	151
7.		R-matrix for the two-sided Hochschild complexes	155
	7.1	Bistructures on $({}^+U_q\mathfrak{n}_-)^{\otimes n}$, $({}^+U_q\mathfrak{n}_-)^{*\otimes n}$	155
	7.2	$U_q\mathfrak{n}_-$-bimodule structure on $(U_q\mathfrak{n}_-)^*$	158
	7.3	$(U_q\mathfrak{n}_-)^*$-bimodule structure on $U_q\mathfrak{n}_-$	158
	7.4	R-matrix	159
	7.5	Symmetrization and R-matrix	165
	7.6	R-matrix for Verma modules	166
	7.7	Connection of R-matrices for two- and one-sided Hochschild complexes	168
8.		Monodromy	171
	8.1	Gauss–Manin connection	171
	8.2	Chain complexes over real points of the base	173
	8.3	Parallel translations along special curves	176
	8.4	Isotopy of the real line	178
	8.5	Factorization properties of cells	180
	8.6	Involution	183
	8.7	Bundle property	183
	8.8	Construction of T_t	185
	8.9	**Lemma.** *The deformation T_t defined on each cell separately is compatible on intersections of cells*	186
	8.10	Example	188
	8.11	Computation of the action $T_1 : C.(Q'/Q'', \mathcal{S}^*(A), z^0) \to C.(Q'/Q'', \mathcal{S}^*(A), z^0)$ for the isotropy T_t constructed in (8.8)	189
	8.12	Proof of Theorem 8.3.4 for T_t^u	190
	8.13	Geometric interpretation of the R-matrix operators acting on the two-sided Hochschild complexes constructed in Sec. 4	191
	8.14	Geometric interpretation of the R-matrix operators on the complexes $C.({}^+U_q\mathfrak{n}_-; M_\delta; {}^+\Delta)_{\lambda'}$ and $C.^*({}^+U_q\mathfrak{n}_-; M_\delta; \mu)_{\lambda'}$ constructed in Sec. 5.	194

9.	R-matrix operator as the canonical element, quantum doubles	197
	9.1 Quantum double	197
	9.2 Quantum doubles $\mathcal{D}((U_q\mathfrak{b}_-)')$ and $\mathcal{D}(U_q\mathfrak{b}_+)$	198
	9.3 The action of the quantum doubles on Verma modules and their duals	202
	9.4 The quotient complex $C.(^+U_q\mathfrak{n}_-; M_\sigma; {}^+\Delta)/\ker S$	207
	9.5 Quantum groups corresponding to Kac–Moody algebras	210
10.	Hypergeometric integrals	215
	10.1 Orlik–Solomon algebra and flags	215
	10.2 Framings and bases	216
	10.3 Quasiclassical contravariant form of a configuration	218
	10.4 Relative complexes	220
	10.5 Integrable connection on the horizontal complexes	223
	10.6 Hypergeometric differential forms	228
	10.7 Hypergeometric integrals	230
	10.8 Resolution of singularities of a configuration of hyperplanes	233
	10.9 Integration pairings and symmetric frames	236
	10.10 Hypergeometric differential equations	239
11.	Kac–Moody Lie algebras without Serre's relations and their doubles	247
	11.1 Kac–Moody Lie algebras without Serre's relations	247
	11.2 Complexes	250
	11.3 The double	252
	11.4 Homology in degree zero	261
	11.5 Knizhnik–Zamolodchikov differential equation	261
12.	Hypergeometric integrals of a discriminantal configuration	265
	12.1 Complexes of a discriminantal configuration	265
	12.2 Hypergeometric pairings	271
	12.3 Hypergeometric integrals and the Knizhnik–Zamolodchikov connection	278
	12.4 Resonances	285
	12.5 Nondegeneracy of the hypergeometric pairing $J(z): H_0(C.(\bar{\mathfrak{n}}_-; L)_\lambda) \otimes H_0(C.(^+U_q\bar{\mathfrak{n}}_-; L(q); \delta)_\lambda) \to \mathbb{C}$ for almost all κ. Asymptotics for $\kappa \to \infty$.	290

13.	Resonances at infinity		295
	13.1	Projective transformations of the complex line and discriminantal configurations	295
	13.2	Complementary weight	296
	13.3	The inversion isomorphism for one Verma module	298
	13.4	An inversion isomorphism for n Verma modules	303
	13.5	Generic sets	309
	13.6	Transformations of flag forms	310
	13.7	The kernel of the hypergeometric pairing for sl_2	312
	13.8	Remarks on the representation theory of the quantum double of the subalgebra $U_q\mathfrak{b}_- \subset U_q sl_2$	317
14.	Degenerations of discriminantal configurations		329
	14.1	Composition of singular vectors	329
	14.2	Asymptotics of the hypergeometric pairing	337
	14.3	Rank of the hypergeometric pairing	341
	14.4	Remarks on the kernel of the hypergeometric pairing	342
	14.5	The Selberg formula	343
	14.6	The hypergeometric pairing for $\mathfrak{g} = sl_2$ and two modules	344
15.	Remarks on homology groups of a configuration with coefficients in local systems more general than complex one-dimensional		347
	15.1	Complexified real configuration	347
	15.2	Universal quantum group	349
	15.3	Discriminantal configuration	352
	15.4	Remarks on homology groups of braid groups	357
	15.5	Local systems of rank greater than 1	359
References			367

1

Introduction

In this book we describe numerous connections of multidimensional hypergeometric functions with Kac–Moody Lie algebras and their quantum deformations. The hypergeometric functions have the following form:

$$I(t_1,\ldots,t_n;\gamma;P;\{a_{ij}\};\kappa)$$
$$= \int_{\gamma(t_1,\ldots,t_n;\{a_{ij}\};\kappa)} \prod_{1\leq i<j\leq N} (t_i - t_j)^{a_{ij}/\kappa} P(t_1,\ldots,t_N)\,dt_{n+1}\wedge\cdots\wedge dt_N.$$

Here $\{a_{ij}\}$, κ are complex parameters, P runs through a suitable space of rational functions, and γ runs through a suitable space of cycles.

There are two main themes in this book.

The first theme: Studying representations of Kac–Moody algebras and the corresponding quantum groups is essentially equivalent to studying the function $\Pi(t_i - t_j)^{a_{ij}/\kappa}$.

The second theme: The hypergeometric functions provide a new and powerful way to compare the representation theory of Kac–Moody algebras and the representation theory of quantum groups.

The differential-geometrical side of the function $\Pi(t_i - t_j)^{a_{ij}/\kappa}$ is connected with the theory of Kac–Moody algebras. Complexes of multivalued differential forms associated with this function have a natural description in terms of Kac–Moody algebras. The differential equation for associated hypergeometric functions has a description in terms of Kac–Moody algebras, and this differential equation is the famous Knizhnik–Zamolodchikov differential equation discovered by physicists in conformal field theory.

The topological side of the function $\Pi(t_i - t_j)^{a_{ij}/\kappa}$ is connected with the theory of quantum groups. Univalued branches of this function define a local system of coefficients over the complement to the union of diagonal hyperplanes. Complexes of chains with coefficients in this local system have a natural description in terms of quantum groups.

It is well-known that the representation theories of Kac–Moody algebras and their quantum deformations are similar and have striking differences at the same time. Studying their interrelations is a fascinating subject.

The hypergeometric functions provide a new approach to this problem. Namely, integration of multivalued differential forms over chains gives a correspondence between the corresponding objects of the representation theory of Kac–Moody algebras and the representation theory of quantum groups.

1.1. Example. Three Points on the Line

Let z_1, z_2, z_3 be points on the complex line. Fix numbers $m_1, m_2, m_3, \kappa \in \mathbb{C}$, and consider a multivalued holomorphic function

$$\ell = \prod_{1 \leq i < j \leq 3} (z_i - z_j)^{m_i m_j / 2\kappa} \prod_{j=1}^{3} (t - z_j)^{-m_j/\kappa}$$

on the complement to the union of the three points, here t is a coordinate on the line.

Set
$$\eta_j = \ell \, dt/(t - z_j)$$

for $j = 1, 2, 3$. $\{\eta_j\}$ are multivalued holomorphic forms on the complex line. They are closed and cohomologically dependent:

$$m_1 \eta_1 + m_2 \eta_2 + m_3 \eta_3 = -\kappa d\ell \,. \tag{1.1.1}$$

Consider a vector

$$I = (I_1, I_2, I_3) = \left(\int_\gamma \eta_1, \int_\gamma \eta_2, \int_\gamma \eta_3 \right), \tag{1.1.2}$$

where γ is a curve shown in Fig. 1.1. Here z_a and z_b are any two of the three points z_1, z_2, z_3. We fix a univalued branch of ℓ over γ, and hence the integrals are well-defined. We have

$$m_1 I_1 + m_2 I_2 + m_3 I_3 = 0 \,. \tag{1.1.3}$$

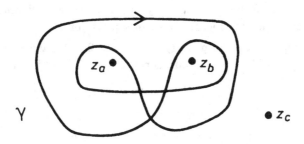

Fig. 1.1

The integrals do not depend on continuous deformations of the curve. Therefore, the vector is a function of z_1, z_2, z_3. The function $I(z_1, z_2, z_3)$ is holomorphic and satisfies the differential equation

$$\frac{\partial I}{\partial z_i} = \frac{1}{\kappa} \sum_{j \neq i} \frac{\Omega_{ij}}{z_i - z_j} I \quad \text{for } i = 1, 2, 3, \quad (1.1.4)$$

where the matrix Ω_{ij} has zero elements with the following exceptions:

$$(\Omega_{ij})_{pp} = m_i m_j / 2 \quad \text{if } p \notin [i, j],$$
$$(\Omega_{ij})_{ii} = m_i m_j / 2 - m_j,$$
$$(\Omega_{ij})_{jj} = m_i m_j / 2 - m_i,$$
$$(\Omega_{ij})_{ij} = m_j, \quad (\Omega_{ij})_{ji} = m_i.$$

This differential equation does not depend on the choice of the curve γ.

Verification of the differential equation is purely combinatorial. For example,

$$\frac{\partial \eta_2}{\partial z_1} = \left[\frac{m_1 m_2}{2\kappa} \frac{1}{z_1 - z_2} + \frac{m_1 m_3}{2\kappa} \frac{1}{z_1 - z_3} + \frac{m_1}{\kappa} \frac{1}{t - z_1} \right] \ell \frac{dt}{t - z_2}$$
$$= \left[\frac{m_1 m_2}{2\kappa} \frac{1}{z_1 - z_2} + \frac{m_1 m_2}{2\kappa} \frac{1}{z_1 - z_2} \right] \eta_2 + \frac{m_1}{\kappa} \frac{1}{z_1 - z_2} (\eta_1 - \eta_2).$$

The key step in the verification is the equality

$$\frac{1}{t - z_1} \frac{1}{t - z_2} = \frac{1}{z_1 - z_2} \left[\frac{1}{t - z_1} - \frac{1}{t - z_2} \right]$$

having deep connections with the Jacobi identity in the theory of Lie algebras.

The differential forms $\{\eta_i\}$, identity (1.1.1), and the differential equation (1.1.4) have the following Lie–algebraic interpretation:

Consider the Lie algebra $\mathfrak{g} = s\ell_2$ of complex 2×2-matrices with the zero trace. $\mathfrak{g}$ is generated by the standard generators e, f, h subject to the relations

$$[e, f] = h, \qquad [h, e] = 2e, \qquad [h, f] = -2f.$$

The tensor

$$\Omega = \frac{1}{2} h \otimes h + e \otimes f + f \otimes e \in \mathfrak{g} \otimes \mathfrak{g}$$

is called the Casimir operator. The Casimir operator is the tensor corresponding to an invariant scalar product on $\mathfrak{g}$.

Let $L_1, \ldots, L_n$ be representations of $\mathfrak{g}$, $L = L_1 \otimes \ldots \otimes L_n$. Let Ω_{ij} be the linear operator on L acting as the Casimir operator on $L_i \otimes L_j$ and as the identity operator on the other factors. The Knizhnik–Zamolodchikov (KZ) equation on an L-valued function $\varphi(z_1, \cdots, z_n)$ is the differential equation

$$\frac{\partial \varphi}{\partial z_i} = \frac{1}{\kappa} \sum_{j \neq i} \frac{\Omega_{ij}}{z_i - z_j} \varphi \quad \text{for } i = 1, \ldots, n,$$

κ is a parameter of the equation.

The KZ equation was discovered in the conformal field theory [**KZ**]. This equation describes conformal blocks in the Wess–Zumino–Witten model of the conformal field theory.

The KZ equation defines an integrable connection on the trivial bundle $L \times \mathbb{C}^n \longrightarrow \mathbb{C}^n$ with singularities over diagonals. Parallel translations for the KZ connection commute with the $\mathfrak{g}$–action on L. Therefore, the bundle $L \times \mathbb{C}^n \longrightarrow \mathbb{C}^n$ has a lot of subbundles invariant with respect to the KZ connection. For example, eigenspaces of the generators of $\mathfrak{g}$ form invariant subbundles.

Consider the following example of the KZ equation.

For $m \in \mathbb{C}$, denote by $M(m)$ the Verma module over $\mathfrak{g}$ with highest weight m. $M(m)$ is the infinite dimensional module generated by one vector v with properties $ev = 0$, $hv = mv$. Set $M = M(m_1) \otimes M(m_2) \otimes M(m_3)$, $m = m_1 + m_2 + m_3$, $\operatorname{Vac} M_{m-2} = \{x \in M \,|\, ex = 0, hx = (m-2)x\}$. Let $v_i \in M(m_i)$ be a generator vector, $i = 1, 2, 3$. Then

$$\begin{aligned}\operatorname{Vac} M_{m-2} = \big\{ x = {}& I_1 f v_1 \otimes v_2 \otimes v_3 + I_2 v_1 \otimes f v_2 \otimes v_3 \\ &+ I_3 v_1 \otimes v_2 \otimes f v_3 \,|\, m_1 I_1 + m_2 I_2 + m_3 I_3 = 0 \big\}.\end{aligned} \quad (1.1.5)$$

Consider the KZ equation with values in $\text{Vac}\, M_{m-2} \subset M$.

The KZ equation on the coordinates I_1, I_2, I_3 in (1.1.5) coincides with the hypergeometric differential equation (1.1.4). \hfill (1.1.6)

Therefore, an M-valued function

$$\int_{\gamma(z_1,z_2,z_3)} \eta_1 f v_1 \otimes v_2 \otimes v_3 + \int_{\gamma(z_1,z_2,z_3)} \eta_2 v_1 \otimes f v_2 \otimes v_3 + \int_{\gamma(z_1,z_2,z_3)} \eta_3 v_1 \otimes v_2 \otimes f v_3$$

takes values in $\text{Vac}\, M_{m-2} \subset M$ and is a solution of the KZ equation.

Set
$$\mathcal{H}^1 = (\mathbb{C}\,\eta_1 \oplus \mathbb{C}\,\eta_2 \oplus \mathbb{C}\eta_3)/\mathbb{C}\,d\ell\,,$$

cf. (1.1.1).

$\mathcal{H}^1$ *is canonically isomorphic to* $(\text{Vac}\, M_{m-2})^*$. \hfill (1.1.7)

Let us return to the hypergeometric differential equation (1.1.4). Its solutions are parametrized by linear combinations of cycles γ shown in Fig. 1.1. Such linear combinations form a suitable homology group of the complement to the three points.

The function ℓ defines the following complex one dimensional local system $\mathcal{S}$ over $\mathbb{C} - \{z_1, z_2, z_3\}$. The sections of $\mathcal{S}$ are complex linear combinations of univalued branches of ℓ. The local system has monodromy around the points z_1, z_2, z_3. The monodromy around the point z_a is multiplication by $\exp(-2\pi i m_a/\kappa)$. Set
$$q = \exp(2\pi i/\kappa)\,,$$
then the monodromy is q^{-m_a}.

Integration defines a pairing

$$I[z_1, z_2, z_3] \,:\, \mathcal{H}^1 \otimes H_1(\mathbb{C} - \{z_1, z_2, z_3\}, \mathcal{S}) \longrightarrow \mathbb{C}\,. \hfill (1.1.8)$$

Each element $\gamma \in H_1(\mathbb{C} - \{z_1, z_2, z_3\}, \mathcal{S})$ generates a solution of the hypergeometric equation (1.1.4).

Now we calculate H_1. Assume that $z_1, z_2, z_3 \in \mathbb{R}$ and $z_1 < z_2 < z_3$. Fix a point b_0 in the upper half plane and three oriented loops b_1, b_2, b_3 as shown in Fig. 1.2.

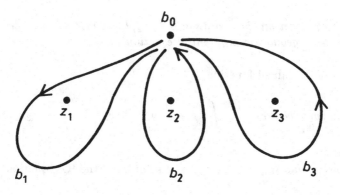

Fig. 1.2

Fix sections s_0, s_1, s_2, s_3 of $\mathcal{S}$ over b_0, b_1, b_2, b_3. For $j = 0, \ldots, 3$, the pair (b_j, s_j) is a singular chain with a coefficient in $\mathcal{S}$. Let

$$d : \bigoplus_{j=1} \mathbb{C}(b_j, s_j) \longrightarrow \mathbb{C}(b_0, s_0) \qquad (1.1.9)$$

be the boundary operator.

Complex (1.1.9) computes $H_\cdot(\mathbb{C} - \{z_1, z_2, z_3\}, \mathcal{S})$. (1.1.10)

There is a natural choice of sections s_0, s_1, s_2, s_3 such that the boundary operator has the following form:

$$\begin{aligned}
(b_1, s_1) &\longmapsto (q^{m_1/2} - q^{-m_1/2})q^{(m_2+m_3)/4}(b_0, s_0), \\
(b_2, s_2) &\longmapsto (q^{m_2/2} - q^{-m_2/2})q^{(-m_1+m_3)/4}(b_0, s_0), \\
(b_3, s_3) &\longmapsto (q^{m_3/2} - q^{-m_3/2})q^{(-m_1-m_2)/4}(b_0, s_0).
\end{aligned} \qquad (1.1.11)$$

Therefore, $H_1(\mathbb{C} - \{z_1, z_3, z_3\}, \mathcal{S})$ is naturally isomorphic to the kernel of the operator d given by formulas (1.1.11).

This computation has the following algebraic interpretation.

The universal enveloping algebra $U\mathfrak{g} = Us\ell_2$ has a one parameter deformation $U_q\mathfrak{g}$, $q = \exp(2\pi i/\kappa) \in \mathbb{C}^*$, called the quantum group. $U_q\mathfrak{g}$ is a $\mathbb{C}$-algebra generated by elements e, f, h subject to the relations

$$\begin{aligned}
[e, f] &= q^{h/2} - q^{-h/2}, \\
[h, e] &= 2e, \quad [h, f] = -2f.
\end{aligned} \qquad (1.1.12)$$

By definition $q^{ah} = \exp(ah \times 2\pi i/\kappa)$.

In the standard definition of $U_q s\ell_2$, the first relation is replaced by

$$[e, f] = (q^{h/2} - q^{-h/2})/(q^{1/2} - q^{-1/2}). \tag{1.1.13}$$

The substitution of $f \mapsto f/(q^{1/2} - q^{-1/2})$ transforms (1.1.13) into (1.1.12).

The quantum group is a Hopf algebra. The comultiplication $\Delta : U_q\mathfrak{g} \to U_q\mathfrak{g} \otimes U_q\mathfrak{g}$ is defined by the formula

$$\Delta(h) = h \otimes 1 + 1 \otimes h,$$
$$\Delta(f) = f \otimes q^{h/4} + q^{-h/4} \otimes f,$$
$$\Delta(e) = e \otimes q^{h/4} + q^{-h/4} \otimes e.$$

For $m \in \mathbb{C}$, denote by $M(m, q)$ the Verma module over $U_q\mathfrak{g}$ with highest weight m. $M(m, q)$ is the infinite dimensional module generated by one vector v with properties $ev = 0$, $hv = mv$. Set $M(q) = M(m_1, q) \otimes M(m_2, q) \otimes M(m_3, q)$, $m = m_1 + m_2 + m_3$, $\text{Vac}\, M(q)_{m-2} = \{x \in M(q) | ex = 0, hx = (m-2)x\}$. Let $v_i \in M(m_i, q)$ be a generating vector, $i = 1, 2, 3$. Then

$$\begin{aligned}\text{Vac}\, M(q)_{m-2} = \{x = &I_1 fv_1 \otimes v_2 \otimes v_3 + I_2 v_1 \otimes fv_2 \otimes v_3 \\&+ I_3 v_1 \otimes v_2 \otimes fv_3 | (q^{m_1/2} - q^{-m_1/2}) q^{(m_2+m_3)/4} I_1 \\&+ (q^{m_2/2} - q^{-m_2/2}) q^{(-m_1+m_3)/4} I_2 \\&+ (q^{m_3/2} - q^{-m_3/2}) q^{(-m_1-m_2)/4} I_3 = 0\}.\end{aligned} \tag{1.1.14}$$

The map

$$\begin{aligned}(b_1, s_1) &\longmapsto fv_1 \otimes v_2 \otimes v_3, \\(b_2, s_2) &\longmapsto v_1 \otimes fv_2 \otimes v_3, \\(b_3, s_3) &\longmapsto v_1 \otimes v_2 \otimes fv_3,\end{aligned} \tag{1.1.15}$$

defines natural isomorphisms

$$\bigoplus_{j=1}^{3} \mathbb{C}(b_j, s_j) \simeq M(q)_{m-2} := \{x \in M(q) | hx = (m-2)x\}$$

$$H_1(\mathbb{C} - \{z_1, z_2, z_3\}, \mathcal{S}) \simeq \text{Vac}\, M(q)_{m-2}.$$

(1.1.16) *Isomorphisms (1.1.7) and (1.1.15) and integration (1.1.8) define the canonical hypergeometric pairing*

$$I[z_1, z_2, z_3] : (\text{Vac}\, M_{m-2})^* \otimes \text{Vac}\, M(q)_{m-2} \longrightarrow \mathbb{C},$$

8   *Introduction*

that is, the canonical map

$$J[z_1, z_2, z_3] : \operatorname{Vac} M(q)_{m-2} \longrightarrow \operatorname{Vac} M_{m-2}.$$

For any $x \in \operatorname{Vac} M(q)_{m-2}$, *the function*

$$(z_1, z_2, z_3) \longmapsto J[z_1, z_2, z_3](x)$$

is a solution of the KZ equation with values in $\operatorname{Vac} M_{m-2}$.

The hypergeometric pairing is nondegenerate for generic κ. It is an interesting problem to describe degenerations of the hypergeometric pairing.

Isomorphism (1.1.15) gives an algebraic description of the monodromy representation of the KZ equation with values in $\operatorname{Vac} M_{m-2}$. To formulate the result we need some definitions.

Let V_1 and V_2 be $U_q\mathfrak{g}$-modules. The comultiplication induces a $U_q\mathfrak{g}$-module structure on $V_1 \otimes V_2$ and $V_2 \otimes V_1$. The modules $V_1 \otimes V_2$ and $V_2 \otimes V_1$ are isomorphic. The isomorphism is given by the formula

$$V_1 \otimes V_2 \xrightarrow{R} V_1 \otimes V_2 \xrightarrow{P} V_2 \otimes V_1,$$

where P is the transposition of the factors and $R \in U_q\mathfrak{g} \otimes U_q\mathfrak{g}$ is the element called the universal R-matrix of the quantum group:

$$R = q^{h \otimes h/4} \sum_{k \geq 0} q^{-k(k+1)/4} \frac{(1)_q^k}{(k)_q!} q^{kh/4} e^k \otimes q^{-kh/4} f^k.$$

Here $(a)_q = q^{a/2} - q^{-a/2}$, $(a)_q! = (a)_q (a-1)_q \ldots (1)_q$.

Let $V_1, \ldots, V_n$ be representations of the quantum group. Then the pure braid group P_n acts on their tensor product. Namely, let $\sigma_1, \ldots, \sigma_{n-1}$ be the elementary braids shown in Fig. 1.3.

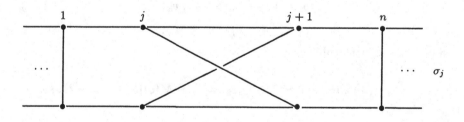

Fig. 1.3

To any braid σ_j, assign a linear operator

$$R_j : V_1 \otimes \ldots V_j \otimes V_{j+1} \cdots \otimes V_n \longrightarrow V_1 \otimes \ldots V_{j+1} \otimes V_j \cdots \otimes V_n$$

acting as PR on the j-th and $(j+1)$-th factors and as the identity on the other factors. These operators define an action of the pure braid group on n strings on $V_1 \otimes \cdots \otimes V_n$. The action of the pure braid group commutes with the action of the quantum group on the tensor product. Therefore, the action of P_n has a lot of invariant subspaces. For example, eigenspaces of the generators of the quantum group form P_n-invariant subspaces.

Let
$$\rho_q : P_3 \longrightarrow \text{Aut}\,(\text{Vac}\,M(q)_{m-2})$$
be the R-matrix representation. Let
$$\tau_\kappa : P_3 \longrightarrow \text{Aut}\,(H_1(\mathbb{C} - \{z_1, z_2, z_3\}, \mathcal{S}))$$
be the natural representation induced by deformations of points z_1, z_2, z_3 on the complex line.

The representations ρ_q and τ_κ are canonically isomorphic for $q = \exp(2\pi i / \kappa)$, the isomorphism is induced by (1.1.15). (1.1.17)

Let $\gamma_\kappa : P_3 \longrightarrow \text{Aut}\,(\text{Vac}\,M(q)_{m-2})$ be the monodromy representation of the KZ equation.

1.1.18. Corollary. *The monodromy representation γ_κ of the KZ equation with values in $\text{Vac}\,M_{m-2}$ is canonically isomorphic to the R-matrix representation ρ_q for $q = \exp(2\pi i/\kappa)$ and generic κ.*

In fact, for generic values of κ, the hypergeometric pairing is nondegenerate, all solutions of the KZ equation are given by hypergeometric integrals.

Remark. The vector valued function $I(z_1, z_2, z_3)$ defined in (1.1.2) is an object of classical mathematics. Each of its components is a classical hypergeometric function. According to one of the numerous definitions [**WW**], the classical hypergeometric function is an integral of a product of powers of three linear functions on the line. Hence, in this section, we have discussed connections between the theory of the classical hypergeometric function and the representation theory of the Lie algebra $s\ell_2$ and the quantum group $U_q s\ell_2$.

1.2. Brief Description of the Contents

The main subject of this book is the comparison between the geometric and analytic constructions in the theory of hypergeometric functions and algebraic constructions in the theory of Lie algebras and quantum groups.

In Secs. 2–5 we discuss the correspondence between the representation theory of quantum groups and geometry of configurations of hyperplanes. Secs. 6–9 are devoted to the correspondence between the universal R-matrix in the theory of quantum groups and the monodromy of configurations of hyperplanes depending on parameters. Secs. 10–12.1, 12.3 are on the connections between the representation theory of Kac–Moody Lie algebras and the geometry of configurations of hyperplanes. In Secs. 12.2, 12.4, 13, 14 we discuss the interrelations between the representation theory of Kac–Moody algebras and the representation theory of quantum groups coming from hypergeometric functions. In Sec. 15 we discuss possible generalizations.

There is another division of this book into parts: geometric, algebraic, and analytic. Secs. 2, 3, 6, 8, 13.1–13.4 are mainly geometric, Secs. 4, 5, 7, 9, 10.1–10.5, 11, 13.8, 15 algebraic, and Secs. 10.6–10.10, 12, 13.5–13.7 are mainly analytic.

In Sec. 2 we consider a configuration of hyperplanes in a complex affine space and a complex one-dimensional local system of coefficients over the complement to the configuration. We assume that the configuration is the complexification of a configuration in a real space and is weighted, that is a number is assigned to any hyperplane of the configuration. We give a combinatorial construction of two finite dimensional complexes and a homomorphism between them. The first complex computes the homology groups of the complement to the configuration with coefficients in the local system. The second complex computes the homology groups of the affine space modulo the configuration. The homomorphism of the first complex to the second induces the natural homomorphism of the homology of the complement of the configuration to the homology of the space modulo the configuration.

This construction can be considered as an analog of the Orlik–Solomon combinatorial description [OS] of the integral cohomology of the complement to the configuration.

In Sec. 3 we introduce discriminantal configurations. These are configurations connected with the representation theory of Lie algebras and quantum groups and the conformal field theory. Consider the configuration of all diagonal hyperplanes in a complex coordinate space and a projection of the space onto another coordinate space along a part of coordinates. Then the configuration of hyperplanes induced in a generic fiber is called a discriminantal configuration. In Sec. 3 we apply the constructions of Sec. 2 to the discriminantal configurations.

In Sec. 4 we introduce the quantum group $U_q\mathfrak{g}$ associated with an arbitrary complex symmetric matrix. Having a collection of highest weights we construct two complexes and a homomorphism between them. The first complex is a two-sided Hochschild complex of the nilpotent subalgebra $U_q\mathfrak{n}_- \subset U_q\mathfrak{g}$ with coefficients in the tensor product of two-sided Verma modules. The second complex is a two-sided Hochschild complex of the dual to $U_q\mathfrak{n}_-$ with coefficients in the tensor product of the contragradient modules corresponding to the Verma modules. The homomorphism between the complexes is given by a suitable contravariant form. We show that these two-sided Hochschild complexes and the homomorphism between them coincide with the combinatorial complexes and the homomorphism between them constructed for a discriminantal configuration in Secs. 2 and 3.

The two-sided Hochschild complexes constructed in Sec. 4 are non-standard for homological algebra. Their spaces are unusually big. Thus, in Sec. 5 we introduce a standard (one-sided) Hochschild complex of $U_q\mathfrak{n}_-$ with coefficients in the tensor product of the standard (one-sided) Verma modules and a standard Hochschild complex of the dual to $U_q\mathfrak{n}_-$ with coefficients in the tensor product of the contragradient representations. We introduce a homomorphism of the first standard complex to the second in terms of a suitable contravariant form. We construct a monomorphism of each of the standard complexes to the corresponding two-sided Hochschild complex and prove that these monomorphisms are quasi-isomorphisms, i.e., they induce isomorphisms of homology groups.

Thus, as a result of Secs. 2–5 we show that the homology groups of a discriminantal configuration with coefficients in a one-dimensional complex local system can be interpreted as the standard Hochschild cohomology of a suitable quantum group. This allows us to identify constructions in representation theory of quantum groups with geometric constructions for discriminantal configurations.

The theorem on quasi-isomorphism suggests that there might be a general construction of a standard (one-sided) Hochschild complex which is quasi-isomorphic to a given two-sided Hochschild complex with coefficients in the tensor product of two-sided modules.

The main body of Sec. 5 is devoted to the proof of the theorem on quasi-isomorphism and could be skipped under first reading. In Sec. 5.11 we give a review of the results of Secs. 2–5.

A discriminantal configuration depends on parameters. The space of parameters has the form of a complex space with deleted diagonal hyperplanes. Thus, the fundamental group of the parameter space is a pure braid group. The fundamental group of the parameter space has the monodromy representation in the homology groups of the discriminantal configuration. The homology groups of a discriminantal configuration were identified with the Hochschild cohomology of a quantum group in Secs. 2–5. We show that under this identification the monodromy representation is identified with the universal R-matrix representation of the braid group in the tensor product of representations of a quantum group. This theorem is proved in Sec. 8. In Sec. 7 we define R-matrix operators and the R-matrix representation of a braid group. In Sec. 6 we prove an auxiliary property of a discriminantal configuration which is basic for the proof of the monodromy theorem in Sec. 8. In Sec. 9 we discuss algebraic properties of the R-matrix, in particular, if the parameter of the quantum group is a root of unity.

In Sec. 10.1–10.3 we consider a weighted configuration of hyperplanes in a complex affine space and give a combinatorial construction of two finite dimensional complexes and a homomorphism between them. The first complex is defined in terms of the flags of the configuration and the spaces of the second are the graded pieces of the Orlik–Solomon algebra. We realize the complexes as subcomplexes of the holomorphic de Rham complex of the complement to the configuration with values in a suitable trivial line bundle with a holomorphic flat connection. The differential forms of these subcomplexes are called the hypergeometric differential forms associated with a configuration.

If the weights of a configuration are in general position, then these finite dimensional complexes compute the cohomology groups of the complement with coefficients in the sheaf of the horizontal sections of the connection.

Assume that a configuration of hyperplanes in a complex affine space is the complexification of a configuration in a real space. Then there are four finite dimensional complexes associated with this configuration. The first two are the combinatorial cell complexes constructed in Sec. 2, the second two are the complexes of hypergeometric forms constructed in Secs. 10.1–10.3. The integration of the differential forms over the cells defines a pairing of the first pair of complexes with the second pair of complexes. This pairing is called the hypergeometric pairing associated with a configuration. We study this pairing in Secs. 10.6–10.9.

Assume that we have a weighted configuration in an affine space and a projection of the space onto another affine space. Then each fiber has a weighted configuration induced by the initial one. The combinatorial structure of the fiber configuration is the same for almost all fibers. The exceptions lie over the discriminant, which is a collection of hyperplanes in the base of the projection. The complexes of hypergeometric forms of a fiber are defined combinatorially and do not depend on a fiber over the complement to the discriminant. In Sec. 10.5 we combinatorially construct a flat connection in the trivial bundle over the complement to the discriminant with the complex of hypergeometric forms as its fiber. We show in Sec. 10.10 that this combinatorial flat connection is realized as the differential equation describing the dependence of the hypergeometric integrals in the fiber on the parameters in the base.

In Sec. 11 we consider a Kac–Moody algebra $\mathfrak{g}$ associated with a complex symmetric matrix. Having a collection of highest weights we construct two complexes and a homomorphism between them. The first complex is the standard Lie algebra chain complex of the nilpotent subalgebra $\mathfrak{n}_- \subset \mathfrak{g}$ with coefficients in the tensor product of Verma modules. We introduce a Lie algebra structure on the dual to $\mathfrak{n}_-$ and an $\mathfrak{n}_-^*$–module structure on the tensor product of the contragradient modules to the Verma modules. The second complex is the standard Lie algebra chain complex of $\mathfrak{n}_-^*$ with coefficients in the tensor product of the contragradient modules. The homomorphism of the first complex to the second is defined by a suitable contravariant form. We introduce the Knizhnik–Zamolodchikov equation with values in these complexes.

We show in Sec. 12 that the two complexes of the hypergeometric differential forms of a weighted discriminantal configuration and a homomorphism between them constructed in Sec. 10 are identical with the two Lie algebra complexes and a homomorphism between them constructed in Sec. 11 for a suitable Kac–Moody Lie algebra. Moreover, under this identification the hypergeometric differential equation is identified with the Knizhnik–Zamolodchikov differential equation.

Thus, results of Secs. 10–12 allow us to identify constructions in the representation theory of Kac–Moody algebras with geometric constructions for discriminantal configurations.

In Secs. 12.2, 12.4, and 12.5 we discuss general properties of the hypergeometric pairing for discriminantal configurations as a pairing between objects of the representation theory of a Kac–Moody algebra and objects of the representation theory of the corresponding quantum group. In particular, we show the nondegeneracy of the pairing for generic values q of the parameter of the quantum group.

Secs. 13 and 14 are devoted to the hypergeometric pairing corresponding to the case of $U_q s\ell_2$ where $q = e^{2\pi i/\kappa}$ and κ is a natural number. The main results are Theorems 13.7.19 and 13.7.27 claiming that the hypergeometric pairing in this case is reduced to a nondegenerate pairing between the space, called the space of conformal blocks in the conformal field theory, and the "path subspace" in the tensor product of the corresponding $U_q s\ell_2$-modules. In particular, the nondegeneracy of this pairing shows that the monodromy representation of the Knizhnik–Zamolodchikov equation in the space of conformal blocks is the R-matrix representation in the "path subspace". To prove the result we discuss in Sec. 13.8 elements of the representation theory of the quantum double of $U_q \mathfrak{n}_- \subset U_q s\ell_2$.

In Sec. 15 we discuss how the constructions of the previous sections could be applied to studying homology groups of configurations with coefficients more general than complex one-dimensional and to studying homology groups of braid groups.

1.3. Acknowledgements

The starting points of this work were articles [**CF, DF, K**]. Interrelations of hypergeometric functions and the representation theory of Lie algebras and quantum groups are studied in interesting works [**BMP, Ch, Do, FW, Ga, L, M, Sc, TK**].

The program of this work and a part of it were developed together with V. Schechtman, see [**SV1–SV4**]. I would like to thank V. Schechtman, V.I. Arnold, I. Cherednik, G. Felder, B. Feigin, E. Frenkel, I. Frenkel, G. Gawedzki, S. Gelfand, T. Kohno, S. Kumar, R. Lawrence, D. Lebedev, P. Orlik, N. Reshetikhin, M. Rosso, M. Salvetti, Y. Soibelman, J. Stasheff, and I.T. Todorov for many valuable discussions.

This work had been started in 1990 when I was a member of the Institute for Advanced Study at Princeton. Parts of this work were written when I visited ETH at Zurich and Max–Planck–Institut für Mathematik at Bonn. I would like to thank the Institutes for their hospitality. This work was partially supported by the National Science Foundation under Grants DMS–8610730 and DMS–9203929.

Moscow, July, 1992

2

Construction of Complexes Calculating Homology of the Complement of Configuration

2.1. Configuration in a Real Space (see [S])

Let $\mathcal{C}$ be a finite set of pairwise distinct affine hyperplanes in $\mathbb{R}^k$. A configuration $\mathcal{C}$ is called *central* if its hyperplanes have a non-empty intersection. An *edge* of a configuration is any non-empty intersection of some subset of the hyperplanes. The connected components of the complement $Y = \mathbb{R}^k \backslash \mathcal{C}$ are called the *domains*. The domains are k-dimensional polyhedra (not necessarily bounded). The open facets of any dimension on these polyhedra are called the *facets* of the configuration. In particular, the domains are n-dimensional facets. Zero-dimensional facets are called the *vertices*. Let $\mathcal{C}'$ be the set of all facets. The support $|F|$ of a facet F is the edge generated by F. Each facet is open in its support.

Given facet F, let $\text{pr}_F : \mathbb{R}^k \to \mathbb{R}^k/|F|$ be the affine projection, which sends hyperplanes containing F into homogeneous hyperplanes of $\mathbb{R}^k/|F|$. These give a central configuration $\mathcal{C}_F$ in $\mathbb{R}^k/|F|$. Let

$$\pi_F : \mathcal{C}' \longrightarrow \mathcal{C}'_F \tag{2.1.1}$$

be the map sending a facet E into the smallest facet containing $\text{pr}_F(E)$. Let $^F\mathcal{C}'$ be the set formed by all facets E having F as a subfacet, that is, $\bar{E} \supset F$. π_F induces a bijection

$$\pi'_F : {}^F\mathcal{C}' \longrightarrow \mathcal{C}'_F. \tag{2.1.2}$$

16 *Construction of Complexes Calculating Homology of Complement of Configuration*

Order the set of all facets of C by the relation: $E \geq F$ iff $\bar{E} \supset F$.

For each facet F choose a point ${}^F w \in F$. For any two subsets $X, Y \subset \mathbb{R}^k$ denote by $X \vee Y$ its join:

$$X \vee Y = \{z = (1 - \lambda)x + \lambda y | \lambda \in [0, 1], x \in X, y \in Y\}.$$

To any flag $F = (F^{j_0} > \cdots > F^{j_i})$ of facets assign an i-dimensional simplex

$$\sigma(F) = {}^{F^{j_0}}w \vee \cdots \vee {}^{F^{j_i}}w. \qquad (2.1.3)$$

2.1.4. Lemma. [S] *Let F^i be a facet of codimension i. Then*

$$D(F^i) = \bigcup_F \sigma(F),$$

where the union is taken over all flags $F = (F^0 > F^1 > \cdots > F^i)$, codim $F^j = j$, is a triangulated i-cell with boundary

$$S^{i-1}(F^0) = \bigcup_F \sigma(F^0 > F^1 > \cdots > F^{i-1}),$$

where the union is being taken over the same flags.

2.1.5. Example. For the configuration shown in Fig. 2.1.a, the simplices $\sigma(F)$ are shown in Fig. 2.1.b.

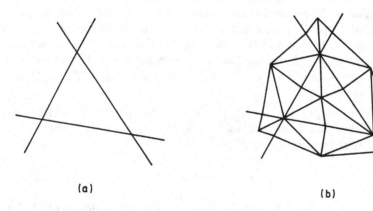

(a)   (b)

Fig. 2.1

Let $\bar{k}$ be the maximum of codimensions of facets. Let

$$Q = \bigcup_{F \in C'} D(F), \qquad (2.1.6)$$

where $D(F)$ is the cell constructed in Lemma (2.1.4). Then Q is a cellular complex endowed with a barycentrical triangulation. One has

$$\partial D(F^j) = \bigcup_{F^{j-1} > F^j} D(F^{j-1}),$$

and each $D(F^{j-1})$ is in the boundary of some $D(F^j)$, $j \leq \bar{k}$. So Q is a regular $\bar{k}$-cell complex, see [S].

For facets $F_1 \geq F_2$, set

$$D(F_1 \geq F_2) = D(F_2) \bigcap \bar{F}_1. \qquad (2.1.7)$$

Then $D(F_1 \geq F_2)$ is a cell of dimension dim F_1 − dim F_2. $D(F_1 \geq F_2)$ is a union of simplices of form (2.1.3). Set

$$Q' = \bigcup_{F_1 \geq F_2} D(F_1 \geq F_2). \qquad (2.1.8)$$

Q' is a regular $\bar{k}$-cell complex, see Lemmas 5 and 6 in [S].

2.1.9. *Example.* For the configuration in Fig. 2.1.a the complexes Q, Q' are shown in Fig. 2.2.

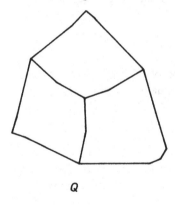

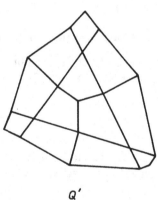

Fig. 2.2

Let

$$Q'' = \bigcup_{\substack{F_1 \geq F_2 \\ \text{codim } F_1 > 0}} D(F_1 \geq F_2). \qquad (2.1.10)$$

Then $Q'' \subset Q'$ is a subcomplex.

$Q' - Q''$ consists of cells $D(F^0 \geq F^i)$, where F^j is a facet of codimension j. We have

$$\partial D(F^0 \geq F^i) = \bigcup_{F^0 \geq F^{i-1} \geq F^i} D(F^0 \geq F^{i-1}) \qquad (2.1.11)$$

modulo cells in Q''.

2.1.12. Lemma. [S] *There is a contraction of the pair $(\mathbb{R}^k, \bigcup_{H \in \mathcal{C}} H)$ onto the pair (Q', Q'').*

Given a facet F^i, let $^{F^i}\mathcal{C}''$ be the set formed by all facets F^0 of codimension 0 having F^i as a facet, that is, $F^0 \geq F^i$.

Given any two facets F^i and F^j, there is a map

$$\pi_{F^i, F^j} : {}^{F^i}\mathcal{C}'' \longrightarrow {}^{F^j}\mathcal{C}'', \qquad (2.1.13)$$

defined by the formula $\pi_{F^i, F^j} = (\pi'_{F^j})^{-1} \circ \pi_{F^j}$, see (2.1.1), (2.1.2).

Let $F^0 \in {}^{F^i}\mathcal{C}''$ and $F^{0'} = \pi_{F^i, F^j}(F^0)$. Set

$$w_{F^0 \geq F^i, F^j} = {}^{F^{0'}}w. \qquad (2.1.14)$$

For example, for every $F^i \geq F^j$ and $F^0 \in {}^{F^i}\mathcal{C}''$, we have $\pi_{F^i, F^j}(F^0) = F^0$, $w_{F^0 \geq F^i, F^j} = {}^{F^0}w$.

2.2. Configuration in a Complex Space (see [S])

Let $\mathbb{C}^k$ be the space obtained by complexification of $\mathbb{R}^k$, with points $x + iy$, where $x, y \in \mathbb{R}^k$, and $\mathcal{R}, \mathcal{I} : \mathbb{C}^k \to \mathbb{R}^k$ the real and imaginary parts. Let

$$Y = \mathbb{C}^k \setminus \bigcup_{H \in \mathcal{C}} H_\mathbb{C},$$

where $H_\mathbb{C}$ is the hyperplane of $\mathbb{C}^k$ which is complexification of $H \in \mathcal{C}$. For each $F^0 \geq F^p$, define a p-cell $E(F^0 \geq F^p)$ in Y as the following embedding

$$\varphi_{F^0 \geq F^p} : D(F^p) \longrightarrow Y,$$

where the p-cell $D(F^p)$ is defined in Lemma 2.1.4.

Each $x \in D(F^p)$ is written as

$$x = \sum_{j=0}^{p} \lambda_j {}^{F^j}w, \quad \sum \lambda_j = 1, \quad \lambda_j \in [0, 1],$$

where $F^0 > \cdots > F^j > \cdots > F^p$, codim $F^j = j$. We set

$$\mathcal{R}\varphi_{F^0 \geq F^p}(x) = x$$
$$I\varphi_{F^0 \geq F^p}(x) = \sum_{j=0}^{p} \lambda_j \left(w_{F^0 \geq F^p, F^j} - {}^{F^j}w\right). \tag{2.2.1}$$

$D(F^0 \geq F^p)$ is a deformation of $D(F^p)$ into the direction of $i({}^{F^0}w - {}^{F^p}w)$. One has

$$\partial E(F^0 \geq F^p) = \bigcup_{F^{p-1} > F^p} E(\pi_{F^p, F^{p-1}}(F^0) \geq F^{p-1}). \tag{2.2.2}$$

Let

$$X = \bigcup_{F^0 \geq F^p} E(F^0 \geq F^p). \tag{2.2.3}$$

Then X is a regular $\bar{k}$-cell complex, where $\bar{k}$ is the maximum of codimensions of facets.

2.2.4. Remark. p-cells of $Q' - Q''$ and p-cells of X are numerated by the same pairs $F^0 \geq F^i$.

2.2.5. Theorem. [S] *There is a homotopy equivalence ϕ_t, $t \in [0,1]$, between Y and X.*

Define a deformation

$$\mathcal{R}_t : X \longrightarrow \mathbb{C}^k, \quad t \in [0,1], \tag{2.2.6}$$

by the formula $\mathcal{R}_t : x + iy \mapsto x + tiy$. Then $\mathcal{R}_0$ defines a projection

$$\mathcal{R}_0 : X \longrightarrow Q, \tag{2.2.7}$$

where Q is the complex defined in (2.1.6).

2.2.8. Remark. Complexes X, Q, Q', Q'' depend on the choice of points $\{{}^F w\}$, but they are canonically piece-wise linear homeomorphic for different choices of points $\{{}^F w\}$. These canonical homeomorphisms depend continuously on $\{{}^F w\}$.

2.3. Homology

Let $\mathcal{S}$ be a local system of coefficients defined on Y. We will consider two sorts of homology.

Homology of the complement $H.(Y, S)$. It may be calculated using the following complex. Let us call an "n-cell" P a convex closed oriented polytope in $\mathbb{R}^n$. A singular n-cell is a continuous map $f : P \to \mathbb{C}^k$. The complex $C.(Y, S)$ has as n-chains finite linear combinations of the pairs (a singular cell $f : P \to Y$; a section $\in \Gamma(P, f^*S)$). We have $H.C.(Y, S) = H.(Y, S)$.

Homology of the pair $H.(\mathbb{C}^k, \cup_{H \in \mathcal{C}} H_\mathbb{C}, S)$. It may be calculated using the following complex $C_!.(Y, S)$. n-chains of this complex are finite linear combinations of the pairs (a singular n-cell $f : P \to Y$ such that $f^{-1}(f(P) \cap \bigcup_{H \in \mathcal{C}} H_\mathbb{C})$ is a union of facets of P; a section $\in \Gamma(f^{-1}(f(P) \cap Y), f^*S)$. We have $H.(\mathbb{C}^k, \cup H_\mathbb{C}, S) = H.C_!.(Y, S)$.

One has an evident map $C.(Y, S) \to C_!.(Y, S)$ inducing the canonical map in homology.

Consider the finite dimensional complex $C.(Q'/Q'', S)$, where the cell complexes $Q'' \subset Q'$ are defined in (2.1). Singular n-cells of $C.(Q'/Q'')$ are pairs (an oriented n-cell D of $Q' - Q''$; a section $\in \Gamma(D, S|_D)$).

There is a natural inclusion

$$C.(Q'/Q'', S) \subset C_!.(Y, S). \tag{2.3.1}$$

2.3.2. Theorem. [S] *The inclusion induces an isomorphism*

$$H.C.(Q'/Q'', S) = H.\Big(\mathbb{C}^k, \bigcup_{H \in \mathcal{C}} H_\mathbb{C}, S\Big).$$

Proof. The pair $(\mathbb{C}^k, \cup H_\mathbb{C})$ is contracted onto the pair $(\mathbb{R}^k, \cup H)$ by the contraction $x + iy \mapsto x + tiy$, $t \in [0, 1]$. By Lemma (2.1.12), there is a contraction of $(R^k, \cup H)$ onto (Q', Q'').

Consider the finite dimensional complex $C.(X, S)$. There is a natural inclusion

$$C.(X, S) \subset C.(Y, S).$$

By Theorem (2.2.5), this inclusion induces an isomorphism

$$H.C.(X, S) = H.(Y, S). \tag{2.3.3}$$

Define an important homomorphism

$$S : C.(X, S) \longrightarrow C.(Q'/Q'', S) \tag{2.3.4}$$

as follows. Let

$$(E(F^0 \geq F^p), \quad s_1 \in \Gamma(E(F^0 \geq F^p), S|_{E(F^0 \geq F^p)}))$$

be any singular cell of $C.(X,\mathcal{S})$. Consider the deformation $\mathcal{R}_t : E(F^0 \geq F^p) \to \mathbb{C}^k$, defined in (2.2.6). Let $s_t \in \Gamma(\mathcal{R}_t(E(F^0 \geq F^p)), \mathcal{S}|_{(\mathcal{R}_t(E(F^0 \geq F^p)))})$, $t \in (0,1]$, be such a section that the union of the sections s_t forms a horizontal section of $\mathcal{S}$ over $\bigcup_t \mathcal{R}_t(E(F^0 \geq F^p))$. Then

$$\left(\mathcal{R}_0(E(F^0 \geq F^p)), \lim_{t \to 0} s_t \right)$$

is an element of the complex $C.(Q'/Q'', \mathcal{S})$. Extended by linearity, this construction defines homomorphism (2.3.4). The homomorphism S will be called *the homomorphism of the projection on real part*. Note that $\mathcal{R}_0(E(F^0 \geq F^p)) = D(F^p)$.

2.3.5. Theorem. *The following diagram is commutative*

$$\begin{array}{ccc} H.C.(X,\mathcal{S}) & \xrightarrow{S} & H.C.(Q'/Q'',\mathcal{S}) \\ \downarrow & & \downarrow \\ H.(Y,\mathcal{S}) & \longrightarrow & H.(\mathbb{C}^k, \cup H_\mathbb{C}, \mathcal{S}), \end{array}$$

where the lower horizontal homomorphism is the canonical homomorphism.

The theorem is a direct corollary of the formulated results.

A bounded domain of $\mathcal{C}$ is a bounded facet of codimension 0. If $\mathcal{C}$ has a bounded domain then $\mathcal{C}$ has a vertex. For each bounded domain D fix its orientation and a basis of sections $\{s_D^i\}$ of the local system $\mathcal{S}$ over D. Each pair (D, s_D^i) determines an element of $H_k(\mathbb{C}^k, \bigcup_{H \in \mathcal{C}} H_\mathbb{C}, \mathcal{S})$.

2.3.6. Theorem. *For a configuration $\mathcal{C}$, having a vertex, $H_n(\mathbb{C}^k, \cup_{H \in \mathcal{C}} H_\mathbb{C}, \mathcal{S}) = 0$ for $n \neq k$. The elements $\{(D, s_D^i) | D$ is any bounded domain and $\{s_D^i\}$ is a basis of local sections$\}$ form a basis of $H_k(\mathbb{C}^k, \bigcup_{H \in \mathcal{C}} H_\mathbb{C}, \mathcal{S})$.*

Proof. The pair $(\mathbb{C}^k, \cup_{H \in \mathcal{C}} H_\mathbb{C})$ is contractible onto $(\mathbb{R}^k, \cup_{H \in \mathcal{C}} H)$. The pair $(\mathbb{R}^k, \cup_{H \in \mathcal{C}} H)$ is contractible onto $((\cup D) \cup (\cup_{H \in \mathcal{C}} H), \cup_{H \in \mathcal{C}} H)$, where D runs through all bounded domains, see [BBR]. These statements imply the theorem.

Let $\bar{k}$ be the maximum of codimensions of facets, and let $\bar{k} < k$. Let $F^{\bar{k}}$ be a facet of maximal codimension and S be any subspace complementary to $|F^{\bar{k}}|$ in $\mathbb{R}^k$. The configuration $\mathcal{C}$ cuts a configuration in S. Denote it by $\mathcal{C}_S$. According to [S], each edge of $\mathcal{C}$ intersects S transversally, therefore, the configuration $\mathcal{C}_S$ has a vertex. There is a natural homotopy equivalence $(\mathbb{C}^k, \cup_{H \in \mathcal{C}} H_\mathbb{C}) \sim (S_\mathbb{C}, \cup_{H \in \mathcal{C}} (H_\mathbb{C} \cap S_\mathbb{C}))$ where $S_\mathbb{C}$ is the complexification of S.

These statements and Theorem (2.3.6) describe the groups $H_n(\mathbb{C}^k, \underset{H\in\mathcal{C}}{\cup} H_\mathbb{C}, \mathcal{S})$ for an arbitrary configuration $\mathcal{C}$.

2.4. Compatible Orientation of Cells of Complexes Q, Q'/Q'', X

A configuration $\mathcal{C}$ will be called *cooriented* if each facet of the configuration is cooriented. Fix a coorientation of $\mathcal{C}$. For each facet F^j of $\mathcal{C}$, a coorientation of F^j determines an orientation of the j-cell $D(F^j)$. Boundary of the oriented j-cell $D(F^i)$ is given by the formula

$$d\, D(F^j) = \sum_{F^{j-1}>F^j} \text{ind}\,(F^{j-1} > F^j)\, D(F^{j-1}), \qquad (2.4.1)$$

where the coefficient $\text{ind}\,(F^{j-1} > F^j)$ is defined as follows. Let a vector v_1 lie in F^{j-1} and have a direction from F^j to F^{j-1}. Let vectors $v_2, \cdots, v_j$ be transversal to F^{j-1} and define the fixed coorientation of F^{j-1}. Then

$$\text{ind}\,(F^{j-1} > F^j) = \pm 1, \qquad (2.4.2)$$

where sign $+$ is taken iff the fixed coorientation of F^j coincides with the coorientation of F^j defined by the ordered sequence $v_1, v_2, \cdots, v_j$.

An orientation of $D(F^j)$ defines an orientation of all j-cells $D(F^0 \geq F^j)$ as these cells are open subsets of $D(F^j)$. Boundary of a singular j-cell $D(F^0 \geq F^j)$ is given by the formula

$$d\, D(F^0 \geq F^j) = \sum_{F^0 \geq F^{j-1} > F^j} \text{ind}\,(F^{j-1} > F^j)\, D(F^0 \geq F^{j-1}). \qquad (2.4.3)$$

An orientation of $D(F^j)$ defines an orientation of all j-cells $E(F^0 \geq F^j)$ as these cells are deformations of $D(F^j)$. Boundary of a singular j-cell $E(F^0 \geq F^j)$ is given by the formula

$$d\, E(F^0 \geq F^j) = \sum_{F^{j-1}>F^j} \text{ind}\,(F^{j-1} > F^j)\, E(\pi_{F^j, F^{j-1}}(F^0) \geq F^{j-1}). \qquad (2.4.4)$$

The orientation of cells $D(F^j)$, $D(F^0 \geq F^j)$, $E(F^0 \geq F^j)$ induced by a coorientation of all facets F^j will be called the *compatible orientation*.

2.5. Local System $\mathcal{S}^*(a)$ and Distinguished Sections Over Cells

A configuration $\mathcal{C}$ will be called *weighted* if a complex number $a(H)$ is assigned to each hyperplane H of $\mathcal{C}$. Fix weights $a = \{a(H), H \in \mathcal{C}\}$. Fix a real affine equation $\ell_H = 0$ for each hyperplane $H \in \mathcal{C}$. Consider the complement

$$Y = \mathbb{C}^k \setminus \bigcup_{H \in \mathcal{C}} H_\mathbb{C}. \qquad (2.5.1)$$

Denote by $\mathcal{L}(a)$ a line bundle over Y with an integrable connection which is trivial as a bundle and whose connection

$$\nabla(a) : \mathcal{O} \longrightarrow \Omega^1 \qquad (2.5.2)$$

is given by

$$d + w(a) = d + \sum_{H \in \mathcal{C}} a(H)\, d\log \ell_H\,,$$

d being the de Rham differential.

Denote by $\Omega^{\cdot}(\mathcal{L}(a))$ the complex of Y-sections of the holomorphic de Rham complex of $\mathcal{L}(a)$. Thus, $\Omega^{\cdot}(\mathcal{L}(a)) = \Omega^{\cdot}(Y)$ as a graded vector space, with the differential $\nabla(a)$:

$$\nabla(a)\, x = dx + w(a) \wedge x\,. \qquad (2.5.3)$$

Denote by $\mathcal{S}(a)$ the sheaf over Y of horizontal sections of $\mathcal{L}(a)$. We have the de Rham isomorphism

$$H^{\cdot}(\Omega^{\cdot}(\mathcal{L}(a))) = H^{\cdot}(Y, \mathcal{S}(a))\,. \qquad (2.5.4)$$

Let $\mathcal{L}^*(a)$, $\mathcal{S}^*(a)$ be the dual line bundle with the dual connection and the sheaf of its horizontal sections, respectively. $\mathcal{S}^*(a)$ is dual to $\mathcal{S}(a)$. $\mathcal{L}^*(a)$ is trivial as a bundle.

$\mathcal{L}(a), \mathcal{L}^*(a), \mathcal{S}(a), \mathcal{S}^*(a)$ will be called *the line bundles and the local systems of coefficients associated with the weights a*.

Consider a multivalued analytic function on Y,

$$\ell_a = \prod_{H \in \mathcal{C}} \ell_H^{a(H)}\,. \qquad (2.5.5)$$

This function is a multivalued horizontal section of $\mathcal{L}^*(a)$. Thus the sections of $\mathcal{L}^*(a)$ have the form

$$s = c \cdot b\,\ell_a \qquad (2.5.6)$$

where $c \in \mathbb{C}$, $b\,\ell_a$ is a univalued branch of ℓ_a.

Below we fix sections of $\mathcal{S}^*(a)$ over all cells $D(F^0 \geq F^j)$, $E(F^0 \geq F^j)$.

Consider any facet F^0 of codimension 0, $F^0 \subset \mathbb{R}^k \subset \mathbb{C}^k$. Define a section of $\mathcal{L}^*(a)$ over F^0 by the formula

$$s_{F^0} = \prod_{H \in \mathcal{C}} |\ell_H|^{a(H)} \qquad (2.5.7)$$

where $|\ell_H|$ is the absolute value of ℓ_H. For given $F^0 \geq F^j$, define a section of $\mathcal{L}^*(a)$ over $D(F^0 \geq F^j) \subset F^0$ by the formula

$$s_{D(F^0 \geq F^j)} = s_{F^0}|_{D(F^0 \geq F^j)}\,. \qquad (2.5.8)$$

2.5.9. Remark. If all weights $a(H)$ are real, then the section s_{F^0} takes positive values.

For each pair $(F^0 \geq F^j)$, we define a section of $\mathcal{S}^*(a)$ over $E(F^0 \geq F^j)$ as follows. We denote this section by $s_{E(F^0 \geq F^j)}$. The cell $E(F^0 \geq F^j)$ contains the point
$$w_0 = {}^{F^j}w + i\left({}^{F^0}w - {}^{F^j}w\right), \tag{2.5.10}$$
see (2.2.1). To define a section it is sufficient to fix its value at w_0. We do this below.

For any $H \in \mathcal{C}$,
$$\ell_H(w_0) = \ell_H({}^{F^j}w) + i\bar{\ell}_H({}^{F^0}w - {}^{F^j}w),$$
where $\bar{\ell}_H = \ell_H - \ell_H(0)$ is the homogeneous linear function corresponding to ℓ_H.

Let $F^j \not\subset H$ and $\ell_H({}^{F^j}w) > 0$, then the complex number $\ell_H(w_0)$ lies in the right half-plane of $\mathbb{C}$. Fix its argument as a number in $[-\pi/2, \pi/2]$. Introduce a number $\alpha(F^0 \geq F^j, H)$ setting $\alpha(F^0 \geq F^j, H) = 0$.

Let $F^j \not\subset H$ and $\ell_H({}^{F^j}w) < 0$, then $\ell_H(w_0)$ lies in the left half-plane of $\mathbb{C}$. Fix its argument as a number in $[\pi/2, 3\pi/2]$. Set $\alpha(F^0 \geq F^j, H) = -\pi$.

Let $F^j \subset H$, then $\ell_H(w_0) = i\bar{\ell}_H({}^{F^0}w - {}^{F^j}w)$. If $\bar{\ell}_H({}^{F^0}w - {}^{F^j}w) > 0$, set $\arg \ell_H(w_0) = \pi/2$, $\alpha(F^0 \geq F^j, H) = -\pi/2$. If $\bar{\ell}_H({}^{F^0}w - {}^{F^j}w) < 0$, set $\arg \ell_H(w_0) = -\pi/2$, $\alpha(F^0 \geq F^j, H) = \pi/2$.

Define $s_{E(F^0 \geq F^j)}(w_0)$ by the formula
$$s_{E(F^0 \geq F^j)}(w_0) = \prod_{H \in \mathcal{C}} \left(\ell_H(w_0)^{a(H)} \exp\left(a(H)\alpha(F^0 \geq F^j, H)\right)\right) \tag{2.5.11}$$
where the argument of all numbers $\ell_H(w_0)$ is chosen above.

The section $s_{E(F^0 \geq F^j)}$ has the following positivity property analogous to the property (2.5.9). Namely, consider the curve $\gamma : [0,1) \to \mathbb{C}^k$, $t \mapsto {}^{F^j}w + (1-t)i\left({}^{F^0}w - {}^{F^j}w\right)$. Consider the section s of $\gamma^*\mathcal{S}^*(a)$ over $[0,1)$ having value $s_{E(F^0 \geq F^j)}(w_0)$ at $t = 0$.

2.5.12. Lemma. *Assume that the numbers $\{a(H), H \in \mathcal{C}\}$ are real. Then*
$$\lim_{t \to 1} \arg s(t) = 0,$$
if the argument is chosen in $[-\pi, \pi)$; and
$$|s(t)| = \prod_{H \in \mathcal{C}} |\ell_H(\gamma(t))|^{a(H)}.$$

The lemma is obvious.

Fix any coorientation Or of the configuration $\mathcal{C}$. The coorientation defines a compatible orientation of all cells $D(F^j)$, $D(F^0 \geq F^j)$, $E(F^0 \geq F^j)$ and the indices in (2.4.1). The compatible orientation is denoted by Or, the indices in (2.4.1) by $\text{ind}(F^{j-1} > F^j, Or)$.

Denote a pair (the oriented $D(F^0 \geq F^j)$, $s_{D(F^0 \geq F^j)}$) by $D(F^0 \geq F^j, a, Or)$. This pair is an element of $C_j(Q'/Q'', \mathcal{S}^*(a))$. The elements $\{D(F^0 \geq F^j, a, Or) \mid F^0 \geq F^j$ runs through all pairs of adjacent facets of codimension 0 and j, respectively$\}$ form a basis of $C_j(Q'/Q'', \mathcal{S}^*(a))$.

2.5.13. Lemma. *The boundary operator in $C.(Q'/Q'', \mathcal{S}^*(a))$ is given by the formula*

$$d\,D(F^0 \geq F^j, a, Or) = \sum_{F^0 \geq F^{j-1} > F^j} \text{ind}(F^{j-1} > F^j, Or) \cdot D(F^0 \geq F^{j-1}, a, Or).$$

The lemma is a corollary of (2.4.3).

Denote a pair (the oriented $E(F^0 \geq F^j)$, $s_{E(F^0 \geq F^j)}$) by $E(F^0 \geq F^j, a, Or)$. This pair is an element of $C_j(X, \mathcal{S}^*(a))$. The elements $\{E(F^0 \geq F^j, a, Or) \mid F^0 \geq F^j$ runs through all pairs of adjacent facets of codimension 0 and j, respectively$\}$ form a basis of $C_j(X, \mathcal{S}^*(a))$.

For each triple (F^0, F^{j-1}, F^j) such that $F^0 \geq F^j$, $F^{j-1} > F^j$, introduce the *twisting number* of the triple, $B(F^0, F^{j-1}; F^j)$, by the formula

$$B(F^0, F^{j-1}; F^j) = \exp\left(\frac{\pi i}{2} \sum_{\substack{H \in \mathcal{C} \\ F^j \subset H, F^{j-1} \not\subset H}} \pm a(H),\right) \qquad (2.5.14)$$

where the sign of $a(H)$ is positive if H separates F^0, F^{j-1}, and is negative otherwise.

2.5.15. Lemma. *The boundary operator in $C.\bigl(X, \mathcal{S}^*(a)\bigr)$ is given by the formula*

$$dE(F^0 \geq F^j, a, Or) = \sum_{F^{j-1} > F^j} \text{ind}(F^{j-1} > F^j, Or)$$

$$\cdot B(F^0, F^{j-1}; F^j) \cdot E(\pi_{F^j, F^{j-1}}(F^0) \geq F^{j-1}, a, Or)$$

cf. (2.2.2).

The proof of Lemma (2.5.15) is given in (2.7).

2.6. Quantum Bilinear Form of a Configuration

Let a configuration $\mathcal{C}$ in $\mathbb{R}^k$ be weighted and let $b = \{b(H) \mid H \in \mathcal{C}\}$ be the weights. Consider the vector space $M_{\mathcal{C}}$ of all $\mathbb{C}$-linear combinations of domains of $\mathcal{C}$. Define a symmetric bilinear form $B_{\mathcal{C}}$ on $M_{\mathcal{C}}$ by the formula

$$B_{\mathcal{C}}(P,Q) = \prod a(H)\,,$$

where P, Q are arbitrary domains of $\mathcal{C}$, the product is taken over all hyperplanes $H \in \mathcal{C}$ separating P and Q, see [V4]. This form is called the *quantum bilinear form of the configuration.*

2.6.1. ***Examples.***

1. For a configuration of two points in the line, the matrix of the form is shown in Fig. 2.3.

(P_i / P_j)

1	b_1	b_1b_2
b_1	1	b_2
b_1b_2	b_2	1

$$\det = (1-b_1^2)(1-b_2^2)$$

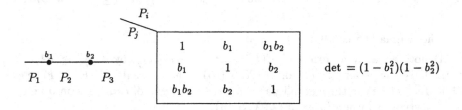

Fig. 2.3

2. For a configuration of two lines in the plane the matrix of the form is shown in Fig. 2.4.

1	b_1	b_1b_2	b_2
b_1	1	b_2	b_1b_2
b_1b_2	b_2	1	b_1
b_2	b_1b_2	b_1	1

$$\det = (1-b_1^2)^2(1-b_2^2)^2.$$

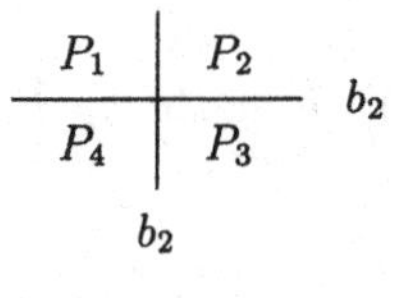

Fig. 2.4

It turns out that the determinant of the matrix of $B_{\mathcal{C}}$ is a product of binomials of the form $1 - (\prod b)^2$. To formulate this result we need some notions of the weight of an edge and the multiplicity of an edge.

Define *the weight of an edge* of $\mathcal{C}$ as the product of weights of all hyperplanes which contain the edge.

Define *the multiplicity of an edge L* as

$$\ell(L) = n(L)\, p(L),$$

where the numbers $n(L)$, $p(L)$ are defined below. Let

$$\mathcal{C}^L = \{H \cap L \mid H \in \mathcal{C}, H \not\supset L\}$$

be the configuration in L cut by $\mathcal{C}$. Let $n(L)$ be the number of domains of the configuration $\mathcal{C}^L$. The number $n(L)$ will be called *the discrete mass* of L. Let

$$\mathcal{C}_L = \{H \mid H \in \mathcal{C}, H \supset L\}$$

be the configuration of all hyperplanes of $\mathcal{C}$ containing L. $\mathcal{C}$ is called the *localization* of $\mathcal{C}$ at L.

For every edge L of codimension r, we construct a configuration in the $(r-1)$-dimensional projective space. Namely, let N be an r-dimensional normal subspace to L. Consider the localization at the edge and its section by the normal subspace. All of the hyperplanes of the resulting configuration $(\mathcal{C}_L)^N$ pass through the point $v = L \cap N$. We consider the configuration which $(\mathcal{C}_L)^N$ induces in the tangent space $T_v N$. It determines a configuration in the projectivization of the tangent space $T_v N$, which we call the *projective localization* and denote by $P\mathcal{C}_L$. The configurations corresponding to different normals are naturally isomorphic.

A domain of a configuration is said to be *bounded with respect to a hyperplane* if the closure of the domain does not intersect the hyperplane. For any configuration $\mathcal{C}$ in a real projective space the number of the domains which are bounded with respect to a hyperplane $H \in \mathcal{C}$ does not depend on H [**V2**]. For example, for the case of a configuration of n points in the projective line, this number is $n - 2$. The number of domains which are bounded with respect to one of the hyperplanes of the configuration will be denoted by $e(\mathcal{C})$. We agree to set $e(\mathcal{C})$ of the empty configuration $\mathcal{C}$ to be equal to 1.

Let L be an edge of a configuration $\mathcal{C}$ in a real affine space. Set

$$p(L) = e(P\mathcal{C}_L).$$

The number $p(L)$ will be called the *discrete density* of L. An edge L will be called *dense* if $p(L) > 0$.

2.6.2. Theorem. [**V4**] *Determinant of the matrix of the bilinear form of a real configuration is given by the formula*

$$\det B_{\mathcal{C}} = \prod_L (1 - b(L)^2)^{\ell(L)},$$

where the product is taken over all edges L of C, $b(L)$ is the weight of L, $\ell(L)$ is the multiplicity of L.

2.6.3. Example. For a configuration of n points in the line having weights $b_1, \ldots, b_n$,
$$\det B_C = (1 - b_1^2) \ldots (1 - b_n^2).$$

2.6.4. Example. For a configuration of n lines in the plane, intersecting at one point and having weights $b_1, \ldots, b_n$,
$$\det B = (1 - b_1^2)^2 \ldots (1 - b_n^2)^2 (1 - (b_1 \ldots b_n)^2)^{n-2}.$$

Assume that C is central and weighted, and let $a = \{a(H) | H \in C\}$ be new weights. In the space M_C of all $\mathbb{C}$-linear combinations of domains of C introduce a new symmetric bilinear form B'_C called the *reduced quantum bilinear form* of a central real configuration:

$$B'_C(P, Q) = \exp\left(\frac{\pi i}{2} \sum_{H \in C} \pm a(H)\right), \qquad (2.6.5)$$

where P, Q are arbitrary domains of C, the sign of $a(H)$ is positive if H separates P, Q and is negative otherwise.

Introduce the *quasiclassical weight* of an edge L of C as the sum of weights $a(H)$ of all hyperplanes $H \in C$ containing L.

2.6.6. Theorem. (see [V4, Theorem (1.1), Lemma (13.1)]) *Determinant of the reduced bilinear form of a central real configuration is given by*

$$\det B'_C = \prod_L (\exp(-\pi i a(L)) - \exp(\pi i a(L)))^{\ell(L)},$$

where the product is taken over all edges of C, $a(L)$ is the quasiclassical weight of L, $\ell(L)$ is the multiplicity of L.

We say that the bilinear form B_C has a resonance at an edge L if

$$b(L)^2 = 1. \qquad (2.6.7)$$

We say that the bilinear form B'_C has a resonance at an edge L if

$$\exp(2\pi i a(L)) = 1. \qquad (2.6.8)$$

According to (2.6.2) and (2.6.6), the matrices of B_C and $B_{C'}$ are invertible if the forms B_C and $B_{C'}$ have no resonances at dense edges.

The forms B_C and B'_C have the following obvious functorial property. Let C_i be a configuration in an affine space V_i, B_{C_i} its quantum bilinear form, $i=1,2$. Let $C_1 \vee C_2$ be the configuration in $V_1 \times V_2$ consisting of the hyperplanes $H \times V_2$, $H \in C_1$, and $V_1 \times H$, $H \in C_2$. Let C_1, C_2 be weighted. Define the weights of $C_1 \vee C_2$ assuming that the weight of $H \times V_2$ is equal to the weight of H, $H \in C_1$, and the weight of $V_1 \times H$ is equal to the weight of H, $H \in C_2$. Then $M_{C_1 \vee C_2} = M_{C_1} \otimes M_{C_2}$, $B_{C_1 \vee C_2} = B_{C_1} \otimes B_{C_2}$,

$$\det B_{C_1 \vee C_2} = (\det B_{C_1})^{n(C_2)} (\det B_{C_2})^{n(C_1)}, \qquad (2.6.9)$$

where $n(C_i)$ is the number of domains of C_i.

The same property holds for the reduced quantum forms B'_{C_i}.

2.6.10. Corollary. *Let L_i be an edge of C_i, $i=1,2$, then the edge $L_1 \times L_2$ of the configuration $C_1 \vee C_2$ has density 0.*

Proof. According to (2.6.9), $\det B_{C_1 \vee C_2}$ does not contain the factor $1 - b(L_1)^2 b(L_2)^2$.

2.7. Homomorphism of the Projection on the Real Part

In (2.3) we defined a homomorphism of complexes

$$S : C.(X, \mathcal{S}^*(a)) \longrightarrow C.(Q'/Q'', \mathcal{S}^*(a)). \qquad (2.7.1)$$

In this section we describe the action of S on the cells $E(F^0 \geq F^j, a, Or)$.

Given a facet F of C, consider the configuration $C_F = \{H \in C | F \subset H\}$.

Let B'_F be the reduced quantum bilinear form of this configuration with respect to the weights $\{a(H)|H \in C_F\}$, see (2.6). Let $^F C''$ be the set formed by all domains G^0 of C having F as a facet, that is, such that $G^0 \geq F$. Let C''_F be the set of all domains of C_F. Denote by

$$\pi''_F : {}^F C'' \longrightarrow C''_F$$

the bijection sending a domain $G^0 \in {}^F C''$ to the domain of C_F containing G^0. For any two domains $G^0, F^0 \in {}^F C''$, set

$$B'_F(G^0, F^0) := B'_F(\pi''_F(G^0), \pi''_F(F^0)).$$

2.7.2. Theorem. *The homomorphism (2.7.1) of the projection on the real part is defined by the formula*

$$S : E(F^0 \geq F^j, a, Or) \mapsto \sum_{G^0 \geq F^j} B'_{F^j}(G^0, F^0) \, D(G^0 \geq F^j, a, Or).$$

2.7.3. Example. For the configuration and facets shown in Fig. 2.5, we have

$$S : E(F_1^0 \geq F^2) \mapsto \exp\left(\pi i(-a_1 - a_2 - a_3)/2\right) D(F_1^0 \geq F^2)$$
$$+ \exp\left(\pi i(a_1 - a_2 - a_3)/2\right) D(F_2^0 \geq F^2)$$
$$+ \exp\left(\pi i(a_1 + a_2 - a_3)/2\right) D(F_3^0 \geq F^2)$$
$$+ \cdots + \exp\left(\pi i(-a_1 - a_2 + a_3)/2\right) D(F_6^0 \geq F^2).$$

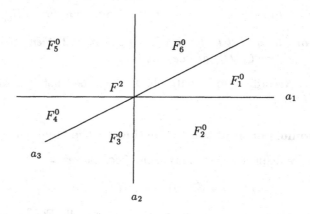

Fig. 2.5

Proof. By definition of S,

$$S(E(F^0 \geq F^j, a, Or)) = \sum_{G^0 \geq F^j} A(G^0, F^0) \cdot D(G^0 \geq F^j, a, Or)$$

where $\{A(G^0, F^0)\}$ are some numbers. The problem is to compute these numbers.

For each F^0, G^0 such that $F^0 \geq F^j$, $G^0 \geq F^j$, the cells $E(F^0 \geq F^j)$, $D(G^0 \geq F^j)$ have a non-empty intersection consisting of one point $^{G^0}w$, see formulas (2.2.1). The number $A(G^0, F^0)$ equals

$$s_{E(F^0 \geq F^j)}(^{G^0}w)/s_{D(G^0 \geq F^j)}(^{G^0}w) \tag{2.7.4}$$

by the definition of the homomorphism S, here s_E, s_D are the sections defined in (2.5). The number $s_{D(G^0 \geq F^j)}(^{G^0}w)$ is defined by formula (2.5.7). The horizontal section $s_{E(F^0 \geq F^j)}$ is defined at the point w_0, see (2.5.10), (2.5.11). The segment I joining points w_0 and $^{G^0}w$ lies in $E(F^0 \geq F^j)$. Therefore, we must continue the section $s_{E(F^0 \geq F^j)}$ horizontally from w_0 to $^{G^0}w$ along I and compute ratio (2.7.4).

In (2.5) we have defined argument of the numbers $\ell_H(w_0)$, $H \in \mathcal{C}$, and, therefore, a branch at w_0 of functions $\ell_H^{a(H)}$, $H \in \mathcal{C}$. We denote this branch of the function $\ell_H^{a(H)}$ by $b_{F^0}\ell_H^{a(H)}$.

2.7.5. Lemma. *The analytic continuation of the branch $b_{F^0}\ell_H^{a(H)}$ from w^0 to $^{G^0}w$ along I equals:*

a) $|\ell_H(^{G^0}w)|^{a(H)}$ *if* $F^j \not\subset H$, $\ell_H(F^j) > 0$;

b) $|\ell_H(^{G^0}w)|^{a(H)} \exp(\pi i a(H))$ *if* $F^j \not\subset H$, $\ell_H(F^j) < 0$;

c) $|\ell_H(^{G^0}w)|^{a(H)} \exp(\pi i a(H))$ *if* $F^j \subset H$, $\ell_H(F^0) > 0$, *and H separates F^0 and G^0;*

d) $|\ell_H(^{G^0}w)|^{a(H)}$ *if* $F^j \subset H$, $\ell_H(F^0) < 0$, *and H separates F^0 and G^0;*

e) $|\ell_H(^{G^0}w)|^{a(H)}$ *if* $F^j \subset H$, $\ell_H(F^0) > 0$, *and H does not separate F^0 and G^0;*

f) $|\ell_H(^{G^0}w)|^{a(H)} \exp(-\pi i a(H))$ *if* $F^j \subset H$, $\ell_H(F^0) < 0$, *and H does not separate F^0 and G^0.*

The lemma is a direct corollary of the definitions in (2.5).

2.7.6. Corollary.

$$s_{E(F^0 \geq F^j)}(^{G^0}w)/s_{D(G^0 \geq F^j)}(^{G^0}w) = B'_{F^j}(G^0, F^0).$$

Theorem (2.7.2) is proved.

2.7.7. Corollary of Theorem 2.7.2. *If a configuration $\mathcal{C}$ has no resonances at its dense edges of codimension $j, j-1, \ldots, 1$, then*

$$S : C_j(X, \mathcal{S}^*(a)) \longrightarrow C_j(Q', Q'', \mathcal{S}^*(a))$$

is an isomorphism, see (2.6.7).

Proof. According to Theorem (2.7.2) the matrix of S with respect to the bases $\{E(F^0 \geq F^j, a, Or)\}$, $\{D(F^0 \geq F^j, a, Or)\}$ has a block form. The blocks are numerated by facets F^j. Namely, for a given F^j, the image of any singular cell $E(F^0 \geq F^j, a, Or)$ is a linear combination of singular cells $D(G^0 \geq F^j, a, Or)$ for different G^0. Moreover, the matrix of this block coincides with the matrix of the reduced quantum bilinear form of the configuration C_{F^j}. The determinant of this matrix is given by (2.6.6). If the configuration C has no resonances at edges of codimension $j, j-1, \ldots, 1$, then all these blocks are invertible.

2.7.8. Corollary of Theorems 2.7.2, 2.2.5, 2.3.2, 2.3.5. *If a configuration C has no resonances at its dense edges of codimension $j, j-1, \ldots, 1$ then the natural homomorphism*

$$H_p(Y, S^*(a)) \longrightarrow H_p\Big(\mathbb{C}^k, \bigcup_{H \in C} H_{\mathbb{C}}, S^*(a)\Big)$$

is an isomorphism for $p < j$ and is an epimorphism for $j = p$.

2.7.9. Proof of Lemma 2.5.15. For general values of weights $\{a(H)\}$, the homomorphism S is an isomorphism. Therefore, the boundary operator in $C.(X, S^*(a))$ is uniquely determined by the boundary operator in $C.(Q'/Q'', S^*(a))$ for general values of the weights. More precisely,

$$E(F^0 \geq F^j, a, Or) \stackrel{S}{\mapsto} \sum_{G^0, G^0 \geq F^j} B'_{F^j}(G^0, F^0) \cdot D(G^0 \geq F^j, a, Or)$$

$$\stackrel{d}{\mapsto} \sum_{G^0, G^0 \geq F^j} B'_{F^j}(G^0, F^0) \sum_{F^{j-1}, G^0 \geq F^{j-1} > F^j} \text{ind}\, (F^{j-1} > F^j, Or)$$

$$\cdot D(G^0 \geq F^{j-1}, a, Or) \qquad (2.7.10)$$

$$= \sum_{F^{j-1}, F^{j-1} > F^j} \text{ind}\, (F^{j-1} > F^j, Or) \sum_{G^0, G^0 \geq F^{j-1}} B'_{F^j}(G^0, F^0)$$

$$\cdot D(G^0 \geq F^{j-1}, a, Or).$$

It is easy to see that the reduced quantum bilinear form has the following multiplicative property:

$$B'_{F^j}(G^0, F^0) = B'_{F^{j-1}}(G^0, \pi_{F^j, F^{j-1}}(F^0)) \cdot B(F^0, F^{j-1}; F^j) \qquad (2.7.11)$$

for every F^0, G^0, F^{j-1}, F^j such that $F^0 \geq F^j$, $G^0 \geq F^{j-1} > F^j$.

According to (2.7.11), the right-hand side can be written as

$$\sum_{F^{j-1}, F^{j-1} > F^j} \operatorname{ind}(F^{j-1}, F^j, Or) B(F^0, F^{j-1}; F^j)$$

$$\times \sum_{G^0, G^0 \geq F^{j-1}} B'_{F^{j-1}}(G^0, \pi_{F^j, F^{j-1}}(F^0)) D(G^0 \geq F^{j-1}, a, Or) \qquad (2.7.12)$$

$$= \sum_{F^{j-1}, F^{j-1} > F^j} \operatorname{ind}(F^{j-1}, F^j, Or) B(F^0, F^{j-1}; F^j)$$

$$\times S\left(E(\pi_{F^j, F^{j-1}}(F^0) \geq F^{j-1}, Or)\right).$$

Formulas (2.7.10) and (2.7.12) prove Lemma (2.5.15) for general values of weights $\{a(H)\}$. Formula (2.5.15) analytically depends on $\{a(H)\}$, therefore, it is proved for arbitrary $\{a(H)\}$.

2.8. Action of a Group Preserving Configuration

Let a finite group G act on $\mathbb{R}^k$ by affine transformations. Assume that G preserves a configuration $\mathcal{C}$ in $\mathbb{R}^k$. Then G acts on the complexes $C.(X, S^*(a))$, $C.(Q'/Q'', S^*(a))$ in the following way.

Fix a real affine equation $\ell_H = 0$ for each $H \in \mathcal{C}$. We say that equations are compatible with the action of G if for every $H \in \mathcal{C}$ and $g \in G$, we have

$$g^* \ell_{gH} = \pm \ell_H. \qquad (2.8.1)$$

2.8.2. Lemma. *There exist compatible equations for all $H \in \mathcal{C}$.*

Proof. Set

$$\bar{\ell}_H = \sum_{g \in G} \operatorname{sign}(g^* \ell_{gH} / \ell_H) g^* \ell_{gH},$$

where sign $= 1$ if the number $g^* \ell_{gH} / \ell_H$ is positive, sign $= -1$ otherwise. It is easy to see that $\bar{\ell}_H = 0$ is an affine equation of H and these equations are compatible.

Fix a system of compatible affine equations $\ell_H = 0$ for all $H \in \mathcal{C}$. As G preserves $\mathcal{C}$, G acts on the set of facets of a given dimension, and this action preserves the partial order on the set of all facets of $\mathcal{C}$.

The construction of X, Q', Q'' depends on the choice of a point ^Fw in each facet of $\mathcal{C}$, see (2.1). Choose and fix a point ^Fw in each facet F in such a way that G preserves the set $\{^Fw | F \text{ is a facet of } \mathcal{C}\}$ of chosen points. It is easy to see that such a choice exists.

Now G acts on the sets of cells $\{D(F^i)\}$, $\{D(F^0 \geq F^i)\}$, $\{E(F^0 \geq F^i)\}$.

For every $g \in G$ and $F^0 \geq F^i$,

$$g : D(F^i) \longrightarrow D(gF^i),$$
$$g : D(F^0 \geq F^i) \longrightarrow D(gF^0 \geq gF^i) \tag{2.8.3}$$

are affine diffeomorphisms;

$$g : E(F^0 \geq F^i) \longrightarrow E(gF^0 \geq gF^i)$$

is a piece-wise linear homeomorphism.

Assume that $\mathcal{C}$ is weighted by weights $a = \{a(H) | H \in \mathcal{C}\}$, and G preserves weights of hyperplanes of $\mathcal{C}$. Then G acts on the line bundle $\mathcal{L}(a)$ preserving the connection $d(a)$ (G acts on the base of the line bundle and trivially acts on its fibers), see (2.5). Therefore, G acts on the sheaf $\mathcal{S}(a)$ of its horizontal sections and on the dual sheaf $\mathcal{S}^*(a)$.

In (2.5) for every $F^0 \geq F^i$, we have fixed the sections $s_{D(F^0 \geq F^i)}$, $s_{E(F^0 \geq F^i)}$ over cells $D(F^0 \geq F^i)$, $E(F^0 \geq F^i)$, respectively.

2.8.4. Lemma. *For every $g \in G$, and $F^0 \geq F^i$,*

$$g : s_{D(F^0 \geq F^i)} \mapsto s_{D(gF^0 \geq gF^i)},$$
$$g : s_{E(F^0 \geq F^i)} \mapsto s_{E(gF^0 \geq gF^i)}.$$

The lemma is a corollary of definitions of the sections s_D, s_E, see (2.5.9), (2.5.12).

Now fix a coorientation Or of the configuration $\mathcal{C}$, see (2.4). Consider singular cells $\{D(F^0 \geq F^j, a, Or)\}$, $\{E(F^0 \geq F^j, a, Or)\}$. According to (2.8.3), (2.8.4), we have

2.8.5. Lemma. *For every $g \in G$ and $F^0 \geq F^j$,*

$$g : E(F^0 \geq F^j, a, Or) \mapsto \deg(F^j, g, Or) \, E(gF^0 \geq gF^j, a, Or),$$
$$g : D(F^0 \geq F^j, a, Or) \mapsto \deg(F^j, g, Or) \, D(gF^0 \geq gF^j, a, Or),$$

where deg *is the degree of the affine diffeomorphism* $g : D(F^j) \mapsto D(gF^j)$ *of the oriented cells* $D(F^j), D(gF^j)$.

2.8.6. Remark. The number $\deg(F^j, g, Or)$ may be defined as follows. Let $|F^j|$ be the affine subspace generated by F^j. A coorientation of F^j defines an orientation of $\mathbb{R}^k/|F^j|$. The transformation $g : \mathbb{R}^k \to \mathbb{R}^k$ induces a map $\mathbb{R}^k/|F^j| \to \mathbb{R}^k/|gF^j|$ of oriented spaces. Then $\deg(F^j, g, Or)$ is the degree of this map.

The lemma implies that for any $g \in G$ the matrix of its action on the complexes $C.(Q'/Q'', \mathcal{S}^*(a))$ or $C.(X, \mathcal{S}^*(a))$ in the bases $\{D(F^0 \geq F^j, a, Or)\}$, $\{E(F^0 \geq F^j, a, Or)\}$ has a single non-zero element in each row and in each column, and this element equals ± 1.

2.8.7. Example. Let $\mathcal{C}$ be a central configuration in $\mathbb{R}^k$, and F^k be the single vertex of $\mathcal{C}$. Then for every $F^0 \geq F^k$, $g : E(F^0 \geq F^k, a, Or) \mapsto \deg g \cdot E(gF^0 \geq F^k, a, Or)$, where $\deg g = 1$ if g preserves an orientation of $\mathbb{R}^k$ and $\deg g = -1$ otherwise.

2.8.8. Lemma. *The homomorphism S of the projection on the real part defined in (2.7.2) commutes with the action of G.*

The lemma is obvious.

2.9. Abstract Complexes of a Real Configuration

In this section we axiomatically describe the complexes $C.(X, \mathcal{S}^*(a))$ and $C.(Q'/Q'', \mathcal{S}^*(a))$.

Let $\mathcal{C}$ be a cooriented and weighted configuration of affine hyperplanes in $\mathbb{R}^k$. Its weights are denoted by a, its coorientation is denoted by Or. Define two complexes $\mathcal{X}.(\mathcal{C})$ and $Q.(\mathcal{C})$ and a homomorphism of complexes $S : \mathcal{X}.(\mathcal{C}) \to Q.(\mathcal{C})$.

Complex $\mathcal{X}.(\mathcal{C})$. Define a complex vector space $\mathcal{X}_j(\mathcal{C})$, $0 \leq j \leq k$, as the space of complex linear combinations of symbols $(F^0 \geq F^j)$, where $F^0 \geq F^j$ runs through all pairs of adjacent facets of $\mathcal{C}$ of codimension 0 and j, respectively. Define a differential

$$d : \mathcal{X}_j(\mathcal{C}) \longrightarrow \mathcal{X}_{j-1}(\mathcal{C})$$

by the formula

$$d : (F^0 \geq F^j) \mapsto \sum_{F^{j-1} > F^j} \text{ind}\,(F^{j-1} > F^j) \qquad (2.9.1)$$
$$\cdot B(F^0, F^{j-1}; F^j) \cdot (\pi_{F^j, F^{j-1}}(F^0) \geq F^{j-1})\,,$$

where the sum is taken over all facets F^{j-1} of codimension $j-1$ adjacent to F^j, the number $\text{ind}\,(F^{j-1} > F^j)$ is defined in (2.4.2), the number $B(F^0, F^{j-1}; F^j)$ is defined in (2.5.14), the facet $\pi_{F^j, F^{j-1}}(F^0)$ is defined in (2.1.13).

Complex $Q.(\mathcal{C})$. Define a complex vector space $Q_j(\mathcal{C})$, $0 \leq j \leq k$, as the space dual to $\mathcal{X}_j(\mathcal{C})$. The basis of $Q_j(\mathcal{C})$ dual to the basis $\{(F^0 \geq F^j)\}$ of $\mathcal{X}_j(\mathcal{C})$ is denoted by $\{(F^0 \geq F^j)^*\}$. Define a differential

$$d : Q_j(\mathcal{C}) \longrightarrow Q_{j-1}(\mathcal{C})$$

by the formula

$$d : (F^0 \geq F^j)^* \mapsto \sum_{F^0 \geq F^{j-1} > F^j} \text{ind}\,(F^{j-1} > F^j) \cdot (F^0, F^{j-1})^*, \qquad (2.9.2)$$

where the sum is taken over all facets F^{j-1}, $F^0 \geq F^{j-1} > F^j$, the number $\text{ind}\,(F^{j-1} > F^j)$ is defined in (2.4.2).

Define a symmetric bilinear form S on $\mathcal{X}_j(\mathcal{C})$, $j = 0, \ldots, k$, by the formulas

For $j = 0$, $S((F^0 \geq F^0), (G^0 \geq G^0))$ equals 1 if $F^0 = G^0$
and equals 0 otherwise. $\qquad (2.9.3)$

For $j > 0$, $S((F^0 \geq F^j), (G^0 \geq G^j)) = 0$ if $F^j \neq G^j$ and
$$S((F^0 \geq F^j), (G^0 \geq F^j)) = B'_{F^j}(F^0, G^0)\,, \qquad (2.9.4)$$

where the right-hand side is defined before (2.7.2).

2.9.5. Theorem. *The spaces $\mathcal{X}_j(\mathcal{C})$, $0 \leq j \leq k$, and the linear maps d defined in (2.9.1) form a complex. The spaces $Q_j(\mathcal{C})$, $0 \leq j \leq k$, and the linear maps d defined in (2.9.2) form a complex. The bilinear forms S, defined in (2.9.3) and (2.9.4), define a homomorphism of complexes*

$$S : \mathcal{X}.(\mathcal{C}) \longrightarrow Q.(\mathcal{C})\,.$$

Proof. The map sending $(F^0 \geq F^j)$ to $E(F^0 \geq F^j, a, Or)$ and $(F^0 \geq F^j)^*$ to $D(F^0 \geq F^j, a, Or)$ identifies the complex $\mathcal{X}.(\mathcal{C})$ with the complex $C.(X, \mathcal{S}^*(a))$, $Q.(\mathcal{C})$ with $C.(Q'/Q'', \mathcal{S}^*(a))$, and homomorphism (2.7.1) of the projection on the real part with the homomorphism $S : \mathcal{X}.(\mathcal{C}) \to Q.(\mathcal{C})$.

The complexes $C.(X, \mathcal{S}^*(a))$, $C.(Q'/Q'', \mathcal{S}^*(a))$, the homomorphism (2.7.1) of the projection on the real part will be called a *realization* of the complexes $\mathcal{X}.(\mathcal{C})$, $Q.(\mathcal{C})$ and the homomorphism S of (2.9.5), respectively. (2.9.6)

A realization is uniquely defined by a choice of the points $^F w \in F$, where F runs through the set of all facets of $\mathcal{C}$, see (2.1), and by a choice of affine equations for hyperplanes of the configuration, see (2.5).

2.9.7. Theorem. *Any realization w defines a monomorphism of complexes*

$$w : \mathcal{X}.(\mathcal{C}) \hookrightarrow C.\Big(\mathbb{C}^k - \bigcup_{H \in \mathcal{C}} H_\mathbb{C}, \mathcal{S}^*(a)\Big),$$
$$w : Q.(\mathcal{C}) \hookrightarrow C_!.\Big(\mathbb{C}^k - \bigcup_{H \in \mathcal{C}} H_\mathbb{C}, \mathcal{S}^*(a)\Big).$$
(2.9.7.1)

The following diagram is commutative:

$$\begin{array}{ccc} \mathcal{X}.(\mathcal{C}) & \xrightarrow{S} & Q.(\mathcal{C}) \\ {\scriptstyle w}\downarrow & & \downarrow {\scriptstyle w} \\ C.\Big(\mathbb{C}^k - \bigcup_{H \in \mathcal{C}} H_\mathbb{C}, \mathcal{S}^*(a)\Big) & \longrightarrow & C_!.\Big(\mathbb{C}^k - \bigcup_{H \in \mathcal{C}} H_\mathbb{C}, \mathcal{S}^*(a)\Big) \end{array}$$

where the lower horizontal homomorphism is the canonical homomorphism. The monomorphisms w induce isomorphisms of homology groups:

$$w : H.\mathcal{X}.(\mathcal{C}) \longrightarrow H.\Big(\mathbb{C}^k - \bigcup_{H \in \mathcal{C}} H_\mathbb{C}, \mathcal{S}^*(a)\Big),$$
$$w : H.Q.(\mathcal{C}) \longrightarrow H.\Big(\mathbb{C}^k - \bigcup_{H \in \mathcal{C}} H_\mathbb{C}, \mathcal{S}^*(a)\Big).$$

The isomorphisms of homology groups do not depend on the choice of a realization w.

The theorem is a direct corollary of the constructions and results of Sec. 2.

2.9.8. Remark. It would be interesting to find a topological meaning for the subcomplex $S(\mathcal{X}.(\mathcal{C})) \subset \mathcal{Q}.(\mathcal{C})$.

2.9.9. Remark. Assume that a configuration $\mathcal{C}$ continuously depends on a parameter $t \in [0,1]$, we assume that the number of hyperplanes of $\mathcal{C}_t$ does not depend on t but their positions continuously depend on t. Assume that the pairs $(\mathbb{R}^k, \mathcal{C}_t)$, $t \in [0,1]$, are isotopic, that is, there exists a homeomorphism $h_t : \mathbb{R}^k \to \mathbb{R}^k$ continuously depending on t such that h_t sends hyperplanes of $\mathcal{C}_0$ onto hyperplanes of $\mathcal{C}_t$ and intersections of hyperplanes of $\mathcal{C}_0$ onto the corresponding intersections of hyperplanes of $\mathcal{C}_t$. In particular, this means that the set of facets of $\mathcal{C}_0$ is identified with the set of facets of $\mathcal{C}_t$, and, moreover, this identification preserves the order on the set of facets. Assume that configurations $\mathcal{C}_t$ are weighted and cooriented, and the weights and the coorientation of $\mathcal{C}_t$ do not depend on t. Then the complexes $\mathcal{X}.(\mathcal{C}_0)$, $\mathcal{Q}.(\mathcal{C}_0)$ and the homomorphism $S : \mathcal{X}.(\mathcal{C}_0) \to \mathcal{Q}.(\mathcal{C}_0)$ are canonically identified with $\mathcal{X}.(\mathcal{C}_t)$, $\mathcal{Q}.(\mathcal{C}_t)$, $S : \mathcal{X}.(\mathcal{C}_t) \to \mathcal{Q}.(\mathcal{C}_t)$, respectively.

Let a group G act on $\mathbb{R}^k$ by affine transformations. Assume that G preserves a weighted configuration $\mathcal{C}$ in $\mathbb{R}^k$. Assume that $\mathcal{C}$ is cooriented. Define an action of G on $\mathcal{X}.(\mathcal{C})$, $\mathcal{Q}.(\mathcal{C})$ by the formula

$$\begin{aligned} g : (F^0 \geq F^j) &\mapsto \deg(F^j, g, Or)(gF^0 \geq gF^j), \\ g : (F^0 \geq F^j)^* &\mapsto \deg(F^j, g, Or)(gF^0 \geq gF^j)^*, \end{aligned} \quad (2.9.10)$$

where the number $\deg(F^j, g, Or)$ is defined in (2.8.6).

(2.9.11) The formula (2.9.10) defines an action of G on $\mathcal{X}.(\mathcal{C})$, $\mathcal{Q}.(\mathcal{C})$. This action commutes with the homomorphism S.

Let
$$\begin{aligned} w : \mathcal{X}.(\mathcal{C}) &\longrightarrow C.(X, \mathcal{S}^*(a)), \\ w : \mathcal{Q}.(\mathcal{C}) &\longrightarrow C.(Q'/Q'', \mathcal{S}^*(a)) \end{aligned} \quad (2.9.12)$$

be a realization such that the set of points $\{{}^F w | F \text{ is a facet of } \mathcal{C}\}$ is invariant under the action of G and the system of affine equations for hyperplanes of $\mathcal{C}$ is compatible with the action of G, see (2.8). Then G acts on $C.(X, \mathcal{S}^*(a))$, $C.(Q'/Q'', \mathcal{S}^*(a))$ as described in (2.8.5).

(2.9.13) A realization w is an isomorphism of complexes of G-modules.

These statements follow from (2.8).

3

Construction of Homology Complexes for Discriminantal Configuration

3.1. Discriminantal Configuration

Let $x_1, \ldots, x_k$ be coordinates in $\mathbb{R}^k$, and let $z = (z_1, \ldots, z_n)$, $z_1 < \cdots < z_n$, $n \geq 0$, be a set of real numbers. Consider in $\mathbb{R}^k$ a configuration of hyperplanes given by the equations:

$$\begin{aligned} x_i &= x_j, \quad 1 \leq i < j \leq k, \\ x_i &= z_j, \quad i = 1, \ldots, k, j = 1, \ldots, n. \end{aligned} \qquad (3.1.1)$$

The configuration is denoted by $\mathcal{C} = \mathcal{C}_{n,k}(z)$ and called *discriminantal*. See an example in Fig. 3.1.

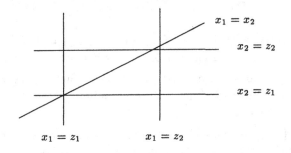

Fig. 3.1

In this section we apply the construction of Sec. 2 to the configuration $C_{n,k}(z)$ and describe the complexes $\mathcal{X}.$, $Q.$ for this configuration.

To apply the construction of Sec. 2 it is necessary to choose a point $^F w \in F$ for any facet F of the configuration. This will be done in the next four sections.

3.2. Facets of Discriminantal Configuration

Consider a set S consisting of the symbols $x_1, \ldots, x_k, z_1, \ldots, z_n$. *Admissible decomposition* of S is any decomposition

$$S = \bigcup_{\alpha \in A} S_\alpha$$

into a union of disjoint non-empty subsets $\{S_\alpha\}$ such that for any subset S_α the number of elements of $S_\alpha \cap \{z_1, \ldots, z_n\}$ is not greater than 1. Admissible decomposition is *ordered* if an order on A is given with the following property. For any $i, j, 1 \leq i < j \leq k$, if $z_i \in S_\alpha$, $z_j \in S_\beta$ then $\alpha < \beta$.

Facets of C are numerated by ordered admissible decompositions $S_A = \{S_\alpha\}_{\alpha \in A}$. The facet $F(S_A)$ corresponding to S_A consists of all $(x_1, \ldots, x_k) \in \mathbb{R}^k$ such that all numbers lying in the same subset S_α of the decomposition are equal, and $x_i < x_j$ for every $x_i \in S_\alpha$, $x_j \in S_\beta$, $\alpha < \beta$.

3.2.1. Example. For the admissible ordered decomposition

$$S_A : \{x_3, x_2\} < \{x_1, z_1\} < \{x_4\} < \{z_2\}$$

of the set $\{x_1, \ldots, x_4, z_1, z_2\}$,

$$F(S_A) = \{(x_1, \ldots, x_4) \in \mathbb{R}^4 | x_3 = x_2 < x_1 = z_1 < x_4 < z_2\}.$$

A subset S_α is called *moving* if $S_\alpha \cap \{z_1, \ldots, z_n\} = \emptyset$. The dimension of the facet $F(S_A)$ equals the number of the moving subsets of S_A.

For example, the facets of dimension k are numerated by the admissible ordered decompositions S_A such that each subset S_α of the decomposition consists of one element.

Let $S_A = \{S_\alpha\}_{\alpha \in A}$, $S_B = \{S_\beta\}_{\beta \in B}$ be two admissible ordered decompositions. S_B is *adjacent* to S_A if the following conditions hold:

For any $\alpha \in A$, there exist $\beta = \beta(\alpha) \in B$ such that $S_\alpha \subset S_\beta$. (3.2.2)

If $\alpha_1, \alpha_2 \in A$, $\alpha_1 < \alpha_2$, then $\beta(\alpha_1) \leq \beta(\alpha_2)$. (3.2.3)

The notation: $S_B \leq S_A$.

3.2.4. Lemma. *A facet $F(S_B)$ lies in the boundary of a facet $F(S_A)$ iff S_B is adjacent to S_A:*
$$F(S_B) \leq F(S_A) \iff S_B \leq S_A.$$

The lemma is obvious.

3.2.5. Example. The 2-dimensional facet $F(S_A)$ of the Example (3.2.1) lies in the boundary of the four 4-dimensional facets $F(S_B)$, $F(S_C)$, where $S_B : \{x_3\} < \{x_2\} < \{z_1\} < \{x_1\} < \{x_4\} < \{z_2\}$, $S_C : \{x_2\} < \{x_3\} < \{z_1\} < \{x_1\} < \{x_4\} < \{z_2\}$.

3.3. Centers of Top-Dimensional Facets

We give a construction of a point $^F w$ in any facet F of a configuration $C_{n,k}(z)$. This point $^F w$ will be called the *center* of a facet F.

3.3.1. Equipment. A set of real numbers T_j^i, $j = 0, \ldots, n$, $i = 1, \ldots, k$, such that

$$T_0^1 < T_0^2 < \cdots < T_0^k < z_1 < T_1^1 < \cdots < T_1^k < z_2 < \cdots$$
$$< z_j < T_j^1 < \cdots < T_j^k < z_{j+1} < \cdots < z_n < T_n^1 < \cdots < T_n^k,$$

will be called an *equipment* of the configuration $C_{n,k}(z)$.

Remark. Any two equipments $\{T_j^i(0)\}$, $\{T_j^i(1)\}$ are connected by a canonical one-parameter family of equipments $\{T_j^i(s)\}$, where

$$s \in [0,1] \quad \text{and} \quad T_j^i(s) := (1-s)T_j^i(0) + sT_j^i(1).$$

Any facet of codimension 0 has the form $F = (S_A)$, where $S_A = \{S_\alpha\}_{\alpha \in A}$ is an admissible ordered decomposition, and each S_α consists of one element. Set

$$^F w = (T_{v_1}^{\sigma_1}, \ldots, T_{v_k}^{\sigma_k}), \tag{3.3.2}$$

where $\sigma_1, \ldots, \sigma_k$, $v_1, \ldots, v_k$ are defined as follows. Let $x_i \in S_\alpha$ then $\sigma_i - 1$ is the number of the indexes β such that $\beta < \alpha$ and $S_\beta \cap \{x_1, \ldots, x_k\} \neq 0$, v_i is the number of the indexes β such that $\beta < \alpha$ and $S_\beta \cap \{z_1, \ldots, z_n\} \neq 0$.

3.3.3. Example. For the facet $F(S_A)$,

$$S_A : \{x_3\} < \{x_2\} < \{z_1\} < \{x_1\} < \{x_4\} < \{z_2\},$$

we have $^F w = (T_1^3, T_0^2, T_0^1, T_1^4)$.

3.4. Basic Polytopes

Fix real numbers $T_0^1 < \cdots < T_0^k$. Let $T = (T_0^1, \ldots, T_0^k)$. For any permutation $\sigma \in \Sigma_k$, let $\sigma T = (T_0^{\sigma(1)}, \ldots, T_0^{\sigma(k)})$. Let $\Delta(T)$ be the convex hull in $\mathbb{R}^k$ of the set $\Sigma T = \{\sigma T \mid \sigma \in \Sigma_k\}$.

The polytope $\Delta(T)$ will be called a *basic polytope of the first type*. It is also called a permutahedra, see [**Ka**].

3.4.1. Lemma. $\Delta(T)$ *is a $(k-1)$-dimensional polytope. The set ΣT forms the set of its vertices.*

Proof. The first statement is obvious. The set ΣT is invariant under the action of Σ. This implies the second statement.

Fix real numbers $T_0^1 < \cdots < T_0^k < T_1^1 < \cdots < T_1^k$. For any $i = 0, 1, \ldots, k$, set
$$T_i = (T_0^1, \ldots, T_0^i, T_1^{i+1}, \ldots, T_1^k).$$
For any $\sigma \in \Sigma_k$, let σT_i be the image of T_i under the permutation σ of the coordinates in $\mathbb{R}^k$.

Let $\Sigma T = \{\sigma T_i \mid \sigma \in \Sigma_k, i = 0, \ldots, k\}$, let $\Delta(T)$ be the convex hull in $\mathbb{R}^k$ of the set ΣT.

The polytope $\Delta(T)$ will be called a *basic polytope of the second type*.

3.4.2. Lemma. $\Delta(T)$ *is a k-dimensional polytope. The set ΣT forms the set of its vertices.*

Proof. The first statement is obvious. The second statement is proved by induction on p. In fact, for any point $x \in \Sigma T$, there exists a hyperplane H containing x such that all points of ΣT lie on one side of H, and the set $\Sigma T \cap H$ generates a basic polytope of dimension $k-1$. For example, for the point T_i, the hyperplane $x_1 = T_0^1$ has such properties.

3.5. Centers of Facets of Arbitrary Codimension

Given a facet $F = F(S_A)$ of dimension i, let X_F be the set of centers of all k-dimensional facets G of the configuration $C_{n,k}(z)$ such that $F \leq G$. Let $\Delta(F)$ be the convex hull in $\mathbb{R}^k$ of the set X_F.

3.5.1. Lemma. $\Delta(F)$ *is a polytope of codimension i. The set X_F forms the set of its vertices. The intersection $\Delta(F) \cap F$ is non-empty and consists of one* *point.*

This point will be called the *center* of F and denoted by $^F w$.

Proof. It is easy to see that $\Delta(F)$ is a direct product of basic polytopes of the first and second types. The factors are numerated by the sets $\{S_\alpha\}$ of the decomposition S_A. Any moving set S_α consisting of p elements gives a $(p-1)$-dimensional basic polytope of the first type. Any non-moving set S_α consisting of p elements gives a $(p-1)$-dimensional basic polytope of the second type. Now the first two statements of Lemma 3.5.1 follow from Lemmas 3.4.1 and 3.4.2.

For example, for the facet $F(S_A)$ of Example 3.2.1, X_F consists of four points and
$$\begin{aligned}\Delta(F) &= [T_0^3, T_1^3] \times \{(x_2, x_3)|x_2 + x_3 \\ &= T_0^1 + T_0^2, T_0^1 < x_2 < T_0^2\} \times T_1^4 \subset \mathbb{R}^4 \,.\end{aligned} \quad (3.5.2)$$

$\Delta(F)$ lies in the $(k-i)$-dimensional subspace given by the following system of equations. The equations are numerated by the moving subsets of the decomposition S_A. The equation corresponding to the moving subset S_α has the form
$$\sum_{j \in S_\alpha} x_j = \sum_{i=\sigma(\alpha)+1}^{\sigma(\alpha)+\#S_\alpha} T_{v(\alpha)}^i\,,$$
where $\#S_\alpha$ is the number of elements of S_α, $v(\alpha) = \#(\{z_1, \ldots, z_n\} \cap \bigcup_{\beta < \alpha} S_\beta)$, $\sigma(\alpha) = \#(\{x_1, \ldots, x_k\} \cap \bigcup_{\beta < \alpha} S_\beta)$, see (3.5.2) as an example. This remark proves the third statement of Lemma 3.5.1.

3.5.3. Example. For the configuration shown in Fig. 3.1, the centers of edges are shown in Fig. 3.2. Some of the centers are marked by the letters $a, b, \ldots, m$. For example, $a = (T_1^2, T_1^1)$, $k = (T_1^1, T_1^2)$, $c = (T_2^2, T_1^1)$, $e = (T_2^2, T_2^1)$.

3.6. Cells of the Construction of Sec. 2 are Convex Polytopes

Apply the construction of Sec. 2 to $C_{n,k}(z)$, choosing the centers of facets as in (3.3)–(3.5).

3.6.1. Lemma. *For a given facet F, the cell $D(F)$ defined in Lemma (2.1.4) coincides with the convex polytope $\Delta(F)$ defined in (3.5).*

Lemma (3.6.1) is a direct corollary of the definition of $D(F)$ and Lemma 3.5.1.

44 Construction of Homology Complexes for Discriminantal Configuration

Consider the cellular complex Q constructed in (2.1.6). According to Lemma 3.6.1 its cells are convex polytopes $\{\Delta(F)\}$. Each of these polytopes is a direct product of basic polytopes of the first and second types.

3.6.2. Example. For the configuration in Fig. 3.1, the complex Q is shown in Fig. 3.3.

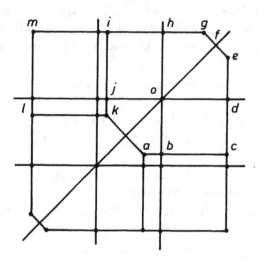

Fig. 3.2

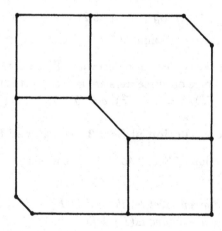

Fig. 3.3

Let X be the cellular complex of $\mathcal{C}_{n,k}(z)$ defined in (2.2.3). Consider its arbitrary p-cell $E(F^0 \geq F^p)$ and the deformation of the cell, $R_t : E(F^0 \geq F^p) \to \mathbb{C}^k$, $t \in [0,1]$, defined in (2.2.6).

3.6.3. Corollary of Lemma 3.6.1. *The set $\bigcup_{t \in [0,1]} R_t E(F^0 \geq F^p)$ lies in a p-dimensional complex affine subspace of $\mathbb{C}^k$.*

3.7. Description of Basic Polytopes by Inequalities

3.7.1. Lemma. *Consider a $(k-1)$-dimensional basic polytope $\Delta(T)$ of the first type defined in (3.4). Then $\Delta(T)$ lies in the hyperplane*

$$x_1 + \cdots + x_k = T_0^1 + \cdots + T_0^k. \tag{3.7.1.1}$$

Its $(k-2)$-dimensional facets are numerated by all decompositions $S_1 \cup S_2 = \{x_1, \ldots, x_k\}$. The corresponding facet lies in the subspace given by the equations

$$\begin{aligned}
\sum_{x_i \in S_1} x_i &= T_0^1 + \cdots + T_0^{\#S_1}, \\
\sum_{x_i \in S_2} x_i &= T_0^{\#S_1+1} + \cdots + T_0^k.
\end{aligned} \tag{3.7.1.2}$$

The polytope $\Delta(T)$ is determined by the inequalities

$$\begin{aligned}
\sum_{x_i \in S_1} x_i &\geq T_0^1 + \cdots + T_0^{\#S_1}, \\
\sum_{x_i \in S_2} x_i &\leq T_0^{\#S_1+1} + \cdots + T_0^k.
\end{aligned} \tag{3.7.1.3}$$

To prove Lemma 3.7.1 consider the configuration $\mathcal{C}$ of all the diagonal hyperplanes in $\mathbb{R}^k$. This configuration is the special case of the discriminantal configuration $\mathcal{C}_{n,k}(z_1, \ldots, z_n)$ for $n = 0$. Fix an equipment $T_0^1 < \cdots < T_0^k$ and apply results of Secs. 2 and 3.

Let F^{k-1} be the only facet of codimension $k-1$. Consider the $(k-1)$-cell $D(F^{k-1})$ constructed in (2.1.4). According to Lemma 3.6.1, $D(F^{k-1})$ coincides with the basic polytope $\Delta(T)$ of the first type. According to (2.1)

$$\partial D(F^{k-1}) = \bigcup_{F^{k-2}} D(F^{k-2}),$$

where F^{k-2} is any facet of codimension $k-2$. Knowing the equations of the polytopes $D(F^{k-2})$ we get the equations of $(k-2)$-dimensional facets of $D(F^{k-1})$. This proves Lemma 3.7.1.

3.7.2. Lemma. *Consider a k-dimensional basic polytope $\Delta(T)$ of the second type defined in (3.4). Then its $(k-1)$-dimensional facets are numerated by all decompositions $S_1 \cup S_2 = \{x_1, \ldots, x_k, z_1\}$. The corresponding facet lies in the subspace given by the equation*

$$\sum_{x_i \in S_1} x_i = T_0^1 + \cdots + T_0^{\#S_1} \qquad (3.7.2.1)$$

if $z_1 \in S_2$, and given by the equation

$$\sum_{x_i \in S_2} x_i = T_1^{\#S_1+1} + \cdots + T_1^k \qquad (3.7.2.2)$$

if $z_1 \in S_1$. The polytope $\Delta(T)$ is determined by the inequalities numerated by decompositions $S_1 \cup S_2 = \{x_1, \ldots, x_k, z_1\}$ and having the form

$$\sum_{x_i \in S_1} x_i \geq T_0^1 + \cdots + T_0^{\#S_1} \qquad (3.7.2.3)$$

if $z_1 \in S_2$ and having the form

$$\sum_{x_i \in S_2} x_i \leq T_1^{\#S_1+1} + \cdots + T_1^k \qquad (3.7.2.4)$$

otherwise.

To prove Lemma 3.7.2 consider the configuration $\mathcal{C}(z_1)$. This is a special case of the configuration $\mathcal{C}(z)$ defined in (3.1) corresponding to $n=1$.

Let F^k be the only facet of condimension k. Consider the k-cell $D(F^k)$ constructed in (2.1.4). Then $D(F^k)$ coincides with the basic polytope $\Delta(T)$ of the second type. Applying the formula $\partial D(F^k) = \bigcup_{F^{k-1}} D(F^{k-1})$, where F^{k-1} is any facet of codimension $k-1$, we get the lemma.

3.8. Admissible Monomials

Let xz be the set $\{x_1, \ldots, x_k, z_1, \ldots, z_n\}$. The permutation group Σ_{k+n} acts on xz. A permutation σxz, $\sigma \in \Sigma_{k+n}$ is called *admissible* if for every i, j, $1 \leq i < j \leq n$, the symbol z_i appears earlier in σxz than the symbol z_j.

Let $k + n = q(1) + \cdots + q(p)$ be a decomposition of the number $k + n$ into a sum of non-negative integers. An *admissible monomial* for the configuration $\mathcal{E}_{n,k}(z)$ or an admissible monomial for the symbols $x_1, \ldots, x_k, z_1, \ldots, z_n$ is any expression of the form

$$f = f_1(1)f_2(1) \ldots f_{q(1)}(1) \mid f_1(2)f_2(2) \\ \ldots f_{q(2)}(2) \mid \cdots \mid f_1(p) \ldots f_{q(p)}(p), \quad (3.8.1)$$

such that

(3.8.2) $f_1(1) \ldots f_{q(1)}(1) f_1(2) \ldots f_{q(2)}(2) \ldots f_1(p) \ldots f_{q(p)}(p)$, is an admissible permutation of xz and for any $i = 1, \ldots, p$, $\#(\{f_1(i) \ldots f_{q(i)}(i)\} \cap \{z_1, \ldots, z_n\}) \leq 1$,

The set of all admissible monomials will be denoted by $\mathrm{Adm}\,(x; z) = \mathrm{Adm}\,(x_1, \ldots, x_k; z_1, \ldots, z_n)$. The monomial $f(i) = f_1(i)f_2(i) \ldots f_{q(i)}(i)$ will be called the *i-the factor* of f. The monomial $f(i)$ will be called an *algebra factor* if $\{f_1(i), \ldots, f_{q(i)}(i)\} \cap \{z_1, \ldots, z_n\} = 0$ and a *module factor* otherwise. Set

$$A(f) = \{i \mid f(i) \text{ is an algebra factor}\}, \\ M(f) = \{i \mid f(i) \text{ is an module factor}\}. \quad (3.8.3)$$

3.8.4. Remark. Let A be the free algebra generated by symbols $x_1, \ldots, x_k$, and M_j be its two-sided Verma module generated by one element z_j. Then an admissible monomial can be viewed as an element $f(1) \otimes \ldots \otimes f(p)$ of the tensor product $X_1 \otimes \cdots \otimes X_p$, where $X_i = A$ if $f(i)$ is an algebra factor and $X_i = M_j$ if $f(i)$ contains z_j.

An admissible monomial f' is *adjacent* to an admissible monomial f, $f' < f$, if f' may be obtained from f by an insertion of one vertical line. Define the *index* $\mathrm{ind}\,(f' < f)$ as $(-1)^N$, where N is the number of factors of f' lying on the right of the inserted line.

For example, for $x_3 x_4 \mid z_1 x_1 \mid x_2 < x_3 x_4 z_1 x_1 \mid x_2$, the index of these monomials equals 1.

An admissible monomial f' is weakly adjacent to an admissible monomial f, $f' <_w f$, if f' may be obtained from f by the following procedure.

Choose a factor $f(j)$ of f. Let $I_1 \cup I_2 = \{1, \ldots, q(j)\}$ be a decomposition into a sum of two non-empty subsets $I_1 = \{s_1 < \cdots < s_a\}$, $I_2 = \{t_1 < \cdots < t_b\}$, $a + b = q(j)$. Set

$$f' = f_1(1)\ldots f_{q(1)}(1) \mid \cdots \mid f_1(j-1)\ldots f_{q(j-1)}(j-1) \mid f_{s_1}(j)f_{s_2}(j)$$
$$\ldots f_{s_a}(j) \mid f_{t_1}(j)f_{t_2}(j)\ldots f_{t_b}(j) \mid f_1(j+1)\ldots f_{q(t+1)}(j+1) \mid \quad (3.8.5)$$
$$\cdots \mid f_1(p)\ldots f_{q(p)}(p) \mid .$$

The monomials f, f' have the same first $j-1$ factors and the last $p-j$ factors.

Define the *index* $\mathrm{ind}\,(f' <_w f)$ as $(-1)^N$, where N is the number of factors of f' lying on the right of the new vertical line in between $f_{s_a}(j)$ and $f_{t_1}(j)$.

3.8.6. Examples. The pairs $x_3z_1x_1x_2 >_w z_1x_2|x_3x_1$, $x_3z_1x_1x_2 >_w x_3x_1|z_1x_2$, $x_3z_1x_1x_2|x_4 >_w x_3x_2|z_1x_1|x_4$ are weakly adjacent. Their indexes are -1, -1, 1, respectively.

For each admissible monomial f, define the numbers $v(1, f), \ldots, v(p, f)$, $\sigma(1, f), \ldots, \sigma(p, f)$ as follows:

$$v(i, f) = \#\left(\{z_1, \ldots, z_n\} \cap \{f_1(1), \ldots, f_{q(1)}(1), f_1(2), \ldots, \right.$$
$$\left. f_1(i-1), \ldots, f_{q(i-1)}(i-1)\}\right),$$
$$\sigma(i, f) = \#\left(\{x_1, \ldots, x_k\} \cap \{f_1(1), \ldots, f_{q(1)}(1), f_1(2), \ldots, \right. \quad (3.8.7)$$
$$\left. f_1(i-1), \ldots, f_{q(i-1)}(i-1)\}\right).$$

3.8.8. Lemma. *The set of all pairs $F^0 \geq F^i$ of facets of a discriminantal configuration $C_{n,k}(z)$ is in one-to-one correspondence with the set of all admissible monomials of the form (3.8.1). The pair $F^0 \geq F^i$ corresponding to f is given by the formulas $F^0 = F(S_A)$, $F^i = F(S_B)$, where S_A, S_B are the following admissible ordered decompositions,*

$$S_A : \{f_1(1)\} < \{f_2(1)\} < \cdots < \{f_{q(1)}(1)\} < \{f_1(2)\} < \cdots < \{f_{q(2)}(2)\} <$$
$$\cdots < \{f_{q(p)}(p)\},$$

$$S_B : \{f_1(1), \ldots, f_{q(1)}(1)\} < \{f_1(2), \ldots, f_{q(2)}(2)\} < \cdots < \{f_1(p),$$
$$\ldots, f_{q(p)}(p)\}.$$

The dimension $k-i$ of the facet F^i is the number of algebraic factors in f.

Lemma 3.8.8 is a direct corollary of Lemma 3.2.4.

3.9. Equalities and Inequalities Defining Cells

In (2.1) a pair of cellular complexes $Q'' \subset Q'$ was constructed. We describe below the cells of $Q' - Q''$. According to (2.1), $Q' - Q''$ consists of the cells $D(F^0 \geq F^i)$. According to (3.8), pairs $F^0 \geq F^i$ are numerated by admissible monomials of the form (3.8.1). Denote by $D(f)$ the cell $D(F^0 \geq F^i)$ if f numerates $F^0 \geq F^i$. According to (3.8.8), the codimension of the polytope $D(f)$ in $\mathbb{R}^k$ is the number of algebraic factors of f. According to (2.1.11), boundary of the cell $D(f)$ in Q'/Q'' is given by

$$\partial D(f) = \bigcup_{f' \leq f} D(f'), \qquad (3.9.1)$$

where the sum is taken over all monomials f' adjacent to f.

3.9.2. Example. Consider the configuration shown in Fig. 3.2. Then $D(z_1|z_2\,x_2x_1)$ is the polygon $defo$,

$$D(z_1 \mid x_2z_2x_1) = dcbo,$$
$$D(z_1 \mid z_2 \mid x_2x_1) = fe,$$
$$D(z_1 \mid z_2x_2 \mid x_1) = ed,$$
$$D(z_1 \mid x_2z_2 \mid x_1) = dc,$$
$$D(z_1 \mid x_2 \mid z_2x_1) = bc,$$
$$\partial D(z_1 \mid z_2x_2x_1) = D(z_1 \mid z_2 \mid x_2x_1) \cup D(z_1 \mid z_2x_2 \mid x_1),$$
$$\partial D(z_1 \mid x_2z_2x_1) = D(z_1 \mid x_2 \mid z_2x_1) \cup D(z_1 \mid x_2z_2|x_1)$$

in $(Q'(\mathcal{C})/Q''(\mathcal{C}))$.

Let f be an admissible monomial of the form (3.8.1). We describe the polytope $D(f)$ by equations and inequalities.

$D(f)$ generates an affine subspace $|D(f)|$ determined by the following equations. The equations are numerated by algebraic factors of f. The equation corresponding to an algebraic factor $f(i)$ has the form

$$f_1(i) + \cdots + f_{q(i)}(i) = T_{v(i,f)}^{\sigma(i,f)+1} + \cdots + T_{v(i,f)}^{\sigma(i,f)+q(i)}. \qquad (3.9.3)$$

The polytope $D(f)$ is determined in the affine subspace $\mid D(f) \mid$ by the system of inequalities

$$f_1(1) < f_2(1) < \cdots < f_{q(1)}(1) < f_1(2) < \cdots < f_{q(2)}(2) < \cdots < f_{q(p)}(p) \quad (3.9.4)$$

and by the following system of inequalities numerated by all the monomials f' adjacent to f. Let a monomial f' be obtained from f by an insertion of a new vertical line into the i-th factor of the monomial f:

$$\cdots \mid f_1(i) \ldots f_{q(i)}(i) \mid \cdots \mapsto \cdots \mid f_1(i) \ldots f_{s-1}(i) \mid f_s(i) \ldots f_{q(i)}(i) \mid \cdots$$

There are three possibilities:

(3.9.5) the i-th factor is an algebraic factor,

(3.9.6) the i-th factor contains a symbol z_j for some $j \in \{1,\ldots,n\}$ and z_j lies on the right of $f_{s-1}(i)$,

(3.9.7) the i-th factor contains a symbol z_j for some j and z_j lies on the left of $f_s(i)$.

The inequality corresponding to (3.9.5) has the form

$$f_1(i) + \cdots + f_{s-1}(i) > T_{v(i,f)}^{\sigma(i,f)+1} + \cdots + T_{v(i,f)}^{\sigma(i,f)+s-1}$$
$$\text{or} \quad f_s(i) + \cdots + f_{q(i)}(i) < T_{v(i,f)}^{\sigma(i,f)+s} + \cdots + T_{v(i,f)}^{\sigma(i,f)+q(i)}. \tag{3.9.8}$$

The inequality corresponding to (3.9.6) has the form

$$f_1(i) + \cdots + f_{s-1}(i) > T_{v(i,f)}^{\sigma(i,f)+1} + \cdots + T_{v(i,f)}^{\sigma(i,f)+s-1}. \tag{3.9.9}$$

The inequality corresponding to (3.9.7) has the form

$$f_s(i) + \cdots + f_{q(i)}(i) < T_{v(i,f)+1}^{\sigma(i,f)+s} + \cdots + T_{v(i,f)+1}^{\sigma(i,f)+q(i)}. \tag{3.9.10}$$

The top dimensional facets of $D(f)$ lying in $Q' - Q''$ are numerated by inequalities (3.9.8)–(3.9.10) and are singled out by the corresponding equalities.

All these statements are easy consequences of Lemmas 3.7.1 and 3.7.2.

3.9.11. Example. Consider the configuration shown in Fig. 3.2. The polygon $D(z_1|z_2 x_2 x_1) = \text{defo}$ is determined by inequalities $z_1 < z_2 < x_2 < x_1$; $x_1 < T_2^2$, $x_2 + x_1 < T_2^1 + T_2^2$. The interval $D(z_1|z_2|x_2 x_1)$ lies in the line $x_2 + x_1 = T_2^1 + T_2^2$ and is determined by inequalities $z_1 < z_2 < x_2 < x_1$; $x_2 > T_2^1$, $x_1 < T_2^2$.

3.10. Distinguished Coorientation of a Discriminant Configuration

Let $F = F(S_B)$ be a facet of $C_{n,k}(z)$ corresponding to the decomposition S_B shown in (3.8.8). Below we define a coorientation of F.

A subset $f(i) = \{f_1(i), \ldots, f_{q(i)}(i)\}$ of S_B will be called an *algebra subset* if $f(i) \cap \{z_1, \ldots, z_n\} = 0$ and a *module subset* otherwise. Set $A(S_B) = \{i \mid f(i)$ is an algebra subset$\}$.

For each algebra subset $f(i)$, denote by $j(i)$ the number $\min\{j \mid x_j \in f(i)\}$ and denote by $m(i)$ the number of module subsets lying in S_B on the right of

$f(i)$. Remove the variables $x_{j(i)}$, $i \in A(S_B)$, from the ordered set $x_1, \ldots, x_k$. The resulting set is denoted by $x(S_B)$.

3.10.1. Lemma. *The ordered set of the coordinates $x(S_B)$ forms a coordinate system on the affine space $\mathbb{R}^k / |F(S_B)|$.*

The lemma follows from (3.9.3).

Set

$$\alpha(S_B) = \sum_{i \in A(S_B)} (m(i) + j(i)) + \#\{\{\ell, m\} \in A(S_B) | \ell < m, j(\ell) < j(m)\}. \quad (3.10.2)$$

(3.10.3) Fix an orientation of $\mathbb{R}^k/(F(S_B)$ as the orientation defined by the ordered set of coordinates $x(S_B)$ multiplied by $(-1)^{\alpha(S_B)}$.

This coorientation of $F(S_B)$ will be called the *distinguished coorientation*. Consider the orientation of all cells $\{D(F^0 \geq F^j), E(F^0 \geq F^j)\}$ compatible with the distinguished coorientation of $\mathcal{C}_{n,k}(z)$, see (2.4). This orientation will be called the *distinguished orientation*.

According to (3.8.8), the pairs $F^0 \geq F^j$ are numerated by admissible monomials. For future purposes we repeat the definition of the distinguished orientation in terms of admissible monomials. Namely, let f be an admissible monomial of the form (3.8.1). For each algebra factor $f(i)$, $i \in A(f)$, denote by $j(i)$ the number $\min\{j \mid x_j \in f(i)\}$ and denote by $m(i)$ the number of module factors lying in f on the right of $f(i)$. Remove the variables $x_{j(i)}$, $i \in A(f)$, from the ordered set $x_1, \ldots, x_k$. The resulting set is denoted by $x(f)$. The ordered set of the variables $x(f)$ forms a coordinate system on cells $D(f)$, $E(f)$.

(3.10.4) We say that the distinguished orientation of $E(f)$ and $D(f)$ is the orientation defined by the ordered set of the variables $x(f)$ multiplied by $(-1)^{\alpha(f)}$, where

$$\alpha(f) = \sum_{i \in A(f)} (m(i) + j(i)) + \#\{\{\ell, m\} \subset A(f) \mid \ell < m, j(\ell) < j(m)\}.$$

3.10.5. Example. For the configuration $\mathcal{C}_{1,2}(z_1)$ in $\mathbb{R}^2$, the cells $D(f)$ and their orientations are shown in Fig. 3.4.

3.10.6. Example. All top dimensional cells $D(f)$, $E(f)$ are oriented by the ordered system of coordinates $x_1, \ldots, x_k$.

Consider the complexes $C.(Q'/Q'', \mathbb{Z})$, $C.(X, \mathbb{Z})$ defined in (2.3). The cells $\{D(f)\}$, $\{E(f)\}$, oriented by the distinguished orientation, define elements of these complexes, respectively.

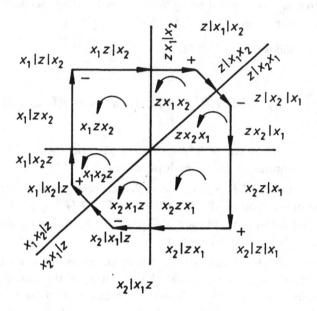

Fig. 3.4

3.10.7. Lemma. *For any admissible monomial f the boundary operator in the above complexes is given by the formulas*

$$d : D(f) \longmapsto \sum_{f > f'} \operatorname{ind}(f > f') D(f'),$$

$$d : E(f) \longmapsto \sum_{f >_w f'} \operatorname{ind}(f >_w f') E(f'),$$

where the first sum is taken over all monomials f' adjacent to f, the second sum is taken over all monomials f' weakly adjacent to f, the indexes $\operatorname{ind}(f > f')$, $\operatorname{ind}(f >_w f')$ are defined in (3.8), all cells are oriented by the distinguished orientation.

The lemma is proved by direct trivial computations.

3.10.8. Example. $dD(x_1 x_3 | x_2 z_1 x_4) = D(x_1 | x_3 | x_2 z_1 x_4) - D(x_1 x_3 | x_2 | z_1 x_4) -$

$D(x_1x_3|x_2z_1|x_4)$. Other examples see in Fig. 3.4.

3.11. Weights of $C_{n,k}(z_1,\ldots,z_n)$ and the Reduced Quantum Bilinear Form for $n \leq 1$

Assume that $C = C_{n,k}(z)$ is weighted. Denote the weight of a hyperplane $x_\ell = x_m$, $1 \leq \ell < m \leq k$, by $a_{\ell,m}$ or $a(x_\ell, x_m)$. Denote the weight of a hyperplane $x_\ell = z_m$, $\ell = 1, \ldots, k$, $m = 1, \ldots, n$, by $a_{\ell,m+k}$ or $a(x_\ell, z_m)$. Denote by a the set of all weights.

Let $n \leq 1$. Then $C_{n,k}$ is central. Let $B' = B'[n,k,a]$ be the reduced quantum bilinear form of the central weighted configuration $C_{n,k}$. In this section we compute the determinant of B' and describe B' in terms of admissible monomials.

Let $n = 0$. Then $C_{0,k}$ is the configuration of all diagonal hyperplanes $x_\ell = x_m$, $1 \leq \ell < m \leq k$. The domains of $C_{0,k}$ are numerated by permutations $\sigma \in \Sigma_k$. The domain corresponding to σ is $F_\sigma = \{(x_i, \ldots, x_k) \in \mathbb{R}^k \mid x_{\sigma(1)} < \cdots < x_{\sigma(k)}\}$.

3.11.1. Lemma.

$$B'(F_\sigma, F_{\sigma'}) = \exp\left(\frac{\pi i}{2} \sum_{\ell < m} \pm a_{\ell m}\right),$$

where the sign of $a_{\ell m}$ is negative if $(\sigma(\ell) - \sigma(m))(\sigma'(\ell) - \sigma'(m)) > 0$, the sign is positive if $(\sigma(\ell) - \sigma(m))(\sigma'(\ell) - \sigma'(m)) < 0$.

The lemma is obvious.

3.11.2. Theorem. *The determinant of $B'[0,k,a]$ is given by the formula*

$$\det B'[0,k,a] = \prod_I \left(\exp(-\pi i a(I)) - \exp(\pi i a(I))\right)^{\ell(I)},$$

where the sum is taken over all the subsets $I \subset \{1,\ldots,k\}$ consisting of more than one element,

$$a(I) = \sum_{\substack{\ell,m \in I \\ \ell < m}} a_{\ell,m}, \qquad \ell(I) = (\#I - 2)!(k+1-\#I)!.$$

3.11.3. Example

$$\det B'[0,3,a] = \bigl(\exp(-\pi i(a_{12}+a_{13}+a_{23})) - \exp(\pi i(a_{12}+a_{13}+a_{23}))\bigr)$$
$$\cdot \prod_{1\le \ell < m \le 3} \bigl(\exp(-\pi i a_{\ell m}) - \exp(\pi i a_{\ell m})\bigr)^2.$$

Proof. A subset $I \subset \{1,\ldots,k\}$ defines an edge

$$L(I) = \{(x_1,\ldots,x_k) \mid x_\ell = x_m \text{ for every } \ell, m \in I\}.$$

It is easy to see that the discrete mass, the discrete density, and the multiplicity of this edge are given by the formulas

$$n(L) = (k+1-\#I)!$$
$$p(L) = (\#I-2)! \tag{3.11.4}$$
$$\ell(L) = (\#I-2)!(k+1-\#I)!$$

see definitions in (2.6).

All other edges of $\mathcal{C}_{0,k}$ have density 0. This fact is a consequence of (2.6.10).

Applying Theorem 2.6.6, we get Theorem 3.11.2.

Let $n=1$. The domains of $\mathcal{C}_{1,k}$ are numerated by permutations $\sigma \in \Sigma_{k+1}$. The domain corresponding to σ is $F_\sigma = \{(x_1,\ldots,x_k) \in \mathbb{R}^k \mid x_{\sigma(1)} < \cdots < x_{\sigma(k)},$ where x_{k+1} should be replaced by $z_1\}$. The value $B'(F_\sigma, F_{\sigma'})$ for σ, $\sigma' \in \Sigma_{k+1}$ is given by formula (3.11.1).

3.11.5. Lemma. $\det B'[1,k,a] = \det B'[0,k+1,a]$.

This lemma and formula (3.11.2) give the formula for $\det B'[1,k,a]$.

Proof. Consider the configuration $\mathcal{C}_{0,k+1}$ in $\mathbb{R}^{k+1}$ having weights a. Consider the hyperplane H given by the equation $x_{k+1} = z_1$. Then $\mathcal{C}_{0,k+1}$ cuts the configuration $\mathcal{C}_{1,k}$ in H. The construction induces natural isomorphisms $M_{\mathcal{C}_{0,k+1}} \cong M_{\mathcal{C}_{1,k}}$, $B'[0,k+1,a] \simeq B'[1,k,a]$.

3.12. Complexes $C.(Q'/Q'', \mathcal{S}^*(a))$, $C.(X, \mathcal{S}^*(a))$ for a Discriminantal Configuration $\mathcal{C}_{n,k}(z)$

Let $\mathcal{C}_{k,n}(z)$ be weighted as in (3.11) and cooriented as in (3.10). Consider the complexes $C.(Q'/Q'', \mathcal{S}^*(a))$, $C.(X, \mathcal{S}^*(a))$ constructed for $\mathcal{C}_{n,k}(z)$ in Sec. 2.

Fix affine equations $\ell_H = 0$ for all $H \in \mathcal{C}_{n,k}(z)$, namely, the equation $x_\ell - x_m = 0$ for a hyperplane $x_\ell = x_m$, $1 \leq \ell < m \leq k$, and the equation $x_\ell - z_m = 0$ for a hyperplane $x_\ell = z_m$, $\ell = 1, \ldots, k$, $m = 1, \ldots, n$.

In Sec. 2 we constructed some bases of the complexes $C.(Q'/Q'', \mathcal{S}^*(a))$, $C.(X, \mathcal{S}^*(a))$ using linear functions ℓ_H, $H \in \mathcal{C}$. The bases are formed by the cells $D(F^0 \geq F^j, a, Or)$, $E(F^0 \geq F^j, a, Or)$. The pairs $F^0 \geq F^j$ are numerated by admissible monomials f. Denote the cells $D(F^0 \geq F^j, a, Or)$, $E(F^0 \geq F^j, a, Or)$ by $D(f, a)$, $E(f, a)$, respectively, if a monomial f corresponds to $F^0 \geq F^j$.

3.12.1. Lemma. *For a configuration $\mathcal{C}_{n,k}(z)$ the boundary operator in $C.(Q'/Q'', \mathcal{S}^*(a))$ is given by the formula*

$$dD(f, a) = \sum_{f > f'} \mathrm{ind}\,(f > f')\, D(f', a),$$

where the sum is taken over all of the monomials f' adjacent to f, $\mathrm{ind}\,(f > f')$ is defined in (3.8).

The lemma is a consequence of (2.5.13) and (3.10.6).

Let $f >_w f'$ be weakly adjacent admissible monomials of the form (3.8.1) and (3.8.5), respectively. Their *twisting number* is defined as the number

$$B(f > f') = \exp\left(\frac{\pi i}{2} \sum_{u=1}^{a} \sum_{v=1}^{b} \pm a(f_{s_u}(j), f_{t_v}(j))\right), \qquad (3.12.2)$$

where $a(f_{s_u}(j), f_{t_v}(j))$ is the weight of the hyperplane $f_{s_u}(j) = f_{t_v}(j)$, the sign of $a(f_{s_u}(j), f_{t_v}(j))$ is positive if $s_u > t_v$, and the sign is negative if $s_u < t_v$.

3.12.3. Lemma. *For the discriminantal configuration $\mathcal{C}_{n,k}(z)$ the boundary operator in $C.(X, \mathcal{S}^*(a))$ is given by the formula*

$$dE(f, a) = \sum_{f >_w f'} \mathrm{ind}\,(f >_w f')\, B(f >_w f')\, E(f', a),$$

where the sum is taken over all the admissible monomials weakly adjacent to

f, ind $(f >_w f')$ is defined in (3.8), $B(f >_w f)$ is the twisting number.

The lemma is a direct corollary of (2.5.15).

3.12.4. Example. For $k = 2$, $n = 1$,

$$\begin{aligned}
dE(x_2 z_1 x_1) = &- \exp\left(\pi i (-a_{23} - a_{12})/2\right) E(x_2 | z_1 x_1) \\
&- \exp\left(\pi i (a_{23} + a_{12})/2\right) E(z_1 x_1 | x_2) \\
&- \exp\left(\pi i (a_{23} - a_{13})/2\right) E(z_1 | x_2 x_1) \\
&- \exp\left(\pi i (-a_{23} + a_{13})/2\right) E(x_2 x_1 | z_1) \\
&- \exp\left(\pi i (a_{12} + a_{13})/2\right) E(x_1 | x_2 z_1) \\
&- \exp\left(\pi i (-a_{12} - a_{13})/2\right) E(x_2 z_1 | x_1),
\end{aligned}$$

where we write $E(f)$ instead of $E(f, a)$.

Let f, f' be admissible monomials of the form (3.8.1). Say that f, f' are *equicomposed*, $f \sim f'$, if f' may be obtained from f by a permutation of the symbols $f_j(t)$ in (3.8.1) preserving the index t, that is, by a permutation preserving the factors of f.

Example. $x_1 z_1 x_2 | x_3 z_2 \sim x_2 z_1 x_1 | z_2 x_3$.

3.12.5. Lemma. *For a discriminantal configuration $C_{n,k}(z)$ the homomorphism of the projection on the real part, $S : C.(X, S^*(a)) \to C.(Q'/Q'', S^*(a))$, defined in (2.3) is given by the formula*

$$S : E(f, a) \longmapsto \sum_{f', f' \sim f} \left(\prod_{j=1}^{p} B'(f(j), f'(j)) \right) D(f', a), \qquad (3.12.5.1)$$

where f is a monomial of the form (3.8.1), p is the number of factors of f, $f(j)$ and $f'(j)$ are the j-th factors of f and f', respectively,

$$B'(f(j), f'(j)) = \exp\left(\frac{\pi i}{2} \sum_{1 \le \ell < m \le q(j)} \pm a(f_\ell(j), f_m(j)) \right), \qquad (3.12.5.2)$$

where $a(f_\ell(j), f_m(j))$ is the weight of the hyperplane $f_\ell(j) = f_m(j)$, the sign of $a(f_\ell(j), f_m(j))$ is positive, if $f_\ell(j)$ lies in $f'(j)$ on the right of $f_m(j)$, the sign *is negative otherwise.*

3.12.6. Example. For $n = 1$, $k = 3$,

$$\begin{aligned}
S : E(x_1 x_2 \mid z_1 x_3) \longmapsto\ & \exp\left(\pi i(a_{12} + a_{34})/2\right) D(x_2 x_1 \mid x_3 z_1) \\
& + \exp\left(\pi i(-a_{12} + a_{34})/2\right) D(x_1 x_2 \mid x_3 z_1) \\
& + \exp\left(\pi i(a_{12} - a_{34})/2\right) D(x_2 x_1 \mid z_1 x_3) \\
& + \exp\left(\pi i(-a_{12} - a_{34})/2\right) D(x_1 x_2 \mid z_1 x_3).
\end{aligned}$$

The lemma is a corollary of Theorem (2.7.2).

3.12.7. Remarks. The right-hand side of (3.12.5.1) consists of $q(1)! \ldots q(p)!$ terms. The number $B'(f(j), f'(j))$ is the value of the reduced quantum bilinear form of the configuration $\mathcal{C}_{0,q(j)}$, if

$$\{f_1(j), \ldots, f_{q(j)}(j)\} \cap \{z_1, \ldots, z_n\}$$

is empty, and of the configuration $\mathcal{C}_{1,q(j)-1}(z_r)$ if this intersection is z_r. The weights of these configurations are given by the numbers $a(f_\ell(j), f_m(j))$. The value $B'(f(j), f'(j))$ of the reduced quantum form is taken at the pair of domains corresponding to $f(j)$, $f'(j)$, see formulas (3.11.1) and (3.12.5.2).

Formula (3.12.5.1) shows that the matrix of S with respect to $\{E(f,a), D(f,a)\}$ has a block form. The blocks are numerated by the equivalence classes of equicomposed monomials. The matrix of the block, corresponding to the class of a monomial f of the form (3.8.1) is the matrix of the bilinear form $B'_1 \otimes \cdots \otimes B'_p$, where B'_j is the reduced quantum bilinear form of the configuration $\mathcal{C}_{0,q(j)}$ or $\mathcal{C}_{1,q(j)-1}$. The determinant of any factor B'_j is given by (3.11.2) and (3.11.5).

Call a subset $I \subset \{x_1, \ldots, x_k, z_1, \ldots, z_k\}$ *admissible* if $\#I > 1$, $\#I \cap \{z_1, \ldots, z_n\} \leq 1$. Call the number

$$a(I) = \sum_{(u,v) \subset I} a(u,v)/2 \qquad (3.12.8)$$

the *weight* of I, here $a(u,v)$ is the weight of the hyperplane $u = v$. Call the subset I a *resonance subset* if I is admissible and $a(I)$ is an integer.

3.12.9. Lemma. *For a discriminantal configuration $\mathcal{C}_{n,k}(z)$, the homomorphism of the projection on the real part, $S : C.(X, \mathcal{S}^*(a)) \to C.(Q'/Q'', \mathcal{S}^*(a))$,* is an isomorphism if there are no resonance subsets. Moreover, the homomorphism $S: C_j(X, \mathcal{S}^*(a)) \to C_j(Q'/Q'', \mathcal{S}^*(a))$ is an isomorphism if there are no resonance subsets I such that $\#I \leq j+1$.

3.13. Action of the Permutation Group Σ_k on a Discriminantal Configuration $\mathcal{C}_{n,k}(z)$

The group Σ_k acts on $\mathbb{R}^k$ permuting $x_1, \ldots, x_k$. This action preserves $\mathcal{C}_{n,k}(z)$. In this section we describe an interaction of this action and constructions of Sec. 3.

3.13.1. Lemma. *For any facet F of $\mathcal{C}_{n,k}(z)$ and for any $\sigma \in \Sigma_k$, $\sigma(^F w) = {}^{\sigma F}w$.*

Proof. According to (3.5), it is sufficient to check Lemma 3.13.1 for top-dimensional facets. In this case the lemma easily follows from Definition 3.3.2.

3.13.2. Corollary. *For all $F^0 \geq F^j$ and $\sigma \in \Sigma_k$,*

$$\sigma : D(F^i) \longrightarrow D(\sigma F^i)$$
$$\sigma : D(F^0 \geq F^i) \longrightarrow D(\sigma F^0 \geq \sigma F^i)$$

are affine diffeomorphisms, and

$$\sigma : E(F^0 \geq F^i) \longrightarrow E(\sigma F^0 \geq \sigma F^i)$$

is a piece-wise linear homeomorphism.

Let f be an admissible monomial of the form (3.8.1) and $\sigma \in \Sigma_k$. Define σf as the admissible monomial which is obtained from f by the permutation σ of the symbols $\{x_1, \ldots, x_k\}$ in f.

Example. For $f = x_2 x_4 z_1 \mid x_3 x_5 \mid z_2 x_1$, $\sigma : \{x_1, \ldots, x_5\} \to \{x_3, x_2, x_1, x_5, x_4\}$, $\sigma f = x_2 x_5 z_1 \mid x_1 x_4 \mid z_2 x_3$.

3.13.3. Corollary. *For an admissible f and $\sigma \in \Sigma_k$, $\sigma : D(f) \to D(\sigma f)$ is an affine diffeomorphisms, and $\sigma : E(f) \to E(\sigma f)$ is a piece-wise linear homeomorphism.*

Let $D(f)$, $D(\sigma f)$, $E(f)$, and $E(\sigma f)$ be oriented by the distinguished orientation.

3.13.4. Lemma. *The degree of the maps $\sigma : D(f) \to D(\sigma f)$, $\sigma : E(f) \to$*

$E(\sigma f)$ is equal to $(-1)^{|\sigma|}$.

As an example see the involution $x_1 \leftrightarrow x_2$ in Fig. 3.4.

Proof. It is sufficient to check the lemma for the case where σ is a transposition of two coordinates x_ℓ, x_m having one of the properties:

(3.13.5) $\ell, m \in \{j(i) \mid i \in A(F)\}$,

(3.13.6) $\ell, m \notin \{j(i) \mid i \in A(F)\}$,

(3.13.7) there exists $i \in A(F)$ such that x_ℓ, x_m belongs to the algebra factor $f(i)$ of f and $\ell = j(i)$.

In each of these cases the lemma easily follows from the definition of the distinguished orientation.

Let

$$\{x_1, \ldots, x_k\} = \coprod_{p=1}^{r} X(p) \qquad (3.13.8)$$

be a decomposition into a union of disjoint subsets $X(p) = \{x_1(p), \ldots, x_{\ell_p}(p)\}$. Let

$$\Sigma' = \Sigma\left(\coprod X(p)\right) := \Sigma_{\ell_1} \times \cdots \times \Sigma_{\ell_r} \subset \Sigma_k \qquad (3.13.9)$$

be the subgroup of all permutations preserving the decomposition. Σ' acts on $\mathbb{R}^k$ by permutations of coordinates.

Assume that $C_{n,k}(z)$ is weighted by the weights $a = \{a_{\ell,m}\}$ such that the action of Σ' preserves the weights. This means that the weights $a(x_\ell, x_m)$, $a(x_\ell, z_m)$ for $x_\ell \in X(a)$ and $x_m \in X(b)$ depend only on a, b and a, m, respectively.

Denote these weights by $a(X(a), X(b))$ and $a(X(a), z_m)$, respectively.

3.13.10. Lemma. *The action of Σ' on the complexes $C.(X, S^*(a))$, $C.(Q'/Q'', S^*(a))$ of the configuration $C_{n,k}(z)$ is given by the formulas:*

$$\sigma : D(f, a) \longmapsto (-1)^{|\sigma|} D(\sigma f, a)$$
$$\sigma : E(f, a) \longmapsto (-1)^{|\sigma|} E(\sigma f, a)$$

for all $\sigma \in \Sigma'$.

The lemma follows from (2.8.5) and (3.13.4).

The homomorphism of the projection on the real part defined in (3.12.5) commutes with the action of Σ' according to (2.8.7).

(3.13.11) Define the subcomplexes $C.(X, \mathcal{S}^*(a))^- \subset C.(X, \mathcal{S}^*(a))$ and $C.(Q'/Q'', \mathcal{S}^*(a))^- \subset C.(Q'/Q'', \mathcal{S}^*(a))$ as the subcomplexes consisting of all elements x such that $\sigma x = (-1)^{|\sigma|} x$ for all $\sigma \in \Sigma'$.

We describe bases of these subcomplexes. Consider an expression

$$F = F_1(1) F_2(1) \ldots F_{q(1)}(1) \mid F_1(2) \ldots F_{q(2)}(2) \mid \ldots \\ \mid F_1(p) \ldots F_{q(p)}(p) \mid, \qquad (3.13.12)$$

where

(3.13.13) $F_j(i) \in \{X(1), \ldots, X(r), z_1, \ldots, z_n\}$ for all j, i;

(3.13.14) For any i, $\#(\{F_1(i), \ldots, F_{q(i)}(i)\} \cap \{z_1, \ldots, z_n\}) \leq 1$;

(3.13.15) $\{z_1, \ldots, z_n\} \subset \{F_1(1), \ldots, F_{q(p)}(p)\}$;

(3.13.16) For every i, j, $1 \leq i < j \leq n$, the symbol z_i appears earlier in F than the symbol z_j.

(3.13.18) For any $p = 1, \ldots, r$, the symbol $X(p)$ stands exactly in ℓ_p positions in F, in particular, $q(1) + \cdots + q(p) = k + n$.

An expression F will be called a *generalized admissible monomial* of weight $(\ell_1, \ldots, \ell_r)$.

Example. $X(2) X(1) z_1 \mid X(2) \mid z_2 X(2)$ is a generalized admissible monomial of weight (1,3).

We say that an admissible monomial f of the form (3.8.1) is *subordinated* to a generalized admissible monomial (3.13.12) if $f_j(i) \in X_j(i)$ for all $f_j(i) \in \{x_1, \ldots, x_n\}$ and $f_j(i) = F_j(i)$ for all $f_j(i) \in \{z_1, \ldots, z_n\}$. Notation: $f \to F$. For any generalized admissible monomial F set

$$D(F, a) = \sum_{f \to F} D(f, a)$$
$$E(F, a) = \sum_{f \to F} E(f, a). \qquad (3.13.18)$$

Each of these sums consists of $\ell_1! \ldots \ell_r!$ terms.

3.13.19. Lemma. *For any generalized admissible monomial F of the form (3.13.12) and for any $\sigma \in \Sigma'$, $\sigma E(F, a) = (-1)^{|\sigma|} E(F, a)$ and $\sigma D(F, a) =*

$(-1)^{|\sigma|} D(F, a)$. The elements $\{E(F,a)|F$ is a generalized admissible monomial of the form (3.13.12)$\}$ form a basis of $C.(X, S^*(a))^-$. The elements $\{D(F, a)|F$ is a generalized admissible monomial of the form (3.13.12)$\}$ form a basis of $C.(Q'/Q'', S^*(a))^-$.

The proof is obvious.

3.14. Complexes $\mathcal{X}.(\mathcal{C}_{n,k})$ and $Q.(\mathcal{C}_{n,k})$

The abstract complexes $\mathcal{X}.$ and $Q.$ and the homomorphism $S : \mathcal{X}. \to Q.$ were introduced in (2.9) for an arbitrary weighted and cooriented configuration. In this section we describe these objects for a weighted discriminantal configuration $\mathcal{C}_{n,k}(z)$ cooriented by the distinguished coorientation.

Complex $\mathcal{X}.(\mathcal{C}_{n,k})$. The space $\mathcal{X}_j(\mathcal{C}_{n,k})$, $0 \leq j \leq k$, is the space of complex linear combination of symbols f, where f runs through the set of all admissible monomials of the form (3.8.1) having $k - j$ algebraic factors. The differential $d : \mathcal{X}_j(\mathcal{C}_{n,k}) \to \mathcal{X}_{j-1}(\mathcal{C}_{n,k})$ is defined by the formula

$$d : (f) \longmapsto \sum_{f', f >_w f'} \mathrm{ind}\,(f >_w f') B(f >_w f')(f'), \qquad (3.14.1)$$

where the sum is taken over all admissible monomials f' weakly adjacent to f, the number $\mathrm{ind}\,(f >_w f')$ is defined in (3.8), the number $B(f >_w f')$ is defined in (3.12).

Complex $Q(\mathcal{C}_{n,k})$. The space $Q_j(\mathcal{C}_{n,k})$, $0 \leq j \leq k$, is the space dual to $\mathcal{X}_j(\mathcal{C}_{n,k})$. It is the space of complex linear combinations of the δ-functions $(f)^*$, where f runs through the set of all admissible monomials of the form (3.8.1) having $k - j$ algebraic factors. The differential $d : Q_j(\mathcal{C}_{n,k}) \to Q_{j-1}(\mathcal{C}_{n,k})$ is defined by the formula

$$d : (f)^* \longmapsto \sum_{f', f > f'} \mathrm{ind}\,(f > f')(f')^*, \qquad (3.14.2)$$

where the sum is taken over all monomials f' adjacent to f, the number $\mathrm{ind}\,(f > f')$ is defined in (3.8).

The homomorphism $S : \mathcal{X}.(\mathcal{C}_{n,k}) \to Q.(\mathcal{C}_{n,k})$ is defined by the formula

$$S : f \longmapsto \sum_{f', f' \sim f} \left(\prod_{j=1}^{p} B'(f(j), f'(j)) \right) (f')^*, \qquad (3.14.3)$$

where the sum is taken over all monomials f' equicomposed with f; p is the number of factors of f; $f(j)$ and $f'(j)$ are the j-th factors of f and f', respectively; the number $B'(f(j), f'(j))$ is defined in (3.12.5.2).

This description of $\mathcal{X}.(\mathcal{C}_{n,k})$ and $Q(\mathcal{C}_{n,k})$ follows from (3.12).

3.14.4. Remark. The complexes $\mathcal{X}.(\mathcal{C}_{n,k}(z))$ and $Q.(\mathcal{C}_{n,k}(z))$ and the homomorphism S do not depend on z and are determined by n, k and the weights of a configuration $\mathcal{C}_{n,k}(z)$, cf. (2.9.9).

3.15. Homology of $Q.(\mathcal{C}_{n,k})$

$H_j Q.(\mathcal{C}_{n,k}) = 0$ for $j \neq k$ or $j = k$, $n \leq 1$, according to (2.3.5) and (2.3.6).

Let $n > 1$. Then $H_k Q.(\mathcal{C}_{n,k})$ has the following basis. The elements of this basis are numerated by the following permutations σxz of the set $xz = \{x_1, \ldots, x_k, z_1, \ldots, z_n\}$: a permutation σxz starts with z_1 and ends with z_n, and for any i, j, $1 \leq i < j \leq n$, the symbol z_i appears earlier in σxz than the symbol z_j. These permutations are in obvious one-to-one correspondence with the bounded domains of $\mathcal{C}_{n,k}(z)$. The element of the basis corresponding to σxz has the form

$$G(\sigma, xz) = \sum (f)^*, \qquad (3.15.1)$$

where the sum is taken over all admissible monomials of the form (3.8.1) such that f has n factors and $f_1(1), f_2(1), \ldots, f_{q(1)}(1), \ldots, f_{q(n)}(n) = \sigma xz$.

This description is a corollary of (2.3.5) and (2.3.6).

3.15.2. Example. $\dim H_2 Q.(\mathcal{C}_{2,2}) = 2$. $G(z_1 x_1 x_2 z_2)$ and $G(z_1 x_2 x_1 z_2)$ form a basis of H_2. $G(z_1 x_1 x_2 z_2) = (z_1 | x_1 x_2 z_2)^* + (z_1 x_1 | x_2 z_2)^* + (z_1 x_1 x_2 | z_2)^*$, $G(z_1 x_2 x_1 z_2) = (z_1 | x_2 x_1 z_2)^* + (z_1 x_2 | x_1 z_2)^* + (z_1 x_2 x_1 | z_2)^*$.

3.16. Action of the Permutation Group on $\mathcal{X}.(\mathcal{C}_{n,k})$, $Q.(\mathcal{C}_{n,k})$

As in (3.8) let $\{x_1, \ldots, x_k\} = \coprod^r X(j)$ be a decomposition into a union of disjoint subsets. Set $\ell_j = \#X(j)$, $\nu = (\ell_1, \ldots, \ell_r)$.

Let $\Sigma' \subset \Sigma_k$ be the subgroup of all permutation of coordinates in $\mathbb{R}^k$ preserving the decomposition. Assume that $\mathcal{C}_{n,k}(z)$ is weighted by the weights such that the action of Σ' preserves the weights. In this situation we denote the discriminant configuration $\mathcal{C}_{n,k}(z)$ by $\mathcal{C}_{n,k,\nu}(z)$. The group Σ' acts on $\mathcal{X}.(\mathcal{C}_{n,k,\nu})$, $Q.(\mathcal{C}_{n,k,\nu})$ as defined in (2.9).

(3.16.1) The action of Σ' on $\mathcal{X}.(\mathcal{C}_{n,k,\nu})$ and $Q.(\mathcal{C}_{n,k,\nu})$ is given by the formulas

$$\sigma: f \longmapsto (-1)^{|\sigma|}\sigma f,$$
$$\sigma: (f)^* \longmapsto (-1)^{|\sigma|}(\sigma f)^*,$$

where $\sigma \in \Sigma'$, f is an admissible monomial of the form (3.8.1), σf is the admissible monomial obtained from f by the permutation σ of the symbols $\{x_1, \ldots, x_k\}$ in f.

This action commutes with the homomorphism S.

Let
$$w: \mathcal{X}.(\mathcal{C}_{n,k,\nu}) \longrightarrow C.(X, S^*(a)),$$
$$w: Q.(\mathcal{C}_{n,k,\nu}) \longrightarrow C.(Q'/Q'', S^*(a)) \qquad (3.16.2)$$

be a realization described in (3.12). Then this realization is an isomorphism of Σ' modules.

Define the subcomplexes
$$\mathcal{X}.(\mathcal{C}_{n,k,\nu})^{-'} \subset \mathcal{X}.(\mathcal{C}_{n,k,\nu}),$$
$$Q.(\mathcal{C}_{n,k,\nu})^{-'} \subset Q.(\mathcal{C}_{n,k,\nu}) \qquad (3.16.3)$$

as the subcomplexes consisting of all elements x such that $\sigma x = (-1)^{|\sigma|} x$ for all $\sigma \in \Sigma'$. A restriction of homomorphism (3.14.3) defines a homomorphism of complexes

$$S: \mathcal{X}.(\mathcal{C}_{n,k,\nu})^{-'} \longrightarrow Q.(\mathcal{C}_{n,k,\nu})^{-'} \qquad (3.16.4)$$

The spaces of the subcomplexes have the following description. Define the space $\mathcal{X}_j(\mathcal{C}_{n,k,\nu})^-$, $0 \leq j \leq k$, as the space of complex linear combinations of symbols F, where F runs through the set of all generalized admissible monomials of the form (3.13.13) having $k - j$ algebra factors.

Define the space $Q_j(\mathcal{C}_{n,k,\nu})^-$, $0 \leq j \leq k$, as the space dual to $\mathcal{X}_j(\mathcal{C}_{n,k,\nu})^-$. $Q_j(\mathcal{C}_{n,k,\nu})^-$ is the space of complex linear combinations of δ-functions $(F)^*$, where F runs through the set of all generalized admissible monomials of the form (3.13.13) having $k - j$ algebra factors.

(3.16.5) The maps $\pi: \mathcal{X}_j(\mathcal{C}_{n,k,\nu})^- \to \mathcal{X}_j(\mathcal{C}_{n,k,\nu})^{-'}$ and $\pi: Q_j(\mathcal{C}_{n,k,\nu})^- \to Q_j(\mathcal{C}_{n,k,\nu})^{-'}$ for $j = 0, \ldots, k$ defined by the formulas

$$\pi: F \longmapsto \sum_{f, f \to F} f,$$

$$\pi: (F)^* \longmapsto \sum_{f, f \to F} (f)^*,$$

where the sums are taken over all admissible monomials f of the form (3.8.1) subordinated to F, are isomorphisms.

These statements follow from (3.13).

The isomorphisms π induce on $\mathcal{X}.(\mathcal{C}_{n,k,\nu})^-$ and $\mathcal{Q}.(\mathcal{C}_{n,k,\nu})^-$ a structure of complexes. Homomorphism (3.16.4) induces a homomorphism of complexes

$$S : \mathcal{X}.(\mathcal{C}_{n,k,\nu})^- \longrightarrow \mathcal{Q}.(\mathcal{C}_{n,k,\nu})^-. \tag{3.16.6}$$

An algebraic description of these complexes and homomorphism (3.16.6), in particular, explicit formulas for the differentials in these complexes and for homomorphism (3.16.6) in terms of generalized monomials are given in Sec. 4, see (4.7.5).

Let w be a realization of $\mathcal{X}.(\mathcal{C}_{n,k,\nu})$, $\mathcal{Q}.(\mathcal{C}_{n,k,\nu})$ by $C.(X, \mathcal{S}^*(a))$, $C.(Q'/Q'', \mathcal{S}^*(a))$, resp., described in (3.12). Then the isomorphisms

$$\begin{aligned} w\pi : \mathcal{X}.(\mathcal{C}_{n,k,\nu})^- &\longrightarrow C.(X, \mathcal{S}^*(a))^-, \\ w\pi : \mathcal{Q}.(\mathcal{C}_{n,k,\nu})^- &\longrightarrow C.(Q'/Q'', \mathcal{S}^*(a))^- \end{aligned} \tag{3.16.7}$$

will be called a *realization* of $\mathcal{X}.(\mathcal{C}_{n,k,\nu})^-$, $\mathcal{Q}.(\mathcal{C}_{n,k,\nu})^-$. Homomorphism (3.12.5) of the projection on the real part restricted to $C.(X, \mathcal{S}^*(a))^-$ will be called a *realization of homomorphism* (3.16.6). The following diagram is commutative

$$\begin{array}{ccc} \mathcal{X}.(\mathcal{C}_{n,k,\nu})^- & \xrightarrow{S} & \mathcal{Q}.(\mathcal{C}_{n,k,\nu})^- \\ {\scriptstyle w\pi}\downarrow & & \downarrow{\scriptstyle w\pi} \\ C.(X, \mathcal{S}^*(a))^- & \xrightarrow{S} & C.(Q'/Q'', \mathcal{S}^*(a))^-. \end{array} \tag{3.16.8}$$

Homology groups of $\mathcal{Q}.(\mathcal{C}_{n,k,\nu})^-$

$$H_j \mathcal{Q}.(\mathcal{C}_{n,k,\nu})^- = 0 \tag{3.16.9}$$

for $j \neq k$ and $j = k$, $n \leq 1$.

(3.16.10) For $n > 1$, the space $H_k \mathcal{Q}.(\mathcal{C}_{n,k,\nu})^-$ has the following basis. The elements of this basis are numerated by ordered sequences $S = S_1, \ldots, S_{n+k}$ such that

(a) $S_1 = z_1$, $S_{n+k} = z_n$, $S_j \in \{X(1), \ldots, X(r), z_1, \ldots, z_n\}$ for all j.

(b) Each $z_1, \ldots, z_n$ appears once in S. Moreover, for every $1 \leq i < j \leq n$

the symbol z_i appears earlier in S than z_j.

(c) For any $j = 1, \ldots, r$, the symbol $X(j)$ stands exactly in ℓ_i positions in S.

The element of the basis corresponding to S has the form

$$G(S) = \sum (F)^*,$$

where the sum is taken over all generalized monomials F of form (3.13.12) such that $p = n$ and $S = F_1(1), F_2(1), \ldots, F_{q(1)}(1), F_1(2), \ldots, F_1(p), \ldots, F_{q(p)}(p)$.

These statements follow from (3.15).

3.16.11. Example. Let $n = 3$, $k = 2$, $r = 1$, $\nu = (2)$. Then $H_2Q.(\mathcal{C}_{n,k,\nu})^-$ is three-dimensional. Its basis is formed by $G(z_1, X(1), X(1), z_2, z_3)$, $G(z_1, X(1), z_2, X(1), z_3)$, $G(z_1, z_2, X(1), X(1), z_3)$, where $G(z_1, X(1), X(1), z_2, z_3) = (z_1 | X(1)X(1)z_2|z_3)^* + (z_1X(1)|X(1)z_2|z_3)^* + (z_1X(1)X(1)|z_2|z_3)^*$, and so on.

4

Algebraic Interpretation of Chain Complexes of a Discriminantal Configuration

4.1. Quantum Groups [D1, SV4]

Let us fix the following data:

(a) a finite-dimensional complex vector space $\mathfrak{h}$;

(b) a nondegenerate symmetric bilinear form $(\,,\,)$ on $\mathfrak{h}$;

(c) linearly independent covectors ("simple roots") $\alpha_1, \ldots, \alpha_r \in \mathfrak{h}^*$;

(d) a nonzero complex number $\varkappa$.

We will denote by $b : \mathfrak{h} \simeq \mathfrak{h}^*$ the isomorphism induced by $(\,,\,)$. We will transfer the form $(\,,\,)$ to $\mathfrak{h}^*$ using b. Set $b_{ij} = (\alpha_i, \alpha_j)$; $B = (b_{ij}) \in \mathrm{Mat}_r(\mathbb{C})$, $h_i = b^{-1}(\alpha_i) \in \mathfrak{h}$. Set $q = \exp(2\pi i/\varkappa)$; for $a \in \mathbb{C}$, set $q^a = \exp(2\pi i a/\varkappa)$.

We shall denote by $U_q\mathfrak{g} = U_q\mathfrak{g}(B)$ the $\mathbb{C}$-algebra generated by the elements $1, e_i, f_i$, for $i = 1, \ldots, r$ and the space $\mathfrak{h}$, subject to the relations

$$[h, e_i] = \langle \alpha_i, h \rangle e_i\,; \qquad [h, f_i] = -\langle \alpha_i, h \rangle f_i\,; \tag{4.1.1}$$

$$[e_i, f_j] = (q^{h_i/2} - q^{-h_i/2}) \cdot \delta_{ij}\,; \tag{4.1.2}$$

$$[h, h'] = 0 \tag{4.1.3}$$

for all i, j and $h, h' \in \mathfrak{h}$. By definition, $q^{ah} = \exp(ah \cdot \frac{2\pi i}{\varkappa})$. (We include in our algebra convergent power series in $h \in \mathfrak{h}$, i.e., such power series that become convergent in any $\mathfrak{h}$-diagonalizable representation.)

Define a comultiplication $\Delta : U_q\mathfrak{g} \to U_q\mathfrak{g} \otimes U_q\mathfrak{g}$ by the rule

$$\Delta(h) = h \otimes 1 + 1 \otimes h, \qquad (4.1.4)$$

$$\Delta(f_i) = f_i \otimes q^{h_i/4} + q^{-h_i/4} \otimes f_i,$$
$$\Delta(e_i) = e_i \otimes q^{h_i/4} + q^{-h_i/4} \otimes e_i. \qquad (4.1.5)$$

This makes $U_q\mathfrak{g}$ a Hopf algebra.

Define the counit $\epsilon : U_q\mathfrak{g} \to \mathbb{C}$ and the antipode $A : U_q\mathfrak{g} \to U_q\mathfrak{g}$ by

$$\epsilon(e_i) = \epsilon(f_i) = \epsilon(h) = 0 \qquad (4.1.6)$$

$$A(h) = -h, \quad A(e_i) = -q^{b_{ii}/4} e_i, \quad A(f_i) = -q^{-b_{ii}/4} f_i. \qquad (4.1.7)$$

We denote by $U_q\mathfrak{n}_-$ (resp. $U_q\mathfrak{n}_+$ and $U_q\mathfrak{h}$) the subalgebra generated by 1, $f_1, \ldots, f_r$ (resp. $1, e_1, \ldots, e_r$ and $1, h \in \mathfrak{h}$). $U_q\mathfrak{n}_\pm$ are free. We have $U_q\mathfrak{g} = U_q\mathfrak{n}_- \cdot U_q\mathfrak{h} \cdot U_q\mathfrak{n}_+$. We set $U_q\mathfrak{b}_\pm = U_q\mathfrak{n}_\pm \cdot U_q\mathfrak{h}$. These are Hopf subalgebras of $U_q\mathfrak{g}$.

Denote by $^+U_q\mathfrak{n}_-$ (resp. $^+U_q\mathfrak{n}_+$ and $^+U_q\mathfrak{h}$) the subalgebra generated by $f_1, \ldots, f_r$ (resp. $e_1, \ldots, e_r$, and $h \in \mathfrak{h}$), $i = 1, \ldots, r$. $U_q\mathfrak{n}_- = \mathbb{C} \oplus {}^+U_q\mathfrak{n}_-, \ldots$

For $\Lambda \in \mathfrak{h}^*$, we denote by $M(\Lambda)$ the Verma module over $U_q\mathfrak{g}$ generated by a vector v subject to the relations $hv = \langle \Lambda, h \rangle v$; $U_q\mathfrak{n}_+ v = 0$.

For $\lambda = (\ell_1, \ldots, \ell_r) \in \mathbb{N}^r$, set

$$(U_q\mathfrak{n}_-)_\lambda = \left\{ x \in U_q\mathfrak{n}_- \;\middle|\; [h, x] = \left\langle -\sum \ell_i \alpha_i; h \right\rangle x \quad \text{for all } h \in \mathfrak{h} \right\},$$

$$(U_q\mathfrak{n}_+)_\lambda = \left\{ x \in U_q\mathfrak{n}_+ \;\middle|\; [h, x] = \left\langle \sum \ell_i \alpha_i; h \right\rangle x \quad \text{for all } h \in \mathfrak{h} \right\},$$

$$M(\Lambda)_\lambda = \left\{ x \in M(\Lambda) \;\middle|\; hx = \left\langle \Lambda - \sum \ell_i \alpha_i; h \right\rangle x \quad \text{for all } h \in \mathfrak{h} \right\}.$$

We have

$$U_q\mathfrak{n}_- = \bigoplus_\lambda (U_q\mathfrak{n}_-)_\lambda, \quad {}^+U_q\mathfrak{n}_- = \bigoplus_{\lambda \neq 0} (U_q\mathfrak{n}_-)_\lambda, \quad M(\Lambda) = \bigoplus_\lambda M(\Lambda)_\lambda.$$

Let

$$\tau : U_q\mathfrak{g} \longrightarrow U_q\mathfrak{g}$$

be the algebra antihomomorphism such that $\tau(e_i) = f_i$, $\tau(f_i) = e_i$, $\tau(h) = h$, $h \in \mathfrak{h}$. Set $M(\Lambda)^* = \oplus_\lambda M(\Lambda)^*_\lambda$. Define a structure of a $U_q\mathfrak{g}$-module on $M(\Lambda)^*$

by the rule $\langle g\varphi, x\rangle = \langle \varphi, \tau(g)x\rangle$ for all $\varphi \in M^*$, $g \in U_q\mathfrak{g}$, $x \in M$. For example, $\langle f_i f_j \varphi, x\rangle = \langle \varphi, e_j e_i x\rangle$. This structure is called contragradient to the module structure of M.

Serre relations. Drinfeld–Jimbo quantized Kac–Moody algebras

Suppose that $b_{ii} \neq 0$ for all i. Set $a_{ij} = 2b_{ij}/b_{ii}$. Suppose that the condition

$$a_{ij} \in \mathbb{Z}, \ a_{ij} \leq 0 \quad \text{for } i \neq j$$

holds. So, $A = (a_{ij})$ is a Generalized Cartan Matrix. Suppose that (a)–(c) is its realization [K, Sec. 1.1]. Introduce the following notation. For $a \in \mathbb{C}$,

$$(a)_q = q^{a/2} - q^{-a/2}. \tag{4.1.8}$$

For $b, a \in \mathbb{N}$,

$$(a)_q! = \prod_{i=1}^{a} (i)_q \tag{4.1.9}$$

$$\binom{a}{b}_q = \frac{(a)_q!}{(b)_q!(a-b)_q!}. \tag{4.1.10}$$

Set $n_{ij} = -a_{ij} + 1$, $q_i = q^{b_{ii}/2}$. Introduce Chevalley–Serre elements

$$R_{ij}(f) = \sum_{v=0}^{n_{ij}} (-1)^v f_i^v \cdot f_j \cdot f_i^{n_{ij}-v} \cdot \binom{n_{ij}}{v}_{q_i} \in U_q\mathfrak{n}_- \tag{4.1.11}$$

$R_{ij}(e) = (R_{ij}(f)$ with $f_{i,j}$ replaced by $e_{i,j}) \in U_q\mathfrak{n}_+$.

Let $U_q\mathfrak{g}^{DJ}$ be the quotient algebra of $U_q\mathfrak{g}$ by the two-sided ideal generated by all $R_{ij}(f)$ and $R_{ij}(e)$, cf. [**D1**], [**D2**].

4.2. Contravariant Form [SV4]

Define a symmetric bilinear form $S = S_-$ on the spaces $(U_q\mathfrak{n}_-)_\lambda$. Define S on $(U_q\mathfrak{n}_-)_{\lambda=0}$ by $S(1,1) = 1$. Suppose that $\lambda = (1, \ldots, 1) \in \mathbb{N}^r$. If $f_I = f_{i_1} \ldots f_{i_r}$, $f_J = f_{j_1} f_{j_2} \ldots f_{j_r} \in (U_q\mathfrak{n}_-)_\lambda$, then there exists a unique $\sigma \in \Sigma_r$ such that $j_\ell = i_{\sigma(\ell)}$ for all ℓ. Set

$$S(f_I, f_J) = q^{\sum_{\ell<m} \pm b_{i_\ell i_m}/4}, \tag{4.2.1}$$

where the sign of $b_{i_\ell i_m}$ is positive if $\sigma(\ell) > \sigma(m)$ and is negative otherwise.

We will give four definitions of S for an arbitrary $\lambda = (\ell_1, \ldots, \ell_r) \in \mathbf{N}^r$.

4.2.2. Definition. Let $f_J = f_{j_1} \ldots f_{j_p}$, $f_I = f_{i_1} \ldots f_{i_p} \in (U_q \mathfrak{n}_-)_\lambda$. Set

$$S(f_I, f_J) = \sum_\sigma q^{\sum_{\ell < m} \pm b_{i_\ell i_m}/4}, \qquad (4.2.2.1)$$

where the sum is taken over all $\sigma \in \Sigma_p$ such that $j_\ell = i_{\sigma(\ell)}$ for all ℓ, the sign in front of $b_{i_\ell i_m}$ is positive if $\sigma(\ell) > \sigma(m)$ and is negative otherwise.

The form S defines a map $S : U_q\mathfrak{n}_- \to (U_q\mathfrak{n}_-)^*$ given by the formula

$$S : f_{i_1} \ldots f_{i_p} \longmapsto \sum_{\sigma \in \Sigma_p} q^{\sum_{\ell < m} \pm b_{i_\ell i_m}/4} \left(f_{i_{\sigma(1)}} \ldots f_{i_{\sigma(p)}}\right)^*, \qquad (4.2.2.2)$$

where the sign of $b_{i_\ell i_m}$ is explained in (4.2.2.1); $\{(f_I)^*\}$ is the basis of $(U_q\mathfrak{n}_-)^*$ dual to the monomial basis $\{f_I\}$ of $U_q\mathfrak{n}_-$.

Introduce operators

$$g_j, g'_j : (U_q\mathfrak{n}_-)_\lambda \longrightarrow (U_q\mathfrak{n}_-)_{\lambda_j}, \quad j = 1, \ldots, r,$$

where $\lambda_j = (\ell_1, \ldots, \ell_j - 1, \ldots, \ell_r)$. Set

$$g_j(f_I) = \sum_{\ell : i_\ell = j} q^{\sum_{m < \ell} b_{ji_m}/4 - \sum_{m > \ell} b_{ji_m}/4} \cdot f_{i_1} \ldots \hat{f}_{i_\ell} \ldots f_{i_p},$$

$$g'_j(f_I) = \sum_{\ell : i_\ell = j} q^{-\sum_{m < \ell} b_{ji_m}/4 + \sum_{m > \ell} b_{ji_m}/4} \cdot f_{i_1} \ldots \hat{f}_{i_\ell} \ldots f_{i_p}. \qquad (4.2.3)$$

4.2.4. Definition. Define a bilinear form S' on $U_q\mathfrak{n}_-$ as the unique bilinear form satisfying the following conditions:

$$S'(f_j, f_j) = 1, \quad S'(1,1) = 1$$
$$S'(f_j x, y) = S'(x, g_j(y))$$

for all $x, y \in U_q\mathfrak{n}_-$, $j = 1, \ldots, r$.

4.2.5. Definition. Define a bilinear form S'' on $U_q\mathfrak{n}_-$ as the unique bilinear form satisfying the following conditions:

$$S''(f_j, f_j) = 1, \quad S''(1,1) = 1$$
$$S''(xf_j, y) = S''(x, g'_j(y))$$

for all $x, y \in U_q\mathfrak{n}_-$, $j = 1, \ldots, r$.

Let $f_I = f_{i_1} \ldots f_{i_p} \in (U_q\mathfrak{n}_-)_\lambda$, let $\Delta_p : U_q\mathfrak{n}_- \to (U_q\mathfrak{n}_-)^{\otimes p}$ be the iterated comultiplication. $\Delta_p f_I$ is a sum of p^p monomials of the form $a = a_1 q^{b_1} | \ldots | a_p q^{b_p}$, where a_j is a monomial in $U_q\mathfrak{n}_-$, $b_j \in \mathfrak{h}$ for $j = 1, \ldots, p$, see (4.1.5).

4.2.6. Example. $\Delta_2 f_1 f_2 = f_1 f_2 | q^{(h_1+h_2)/4} + q^{-(h_1+h_2)/4} | f_1 f_2 + q^{-b_{12}/4} f_1 q^{-h_2/4} | f_2 q^{h_1/4} + q^{b_{12}/4} f_2 q^{-h_1/4} | f_1 q^{h_2/4}$. Here we use the notation $a|b$ for $a \otimes b$.

The sum of $p!$ monomials of $\Delta_p f_I$ such that $a_1, \ldots, a_p$ is just a transposition of $f_{i_1} \ldots f_{i_p}$ will be called the *middle part* of $\Delta_p f_I$ and denoted by $md\ell\Delta_p f_I$. In (4.2.6) the last two monomials form the middle part.

The middle part has the form

$$md\ell\Delta_p f_I = \sum_J C_{IJ} f_{j_1} q^{b_1} | \ldots | f_{j_p} q^{b_p}, \qquad (4.2.7)$$

where $b_i \in \mathfrak{h}$ are suitable elements, the sum is taken over all monomials $f_J \in (U_q\mathfrak{n}_-)_\lambda$.

4.2.8. Definition. Set

$$S'''(f_I, f_J) = C_{IJ}, \qquad S'''(1,1) = 1.$$

Different subspaces of $(U_q\mathfrak{n}_-)_\lambda$ are pairwise orthogonal with respect to S, S', S'', S'''.

4.2.9. Lemma. $S = S' = S'' = S'''$. The form S is symmetric. $S(f_{i_1} \ldots f_{i_p}, f_{j_1} \ldots f_{j_p}) = S(f_{i_p} \ldots f_{i_1}, f_{j_p} \ldots f_{j_1})$ for all f_I, f_J.

The form S is called the *contravariant form*.

Examples. $S(f_2 f_1 f_3, f_1 f_3 f_2) = q^{(b_{12}+b_{23}-b_{13})/4}$,

$$S(f_j^p, f_j^p) = (p)_{q_j}!/(1)_{q_j}^p,$$

where $(p)_q!$ is defined in (4.1.9), $q_j := q^{b_{jj}/2}$.

Define a symmetric bilinear form $S = S_+$ on $U_q\mathfrak{n}_+$ by the same formulas (4.2.1)–(4.2.8), replacing $f_1, \ldots, f_r$ by $e_1, \ldots e_r$, respectively.

Example. $S(e_j^p, e_j^p) = (p)_{q_j}!/(1)_{q_j}^p$.

Set $a_{ij} = 2b_{ij}/b_{ii}$.

4.2.10. Lemma. [SV4]. *Assume that for given i, j, $1 \leq i < j \leq r$, we have*

$b_{ii} \neq 0$, $a_{ij} \in \mathbb{Z}$, $a_{ij} \leq 0$. Then the two-sided ideal of $U_q\mathfrak{n}_-$ generated by the element $R_{ij}(f)$ of (4.1.11) lies in the kernel of $S : U_q\mathfrak{n}_- \to (U_q\mathfrak{n}_-)^*$.

Proof. According to (4.2.4) and (4.2.5), $S(xR_{ij}(f)y, z) = S(R_{ij}(f), v)$ for a suitable $v = v(x, y, z)$. Therefore, it is sufficient to prove the equality

$$S(R_{ij}(f), x) = 0 \qquad (4.2.11)$$

for all $x \in U_q\mathfrak{n}_-$. To check (4.2.11), it is sufficient to check that $g_j R_{ij}(f) = 0$ and $g_i R_{ij}(f) = 0$.

$$\begin{aligned} g_j R_{ij}(f) &= \sum_{v=0}^{n_{ij}} (-1)^v \binom{n_{ij}}{v}_{q_i} q^{(2v - n_{ij}) b_{ij}/4} f_i^{n_{ij}} \\ &= \sum_{v=0}^{n_{ij}} (-1)^v \binom{n_{ij}}{v}_{q_i} q_i^{(2v - n_{ij})(1 - n_{ij})/4} f_i^{n_{ij}} . \end{aligned} \qquad (4.2.12)$$

The formula

$$(1 + t)(1 + qt) \ldots (1 + q^{n-1} t) = \sum_{v=0}^{n} \binom{n}{v}_q q^{\frac{v(n-1)}{2}} t^v \qquad (4.2.13)$$

after a substitution $q_i \mapsto q$, $n_{ij} \mapsto n$, $-q^{1-n} \mapsto t$ proves that (4.2.12) is zero.

To prove $g_i R_{ij}(f) = 0$, it is sufficient to check that

$$\begin{aligned} g_i f_i^v f_j f_i^{n_{ij}-v} &= f_i^{v-1} f_j f_i^{n_{ij}-v} q_i^{-(n_{ij}-1)/4 + (v-1)/2} (v)_{q_i} / (1)_{q_i} \\ &\quad + f_i^v f_j f_i^{n_{ij}-v-1} q_i^{-(n_{ij}-1)/4 + v/2} (n_{ij} - v)_{q_i} / (1)_{q_i} \end{aligned}$$

and apply the formula

$$\binom{n}{v}_q (v)_q = \binom{n}{v-1}_q (n - v + 1)_q. \qquad (4.2.14)$$

Let $M(\Lambda)$ be the Verma module over $U_q\mathfrak{g}$ with a highest weight $\Lambda \in \mathfrak{h}^*$ generated by a vacuum vector v. There exists the unique bilinear form S on $M(\Lambda)$ such that

$$\begin{aligned} S(v, v) &= 1, \\ S(f_j x; y) &= S(x; e_j y) \end{aligned} \qquad (4.2.15)$$

for all $x, y \in M(\Lambda)$ and $j = 1, \ldots, r$. The form S is symmetric and called the *contravariant form*. Different subspaces $M(\Lambda)_\lambda$ are pairwise orthogonal with respect to S.

4.2.16. Examples. $S(f_2 f_1 f_3 v; f_1 f_3 f_2 v) = (\langle h_2, \Lambda \rangle)_q (\langle h_1, \Lambda - \alpha_3 \rangle)_q (\langle h_3, \Lambda \rangle)_q$.
$S(f_j^p v; f_j^p v) = (\langle h_j, \Lambda \rangle)_q (\langle h_j, \Lambda - \frac{1}{2}\alpha_j \rangle)_q \ldots (\langle h_j, \Lambda - \frac{p-1}{2}\alpha_j \rangle)_q (p)_{q_j}!/(1)_{q_j}^p$.

The form S defines a homomorphism of $U_q\mathfrak{g}$-modules

$$S : M(\Lambda) \longrightarrow M(\Lambda)^*. \qquad (4.2.17)$$

4.2.18. Lemma. *Assume that for given i, j, $1 \leq i < j \leq r$, we have $b_{ii} \neq 0$, $a_{ij} = 2b_{ij}/b_{ii} \in \mathbb{Z}$, $a_{ij} \leq 0$. Let I_{ij} be the two-sided ideal of $U_q\mathfrak{g}$ generated by the elements $R_{ij}(f)$ and $R_{ij}(e)$ of (4.1.11). Then $I_{ij}v \subset M(\Lambda)$ lies in the kernel of $S : M(\Lambda) \to M(\Lambda)^*$.*

Proof. According to properties (4.2.15), it is sufficient to check that

$$e_k R_{ij}(f) v = 0, \quad k = 1, \ldots, r.$$

It is obvious for $k \neq i, j$. We have

$$e_j R_{ij}(f) v = \sum_{v=0}^{n_{ij}} (-1)^v \binom{n_{ij}}{n}_{q_i} (q^{\langle h_j, \Lambda - (n_{ij}-v)\alpha_i \rangle/2} - q^{-\langle h_j, \Lambda - (n_{ij}-v) \rangle/2}) f_i^{n_{ij}} v.$$

This is zero according to formula (4.2.13), cf. the proof of the equality $g_j R_{ij}(f) = 0$ in Lemma (4.2.10).

$$e_i R_{ij}(f) v = \sum_{v=0}^{n_{ij}} (-1)^v \binom{n_{ij}}{v}_{q_i} \left(\left(\left\langle h_i, \Lambda - (n_{ij}-v)\alpha_i - \alpha_j - \frac{v-1}{2}\alpha_i \right\rangle \right)_q \right.$$
$$\left. \cdot \frac{(v)_{q_i}}{(1)_{q_i}} f_i^{v-1} f_j f_i^{n_{ij}-v} + \left(\left\langle h_j, \Lambda - \frac{n-v-1}{2}\alpha_i \right\rangle \right)_q \frac{(n-v)_{q_i}}{(1)_{q_i}} f_i^v f_j f_i^{n_{ij}-v-1} \right).$$

This is zero according to formula (4.2.10).

Now let $(a_{ij}) = (2b_{ij}/b_{ii})$ be a Generalized Cartan Matrix, and let $U_q\mathfrak{g}^{DJ}$ be the corresponding Drinfeld–Jimbo quantum group, see (4.1).

Let $\overline{U_q\mathfrak{g}}$ be the quotient algebra of $U_q\mathfrak{g}$ factored by the two-sided ideal generated by ker $(S_- : U_q\mathfrak{n}_- \to (U_q\mathfrak{n}_-)^*)$ and by ker $(S_+ : U_q\mathfrak{n}_+ \to (U_q\mathfrak{n}_+)^*)$.

By (4.2.10) we have a natural epimorphism

$$U_q\mathfrak{g}^{DJ} \longrightarrow \overline{U_q\mathfrak{g}}. \qquad (4.2.19)$$

4.2.20. Remark. In (9.5.7), we will prove that (4.2.19) is an isomorphism if $\varkappa$ is not a rational number. This statement was conjectured in [**SV**].

Consider a Verma module $M(\Lambda)$ and set $\overline{M}(\Lambda) = M(\Lambda)/\ker S$, where S is the contravariant form on $M(\Lambda)$. $\overline{M}(\Lambda)$ has a natural $U_q\mathfrak{g}^{DJ}$-module structure:

$$U_q\mathfrak{g}^{DJ} \otimes \overline{M}(\Lambda) \longrightarrow \overline{M}(\Lambda). \qquad (4.2.21)$$

An easy standard reason shows

(4.2.22) $\overline{M}(\Lambda)$ is an irreducible $U_q\mathfrak{g}^{DJ}$-module with highest weight Λ.

4.2.23. Remarks. Let $f_I v, f_J v \in M(\Lambda)_\lambda$, where $\lambda = (\ell_1, \ldots, \ell_r) \in \mathbf{N}^r$, $f_I = f_{i_1} \ldots f_{i_k}$, $f_J = f_{j_1} \ldots f_{j_k}$. Compute $S(f_I v, f_J v)$ as a function of f_I, f_J, Λ:

$$S(f_I v, f_J v) = \sum_L C_L(f_I, f_J) q^{L(\Lambda)},$$

where the sum is finite, $\{L : \mathfrak{h}^* \to \mathbf{C}\}$ are suitable linear functions, and $\{C_L\}$ are suitable bilinear forms on $(U_q\mathfrak{n}_-)_\lambda$. Let $L_0(\Lambda) = \langle \sum \ell_i \alpha_i, \Lambda \rangle/2$. Then

$$C_{L_0}(f_I, f_J) = q^{A_\lambda} S(f_I, f_J),$$

where S is the contravariant form on $(U_q\mathfrak{n}_-)_\lambda$,

$$A_\lambda = \sum_{j=1}^r b_{jj}\ell_j(\ell_j - 1)/8 + \sum_{i<j} b_{ij}\ell_i\ell_j/4.$$

The other linear functions L have the form $\langle \sum \ell'_i \alpha_i, \Lambda \rangle/2$, where $-\ell_i \leq \ell'_i \leq \ell_i$, $\ell_i - \ell'_i \equiv 0 \bmod 2$, for all $i = 1, \ldots, r$.

Corollaries. 1. If $x \in (U_q\mathfrak{n}_-)_\lambda$ and $xv \in \ker(S : M(\Lambda) \to M(\Lambda)^*)$ for all Λ, then $x \in \ker(S : U_q\mathfrak{n}_- \to (U_q\mathfrak{n}_-)^*)$.

2. If $S : (U_q\mathfrak{n}_-)_\lambda \to (U_q\mathfrak{n}_-)^*_\lambda$ is nondegenerate, then $S : M(\Lambda)_\lambda \to M(\Lambda)^*_\lambda$ is nondegenerate for general values of $\Lambda \in \mathfrak{h}^*$.

The connections between the contravariant forms on $U_q\mathfrak{n}_-$ and on $M(\Lambda)$ are discussed in Secs. 5 and 9. In particular, see (5.3.14) and (9.3.8). Also, in (9.3.8), we prove that if $x \in \ker(S : (U_q\mathfrak{n}_-)_\lambda \to (U_q\mathfrak{n}_-)^*_\lambda)$, then $xM(\Lambda) \subset \ker(S : M(\Lambda) \to M(\Lambda)^*)$ for all Λ.

4.3. Coalgebra Structure on $U_q\mathfrak{n}_-$, Algebra Structure on $(U_q\mathfrak{n}_-)^*$

Define $\mu : U_q\mathfrak{n}_- \to U_q\mathfrak{n}_- \otimes U_q\mathfrak{n}_-$ as follows. Let $f_I = f_{i_1} \ldots f_{i_p}$. Then $\Delta(f)$ is the sum of 2^p monomials $a \otimes b$ of the form $a_1 \ldots a_p \otimes b_1 \ldots b_p$, where $a_j = f_{i_j}$, $b_j = q^{h_{i_j}/4}$ or $a_j = q^{-h_{i_j}/4}$, $b_j = f_{i_j}$, $j = 1, \ldots, p$. Each of these monomials $a \otimes b$ can be written in the form

$$Bf_{i_{\sigma(1)}} \ldots f_{i_{\sigma(\ell)}} q^{-(h_{i_{\sigma(\ell+1)}} + \cdots + h_{i_{\sigma(p)}})/4} \otimes f_{i_{\sigma(\ell+1)}} \ldots f_{i_{\sigma(p)}} q^{(h_{i_{\sigma(1)}} + \cdots + h_{i_{\sigma(\ell)}})/4}, \quad (4.3.1)$$

where $1 \leq \ell \leq p$; σ is a permutation of $\{1, \ldots, p\}$ satisfying the following inequalities: $\sigma(u) < \sigma(v)$ for all $1 \leq u < v \leq \ell$, and for all $\ell + 1 \leq u < v \leq p$; $B = B_\sigma$ is some complex number.

Set

$$\text{proj} : a \otimes b \longmapsto B f_{i_{\sigma(1)}} \ldots f_{i_{\sigma(\ell)}} \otimes f_{i_{\sigma(\ell+1)}} \ldots f_{i_{\sigma(p)}}, \quad (4.3.2)$$

$$\overline{\Delta} = \text{proj} \circ \Delta : f_I \longmapsto \sum_{a \otimes b} \text{proj}(a \otimes b). \quad (4.3.3)$$

Set $f_{I_1} = f_{i_{\sigma(1)}} \ldots f_{i_{\sigma(\ell)}}$, $f_{I_2} = f_{i_{\sigma(\ell+1)}} \ldots f_{i_{\sigma(p)}}$, $f_{I_1}|f_{I_2} := f_{I_1} \otimes f_{I_2}$. We say that $f_{I_1}|f_{I_2}$ is weakly adjacent to f_I if $f_{I_1}|f_{I_2}$ appears in (4.3.3) and $\ell > 0$, $p - \ell > 0$. Notation: $f_I >_w f_{I_1}|f_{I_2}$. The number B in (4.3.2) will be called the *twisting number* and denoted by $B(f_I >_w f_{I_1}|f_{I_2})$.

4.3.4. Examples. $\overline{\Delta}(f_i) = f_i \otimes 1 + 1 \otimes f_i$,

$$\overline{\Delta}(f_1 f_2) = q^{-b_{12}/4} f_1 \otimes f_2 + q^{b_{12}/4} f_2 \otimes f_1 + f_1 f_2 \otimes 1 + 1 \otimes f_1 f_2,$$

$$\overline{\Delta}(f_i^p) = \sum_{j=1}^{p-1} \binom{p}{j}_{q_i} f_i^j \otimes f_i^{p-j} + f_i^p \otimes 1 + 1 \otimes f_i^p,$$

where $q_i = q^{b_{ii}/2}$.

4.3.5. Lemma. $\overline{\Delta} : U_q\mathfrak{n}_- \to U_q\mathfrak{n}_- \otimes U_q\mathfrak{n}_-$ *defines a coalgebra structure on* $U_q\mathfrak{n}_-$.

The lemma is a direct corollary of the fact that $U_q\mathfrak{g}$ is a Hopf algebra.

Let $\mu : U_q\mathfrak{n}_- \otimes U_q\mathfrak{n}_- \to U_q\mathfrak{n}_-$ be the multiplication in $U_q\mathfrak{n}_-$, and

$$\mu^* : (U_q\mathfrak{n}_-)^* \longrightarrow (U_q\mathfrak{n}_-)^* \otimes (U_q\mathfrak{n}_-)^*$$

be its dual map.

4.3.6. Lemma. *The following diagram is commutative:*

$$\begin{array}{ccc} (U_q\mathfrak{n}_-)^* & \xrightarrow{\mu^*} & (U_q\mathfrak{n}_-)^* \otimes (U_q\mathfrak{n}_-)^* \\ S\uparrow & & \uparrow S\otimes S \\ U_q\mathfrak{n}_- & \xrightarrow{\overline{\Delta}} & U_q\mathfrak{n}_- \otimes U_q\mathfrak{n}_- \end{array}$$

Proof. To prove the lemma, it is sufficient to check the commutativity for each monomial $f_I = f_{i_1}\ldots f_{i_p} \in (U_q\mathfrak{n}_-)_\lambda$, $\lambda = (\ell_1,\ldots,\ell_r)$.

$$\mu^* S(f_I) = \mu^* \left(\sum_{\sigma\in\Sigma_p} q^{D_\sigma} (f_{i_{\sigma(1)}}\ldots f_{i_{\sigma(p)}})^* \right) \tag{4.3.7}$$
$$= \sum_{\sigma\in\Sigma_p} q^{D_\sigma} \sum_{s=1}^{p-1} (f_{i_{\sigma(1)}}\ldots f_{i_{\sigma(s)}})^* \mid (f_{i_{\sigma(s+1)}}\ldots f_{i_{\sigma(p)}})^*,$$

where D_σ can be deduced by comparison with (4.2.2).

$$(S\otimes S) \circ \overline{\Delta}(f_I) = S\otimes S\left(\sum_{s=1}^{p-1}\sum_{\tau_s} q^{C_\tau} f_{i_{\tau_s(1)}}\ldots f_{i_{\tau_s(s)}} \Big| f_{i_{\tau_s(s+1)}}\ldots f_{i_{\tau_s(p)}} \right)$$
$$= \sum_{s=1}^{p-1}\sum_{\tau_s} q^{C_\tau} \left(\sum_{\nu\in\Sigma_s} q^{C_{\tau,\nu,1}} (f_{i_{\tau_s(\nu(1))}}\ldots f_{i_{\tau_s(\nu(s))}})^* \right| \tag{4.3.8}$$
$$\sum_{\mu\in\Sigma_{p-s}} q^{C_{\tau,\mu,2}} (f_{i_{\tau_s(\mu(1)+s)}}\ldots f_{i_{\tau_s(\mu(p-s)+s)}})^* \Big),$$

where τ_s runs through the set of all permutations of $\{1,\ldots,p\}$ satisfying the inequalities $\tau(u) < \tau(v)$ for all $1 \leq u < v \leq s$ and for all $s+1 \leq u < v \leq p$; q^{C_τ} is the twisting number of the pair $f_I >_w f_{i_{\tau_s(1)}}\ldots f_{i_{\tau_s(s)}} | f_{i_{\tau_s(s+1)}}\ldots f_{i_{\tau_s(p)}}$ and is given by the formula

$$C_\tau = \sum_{\substack{u\leq s, v>s \\ \tau_s(v)<\tau_s(u)}} b_{i_{\tau(u)} i_{\tau(v)}}/4 - \sum_{\substack{u\leq s, v>s \\ \tau_s(v)>\tau_s(u)}} b_{i_{\tau(u)} i_{\tau(v)}}/4 ; \tag{4.3.9}$$

the numbers $q^{C_{\tau,\nu,1}}$, $q^{C_{\tau,\mu,2}}$ are defined in (4.2.2).

The terms on the right-hand side of (4.3.7) are numerated by the pairs (σ, s). The terms on the right-hand side of (4.3.8) are numerated by the four-tuples (τ, s, ν, μ). The pairs (σ, s) and the four-tuples are in one-to-one natural correspondence: $\sigma = \tau(\nu \oplus \mu)$. The corresponding numbers are connected by the formula
$$D_\sigma = C_\tau + C_{\tau,\nu,1} + C_{\tau,\mu,2}.$$
This proves the lemma.

Define a map
$$^+\Delta : {}^+U_q\mathfrak{n}_- \longrightarrow {}^+U_q\mathfrak{n}_- \otimes {}^+U_q\mathfrak{n}_- \qquad (4.3.10)$$
by the formula
$$^+\Delta(x) = \overline{\Delta}(x) - x \otimes 1 - 1 \otimes x \qquad (4.3.11)$$
for all $x \in {}^+U_q\mathfrak{n}_-$.

4.3.12. Lemma. $^+\Delta$ *is a coalgebra structure on* $^+U_q\mathfrak{n}_-$. *Moreover, the following diagram is commutative:*

$$\begin{array}{ccc} ({}^+U_q\mathfrak{n}_-)^* & \xrightarrow{\mu^*} & ({}^+U_q\mathfrak{n}_-)^* \otimes ({}^+U_q\mathfrak{n}_-)^* \\ S\uparrow & & \uparrow S\otimes S \\ {}^+U_q\mathfrak{n}_- & \xrightarrow{{}^+\Delta} & ({}^+U_q\mathfrak{n}_-) \otimes ({}^+U_q\mathfrak{n}_-) \end{array}$$

where μ^* is the map dual to the multiplication.

The lemma is a corollary of (4.3.5) and (4.3.6).

Define an algebra structure on $(U_q\mathfrak{n}_-)^*$ by the map
$$\overline{\Delta}^* : (U_q\mathfrak{n}_-)^* \otimes (U_q\mathfrak{n}_-)^* \longrightarrow (U_q\mathfrak{n}_-)^* \qquad (4.3.13)$$
dual to the map $\overline{\Delta}$. According to Lemma (4.3.6), the contravariant form defines a homomorphism of algebras $S : U_q\mathfrak{n}_- \to (U_q\mathfrak{n}_-)^*$.

Define an algebra structure on $({}^+U_q\mathfrak{n}_-)^*$ by the map
$$^+\Delta^* : ({}^+U_q\mathfrak{n}_-)^* \otimes ({}^+U_q\mathfrak{n}_-)^* \longrightarrow ({}^+U_q\mathfrak{n}_-)^* \qquad (4.3.14)$$
dual to the map $^+\Delta$. According to Lemma (4.3.12), $S : {}^+U_q\mathfrak{n}_- \to ({}^+U_q\mathfrak{n}_-)^*$ is a homomorphism of algebras.

4.3.15. Examples. $^+\Delta^*(f_1^* \otimes f_2^*) = q^{-b_{12}/4}(f_1 f_2)^* + q^{b_{12}/4}(f_2 f_1)^*$, $^+\Delta^*((f_1 f_2)^* \otimes (f_3)^*) = q^{-b_{13}/4 - b_{23}/4}((f_1 f_2 f_3)^* + q^{b_{23}/2}(f_1 f_3 f_2)^* + q^{b_{13}/2 + b_{23}/2}(f_3 f_1 f_2)^*)$.

At the end of this section, we will discuss properties of the subspaces $\ker S_- \subset U_q \mathfrak{n}_-$ and $\ker S_+ \subset U_q \mathfrak{n}_+$.

4.3.16. Lemma. $\overline{\Delta}(\ker S_-) \subset \ker S_- \otimes U_q \mathfrak{n}_- + U_q \mathfrak{n}_- \otimes \ker S_-$.

The lemma follows from (4.3.6).

Let $f_I = f_{i_1} \ldots f_{i_p} \in (U_q \mathfrak{n}_-)_\lambda$ be any monomial, $p > 2$. Let

$$\Delta f_I = f_I \otimes q^{(h_{i_1} + \cdots + h_{i_p})/4} + q^{-(h_{i_1} + \cdots + h_{i_p})/4} \otimes f_I$$

$$+ \sum_{j=1}^{r} \Big(x_j q^{-h_j/4} \otimes f_j q^{(h_{i_1} + \cdots + h_{i_p} - h_j)/4}$$

$$+ f_j q^{-(h_{i_1} + \cdots + h_{i_p} - h_j)/4} \otimes y_j q^{h_j/4} \Big) + \sum_s a_s \otimes b_s,$$

where $\{x_j, y_j\}$ are suitable elements of $U_q \mathfrak{n}_-$, and for every s, both a_s and b_s contain at least two f's.

4.3.17. Lemma. For all j, we have $x_j = g_j(f_I)$ and $y_j = g'_j(f_I)$, where g_j, g'_j are defined in (4.2.8).

The lemma is obvious. The lemma has obvious reformulation for $p = 2$.

4.3.18. Corollary. Let $x \in (U_q \mathfrak{n}_-)_\lambda$ where $|\lambda| > 1$. Then $x \in \ker S_-$ if and only if $g_1(x), \ldots, g_r(x) \in \ker S_-$ and if and only if $g'_1(x), \ldots, g'_r(x) \in \ker S_-$.

Set $I_- = U_q \mathfrak{n}_+ \cdot U_q \mathfrak{h} \cdot \ker S \subset U_q \mathfrak{g}$.

4.3.19. Lemma. $I_- \subset U_q \mathfrak{g}$ is an ideal.

Proof. $\ker S_-$ is graded. Let $x \in (\ker S_-)_\lambda$, where $\lambda = (\ell_1, \ldots, \ell_r) \in \mathbb{N}^r$. Then

$$xh = hx + \Big\langle h, \sum \ell_j \alpha_j \Big\rangle x, \tag{4.3.20}$$

$$xe_j = e_j x + q^{-A_\lambda} q^{-h_j/2} g'_j(x) - q^{A_\lambda} q^{h_j/2} g_j(x), \tag{4.3.21}$$

where A_λ is given in (4.2.23). These formulas prove the lemma.

4.3.22. Lemma. I_- is the maximal ideal of the form $U_q \mathfrak{n}_+ \cdot U_q \mathfrak{h} \cdot Y$, where $Y \subset U_q \mathfrak{n}_-$.

Proof. By (4.3.20), $Y = \oplus_\lambda Y \cap (U_q \mathfrak{n}_-)_\lambda$. Let $(\ker S_-)_{\lambda^0} \not\supset Y_{\lambda^0}$ for some $\lambda^0 = (\ell^0_1, \ldots, \ell^0_r) \in \mathbb{N}^r$ and $(\ker S_-)_\lambda \supset Y_\lambda$ for all $\lambda = (\ell_1, \ldots, \ell_r)$ such that

$\ell_1 + \cdots + \ell_r < \ell_1^0 + \cdots + \ell_r^0$. Let $x \in Y_{\lambda^0} - (\ker S_-)_{\lambda^0}$. It follows from (4.3.21) that $g_j(x) \in \ker S_-$ for all j. By (4.3.18), $x \in \ker S_-$ and we get a contradiction.

(4.3.23) Analogously, one can prove that $I_+ = (\ker S_+) \cdot U_q \mathfrak{h} \cdot U_q \mathfrak{n}_- \subset U_q \mathfrak{g}$ is an ideal, and I_+ is the maximal ideal of the form $Y \cdot U_q \mathfrak{h} \cdot U_q \mathfrak{n}_-, Y \subset U_q \mathfrak{n}_+$.

Cf. these statements with [**K**, Theorem 1.2].

4.4. Hochschild Homology

(4.4.1) For an algebra A and a left A-module M we denote by $C.(A; M)$ the *standard Hochschild complex* of A with coefficients in M : $C_i(A; M) = A^{\otimes i} \otimes M$; $d : C_i(A; M) \to C_{i-1}(A; M)$ maps $a_i \otimes \cdots \otimes a_1 \otimes m$ to

$$-a_i \otimes \cdots \otimes a_2 \otimes a_1 m + \sum_{p=2}^{i} (-1)^p a_i \otimes \cdots \otimes a_p a_{p-1} \otimes \cdots \otimes a_1 \otimes m.$$

$dC_0 = 0$. We will use the notation $a_1 | a_2 | \ldots a_i | m$ for $a_1 \otimes \cdots \otimes a_i \otimes m$. We set $H.(A; M) = H.(C.(A; M))$.

(4.4.2) For an algebra A, M is a two-sided A-module (or A-bimodule) if M is a left and right A-module and left and right multiplications commute: $(a_1 m)a_2 = a_1(ma_2)$.

Let $M_1, \ldots, M_n$ be two-sided A-modules. Introduce the *two-sided Hochschild complex* $C.(A; M_1, \ldots, M_n; 2)$ as follows:

Define the *j-admissible tensor product for the sequence* $M_1, \ldots, M_n$ as the tensor product of the form $X = X_1 \otimes \cdots \otimes X_{j+n}$, where

a) $X_\ell \in \{A, M_1, \ldots, M_n\}$ for all ℓ,

b) A appears in exactly j places of X,

c) for any ℓ, M_ℓ is one of the factors of X, moreover, for all $1 \leq \ell < m \leq n$, M_ℓ appears earlier in X than M_m.

Set
$$C_j(A; M_1, \ldots, M_n; 2) = \oplus X, \qquad (4.4.2.1)$$

where the sum is taken over all pairwise different j-admissible tensor products X. Let $d : C_j(A; M_1, \ldots, M_n; 2) \to C_{j-1}(A; M_1, \ldots, M_n; 2)$ map $a_{j+n} | a_{j+n-1} | \ldots | a_1$ to

$$\sum (-1)^p a_{j+n} | \ldots | a_{p+1} a_p | \ldots | a_1,$$

where the sum is taken over all p, $1 \leq p < j+n$, such that at least one element of a_p, a_{p+1} belongs to A. $dC_0 = 0$.

Example. $C.(A; M_1, M_2; 2)$ has the form $0 \leftarrow M_1 \otimes M_2 \leftarrow A \otimes M_1 \otimes M_2 \oplus M_1 \otimes A \otimes M_2 \oplus M_1 \otimes M_2 \otimes A \leftarrow A^{\otimes 2} \otimes M_1 \otimes M_2 \oplus A \otimes M_1 \otimes A \otimes M_2 \oplus \cdots \oplus M_1 \otimes M_2 \otimes A^{\otimes 2} \leftarrow \ldots$, $d : a_1|m_1|m_2 + m_3|a_2|m_4 + m_5|m_6|a_3 \mapsto a_1 m_1|m_2 + m_3 a_2|m_4 - m_3|a_2 m_4 - m_5|m_6 a_3$.

(4.4.2.2) For the special case $n = 0$, we set $C_0(A; 2) = 0$, $C_j(A; 2) = A^{\otimes j}$ for $j > 0$, $d : a_j| \ldots |a_1 \mapsto \sum_{p=1}^{j-1}(-1)^p a_j| \ldots |a_{p+1} a_p| \ldots |a_1$.

(4.4.3) For a coalgebra A and a left A-comodule M, we denote by $C.(A; M)$ the standard Hochschild complex of A with coefficients in M : $C_i(A; M) = A^{\otimes j} \otimes M$; $d : C_j(A; M) \to C_{j+1}(A; M)$ maps $a_j \otimes \cdots \otimes a_1 \otimes m$ to

$$-a_j \otimes \cdots \otimes a_1 \otimes \Delta m + \sum_{p=1}^{j}(-1)^{p+1} a_j \otimes \cdots \otimes \Delta a_p \otimes \cdots \otimes a_1 \otimes m,$$

where $\Delta : M \to A \otimes M$, $\Delta : A \to A \otimes A$ are comodule and coalgebra structures.

(4.4.3.1) For a coalgebra A, M is a two-sided A-comodule (or A-bicomodule) if M is a left and right comodule and the left and right comultiplications commute, that is, the comodule structures $\Delta_\ell : M \to A \otimes M$, $\Delta_r : M \to M \otimes A$ are given and $(1 \otimes \Delta_r) \circ \Delta_\ell = (\Delta_\ell \otimes 1) \circ \Delta_r$.

Let $M_1, \ldots, M_n$ be two-sided A-comodules. Introduce the two-sided Hochschild complex $C.(A; M_1, \ldots, M_n; 2)$ as follows. Define $C_j(A; M_1, \ldots, M_n; 2)$ by formula (4.4.2.1). Let $d : C_j(A; M_1, \ldots, M_n; 2) \to C_{j+1}(A; M_1, \ldots, M_n; 2)$ map $a_{j+n}| \ldots |a_1$ to

$$\sum_{p=1}^{j+n}(-1)^p a_{j+n}| \ldots |\Delta a_p| \ldots |a_1,$$

where $\Delta : A \to A \otimes A$ is the comultiplication in A if $a_p \in A$, $\Delta = \Delta_\ell \oplus \Delta_r : M_i \to A \otimes M_i \oplus M_i \otimes A$ if $a_p \in M_i$, $i = 1, \ldots, n$.

(4.4.3.2) For the special case $n = 0$, we set $C_0(A; 2) = 0$, $C_j(A; 2) = A^{\otimes j}$ and $d : a_j| \ldots |a_1 \mapsto \sum_p^{j-1}(-1)^p a_j| \ldots |\Delta a_p| \ldots |a_1$.

4.5. Two-Sided Hochschild Complexes Connected with a Discriminantal Configuration

Let $U_q \bar{\mathfrak{g}}$ be a quantum group of the form (4.1) such that $U_q \bar{\mathfrak{g}}$ is generated by $1, \bar{e}_i, \bar{f}_i, i = 1, \ldots, s$, and the space $\bar{\mathfrak{h}}$, and let $\bar{\alpha}_i, \ldots, \bar{\alpha}_s$ be its simple roots.

Let $s = n + k$, where $n, k \in \mathbb{N}$; let $U_q\bar{\mathfrak{n}}'_- \subset U_q\bar{\mathfrak{g}}$ (resp. $^+U_q\bar{\mathfrak{n}}'_-$) be the subalgebra generated by $1, f_i$ (resp. f_i) for $i = 1,\ldots,k$; Let $U_q\bar{\mathfrak{n}}_- \subset U_q\bar{\mathfrak{g}}$ (resp. $^+U_q\bar{\mathfrak{n}}$) be the subalgebra generated by 1, f_i (resp. f_i) for $i = 1,\ldots,s$; $\overline{M}_i \subset U_q\bar{\mathfrak{g}}$ the two-sided submodule over $U_q\bar{\mathfrak{n}}'_-$ generated by f_{k+i} for $i = 1,\ldots,n$.

Set $\overline{M} = (\overline{M}_1,\ldots,\overline{M}_n)$, $b_{ij} = (\bar{\alpha}_i, \bar{\alpha}_j)$ for all i, j.

$^+U_q\bar{\mathfrak{n}}_-$ has the coalgebra structure defined in (4.3). The restriction of this coalgebra structure to $^+U_q\bar{\mathfrak{n}}'_-$, $\overline{M}_1, \ldots, \overline{M}_n$ defines a coalgebra structure on $^+U_q\bar{\mathfrak{n}}'_-$ and a bicomodule $^+U_q\bar{\mathfrak{n}}'_-$-structure on $\overline{M}_1, \ldots, \overline{M}_n$:

$$\begin{aligned}^+\Delta : {}^+U_q\bar{\mathfrak{n}}'_- &\longrightarrow {}^+U_q\bar{\mathfrak{n}}'_- \otimes {}^+U_q\bar{\mathfrak{n}}'_- \\ ^+\Delta : \overline{M}_i &\longrightarrow \overline{M}_i \otimes {}^+U_q\bar{\mathfrak{n}}'_- \oplus {}^+U_q\bar{\mathfrak{n}}'_- \otimes \overline{M}_i .\end{aligned} \quad (4.5.1)$$

There is the weight decomposition

$$^+U_q\bar{\mathfrak{n}}'_- = \oplus'(U_q\bar{\mathfrak{n}}_-)_\lambda, \quad \overline{M}_i = \oplus''(U_q\bar{\mathfrak{n}}_-)_\lambda \quad \text{for } i = 1,\ldots,n$$

where the first sum is taken over all $\lambda = (\ell_1,\ldots,\ell_k,0,\ldots,0) \in \mathbb{N}^s$, $\lambda \neq 0$, and the second sum is taken over all $\lambda = (\ell_1,\ldots,\ell_k,0,\ldots,0,1_{k+i},0,\ldots,0) \in \mathbb{N}^s$.

The multiplication $\mu : {}^+U_q\bar{\mathfrak{n}}'_- \otimes {}^+U_q\bar{\mathfrak{n}}'_- \to {}^+U_q\bar{\mathfrak{n}}'_-$ and the comultiplications (4.5.1) preserve the natural weight decomposition of the tensor products.

Set $(^+U_q\bar{\mathfrak{n}}'_-)^* = \oplus'(U_q\bar{\mathfrak{n}}_-)^*_\lambda$, $\overline{M}_i^* = \oplus''(U_q\bar{\mathfrak{n}}_-)^*_\lambda$. In particular, we have $(^+U_q\bar{\mathfrak{n}}'_-)^*$, $\overline{M}_i^* \subset (U_q\bar{\mathfrak{n}}_-)^*$. Let $S : U_q\bar{\mathfrak{n}}_- \to (U_q\bar{\mathfrak{n}}_-)^*$ be the contravariant form. Then $S(\overline{M}_i) \subset \overline{M}_i^*$, $S(^+U_q\bar{\mathfrak{n}}'_-) \subset (^+U_q\bar{\mathfrak{n}}'_-)^*$.

Let $C.(^+U_q\bar{\mathfrak{n}}'_-; \overline{M}; 2; \mu)$ be the two-sided Hochschild complex of the algebra $^+U_q\bar{\mathfrak{n}}'_-$ with coefficients in the sequence of the modules $\overline{M} = (\overline{M}_1, \ldots, \overline{M}_n)$, let $C.(^+U_q\bar{\mathfrak{n}}'_-; \overline{M}; 2; {}^+\Delta)$ be the two-sided Hochschild complex of the coalgebra $U_q\bar{\mathfrak{n}}'_-$ with coefficients in the sequence of the comodules $\overline{M}$.

These complexes have the same spaces of j-chains for all j.

The differentials of these complexes preserve the weight decomposition. Therefore the complexes are direct sums of their weight subcomplexes:

$$C.(^+U_q\bar{\mathfrak{n}}'_-; \overline{M}; 2; \mu)_\lambda, \quad C.(^+U_q\bar{\mathfrak{n}}'_-; \overline{M}; 2; {}^+\Delta)_\lambda \quad (4.5.2)$$

where the weights have the form $\lambda = (\ell_1,\ldots,\ell_k,1,\ldots,1) \in \mathbb{N}^s$.

Let

$$C.^*(^+U_q\bar{\mathfrak{n}}'_-; \overline{M}; 2; \mu), \quad C.^*(^+U_q\bar{\mathfrak{n}}'_-; \overline{M}; 2; \mu)_\lambda \quad (4.5.3)$$

be the complexes dual to the complexes $C.(^+U_q\bar{\mathfrak{n}}'_-;\overline{M};2;\mu)$ and $C.(^+U_q\bar{\mathfrak{n}}'_-;\overline{M};2;\mu)_\lambda$, respectively.

For any j and λ, consider the space $C_j(^+U_q\bar{\mathfrak{n}}'_-;\overline{M};2)_\lambda$ of j-chains of complexes (4.5.2) and define the contravariant form

$$S: C_j(^+U_q\bar{\mathfrak{n}}'_-;\overline{M};2)_\lambda \longrightarrow C_j(^+U_q\bar{\mathfrak{n}}'_-;\overline{M};2)^*_\lambda \qquad (4.5.4)$$

as the tensor product of the contravariant forms of factors:

$$S: a_{j+n}|\ldots|a_1 \mapsto Sa_{j+n}|\ldots|Sa_1.$$

4.5.5. Lemma. *The contravariant form S defines a homomorphism of complexes*

$$S: C.(^+U_q\bar{\mathfrak{n}}'_-;\overline{M};2;{}^+\Delta)_\lambda \longrightarrow C.^*(^+U_q\bar{\mathfrak{n}}'_-;\overline{M};2;\mu)_\lambda.$$

The lemma is a direct corollary of Lemma (4.3.6).

Let $\sigma \in \Sigma_n$ be a permutation. Set $\overline{M}_\sigma = (\overline{M}_{\sigma(1)},\ldots,\overline{M}_{\sigma(n)})$. Consider the two-sided Hochschild complex $C.(^+U_q\bar{\mathfrak{n}}'_-;\overline{M}_\sigma;2;\mu)$ of the algebra $^+U_q\bar{\mathfrak{n}}'_-$ with coefficients in the sequence of the modules $\overline{M}_{\sigma(1)},\ldots,\overline{M}_{\sigma(n)}$ and the two-sided Hochschild complex $C.(^+U_q\bar{\mathfrak{n}}'_-;\overline{M}_\sigma;2;{}^+\Delta)$ of the coalgebra $^+U_q\bar{\mathfrak{n}}'_-$ with coefficients in the sequence of the comodules $\overline{M}_{\sigma(1)},\ldots,\overline{M}_{\sigma(n)}$.

The tensor product of the contravariant forms defines a homomorphism of the complexes

$$S: C.(^+U_q\bar{\mathfrak{n}}'_-;\overline{M}_\sigma;2;{}^+\Delta)_\lambda \longrightarrow C.^*(^+U_q\bar{\mathfrak{n}}'_-;\overline{M}_\sigma;2;\mu)_\lambda \qquad (4.5.6)$$

for any λ.

4.6. Algebraic Interpretation of the Abstract Complexes $\mathcal{X}.(\mathcal{C}_{n,k})$ and $\mathcal{Q}.(\mathcal{C}_{n,k})$ of a Discriminantal Configuration

Consider the quantum group $U_q\bar{\mathfrak{g}}$ of (4.5). Introduce the set of numbers $(a_{\ell,m})_{1\leq \ell,m\leq s}$ by

$$a_{\ell,m} = b_{\ell,m}/\varkappa. \qquad (4.6.1)$$

Here $\varkappa$ is the parameter of the quantum group, $q = \exp(2\pi i/\varkappa)$, $b_{\ell,m} = (\bar{\alpha}_\ell, \bar{\alpha}_m)$.

Consider the discriminantal configuration $\mathcal{C}_{n,k}(z)$ of (3.1) weighted by the weights $a = (a_{\ell,m})$ and cooriented by the distinguished coorientation of (3.10). Let $\mathcal{X}.(\mathcal{C}_{n,k})$, $\mathcal{Q}.(\mathcal{C}_{n,k})$ be its abstract complexes as in (3.14).

We define an isomorphism of these complexes and the complexes

$$C^{\cdot *}(^+U_q\bar{\mathfrak{n}}'_-; \overline{M}; 2; {}^+\Delta)_\lambda, C.(^+U_q\bar{\mathfrak{n}}'_-; \overline{M}; 2; \mu)_\lambda \quad \text{for} \quad \lambda = (1,\ldots,1) \in \mathbf{N}^s.$$

Define $p : \{\bar{f}_1,\ldots,\bar{f}_s\} \to \{x_1,\ldots,x_k,z_1,\ldots,z_n\}$ as $p(\bar{f}_j) = x_j$ for $j = 1,\ldots,k$, $p(\bar{f}_j) = z_{j-k}$ for $j = k+1,\ldots,s$.

Let

$$\bar{f} = \bar{f}_1(1)\ldots\bar{f}_{q(1)}(1)|\bar{f}_1(2)\ldots\bar{f}_{q(2)}(2)|\ldots|\bar{f}_1(n+j)\ldots\bar{f}_{q(n+j)}(n+j) \quad (4.6.2)$$

be a monomial of $C_j(^+U_q\bar{\mathfrak{n}}'_-; \overline{M}; 2)_\lambda$, $\lambda = (1,\ldots,1) \in \mathbf{N}^s$. Here $\bar{f}_1(1),\ldots,\bar{f}_{q(1)}(1), \bar{f}_1(2),\ldots,\bar{f}_{q(n+j)}(n+j)$ is a permutation of $\bar{f}_1,\ldots,\bar{f}_s$, and each factor $\bar{f}(\ell) = \bar{f}_1(\ell)\ldots\bar{f}_{q(\ell)}(\ell)$ of the tensor product $\bar{f}$ belongs to one of $^+U_q\bar{\mathfrak{n}}'_-$, $\overline{M}_1,\ldots\overline{M}_n$. Set

$$p(\bar{f}) = p(\bar{f}_1(1))\ldots p(\bar{f}_{q(1)}(1))|\ldots|p(\bar{f}_1(n+j))\ldots p(\bar{f}_{q(n+j)}(n+j)). \quad (4.6.3)$$

Then $p(\bar{f})$ is an admissible monomial of the form (3.8.1). Formula (4.6.3) defines an isomorphism

$$p : C_j(^+U_q\bar{\mathfrak{n}}'_-; \overline{M}; 2)_\lambda \longrightarrow \mathcal{X}_{k-j}(\mathcal{C}_{n,k}) \quad \text{for all } j \geq 0. \quad (4.6.4)$$

Let $\bar{f}$ be given by (4.6.2) and let $(\bar{f})^*$ be the corresponding element of $C_j(^+U_q\bar{\mathfrak{n}}'_-; \overline{M}; 2)^*_\lambda$, $(\bar{f})^*$ is an element of the basis dual to the monomial basis. The formula $(\bar{f})^* \mapsto (p(\bar{f}))^*$ defines an isomorphism

$$p : C_j(^+U_q\bar{\mathfrak{n}}'_-; \overline{M}; 2)^*_\lambda \longrightarrow \mathcal{Q}_{k-j}(\mathcal{C}_{n,k}) \quad \text{for all } j \geq 0. \quad (4.6.5)$$

4.6.6. Theorem. *Maps (4.6.4) and (4.6.5) define isomorphisms of complexes*

$$p : C.(^+U_q\bar{\mathfrak{n}}'_-; \overline{M}; 2; {}^+\Delta)_\lambda \longrightarrow \mathcal{X}_{k-.}(\mathcal{C}_{n,k}),$$
$$p : C^{\cdot *}(^+U_q\bar{\mathfrak{n}}'_-; \overline{M}; 2; \mu)_\lambda \longrightarrow \mathcal{Q}_{k-.}(\mathcal{C}_{n,k}),$$

where $\lambda = (1,\ldots,1) \in \mathbf{N}^s$. *The following diagram is commutative:*

$$\begin{array}{ccc} C.(^+U_q\bar{\mathfrak{n}}'_-; \overline{M}; 2; {}^+\Delta)_\lambda & \xrightarrow{S} & C^{\cdot *}(^+U_q\bar{\mathfrak{n}}'_-; \overline{M}; 2; \mu)_\lambda \\ {\scriptstyle p}\downarrow & & \downarrow{\scriptstyle p} \\ \mathcal{X}_{k-.}(\mathcal{C}_{n,k}) & \xrightarrow{S} & \mathcal{Q}_{k-.}(\mathcal{C}_{n,k}). \end{array}$$

The upper S is the contravariant form defined in (4.5.4), the lower S is the form defined in (2.9).

Proof. The differentials of the corresponding complexes are defined by the same formulas. The maps S are defined by the same formulas.

4.6.7. Corollary. *The following diagrams are commutative:*

$$\begin{array}{ccc}
(^+U_q\bar{\mathfrak{n}}'_-)_{\lambda_0} & \xrightarrow{S} & (^+U_q\bar{\mathfrak{n}}'_-)_{\lambda_0} \\
{\scriptstyle p}\downarrow & & \downarrow{\scriptstyle p} \\
\mathcal{X}_{k-1}(\mathcal{C}_{0,k}) & \xrightarrow{S} & Q_{k-1}(\mathcal{C}_{0,k}) ,
\end{array}$$

$$\begin{array}{ccc}
(\overline{M}_i)_{\lambda_i} & \xrightarrow{S} & (\overline{M}_i)_{\lambda_i} \\
{\scriptstyle p}\downarrow & & \downarrow{\scriptstyle p} \\
\mathcal{X}_k(\mathcal{C}_{1,k}(z_i)) & \xrightarrow{S} & Q_k(\mathcal{C}_{1,k}(z_i)) .
\end{array}$$

Here $\lambda_0 = (1_1, \ldots, 1_k, 0, \ldots, 0) \in \mathbf{N}^s$, $i = 1, \ldots, n$, $\lambda_i = (1_1, \ldots, 1_k, 0, \ldots, 0, 1_{k+i}, 0, \ldots, 0) \in \mathbf{N}^s$. The weights of $\mathcal{C}_{0,k}$ are $\{a_{\ell,m} | 1 \leq \ell < m \leq k\}$, the weights of $\mathcal{C}_{1,k}(z_i)$ are $\{a_{\ell,m} | 1 \leq \ell < m \leq k \text{ or } 1 \leq \ell \leq k, m = k+i\}$, where $a_{\ell,m}$ are given by (4.6.1). The lower S in the diagrams is defined in (2.9) and in these cases coincides with the reduced quantum bilinear form of the corresponding configuration, see (2.7.2).

The theorem and its corollary give a geometric interpretation for the quantum group complexes of (4.5) and an algebraic interpretation for the geometric complexes $\mathcal{X}.$, $Q.$ of a discriminantal configuration.

In particular, Theorem (3.11.2) and Lemma (3.11.5) give a formula for the determinant of the forms $S : (^+U_q\bar{\mathfrak{n}}_-)_\lambda \to (^+U_q\bar{\mathfrak{n}}_-)^*_\lambda$ for $\lambda = (1_1, \ldots, 1_k, 0, \ldots, 0)$ and $S : (\overline{M}_i)_{\lambda_i} \to (\overline{M}_i)^*_{\lambda_i}$. The formula shows nondegeneracy of these forms for generic values of the weights $a = \{a_{\ell,m}\}$.

4.6.8. Corollary. *Let $\lambda = (1, \ldots, 1) \in \mathbf{N}^s$. Then $H_j C^{\cdot*}(^+U_q\bar{\mathfrak{n}}_-; \overline{M}; 2; \mu)_\lambda = 0$ for $j > 0$ or for $j = 0$, and $n \leq 1$. There is a basis of $H_0 C^{\cdot*}(^+U_q\bar{\mathfrak{n}}_-; \overline{M}; 2; \mu)_\lambda$ numerated by bounded domains of $\mathcal{C}_{n,k}(z_1, \ldots, z_n)$, it is described in (3.15).*

Let $F \subset \{\bar{f}_1, \ldots, \bar{f}_s\}$ be a subset such that $\#\{F \cap \{\bar{f}_{k+1}, \bar{f}_{k+2}, \ldots, \bar{f}_s\}\} \leq 1, \#F > 1$. F is called a *resonance subset* if the number

$$a(F) = \sum_{\substack{\ell,m \in F \\ \ell < m}} a_{\ell,m} = \sum_{\substack{\ell,m \in F \\ \ell < m}} b_{\ell,m}/\varkappa \qquad (4.6.9)$$

is an integer.

4.6.10. Corollary. *For $\lambda = (1,\ldots,1) \in \mathbb{N}^s$, the homomorphism $S: C_j({}^+U_q\bar{\mathfrak{n}}_-; \overline{M}; 2)_\lambda \to C_j({}^+U_q\bar{\mathfrak{n}}_-; \overline{M}; 2)^*_\lambda$ is an isomorphism if there is no resonance subset F such that $\#F \leq k - j + 1$, see (3.12.9).*

Remark. If the number

$$b(F) = \sum_{\substack{\ell,m \in F \\ \ell < m}} b_{\ell,m}$$

is not zero, then F is not a resonance subset for generic values of the parameter $\varkappa$.

4.7. Geometric Interpretation of the Complexes $C.({}^+U_q\bar{\mathfrak{n}}'_-; \overline{M}; 2; {}^+\Delta)_\lambda$ and $C.^*({}^+U_q\bar{\mathfrak{n}}'_-; \overline{M}; 2; \mu)_\lambda$ Defined in (4.5)

Let $U_q\bar{\mathfrak{g}}$ be the quantum group considered in (4.5), and let $C.({}^+U_q\bar{\mathfrak{n}}'_-; \overline{M}; 2; {}^+\Delta)_\lambda$ and $C.^*({}^+U_q\bar{\mathfrak{n}}'_-; \overline{M}; 2; \mu)_\lambda$ be the complexes defined in (4.5). Fix $\lambda = (\ell_1, \ldots, \ell_k, 1, \ldots, 1) \in \mathbb{N}^s$. Set $\ell = \ell_1 + \cdots + \ell_k$, $\nu = (\ell_1, \ldots, \ell_k) \in \mathbb{N}^k$.

Consider a discriminantal configuration $\mathcal{C}_{n,\ell}(z)$, $z = (z_1, \ldots, z_n)$, $z_1 < \cdots < z_n$. $\mathcal{C}_{n,\ell}$ is a configuration of hyperplanes in $\mathbb{R}^\ell$. Let $x_1(1), \ldots, x_{\ell_1}(1)$, $x_1(2), \ldots, x_{\ell_2}(2), \ldots, x_1(k), \ldots, x_{\ell_k}(k)$ be coordinates in $\mathbb{R}^\ell$.

Define the weights of $\mathcal{C}_{n,\ell}$:

$$\begin{aligned} a(x_i(j), x_u(v)) &= (\bar{\alpha}_j, \bar{\alpha}_v)/\varkappa, \\ a(x_i(j), z_u) &= (\bar{\alpha}_j, \bar{\alpha}_{k+u})/\varkappa \end{aligned} \qquad (4.7.1)$$

for all i, j, u, v. Here $\bar{\alpha}_1, \ldots, \bar{\alpha}_s$ are the simple roots of $U_q\bar{\mathfrak{g}}$.

Let $\Sigma' = \Sigma_{\ell_1} \times \cdots \times \Sigma_{\ell_k}$. Σ' acts on $\mathcal{C}_{n,\ell}$ preserving the weights of the configuration. This action was considered in (3.13) and (3.16). In this situation, the configuration $\mathcal{C}_{n,\ell}(z)$ was denoted by $\mathcal{C}_{n,\ell,\nu}(z)$, see (3.16). Let $\mathcal{X}.(\mathcal{C}_{n,\ell,\nu})^-$ and $\mathcal{Q}.(\mathcal{C}_{n,\ell,\nu})^-$ be the complexes introduced in (3.16).

Below we define isomorphisms

$$\begin{aligned} p &: C_j({}^+U_q\bar{\mathfrak{n}}'_-; \overline{M}; 2)_\lambda \longrightarrow \mathcal{X}_{\ell-j}(\mathcal{C}_{n,\ell,\nu})^-, \\ p &: C_j^*({}^+U_q\bar{\mathfrak{n}}'_-; \overline{M}; 2)_\lambda \longrightarrow \mathcal{Q}_{\ell-j}(\mathcal{C}_{n,\ell,\nu})^-, \end{aligned} \qquad (4.7.2)$$

for all j.

Namely, set $X(j) = \{x_1(j), \ldots, x_{\ell_j}(j)\}$ for $j = 1, \ldots, k$. Define $p: \{\bar{f}_1, \ldots, \bar{f}_s\} \to \{X(1), \ldots, X(k), z_1, \ldots, z_n\}$ as $p(\bar{f}_j) = X(j)$ for $j = 1, \ldots, k$, $p(\bar{f}_j) = z_{j-k}$ for $j = k+1, \ldots, s$.

Let
$$\bar{f} = \bar{f}_1(1) \ldots \bar{f}_{q(1)}(1) | \bar{f}_1(2) \ldots \bar{f}_{q(2)}(2) | \ldots | \bar{f}_1(n+j) \ldots \bar{f}_{q(n+j)}(n+j) \quad (4.7.3)$$

be a monomial in $C_j(^+U_q \bar{\mathfrak{n}}'_-; \overline{M}; 2)_\lambda$. Here $\bar{f}_a(b) \in \{\bar{f}_1, \ldots, \bar{f}_s\}$ for all a, b. Set
$$\begin{aligned} p(\bar{f}) = p(\bar{f}_1(1)) \ldots p(\bar{f}_{q(1)}(1)) | \ldots | \\ p(\bar{f}_1(n+j)) \ldots p(\bar{f}_{q(n+j)}(n+j)). \end{aligned} \quad (4.7.4)$$

Then $p(\bar{f})$ is a generalized admissible monomial of the form (3.13.12) and, therefore, is an element of $\mathcal{X}_{\ell-j}(C_{n,\ell,\nu})^-$, see (3.16). Formula (4.7.4) defines the first isomorphism of (4.7.2).

Let $\bar{f}$ be given by (4.7.3) and $(\bar{f})^*$ be the corresponding element of $C_j^*(^+U_q \bar{\mathfrak{n}}'_-; \overline{M}; 2)_\lambda$. The formula $(\bar{f})^* \mapsto (p(\bar{f}))^*$ defines the second isomorphisms of (4.7.2).

4.7.5. Theorem. *Isomorphism (4.7.2) define isomorphisms of complexes*
$$p : C_{\cdot}(^+U_q \bar{\mathfrak{n}}'_-; \overline{M}; 2; {}^+\Delta)_\lambda \longrightarrow \mathcal{X}_{\ell-\cdot}(C_{n,\ell,\nu})^-,$$
$$p : C^*_{\cdot}(^+U_q \bar{\mathfrak{n}}'_-; \overline{M}; 2; \mu)_\lambda \longrightarrow Q_{\ell-\cdot}(C_{n,\ell,\nu})^-$$

for an arbitrary $\lambda = (\ell_1, \ldots, \ell_k, 1, \ldots, 1) \in \mathbb{N}^s$. *The following diagram is commutative:*

$$\begin{array}{ccc} C_{\cdot}(^+U_q \bar{\mathfrak{n}}'_-; \overline{M}; 2; {}^+\Delta)_\lambda & \xrightarrow{S} & C^*_{\cdot}(^+U_q \bar{\mathfrak{n}}'_-; \overline{M}; 2; \mu)_\lambda \\ {\scriptstyle p}\downarrow & & \downarrow {\scriptstyle p} \\ \mathcal{X}_{\ell-\cdot}(C_{n,\ell,\nu})^- & \longrightarrow & Q_{k-\cdot}(C_{n,\ell,\nu})^-. \end{array}$$

The upper S is the contravariant form defined in (4.5.4), the lower S is the geometric form defined in (3.16.6).

The theorem is proved in (4.9).

4.7.6. Corollary. *The following diagrams are commutative:*

$$\begin{array}{ccc} (^+U_q \bar{\mathfrak{n}}'_-)_{\lambda_0} & \xrightarrow{S} & (^+U_q \bar{\mathfrak{n}}'_-)^*_{\lambda_0} \\ {\scriptstyle p}\downarrow & & \downarrow {\scriptstyle p} \\ \mathcal{X}_{\ell-1}(C_{0,\ell,\nu})^- & \xrightarrow{S} & Q_{\ell-1}(C_{0,\ell,\nu})^-, \end{array}$$

$$(\overline{M}_i)_{\lambda_i} \xrightarrow{S} (\overline{M}_i)^*_{\lambda_i}$$
$$p \downarrow \qquad \qquad \downarrow p$$
$$\mathcal{X}_\ell(\mathcal{C}_{1,\ell,\nu}(z_i))^- \xrightarrow{S} Q_\ell(\mathcal{C}_{1,\ell,\nu}(z_i))^-.$$

Here $\lambda_0 = (\ell_1, \ldots, \ell_k, 0, \ldots, 0) \in \mathbb{N}^s$. The weights of $\mathcal{C}_{0,\ell,\nu}$ are the numbers $\{a(x_a(b), x_u(v))\}$ of (4.7.1), $i = 1, \ldots, n$. $\lambda_i = (\ell_1, \ldots, \ell_k, 0, \ldots, 0, 1_{k+i}, 0, \ldots, 0)$. The weights of $\mathcal{C}_{1,\ell,\nu}(z_i)$ are the numbers $\{a(x_a(b), x_u(v)), a(x_a(b), z_i)\}$ of (4.7.1).

The lower S of the diagrams is the geometric form defined in (3.16.6). In these cases, the lower S coincides with the reduced quantum bilinear form of the corresponding configuration restricted to an invariant subspace of the action of Σ' on the space of linear combination of domains of the configuration.

4.7.7. Corollary Let $\lambda = (\ell_1, \ldots, \ell_k, 1, \ldots, 1) \in \mathbb{N}^s$. Then $H_j C^{\cdot}(^+U_q \bar{\mathfrak{n}}_-; \overline{M}; 2; \mu)_\lambda = 0$ for $j > 0$ or $j = 0$, $n \leq 1$. The dimension of $H_0 C^{\cdot}(^+U_q \bar{\mathfrak{n}}_-; \overline{M}; 2; \mu)_\lambda$ for $n > 1$ is equal to the number of the orbits of the action of Σ' on the set of bounded domains of $\mathcal{C}_{n,\ell,\nu}$. The space $H_0 C^{\cdot}(^+U_q \bar{\mathfrak{n}}_-; \overline{M}; 2; \mu)_\lambda$ has a basis described below in (4.7.10).

Let
$$S = \bar{f}_1(1), \ldots, \bar{f}_{s(1)}(1), \bar{f}_{k+1}, \bar{f}_1(2), \ldots, \bar{f}_{s(2)}(2), \\ \bar{f}_{k+2}, \bar{f}_1(3), \ldots, \bar{f}_1(n), \ldots, \bar{f}_{s(n)}(n), \bar{f}_{k+n} \quad (4.7.8)$$

be a sequence such that $\bar{f}_a(b) \in \{\bar{f}_1, \ldots, \bar{f}_k\}$ for all a, b and each $\bar{f}_j$ for $j = 1, \ldots, k$, appears in S in exactly ℓ_j positions. Such a sequence will be called an *admissible sequence* of weight $\lambda = (\ell_1, \ldots, \ell_k, 1, \ldots, 1) \in \mathbb{N}^s$.

Set
$$G(S) = \sum (g)^*, \quad (4.7.9)$$
where the sum is taken over all monomials,
$$g = g_1(1) \ldots g_{u(1)}(1) \bar{f}_{k+1} g_{u(1)+1}(1) \ldots g_{\nu(1)}(1) | \ldots | g_1(n) \ldots \\ g_{u(n)}(n) \bar{f}_{n+k} \in (\overline{M}_1 \otimes \cdots \otimes \overline{M}_n)_\lambda,$$
such that $S = g_1(1), \cdots, g_{u(1)}(1), \bar{f}_{k+1}, g_{u(1)+1}(1), \ldots, g_{\nu(1)}(1), \ldots, g_1(n), \ldots, g_{u(n)}(n), \bar{f}_{k+n}$.

Example. Let $n = 2$. Then $G(\bar{f}_1 \bar{f}_{k+1} \bar{f}_1 \bar{f}_1 \bar{f}_{k+2}) = (\bar{f}_1 \bar{f}_{k+1} \bar{f}_1 \bar{f}_1 | \bar{f}_{k+2})^* + (\bar{f}_1 \bar{f}_{k+1} \bar{f}_1 | \bar{f}_1 \bar{f}_{k+2})^* + (\bar{f}_1 \bar{f}_{k+1} | \bar{f}_1 \bar{f}_1 \bar{f}_{k+2})^*$.

(4.7.10) A basis of $H_0 C^{\cdot}(^+U_q \bar{\mathfrak{n}}_-; \overline{M}; 2; \mu)_\lambda$ is generated by the elements $G(S)$, where S runs through the set of all admissible sequences of weight λ such that S starts with $\bar{f}_{k+1}$, that is, $s(1) = 0$.

4.7.11. Example. Let $k = 1$, $n = 3$, $\lambda = (2,1,1,1)$. Then a basis of H_0 is generated by the elements $G(\bar{f}_2\bar{f}_3\bar{f}_1\bar{f}_1\bar{f}_4)$, $G(\bar{f}_2\bar{f}_1\bar{f}_3\bar{f}_1\bar{f}_4)$, $G(\bar{f}_2\bar{f}_1\bar{f}_1\bar{f}_3\bar{f}_4)$.

For an admissible sequence S of form (4.7.8), the monomial

$$\bar{f}(S) = \bar{f}_1(1)\ldots \bar{f}_{s(1)}\bar{f}_{k+1}|\ldots|f_1(i)\ldots f_{s(i)}(i)\bar{f}_{k+i}|\ldots| \\ \bar{f}_1(n)\ldots \bar{f}_{s(n)}(n)\bar{f}_{k+n} \qquad (4.7.12)$$

in $\overline{M}_1 \otimes \cdots \otimes \overline{M}_n$ will be called the right monomial associated with S.

Let $\lambda = (\ell_1,\ldots,\ell_k,1,\ldots,1) \in \mathbb{N}^s$, $\lambda' = (\ell'_1,\ldots,\ell'_s,) \in \mathbb{N}^s$, let $\lambda - \lambda'$ be a vector with non-negative coordinates, $\ell'_1 + \cdots + \ell'_s > 1, \ell'_{k+1} + \cdots + \ell'_s \leq 1$. Such a vector λ' will be called a *resonance vector* if the number

$$a(\lambda') = \sum_{i<j} \ell'_i\ell'_j b_{ij}/\varkappa + \sum_{i=1}^{k} \ell'_i(\ell'_i - 1)b_{ii}/2\varkappa \qquad (4.7.13)$$

is an integer.

4.7.14. Corollary. *The contravariant form*

$$S : C_j({}^+U_q\bar{\mathfrak{n}}'_-;\overline{M};2)_\lambda \longrightarrow C_j({}^+U_q\bar{\mathfrak{n}}'_-;\overline{M};2)^*_\lambda$$

is an isomorphism if there is no resonance vector λ' *such that* $\ell'_1 + \cdots + \ell'_s \leq \ell - j + 1$, *see (3.12.9).*

4.8. Symmetrization

Fix a quantum group $U_q\mathfrak{g}$ given by the data in (4.1.a)–(4.1.d). Fix a weight $\lambda = (\ell_1,\ldots,\ell_r)$. Set $\ell = \ell_1 + \cdots + \ell_r$.

Consider a quantum group $U_q\tilde{\mathfrak{g}}$ given by the data

(a) a finite-dimensional complex vector space $\tilde{\mathfrak{h}}$,

(b) a nondegenerate symmetric bilinear form $(\,,\,)$ on $\tilde{\mathfrak{h}}$,

(c) linearly independent covectors $\alpha_1(1),\ldots,\alpha_{\ell_1}(1), \alpha_1(2),\ldots,\alpha_{\ell_2}(2),\ldots,$
    $\alpha_1(r),\ldots,\alpha_{\ell_r}(r) \in \tilde{\mathfrak{h}}^*$;

(d) a nonzero complex number $\varkappa$, the same as in the definition of $U_q\mathfrak{g}$.

Such a quantum group is generated by elements 1, $e_i(j)$, $f_i(j)$ for $j = 1,\ldots,r$, for $i = 1,\ldots,\ell_j$, and the space $\tilde{\mathfrak{h}}$.

The quantum group $U_q\tilde{\mathfrak{g}}$ will be called *adjacent* to the quantum group $U_q\mathfrak{g}$ at weight λ if

$$(\alpha_\ell(i), \alpha_m(j)) = (\alpha_i, \alpha_j) \qquad (4.8.1)$$

for all ℓ, m, i, j; where the first scalar product is taken with respect to the bilinear form on $\tilde{\mathfrak{h}}^*$ induced by $(\,,\,) : \tilde{\mathfrak{h}} \to \tilde{\mathfrak{h}}^*$, and the second scalar product is the scalar product on $\mathfrak{h}^*$ of $U_q\mathfrak{g}$.

An adjacent quantum group exists according to [K]. Fix $U_q\tilde{\mathfrak{g}}$. Set $\tilde{\lambda} = (1,\ldots,1) \in \mathbb{N}^\ell$. Let $^+U_q\tilde{\mathfrak{n}}_- \subset U_q\tilde{\mathfrak{g}}$ be the subalgebra generated by all $f_i(j)$. Denote by μ its multiplication. Denote by $^+\Delta$ the coalgebra structure on $U_q\tilde{\mathfrak{n}}$ introduced in (4.3). Consider the complexes

$$C_\cdot(^+U_q\tilde{\mathfrak{n}}_-; 2; \mu)_{\tilde{\lambda}}, \qquad C_\cdot(^+U_q\tilde{\mathfrak{n}}_-; 2; {}^+\Delta)_{\tilde{\lambda}}, \qquad (4.8.2)$$

see (4.4.2.2) and (4.4.3.2), respectively. By definition, $C_j(^+U_q\tilde{\mathfrak{n}}_-; 2; \mu)_{\tilde{\lambda}} = C_j(^+U_q\tilde{\mathfrak{n}}_-; 2; {}^+\Delta)_{\tilde{\lambda}} = ((^+U_q\tilde{\mathfrak{n}}_-)^{\otimes j})_{\tilde{\lambda}}$.

Let $\Sigma' = \Sigma_{\ell_1} \times \cdots \times \Sigma_{\ell_r}$ be the direct product of the permutation groups. Σ' acts on these complexes as follows. Let $a = (a_1, \ldots, a_r) \in \Sigma'$, let f be a monomial in $((^+U_q\tilde{\mathfrak{n}}_-)^{\otimes j})_{\tilde{\lambda}}$, define $a(f) \in ((^+U_q\tilde{\mathfrak{n}}_-)^{\otimes j})_{\tilde{\lambda}}$ as the monomial obtained from f by the permutation sending $f_i(j)$ to $f_{a_j(i)}(j)$. This action of Σ' commutes with the differentials of the complexes by virtue of (4.8.1).

Example. Let $r = 2$, $\ell_1 = \ell_2 = 2$, $a = ((2,1),(2,1)) \in \Sigma_2 \times \Sigma_2$, $j = 3$, and $f = f_1(2)|f_2(1)f_2(2)|f_1(1)$, then $a(f) = f_2(2)|f_1(1)f_1(2)|f_2(1)$.

Denote by

$$\begin{aligned} C_\cdot(^+U_q\tilde{\mathfrak{n}}_-; 2; \mu)_{\tilde{\lambda}}^+ &\subset C_\cdot(^+U_q\tilde{\mathfrak{n}}_-; 2; \mu)_{\tilde{\lambda}}, \\ C_\cdot(^+U_q\tilde{\mathfrak{n}}_-; 2; {}^+\Delta)_{\tilde{\lambda}}^+ &\subset C_\cdot(^+U_q\tilde{\mathfrak{n}}_-; 2; {}^+\Delta)_{\tilde{\lambda}} \end{aligned} \qquad (4.8.3)$$

the subcomplexes consisting of all elements invariant under the action of Σ'.

According to (4.5.5), the contravariant form S on $^+U_q\tilde{\mathfrak{n}}_-$ defines a homomorphism of complexes:

$$S : C_\cdot(^+U_q\tilde{\mathfrak{n}}_-; 2; {}^+\Delta)_{\tilde{\lambda}}^+ \longrightarrow C_\cdot^*(^+U_q\tilde{\mathfrak{n}}_-; 2; \mu)_{\tilde{\lambda}}^+. \qquad (4.8.4)$$

Let $f_1 \ldots f_r$ be the generators of $^+U_q\mathfrak{n}_-$, let $\{f_i(j)\}$ be the generators of $^+U_q\tilde{\mathfrak{n}}_-$. Define a map $p : \{f_i(j)\} \to \{f_j\}$ where $f_i(j) \mapsto f_j$. The map p induces an isomorphism

$$p : ((^+U_q\tilde{\mathfrak{n}}_-)^{\otimes j})_{\tilde{\lambda}}^+ \longrightarrow ((^+U_q\mathfrak{n}_-)^{\otimes j})_\lambda \qquad (4.8.5)$$

for all j. Set

$$\pi = (\ell_1!\ldots\ell_r!)\cdot p^{-1} : ((^+U_q\mathfrak{n}_-)^{\otimes j})_\lambda \longrightarrow ((^+U_q\tilde{\mathfrak{n}}_-)^{\otimes j})^+_{\tilde{\lambda}},$$
$$\pi = p^* : ((^+U_q\mathfrak{n}_-)^{\otimes j})^*_\lambda \longrightarrow (((^+U_q\tilde{\mathfrak{n}}_-)^{\otimes j})^+_{\tilde{\lambda}})^*. \quad (4.8.6)$$

4.8.7. Example. $\pi : f_1^2 \mapsto f_1(1)f_2(1) + f_2(1)f_1(1)$, $\pi : f_1|f_1 \mapsto f_1(1)|f_2(1) + f_2(1)|f_1(1)$, $\pi : (f_1^2)^* \mapsto (f_1(1)f_2(1))^* + (f_2(1)f_1(1))^*$, $\pi : (f_1|f_1)^* \mapsto (f_1(1)|f_2(1))^* + (f_2(1)|f_1(1))^*$.

4.8.8. Theorem. *Isomorphisms (4.8.6) define isomorphisms of complexes:*

$$\pi : C.(^+U_q\mathfrak{n}_-; 2; {}^+\Delta)_\lambda \longrightarrow C.(^+U_q\tilde{\mathfrak{n}}_-; 2; {}^+\Delta)^+_{\tilde{\lambda}},$$
$$\pi : C^*_\cdot(^+U_q\mathfrak{n}_-; 2; \mu)_\lambda \longrightarrow C^*_\cdot(^+U_q\tilde{\mathfrak{n}}_-; 2; \mu)^+_{\tilde{\lambda}}.$$

Moreover, the following diagram is commutative:

$$\begin{array}{ccc} C.(^+U_q\mathfrak{n}_-; 2; {}^+\Delta)_\lambda & \xrightarrow{S} & C^*_\cdot(^+U_q\mathfrak{n}_-; 2; \mu)_\lambda \\ \pi\downarrow & & \downarrow\pi \\ C.(^+U_q\tilde{\mathfrak{n}}_-; 2; {}^+\Delta)^+_{\tilde{\lambda}} & \xrightarrow{S} & C^*_\cdot(^+U_q\tilde{\mathfrak{n}}_-; 2; \mu)^+_{\tilde{\lambda}}. \end{array}$$

Proof. Let

$$g = g_1(1)\ldots g_{q(1)}(1)|\ldots|g_1(p)\ldots g_{q(p)}(p)$$

be a monomial in $((^+U_q\mathfrak{n}_-)^{\otimes p})_\lambda$, where $g_i(j) \in \{f_1\ldots f_r\}$ for all i,j. To prove the theorem, it is sufficient to check the equalities:

$$(\pi{}^+\Delta_j - {}^+\Delta_j\pi)g = 0 \quad \text{for } j = 1,\ldots,p, \quad (4.8.9)$$

where ${}^+\Delta_j$ is the operator ${}^+\Delta$ applied to the jth factor of the tensor product,

$$(\pi\mu_j^* - \mu_j^*\pi)(g)^* = 0, \quad (4.8.10)$$

where μ_j^* is the operator μ^* applied to the jth factor of the tensor product,

$$(\pi S - S\pi)g = 0. \quad (4.8.11)$$

To prove equalities (4.8.9)–(4.8.11), it is sufficient to prove the equalities:

$$(\pi{}^+\Delta - {}^+\Delta\pi)g = 0, \quad (\pi\mu^* - \mu^*\pi)(g)^* = 0, \quad (\pi S - S\pi)g = 0 \quad (4.8.12)$$

for all monomials $g \in (U_q n_-)_\lambda$. These equalities are proved by a simple direct verification.

Examples. $(\pi^+\Delta - {}^+\Delta\pi)f_1^2 = \pi(q^{b_{11}/4} + q^{-b_{11}/4})f_1|f_1 - {}^+\Delta(f_1(1)f_2(1) + f_2(1)f_1(1)) = (1-1)(q^{b_{11}/4} + q^{-b_{11}/4})(f_1(1)|f_2(1) + f_2(1)|f_1(1)) = 0$. $(\pi\mu^* - \mu^*\pi)(f_1^2)^* = \pi(f_1|f_1)^* - \mu^*((f_1(1)f_2(1))^* + (f_2(1)f_1(1))^*) = (1-1)((f_1(1)|f_2(1))^* + (f_2(1)|f_1(1))^*) = 0$. $(\pi S - S\pi)f_1^2 = \pi(q^{b_{11}/4} + q^{-b_{11}/4})(f_1^2)^* - S(f_1(1)f_2(1) + f_2(1)f_1(1)) = 0$.

4.9. Proof of Theorem (4.7.5)

Let $U_q \tilde{\mathfrak{g}}$ be a quantum group adjacent to $U_q \tilde{\mathfrak{g}}$ at weight $\lambda = (\ell_1, \ldots, \ell_k, 1, \ldots, 1) \in \mathbb{N}^s$. $U_q \tilde{\mathfrak{g}}$ is generated by elements 1, $e_i(j)$, $f_i(j)$ for $j = 1, \ldots, k$, $i = 1, \ldots, \ell_j$, elements e_j, f_j, for $j = k+1, \ldots, s$, and the space $\tilde{\mathfrak{h}}$, see (4.8).

Let $U_q \tilde{\mathfrak{n}}'_- \subset U_q \tilde{\mathfrak{g}}$ (resp. ${}^+U_q \tilde{\mathfrak{n}}'_-$) be the subalgebra generated by 1, $f_i(j)$ (resp. $f_i(j)$) for all i, j; let $\widetilde{M}_i \subset U_q \tilde{\mathfrak{g}}$ be the two-sided submodule over $U_q \tilde{\mathfrak{n}}'_-$ generated by f_{k+i}, $i = 1, \ldots, n$. Set $\widetilde{M} = (\widetilde{M}_1, \ldots, \widetilde{M}_n)$.

Consider the complexes $C_\cdot({}^+U_q \tilde{\mathfrak{n}}'_-; \widetilde{M}; 2; {}^+\Delta)_{\tilde{\lambda}}$, $C^*_\cdot({}^+U_q \tilde{\mathfrak{n}}'_-; \widetilde{M}; 2; \mu)_{\tilde{\lambda}}$, where $\tilde{\lambda} = (1, \ldots, 1) \in \mathbb{N}^{n+\ell}$. According to (4.6.6), there are isomorphisms

$$p : C_\cdot({}^+U_q \tilde{\mathfrak{n}}'_-; \widetilde{M}; 2; {}^+\Delta)_{\tilde{\lambda}} \longrightarrow \mathcal{X}_{\ell-\cdot}(\mathcal{C}_{n,\ell,\nu})$$
$$p : C^*_\cdot({}^+U_q \tilde{\mathfrak{n}}'_-; \widetilde{M}; 2; \mu)_{\tilde{\lambda}} \longrightarrow \mathcal{Q}_{\ell-\cdot}(\mathcal{C}_{n,\ell,\nu}),$$
(4.9.1)

and the following diagram is commutative

$$\begin{array}{ccc} C_\cdot({}^+U_q \tilde{\mathfrak{n}}'_-; \widetilde{M}; 2; {}^+\Delta)_{\tilde{\lambda}} & \xrightarrow{S} & C^*_\cdot({}^+U_q \tilde{\mathfrak{n}}'_-; \widetilde{M}; 2; \mu)_{\tilde{\lambda}} \\ {\scriptstyle p}\downarrow & & \downarrow{\scriptstyle p} \\ \mathcal{X}_{\ell-\cdot}(\mathcal{C}_{n,\ell,\nu}) & \xrightarrow{S} & \mathcal{Q}_{\ell-\cdot}(\mathcal{C}_{n,\ell,\nu}). \end{array}$$
(4.9.2)

Let
$$\pi : \mathcal{X}_\cdot(\mathcal{C}_{n,\ell,\nu})^- \hookrightarrow \mathcal{X}_\cdot(\mathcal{C}_{n,\ell,\nu}),$$
$$\pi : \mathcal{Q}_\cdot(\mathcal{C}_{n,\ell,\nu})^- \hookrightarrow \mathcal{Q}_\cdot(\mathcal{C}_{n,\ell,\nu})$$
(4.9.3)

be the monomorphisms defined in (3.16.5). The following diagram is commutative:

$$\begin{array}{ccc} \mathcal{X}_\cdot(\mathcal{C}_{n,\ell,\nu}) & \xrightarrow{S} & \mathcal{Q}_\cdot(\mathcal{C}_{n,\ell,\nu}) \\ {\scriptstyle \pi}\uparrow & & \uparrow{\scriptstyle \pi} \\ \mathcal{X}_\cdot(\mathcal{C}_{n,\ell,\nu})^- & \xrightarrow{S} & \mathcal{Q}_\cdot(\mathcal{C}_{n,\ell,\nu})^-. \end{array}$$
(4.9.4)

Let
$$\pi : C.(^+U_q\bar{\mathfrak{n}}'_-; \overline{M}; 2; {}^+\Delta)_\lambda \hookrightarrow C.(^+U_q\tilde{\mathfrak{n}}'_-; \widetilde{M}; 2; {}^+\Delta)_{\tilde\lambda},$$
$$\pi : C^*(^+U_q\bar{\mathfrak{n}}'_-; \overline{M}; 2; \mu)_\lambda \hookrightarrow C^*(^+U_q\tilde{\mathfrak{n}}'_-; \widetilde{M}; 2; \mu)_{\tilde\lambda}$$
(4.9.5)

be the monomorphisms defined in (4.8.6) and (4.8.8). The following diagram is commutative

$$\begin{array}{ccc} C.(^+U_q\bar{\mathfrak{n}}'_-; \overline{M}; 2; {}^+\Delta)_\lambda & \xrightarrow{S} & C^*(^+U_q\bar{\mathfrak{n}}'_-; \overline{M}; 2; \mu)_\lambda \\ \pi \downarrow & & \downarrow \pi \\ C.(^+U_q\tilde{\mathfrak{n}}'_-; \widetilde{M}; 2; {}^+\Delta)_{\tilde\lambda} & \xrightarrow{S} & C^*(^+U_q\tilde{\mathfrak{n}}'_-; \widetilde{M}; 2; \mu)_{\tilde\lambda} \end{array} \quad (4.9.6)$$

According to formulas (3.16.5) and (4.8.6), the following diagrams are commutative for all j:

$$\begin{array}{ccc} C_j(^+U_q\bar{\mathfrak{n}}'_-; \overline{M}; 2; {}^+\Delta)_\lambda & \xrightarrow{p} & \mathcal{X}_{\ell-j}(\mathcal{C}_{n,\ell,\nu})^- \\ \pi \downarrow & & \downarrow \pi \\ C_j(^+U_q\tilde{\mathfrak{n}}'_-; \widetilde{M}; 2; {}^+\Delta)_{\tilde\lambda} & \xrightarrow{p} & \mathcal{X}_{\ell-j}(\mathcal{C}_{n,\ell,\nu}), \end{array} \quad (4.9.7)$$

$$\begin{array}{ccc} C^*_j(^+U_q\bar{\mathfrak{n}}'_-; \overline{M}; 2; \mu)_\lambda & \xrightarrow{p} & \mathcal{Q}_{\ell-j}(\mathcal{C}_{n,\ell,\nu})^- \\ \pi \downarrow & & \downarrow \pi \\ C^*_j(^+U_q\tilde{\mathfrak{n}}'_-; \widetilde{M}; 2; \mu)_{\tilde\lambda} & \xrightarrow{p} & \mathcal{Q}_{\ell-j}(\mathcal{C}_{n,\ell,\nu}). \end{array}$$

Commutativity of diagrams (4.9.2), (4.9.4), (4.9.6) and (4.9.7) implies Theorem (4.7.5).

5

Quasiisomorphism of Two-sided Hochschild Complexes to Suitable One-sided Hochschild Complexes

In Sec. 4, we have introduced special two-sided Hochschild complexes (4.5.5) and interpreted these complexes as the abstract complexes of a discriminantal configuration. In this section, we prove that these two-sided Hochschild complexes are quasiisomorphic to suitable one-sided Hochschild complexes. These one-sided Hochschild complexes are the Hochschild complexes of a nilpotent subalgebra of a quantum group with coefficients in the tensor product of Verma modules and their duals. As a result, we get an isomorphism of the homology groups of the nilpotent subalgebra and the homology groups of the discriminantal configuration.

5.1. One-sided Hochschild Complexes Connected with a Discriminantal Configuration

Let $U_q\mathfrak{g}$ be a quantum group of type (4.1.a)–(4.1.d), $M_1,\ldots,M_n$ Verma modules over $U_q\mathfrak{g}$ with highest weights $\Lambda_1,\ldots,\Lambda_n \in \mathfrak{h}^*$ and singular generating vectors $v_1 \in M_1,\ldots,v_n \in M_n$, respectively. Set $M = M_1 \otimes \cdots \otimes M_n$. Let $C.(^+U_q\mathfrak{n}_-; M; \mu)$ be the standard Hochschild complex of the algebra $^+U_q\mathfrak{n}_- \subset U_q\mathfrak{g}$ with the standard multiplication μ and with coefficients in the $(^+U_q\mathfrak{n}_-)$-module M. Let

(5.1.1) $C^{\cdot}(^+U_q\mathfrak{n}_-; M; \mu)$ be its dual complex, and let $C^{\cdot}(^+U_q\mathfrak{n}_-; M; \mu)_\lambda$ be its weight component corresponding to a weight $\lambda = (k_1, \ldots, k_r) \in \mathbf{N}^r$.

5.1.2. Example. Let $U_q\mathfrak{g} = U_q s\ell_2$, $M = M_1 \otimes M_2$, $\lambda = 1 \in \mathbf{N}$. Then $C^*(^+U_q\mathfrak{n}_-; M; \mu)_\lambda$ has the form

$$0 \to C_0^* \to C_1^* \to 0,$$

where C_0^* has a basis $(fv_1|v_2)^*$, $(v_1|fv_2)^*$, C_1^* has a basis $(f|v_1|v_2)^*$, $d(fv_1|v_2)^* = q^{(\Lambda_2,\alpha_1)/4}(f|v_1|v_2)^*$, $d(v_1|fv_2)^* = q^{-(\Lambda_1,\alpha_1)/4}(f|v_1|v_2)^*$.

$C_j^*(^+U_q\mathfrak{n}_-; M; \mu) = (^+U_q\mathfrak{n}_-)^{*\otimes j} \otimes M^*$. $^+U_q\mathfrak{n}_-$ has the natural monomial basis $\{ f_I = f_{i_1} \ldots f_{i_p} \}$ where $i_j \in \{1, \ldots, r\}$. M has the natural monomial basis $\{ f_{I_1}v_1 | \ldots | f_{I_n}v_n \}$. Thus $(^+U_q\mathfrak{n}_-)^*$ has the natural dual basis $\{ f_I^* \}$, and M^* has the natural dual basis $\{(f_{I_1}v_1)^* | \ldots | (f_{I_n}v_n)^*\}$.

The differential of $C^*(^+U_q\mathfrak{n}_-; M)$ sends $a_j | \ldots | a_1 | m$ to

$$-a_j | \ldots | a_1 | \mu^* m + \sum_{p=1}^{j}(-1)^{p+1} a_j | \ldots | \mu^* a_p | \ldots | a_1 | m,$$

where $\mu^* : (^+U_q\mathfrak{n}_-)^* \to (^+U_q\mathfrak{n}_-)^{*\otimes 2}$, $\mu^* : (M)^* \to (^+U_q\mathfrak{n}_-)^* \otimes M^*$ are duals to the multiplication. In the basis $\{f_I^*\}$ of $(^+U_q\mathfrak{n}_-)^*$, the map $\mu^* : M^* \to (^+U_q\mathfrak{n}_-)^* \otimes M^*$ is given by the formula

$$\mu^* : m \mapsto \sum_I f_I^* \otimes e_{\bar{I}} m, \qquad (5.1.3)$$

where the sum is taken over all monomials f_I, $f_I = f_{i_1} \ldots f_{i_p}$, $e_{\bar{I}} = e_{i_p} \ldots e_{i_1}$, and the $^+U_q\mathfrak{g}$-module structure on M^* is defined in (4.1). Therefore, $\mu^* = \sum f_I^* \otimes e_{\bar{I}}$.

Following [**SV4**], define a map

$$^+\Delta : M \to {^+U_q\mathfrak{n}_-} \otimes M \qquad (5.1.4)$$

by the condition of commutativity of the diagram

$$\begin{array}{ccc} M^* & \xrightarrow{\mu^*} & (^+U_q\mathfrak{n}_-)^* \otimes M^* \\ S \uparrow & & \uparrow S \\ M & \xrightarrow{^+\Delta} & {^+U_q\mathfrak{n}_-} \otimes M, \end{array} \qquad (5.1.5)$$

where S are the contravariant forms.

According to (4.6) and (4.2.23), S are isomorphisms for general values of q, $\Lambda_1,\ldots,\Lambda_n$, $b_{ij} = (\alpha_i,\alpha_j)$. Therefore the commutativity condition defines $^+\!\Delta$ for such general values. We show below that the map $^+\!\Delta$ is some universal function in q, b_{ij}, Λ_i having no singularities, so, the map $^+\!\Delta$ may be defined by continuity for all q, b_{ij}, Λ_i.

According to (4.1) and (4.2), $S : M \to M^*$ is a homomorphism of $U_q\mathfrak{g}$-modules. Therefore

$$^+\!\Delta = \sum_I \sum_J S_{IJ}^{-1} f_J \otimes e_I = \sum_J f_J \otimes \sum_I S_{IJ}^{-1} e_I = \sum_J f_J \otimes g_J,$$

where (S_{IJ}^{-1}) is the matrix inverse to the matrix $(S_{IJ}) = (S(f_I, f_J)) = (S(e_I, e_J))$ and

$$g_J := \sum_I S_{IJ}^{-1} e_I. \tag{5.1.6}$$

Our problem is to prove that the operators $g_J : M \to M$ are well-defined for all values of q, b_{ij}, Λ_i.

5.1.7. Examples. For $J = (j,\ldots,j) \in \mathbf{N}^p$, we have $g_J = ((1)_{q_j}^p/(p)_{q_j}!)\, e_j^p$ and

$$g_{1\,2} = (q^{b_{1\,2}/4} e_1 e_2 - q^{-b_{1\,2}/4} e_2 e_1)/(q^{b_{1\,2}/2} - q^{-b_{1\,2}/2}).$$

For $f_1 f_2 v \in M(\Lambda)$, $g_{1\,2}(f_1 f_2 v) = (\Lambda,\alpha_2)_q\, q^{(\Lambda,\alpha_1)/2} q^{-b_{1\,2}/4} v$.

5.1.8. Lemma. *Assume that $S : {}^+U_q\mathfrak{n}_+ \to ({}^+U_q\mathfrak{n}_+)^*$ is nondegenerate. Then for any $J = (j_1,\ldots,j_p)$, the operator g_J has the property*

$$\Delta\, g_J = \sum_{i=0}^p g_{j_1\ldots j_i} q^{-(h_{j_{i+1}}+\cdots+h_{j_p})/4} \otimes g_{j_{i+1}\ldots j_p} q^{(h_{j_1}+\cdots+h_{j_i})/4}. \tag{5.1.8.1}$$

Proof. Consider the middle part of the $\Delta_p g_J$, see (4.2.7). By definition of g_J and the middle part, we have

$$\mathrm{mdl}\, \Delta_p g_J = e_{j_1} q^{b_1} |\ldots| e_{j_p} q^{b_p}, \tag{5.1.9}$$

where $b_i = (h_{j_1} + \cdots + h_{j_{i-1}} - h_{j_{i+1}} - \cdots - h_{j_p})/4$. This property uniquely defines the element $g_J \in {}^+U_q\mathfrak{n}_+$ and implies (5.1.8.1). In fact, let

$$\Delta\, g_J = \sum_{K,L} d_{K,L}\, g_K q^{b_{K,L}} \otimes g_L q^{c_{K,L}},$$

where $K, L \subset \{j_1, \ldots, j_p\}$, $b_{K,L}, c_{K,L} \in \mathfrak{h}$, $d_{K,L} \in \mathbb{C}$. For any $i = 1, \ldots, p-1$, $\triangle_p = (\triangle_i \otimes \triangle_{p-i}) \circ \triangle$. Thus

$$\mathrm{mdl}\,(g_J) = \sum_{K,L,|K|=i} c_{K,L}\,\mathrm{mdl}\,\triangle_i(g_K q^{b_{K,L}}) \otimes \mathrm{mdl}\,\triangle_{p-i}(g_L q^{c_{K,L}}).$$

The elements
$$\mathrm{mdl}\,\triangle_i(q_K q^{b_{K,L}}) \otimes \mathrm{mdl}\,\triangle_{p-i}(g_L q^{c_{K,L}})$$
are linearly independent. Therefore (5.1.9) implies $c_{K,L} = 1$ for $K = (j_1, \ldots, j_i)$, $L = (j_{i+1}, \ldots, j_p)$, $c_{K,L} = 0$ otherwise.

5.1.10. Theorem. *The operators g_J defined by (5.1.6) for general values of q, b_{ij} are analytic functions of q, b_{ij} and are well defined by continuity on $M = M_1 \otimes \cdots \otimes M_n$ for all values of q, b_{ij}.*

5.1.11. Corollary. *The operators g_J satisfy (5.1.8.1).*

5.1.12. Corollary. *The operator $^+\triangle = \sum_J f_J \otimes g_J$ makes diagram (5.1.5) commutative and preserves the weight of elements:*

$$^+\triangle : (M)_\lambda \longrightarrow (^+U_q \mathfrak{n}_- \otimes M)_\lambda \quad \text{for all } \lambda.$$

To prove the theorem, it is sufficient to prove it for the case $M = M_1 = M(\Lambda)$. Then formula (5.1.8.1) extends it to the general case $M = M_1 \otimes \cdots \otimes M_n$.

To prove the theorem for the case $M = M(\Lambda)$, we give below a direct definition of the operators $g_J : M(\Lambda) \to M(\Lambda)$.

Let $0 \leq k \leq n \leq p$ be integers. A map $\pi : \{1, \ldots, n\} \to \{1, \ldots, p\}$ is called a (k, n, p)-*hook* if

$$\begin{aligned}
\pi(i) &\neq \pi(j) \quad \text{for all } i \neq j; \\
\pi(i) &< \pi(j) \quad \text{for } 1 \leq i < j \leq k; \\
\pi(i) &> \pi(j) \quad \text{for } k < i < j \leq n.
\end{aligned} \qquad (5.1.13)$$

A hook will be denoted by $(\pi(1), \ldots, \pi(k)|\pi(k+1), \ldots, \pi(n))$.

Examples. $(1|3, 2)$, $(1, 3|2)$, $(2, 6|8, 3)$.

Let $I = (i_1, \ldots, i_n)$, $f_J v = f_{j_1} \ldots f_{j_p} v$. A (k, n, p)-hook π will be called *adjacent* to $(I, f_J v)$ if $i_s = j_{\pi(s)}$ for all s. Notation: $\pi < (I, f_J v)$.

Example. $(1|2)$, $(1, 2\,|)$, $(3|2)$, $(|3, 2)$ are adjacent to $((1, 2), f_1 f_2 f_1 v)$.

Let π be adjacent to $(I, f_J v)$. Remove the set $(\pi(1), \ldots, \pi(n))$ from the ordered set $(1, \ldots, p)$. The resulting ordered set is denoted by P. Remove the elements $f_{j_{\pi(1)}}, \ldots, f_{j_{\pi(n)}}$ from the monomial $f_J v$. Let the resulting monomial be denoted by $f_{J, \pi} v$.

Define an *index* of the (k, n, p)-hook π adjacent to $(I, f_J v)$ as the number

$$\text{ind}\,(\pi, I, f_J v) = (-1)^{n-k} q^{A_0 + \cdots + A_n}\,. \tag{5.1.14}$$

Here

$$A_0 = \sum_{1 \le \ell < m \le n} \pm b_{j_{\pi(\ell)}, j_{\pi(m)}}/4\,, \tag{5.1.15}$$

where the sign of $b_{j_{\pi(\ell)}, j_{\pi(m)}}$ is positive if $\pi(\ell) < \pi(m)$ and the sign is negative otherwise.

$$A_i = \pm \langle \alpha_{j_{\pi(i)}}, \Lambda - \sum_{\substack{p \in P \\ p > \pi(i)}} \alpha_{j_p} \rangle / 2\,, \tag{5.1.16}$$

where the sign of the right-hand side is positive for $i = 1, \ldots, k$, and the sign is negative for $i = k+1, \ldots, n$.

Define an operator $g_I : M(\Lambda) \to M(\Lambda)$, $I = (i_1, \ldots, i_n)$, by the formula

$$g_I : f_J v \mapsto \sum_\pi \text{ind}\,(\pi, I, f_J v)\, f_{J, \pi} v\,, \tag{5.1.17}$$

where $f_J v$ is an arbitrary monomial in $M(\Lambda)$ and the sum is taken over all hooks adjacent to $(I, f_J v)$.

Example. $g_{1\,2}(f_1 f_2 v)$ is given in (5.1.7) and consists of two terms corresponding to the hooks $(1|2)$, $(1, 2\,|\,)$.

Remark. There is a useful formula

$$g_I f_J v = f_{j_1} g_I f_{j_2} \cdots f_{j_p} v + q^A \delta_{i_1, j_1} g_{i_2 \ldots i_n} f_{j_2} \cdots f_{j_p} v$$
$$- q^B \delta_{i_p, j_1} g_{i_1 \ldots i_{n-1}} f_{j_2} \cdots f_{j_p} v\,,$$

where $A = (\alpha_{i_1}, \Lambda - \alpha_{j_2} - \cdots - \alpha_{j_p})/2 + (\alpha_{i_1}, \alpha_{i_2} + \cdots + \alpha_{i_n})/4$, and $B = -(\alpha_{i_n}, \Lambda - \alpha_{j_2} - \cdots - \alpha_{j_p})/2 - (\alpha_{i_n}, \alpha_{i_1} + \cdots + \alpha_{i_{n-1}})/4$.

5.1.18. Theorem. *The operators g_I defined by formula (5.1.17) satisfy the relation*

$$e_{\bar{L}} = \sum_I S_{LI}\, g_I$$

for all L.

As a corollary, we get Theorem (5.1.10).

Proof of Theorem 5.1.18. We have to prove

$$e_L f_J v = \sum_I S_{L,I} \sum_{\pi < (I, f_J v)} \text{ind}\,(\pi, I, f_J v)\, f_{J,\pi} v \qquad (5.1.19)$$

for all L, J.

We prove (5.1.19) for the case $n = p$, $f_J v = f_1 \ldots f_n v$, $e_L = e_{\ell_n} \ldots e_{\ell_1}$, and $L = (\ell_1, \ldots, \ell_n)$ is a permutation of $(1, \ldots, n)$. The general case is proved analogously.

Calculate the left-hand side of (5.1.19):

$$e_L f_J v = \left(\prod_{s=1}^n \left(\alpha_{\ell_s}, \Lambda - \sum_{\substack{t > s \\ \ell_t > \ell_s}} \alpha_{\ell_t} \right)_q \right) v. \qquad (5.1.20)$$

Since $(a)_q = q^{a/2} - q^{-a/2}$, the product in (5.1.20) consists of 2^n terms numerated by all hooks $\pi : (1, \ldots, n) \to (1, \ldots, n)$. To a (k, n, n)-hook $(\pi(1), \ldots, \pi(k) | \pi(k+1), \ldots, \pi(n))$, we assign the term of the product in which $q^{a/2}$ is taken in the factors $\langle \alpha_{\pi(t)}, \Lambda - \sum \ldots \rangle_q$, $t = 1, \ldots, k$, and $q^{-a/2}$ is taken in the other factors. This term is equal to $(-1)^{n-k} q^{B+C}$, where B, C are given below

$$B = (\alpha_{\pi(1)} + \cdots + \alpha_{\pi(k)} - \alpha_{\pi(k+1)} - \cdots - \alpha_{\pi(n)}, \Lambda)/2. \qquad (5.1.21)$$

To describe C, we denote by $\sigma \in \Sigma_n$ the permutation such that $\sigma(\ell) = m$, if $\ell_m = \ell$. Then

$$C = \sum \pm b_{\ell m}/2,$$

where the sum is taken over all $\ell < m$ such that $\sigma(\ell) < \sigma(m)$, the sign of $b_{\ell m}$ is negative, if $\ell \in \{\pi(1), \ldots, \pi(k)\}$, and is positive otherwise.

The right-hand side of (5.1.19) consists of 2^n terms numerated by all hooks $\pi : (1, \ldots, n) \to (1, \ldots, n)$. To a (k, n, n)-hook $(\pi(1), \ldots, \pi(k) | \pi(k+1), \ldots, \pi(n))$ assign the set $I = (\pi(1), \ldots, \pi(n))$. The hook π is adjacent to $(I, f_J v)$. We have $f_{J,\pi} v = v$. $\text{Ind}\,(\pi, I, f_J v) = (-1)^{n-k} q^{A_0 + A_1 + \cdots + A_n}$, where $A_1 + \cdots + A_n$ coincides with B in (5.1.21).

$$A_0 = \sum_{\ell < m} \pm b_{\pi(\ell), \pi(m)}/4$$

where the sign of $b_{\pi(\ell),\pi(m)}$ is positive, if $\pi(\ell) < \pi(m)$, and is negative otherwise. In other words, $q^{A_0} = S_{I,Id}$, where $Id = (1,\ldots,n)$.

Let $\tau \in \Sigma_n$ be a permutation such that $\pi(i) = \ell_{\tau(i)}$ for all i. Then $S_{L,I} = q^D$, where
$$D = \sum_{\ell<m} \pm b_{\pi(\ell),\pi(m)}/4,$$
where the sign of $b_{\pi(\ell),\pi(m)}$ is positive, if $\tau(\ell) > \tau(m)$ and is negative otherwise.

Therefore to prove (5.1.19), it is sufficient to check that $q^C = S_{L,I} S_{I,Id}$, or $C = A_0 + D$.

5.1.22. Lemma. *Let $\pi \in \Sigma_n$ be a hook, let $\tau \in \Sigma_n$ be any permutation, $\sigma = \tau\pi$. Define the numbers A_0, C, D as before. Then $C = A_0 + D$.*

This lemma proves equality (2.1.19).

Proof of the lemma. The conditions $\ell < m$, $\sigma(\ell) < \sigma(m)$ are equivalent to

(a) $\ell < m$, $\pi(\ell) < \pi(m)$, $\tau(\pi(\ell)) < \tau(\pi(m))$,

or

(b) $\ell < m$, $\pi(\ell) > \pi(m)$, $\tau(\pi(\ell)) < \tau(\pi(m))$.

The impact in $A_0 + D$ of a pair (ℓ,m) of type (a) is $-b_{\ell,m}/2$. The impact of a pair of type (b) is $b_{\ell,m}/2$. All other pairs give zero impact. Conditions (a) are equivalent to the conditions

(c) $\ell < m$, $\sigma(\ell) < \sigma(m)$, $\ell \in \{\pi(1),\ldots,\pi(k)\}$.

Conditions (b) are equivalent to the conditions

(d) $\ell < m$, $\sigma(\ell) < \sigma(m)$, $\ell \in \{\pi(k+1),\ldots,\pi(n)\}$.

The lemma is proved.

Consider $^+U_q\mathfrak{n}_-$ as a coalgebra with respect to the coalgebra structure $^+\Delta : {^+U_q\mathfrak{n}_-} \to {^+U_q\mathfrak{n}_-} \otimes {^+U_q\mathfrak{n}_-}$ defined in (4.3).

5.1.23. Lemma. *The operator $^+\Delta = \sum_J f_J \otimes g_J : M \to {^+U_q\mathfrak{n}_-} \otimes M$ defines a $^+U_q\mathfrak{n}_-$-comodule structure on $M = M_1 \otimes \cdots \otimes M_n$.*

Proof. Diagrams (4.3.6) and (5.1.5) are commutative. The form S in the diagrams is an isomorphism for general values of q, b_{ij}, Λ_j. $\mu^* : M^* \to (^+U_q\mathfrak{n}_-)^* \otimes M^*$ is a $(^+U_q\mathfrak{n}_-)^*$-comodule structure. Therefore $^+\Delta$ is a $^+U_q\mathfrak{n}_-$-

comodule structure for general values of q, b_{ij}, Λ_j. $^+\Delta$ continuously depends on the parameters. Thus $^+\Delta$ is a $^+U_q\mathfrak{n}_-$-comodule structure for all values of the parameters.

(5.1.24) Let $C.(^+U_q\mathfrak{n}_-; M; ^+\Delta)$ be the standard Hochschild complex of the coalgebra $^+U_q\mathfrak{n}$ with the comultiplication $^+\Delta$ and with coefficients in the comodule M, and let $C.(^+U_q\mathfrak{n}_-; M; ^+\Delta)_\lambda$ be its weight component corresponding to a weight $\lambda = (k_1, \ldots, k_r) \in \mathbf{N}^r$.

5.1.25. Example. Let $U_q\mathfrak{g} = U_q s\ell_2$, $M = M_1 \otimes M_2$, $\lambda = 1 \in \mathbf{N}$. $C.(^+U_q\mathfrak{n}_-; M; ^+\Delta)_\lambda$ has the form
$$0 \to C_0 \to C_1 \to 0,$$
where C_0 has a basis $fv_1|v_2$, $v_1|fv_2$, C_1 has a basis $f|v_1|v_2$,
$$d(fv_1|v_2) = (\Lambda_1, \alpha_1)_q \, q^{(\Lambda_2, \alpha_1)/4}(f|v_1|v_2)$$
$$d(v_1|fv_2) = (\Lambda_2, \alpha_1)_q \, q^{-(\Lambda_1, \alpha_1)/4}(f|v_1|v_2),$$
cf. (5.1.2).

5.1.26. Theorem. *For all λ the contravariant form S defines a homomorphism of complexes:*
$$S : C.(^+U_q\mathfrak{n}_-; M; ^+\Delta)_\lambda \longrightarrow C^*(^+U_q\mathfrak{n}_-; M; \mu)_\lambda . \quad (5.1.26.1)$$

The theorem is a corollary of commutativity of diagrams (4.3.6) and (5.1.5).

5.1.27. Example. For the complexes of examples (5.1.2) and (5.1.25), the homomorphism S is defined by the formulas $f | v_1|v_2 \mapsto (f|v_1|v_2)^*$, $fv_1|v_2 \mapsto (\Lambda_1, \alpha_1)_q(fv_1|v_2)^*$, and $v_1|fv_2 \mapsto (\Lambda_2, \alpha_1)_q(v_1|fv_2)^*$. Obviously, S commutes with the differentials.

5.2. Complexes (5.1.1) and (5.1.24) as Subcomplexes of Complexes (4.5.5)

Fix a quantum group $U_q\mathfrak{g}$ of type (4.1.a)–(4.1.d), Verma modules over $U_q\mathfrak{g}$ with highest weights $\Lambda_1, \ldots, \Lambda_n \in \mathfrak{h}^*$ and singular generating vectors $v_1 \in M_1, \ldots, v_n \in M_n$.

Consider a quantum group $U_q\bar{\mathfrak{g}}$ given by the data

(a) a finite-dimensional complex vector space $\bar{\mathfrak{h}}$;

(b) a nondegenerate symmetric bilinear form $(\,,\,)$ on $\bar{\mathfrak{h}}$;

(c) a linearly independent covector $\bar{\alpha}_1, \ldots, \bar{\alpha}_{n+r} \in \bar{\mathfrak{h}}^*$, where r is the same as in the definition of $U_q\mathfrak{g}$;

(d) a nonzero complex number $\varkappa$ the same as in the definition of $U_q\mathfrak{g}$.

$U_q\bar{\mathfrak{g}}$ will be called a *quantum group associated with* $(U_q\mathfrak{g}, M_1, \ldots, M_n)$ if

$$(\bar{\alpha}_\ell, \bar{\alpha}_m) = (\alpha_\ell, \alpha_m), \qquad 1 \leq \ell < m \leq r, \qquad (5.2.1)$$

where α_ℓ, α_m are the simple roots of $U_q\mathfrak{g}$;

$$\begin{aligned}(\bar{\alpha}_\ell, \bar{\alpha}_{r+m}) &= -(\alpha_\ell, \Lambda_m), \\ (\bar{\alpha}_{r+\ell}, \bar{\alpha}_{r+m}) &= (\Lambda_\ell, \Lambda_m)\end{aligned} \qquad (5.2.2)$$

for all ℓ, m. Here the left scalar products are scalar products in $\bar{\mathfrak{h}}^*$, the right scalar products are scalar products in $\mathfrak{h}^*$.

An associated quantum group exists according to [**K**].

Fix an associated group $U_q\bar{\mathfrak{g}}$. $U_q\bar{\mathfrak{g}}$ is generated by elements 1, $\bar{e}_i$, $\bar{f}_i$ for $i = 1, \ldots, r + n$ and the space $\bar{\mathfrak{h}}$. Let $U_q\bar{\mathfrak{n}}'_- \subset U_q\bar{\mathfrak{g}}$ (resp. $^+U_q\bar{\mathfrak{n}}'_-$) be the subaglebra generated by $1, \bar{f}_1, \ldots, \bar{f}_r$ (resp. $\bar{f}_1, \ldots, \bar{f}_r$). For any $i = 1, \ldots, n$, let $\overline{M}_i \subset U_q\bar{\mathfrak{g}}$ be the two-sided submodule over $U_q\bar{\mathfrak{n}}'_-$ generated by $\bar{f}_{r+i}$. Set $\overline{M} = (\overline{M}_1, \ldots, \overline{M}_n)$.

In (4.5) we have defined two-sided Hochschild complexes $C.(^+U_q\bar{\mathfrak{n}}'_-; \overline{M}; 2; {}^+\Delta)$ and $C.^*(^+U_q\bar{\mathfrak{n}}'_-; \overline{M}; 2; \mu)$ and a homomorphism

$$S : C.(^+U_q\bar{\mathfrak{n}}'_-; \overline{M}; 2; \mu)_{\bar{\lambda}} \longrightarrow C.^*(^+U_q\bar{\mathfrak{n}}'_-; \overline{M}; 2; {}^+\Delta)_{\bar{\lambda}} \qquad (5.2.3)$$

for all $\bar{\lambda} = (\ell_1, \ldots, \ell_r, 1, \ldots, 1) \in \mathbf{N}^{r+n}$.

In (5.2) and (5.3) for arbitrary $\lambda = (\ell_1, \ldots, \ell_r) \in \mathbf{N}^r$, we will define monomorphisms

$$\begin{aligned}\varphi &: C.(^+U_q\mathfrak{n}_-; M; {}^+\Delta)_\lambda \longrightarrow C.(^+U_q\bar{\mathfrak{n}}'_-; \overline{M}; 2; {}^+\Delta)_{\bar{\lambda}}, \\ \varphi &: C.^*(^+U_q\mathfrak{n}_-; M; \mu)_\lambda \longrightarrow C.^*(^+U_q\bar{\mathfrak{n}}'_-; \overline{M}; 2; \mu)_{\bar{\lambda}},\end{aligned} \qquad (5.2.4)$$

commuting with (5.2.3) and (5.1.26.1). We will prove that these monomorphisms are quasiisomorphisms. Since the right-hand side complexes of (5.2.4) compute the homology groups of the discriminantal configuration, we will get a theorem that the left-hand side complexes of (5.2.4) compute the homology groups of the discriminantal configuration.

Fix $\bar{\lambda} = (\ell_1, \ldots, \ell_r, 1, \ldots, 1) \in \mathbf{N}^{r+n}$. Let

$$f^* = \big(f_1(1) \ldots f_{s(1)}(1) \mid \ldots \mid f_1(j) \ldots f_{s(j)}(j) \mid f_1(j+1) \ldots$$
$$f_{s(j+1)}(j+1)\bar{f}_{r+1}f_{s(j+1)+1}(j+1) \ldots f_{t(j+1)}(j+1)$$
$$\mid \ldots \mid f_1(j+n) \ldots f_{s(j+n)}(j+n)\bar{f}_{r+n}f_{s(j+n)+1}(j+n)$$
$$\ldots f_{t(j+n)}(j+n)\big)^* \quad (5.2.5)$$

be a monomial in $\big((^+U_q\bar{\mathfrak{n}}'_-)^{\otimes j} \otimes \overline{M}_1 \otimes \cdots \otimes \overline{M}_n\big)^*_{\bar{\lambda}}$. Here $f_\ell(m) \in \{\bar{f}_1, \ldots, \bar{f}_r\}$ for all ℓ, m. Let

$$g^* = \big(f_1(1) \ldots f_{s(1)}(1) \mid \ldots \mid f_1(j) \ldots f_{s(j)}(j) \mid g_1(j+1) \ldots$$
$$g_{u(j+1)}(j+1)\bar{f}_{r+1}g_{u(j+1)+1}(j+1) \ldots g_{v(j+1)}(j+1)$$
$$\mid \ldots \mid g_1(j+n) \ldots g_{u(j+n)}(j+n)\bar{f}_{r+n}g_{u(j+n)+1}(j+n)$$
$$\ldots g_{v(j+n)}(j+n)\big)^* \quad (5.2.6)$$

be another monomial in $\big((^+U_q\bar{\mathfrak{n}}'_-)^{\otimes j} \otimes \overline{M}\big)^*_{\bar{\lambda}}$ with the same first j factors; here $g_\ell(m) \in \{\bar{f}_1, \ldots, \bar{f}_r\}$ for all ℓ, m.

Let $t = t(j+1) + \cdots + t(j+n) + n$. Then $t = v(j+1) + \cdots + v(j+n) + n$ since f^*, g^* have the same weight $\bar{\lambda}$. Let $\sigma \in \Sigma_t$ be any permutation sending the ordered sequence

$$f_1(j+1), \ldots, f_{s(j+1)}(j+1), \bar{f}_{r+1}, f_{s(j+1)+1}(j+1), \ldots,$$
$$f_{t(j+1)}(j+1), \ldots, f_1(j+n), \ldots, f_{s(j+n)}(j+n), \quad (5.2.7)$$
$$\bar{f}_{r+n}, f_{s(j+n)+1}(j+n), \ldots, f_{t(j+n)}(j+n)$$

onto the ordered sequence

$$g_1(j+1), \ldots, g_{u(j+1)}(j+1), \bar{f}_{r+1}, g_{u(j+1)+1}(j+1), \ldots,$$
$$g_{v(j+1)}(j+1), \ldots, g_1(j+n), \ldots, g_{u(j+n)}(j+n), \quad (5.2.8)$$
$$\bar{f}_{r+n}, g_{u(j+n)+1}(j+n), \ldots, g_{v(j+n)}(j+n).$$

The triple (f^*, g^*, σ) is called a *left mover* associated with f^* if the following condition is satisfied.

(5.2.9) The permutation σ restricted to each of the following $(n+1)$ subsets of (5.2.7) preserves the order of elements of the subset. These subsets are the subset

$$\bar{f}_{r+1}, f_{s(j+1)+1}(j+1), \ldots, f_{t(j+1)}(j+1), \bar{f}_{r+2}, f_{s(j+2)+1}(j+2), \ldots,$$
$$f_{t(j+2)}(j+2), \bar{f}_{r+3}, \ldots, \bar{f}_{r+n}f_{s(r+n)+1}(r+n), \ldots, f_{t(r+n)}(r+n)$$

and the subsets
$$f_1(j+p), \ldots, f_{s(j+p)}(j+p)\bar{f}_{r+p},$$
for $p = 1, \ldots, n$.

The monomial g^* will be called the *resulting monomial* of the left mover. Define the index of a left mover associated with f^* as the number
$$\text{ind}(f^*, g^*, \sigma) = q^{A+B}, \qquad (5.2.10)$$
where A, B are defined below.
$$A = -\sum (\Lambda_i, \alpha_{f_\ell(j+m)})/2, \qquad (5.2.11)$$
where $\alpha_{f_\ell(j+m)}$ denotes α_p if $f_\ell(j+m) = \bar{f}_p$; the sum is taken over all $\{\Lambda_i, f_\ell(j+m)\}$, $m = 1, \ldots, n$, $\ell = 1, \ldots, s(j+m)$, $i = 1, \ldots, m-1$ such that the element $\sigma f_\ell(j+m)$ stands from the left of $\bar{f}_{r+i}$ in g^*.
$$B = \sum (\alpha_{f_\ell(j+m)}, \alpha_{f_a(j+b)})/2, \qquad (5.2.12)$$
where the sum is taken over all $f_\ell(j+m)$, $f_a(j+b)$ such that $f_\ell(j+m)$ stands from the left of $f_a(j+b)$ in f^*, and $\sigma f_\ell(j+m)$ stands from the right of $\sigma f_a(j+b)$ in g^*.

5.2.13. Example. The triple (f^*, f^*, id), where id is the identity permutation, is a left mover. Its index equals 1.

5.2.14. Example. Let $r = 3$, $n = 2$, $\lambda = (1,1,1,1,1)$, $f^* = (\bar{f}_1|\bar{f}_2\bar{f}_4|\bar{f}_3\bar{f}_5)^*$. There are four left movers associated with f^*. The resulting monomials are f^*, $(\bar{f}_1|\bar{f}_2\bar{f}_4\bar{f}_3|\bar{f}_5)^*$, $(\bar{f}_1|\bar{f}_2\bar{f}_3\bar{f}_4|\bar{f}_5)^*$, $(\bar{f}_1|\bar{f}_3\bar{f}_2\bar{f}_4|\bar{f}_5)^*$. Their indexes are 1, 1, $q^{-(\Lambda_1, \alpha_3)/2}$, $q^{-(\Lambda_1 - \alpha_2, \alpha_3)/2}$, respectively.

5.2.15. Example. Let $r = 1$, $n = 2$, $\lambda = (3,1,1)$, $f^* = (\bar{f}_1|\bar{f}_1\bar{f}_2|\bar{f}_1\bar{f}_3)^*$. There are four left movers associated with f^*, cf. (5.2.14). The resulting monomials are f^*, $(\bar{f}_1|\bar{f}_1\bar{f}_2\bar{f}_1|\bar{f}_3)^*$, $(\bar{f}_1|\bar{f}_1\bar{f}_1\bar{f}_2|\bar{f}_3)^*$, $(\bar{f}_1|\bar{f}_1\bar{f}_1\bar{f}_2|\bar{f}_3)^*$. Their indexes are 1, 1, $q^{-(\Lambda_1, \alpha_1)/2}$, $q^{-(\Lambda_1, -\alpha_1, \alpha_1)/2}$.

5.2.16. Example. Let $r = 1$, $n = 2$, $\lambda = (3,1,1)$, $f^* = (\bar{f}_1|\bar{f}_2|\bar{f}_1\bar{f}_3\bar{f}_1)^*$. There are three left movers. The resulting monomials are f^*, $(\bar{f}_1|\bar{f}_2\bar{f}_1|\bar{f}_3\bar{f}_1)^*$, $(\bar{f}_1|\bar{f}_1\bar{f}_2|\bar{f}_3\bar{f}_1)^*$. Their indexes are 1, 1, $q^{-(\Lambda_1, \alpha_1)/2}$.

Let $f^* \in \left((^+U_q\bar{\mathfrak{n}}'_-)^{\otimes j} \otimes \overline{M} \right)^*_\lambda$ be a monomial. The element
$$\text{left } f^* = \sum \text{ind}(f^*, g^*, \sigma) g^*, \qquad (5.2.17)$$

where the sum is taken over all left movers (f^*, g^*, σ) associated with f^*, is called the *left smash* of f^*. Monomials $\{f^*\}$ form a basis of C_j^*. Therefore (5.2.17) defines a linear map

$$\text{left} : ((^+U_q\bar{\mathfrak{n}}'_-)^{\otimes j} \otimes \overline{M})_{\bar\lambda}^* \longrightarrow ((^+U_q\bar{\mathfrak{n}}'_-)^{\otimes j} \otimes \overline{M})_{\bar\lambda}^*. \qquad (5.2.18)$$

Let

$$f^* = (f_1(1)\ldots f_{s(1)}(1) | \ldots | f_1(j)\ldots f_{s(j)}(j) | f_1(j+1)\ldots \\ f_{s(j+1)}(j+1)v_1 | \ldots | f_1(j+n)\ldots f_{s(j+n)}(j+n)v_n)^* \qquad (5.2.19)$$

be a monomial in $C_j^*(^+U_q\mathfrak{n}_-; M; \mu)_\lambda$, $\lambda = (\ell_1, \ldots, \ell_r) \in \mathbf{N}^r$. Here $f_\ell(m) \in \{f_1, \ldots, f_r\}$. Set

$$\text{inj}(f^*) = (\bar{f}_1(1)\ldots\bar{f}_{s(1)}(1)| \cdots | \bar{f}_1(j)\ldots \bar{f}_{s(j)}(j) | \bar{f}_1(j+1)\ldots \\ \bar{f}_{s(j+1)}(j+1)\bar{f}_{r+1} | \cdots | \bar{f}_1(j+n)\ldots \bar{f}_{s(j+n)}(j+n)\bar{f}_{r+n})^*, \qquad (5.2.20)$$

where $\bar{f}_\ell(m) = \bar{f}_i$ if $f_\ell(m) = f_i$ for all ℓ, m. This formula defines a linear map

$$\text{inj} : C_j^*(^+U_q\mathfrak{n}_-; M; \mu)_\lambda \longrightarrow ((^+U_q\bar{\mathfrak{n}}'_-)^{\otimes j} \otimes \overline{M})_{\bar\lambda}^*. \qquad (5.2.21)$$

For a monomial f^* of (5.2.19), define a number $C = C(f^*)$ as follows:

$$C = C_{\text{alg,alg}} + C_{\text{alg,mod}} + C_{\text{mod,mod}}. \\ C_{\text{alg,alg}} = \sum (\alpha_{f_\ell(m)}, \alpha_{f_a(b)})/4, \qquad (5.2.22)$$

where $\alpha_{f_\ell(m)}$ denotes α_i if $f_\ell(m) = f_i$, the sum is taken over all $f_\ell(m), f_a(b)$, $m \leq j, b \leq j$, such that $f_\ell(m)$ stands from the left of $f_a(b)$ in f^*

$$C_{\text{alg,mod}} = \left(\sum_{m=1}^{j} \sum_{\ell=1}^{s(m)} \alpha_{f_\ell(m)}, \sum_{m=1}^{n} \left(-\Lambda_m + \sum_{\ell=1}^{s(j+m)} \alpha_{f_\ell(m)}\right)\right) / 4$$

$$C_{\text{mod,mod}} = \sum_{m=1}^{n} \sum_{\ell=1}^{s(j+m)} \left(\alpha_{f_\ell(j+m)}, -\Lambda_m + \sum_{k=1}^{\ell-1} \alpha_{f_k(j+m)} \right) / 4.$$

Denote by $a(j)$ the tensor factors of f^*. Therefore $f^* = a(1) | \ldots | a(j+n)$. It is obvious that

(5.2.23) The number $C(f^*)$ depends only on the weights of the elements $a(1)|\ldots|a(j), a(j+1), \ldots, a(j+n)$.

Define a map

$$\text{twist} : C_j^*(^+U_q\mathfrak{n}_-; M; \mu)_\lambda \longrightarrow C_j^*(^+U_q\mathfrak{n}_-; M; \mu)_\lambda \qquad (5.2.24)$$

by the formula $f^* \mapsto q^{C(f^*)} f^*$.

Define a map

$$\varphi : C_j^*(^+U_q\mathfrak{n}_-; M; \mu)_\lambda \longrightarrow C_j^*(^+U_q\bar{\mathfrak{n}}'_-; \overline{M}; 2; \mu)_{\bar\lambda} \qquad (5.2.25)$$

as the composition left ∘ inj ∘ twist. Note that $\left((^+U_q\bar{\mathfrak{n}}'_-)^{\otimes j} \otimes \overline{M}\right)^* \subset C_j^*(^+U_q\bar{\mathfrak{n}}'_-; \overline{M}; 2; \mu)$.

5.2.26. Lemma. *The map φ gives a well-defined monomorphism of complexes*

$$\varphi : C.^*(^+U_q\mathfrak{n}'_-; M; \mu)_\lambda \longrightarrow C.^*(^+U_q\mathfrak{n}_-; \overline{M}; 2; \mu)_{\bar\lambda} .$$

5.2.27. Example. Let $U_q\mathfrak{g} = U_q s\ell_2$, $M = M_1 \otimes M_2$, $\lambda = 1 \in \mathbb{N}$. The complex $C.^*(^+U_q\mathfrak{n}_-; M)_\lambda$ is described in (5.1.2). The monomorphism is given by the formulas

$$(f_1|v_1|v_2)^* \mapsto q^{-(\alpha_1, \Lambda_1+\Lambda_2)/4}(\bar f_1|\bar f_2|\bar f_3)^*,$$

$$(f_1v_1|v_2)^* \mapsto q^{-(\alpha_1, \Lambda_1)/4}(\bar f_1\bar f_2|\bar f_3)^*,$$

$$(v_1|f_1v_2)^* \mapsto q^{-(\alpha_1, \Lambda_2)/4}\left((\bar f_2|\bar f_1\bar f_3)^* + (\bar f_2\bar f_1|\bar f_3)^* + q^{-(\alpha_1, \Lambda_1)/2}(\bar f_1\bar f_2|\bar f_3)^*\right).$$

The differential on the image is given by the formulas: $(\bar f_a|\bar f_b|\bar f_c)^* \mapsto 0$, $(\bar f_a\bar f_b|\bar f_c)^* \mapsto (\bar f_a|\bar f_b|\bar f_c)^*$, $(\bar f_a|\bar f_b\bar f_c)^* \mapsto -(\bar f_a|\bar f_b|\bar f_c)^*$. The differential on the preimage is as seen in (5.1.2). The monomorphism commutes with the differentials.

Proof of Lemma **5.2.26.** Evidently, φ is a monomorphism because it has a triangular matrix with respect to the monomial bases and its diagonal part, $f^* \mapsto q^C \bar f^*$, has no kernel. To prove the lemma, it is sufficient to check that φ commutes with the differentials. This is done by trivial direct computations, see examples (5.1.2) and (5.2.27).

In conclusion of this section, we list some useful properties of left movers.

(5.2.28) If (f^*, g^*, σ_1) is a left mover associated with f^*, and (g^*, h^*, σ_2) is a left mover associated with g^*, then $(f^*, h^*, \sigma_2\sigma_1)$ is a left mover associated with f^*. Moreover, if at least one of σ_1, σ_2 is nontrivial, then $\sigma_2\sigma_1$ is nontrivial, that is, $f^* \neq h^*$.

For any monomial

$$f^* = \big(f_1(1)\ldots f_{s(1)}(1)\bar{f}_{r+1} f_{s(1)+1}(1)\ldots f_{t(1)}(1)|\ldots$$
$$|f_1(n)\ldots f_{s(n)}(n)\bar{f}_{r+n} f_{s(n)+1}(n)\ldots f_{t(n)}(n)\big)^*$$

in $\overline{M}_1 \otimes \cdots \otimes \overline{M}_n$, set

$$S(f^*) = f_1(1),\ldots,f_{s(1)}(1),\bar{f}_{r+1},f_{s(1)+1}(1),\ldots, \\ f_{t(1)}(1),\ldots,f_1(n),\ldots,f_{s(n)}(n)\,. \tag{5.2.29}$$

5.2.30. Lemma. Let $f^* \in \overline{M} = \overline{M}_1 \otimes \cdots \otimes \overline{M}_n$ be a monomial of the form

$$f^* = \big(f_1(1)\ldots f_{s(1)}(1)\bar{f}_{r+1}\,|\,\ldots\,|\,f_1(n)\ldots f_{s(n)}(n)\bar{f}_{r+n}\big)^*\,.$$

Then

$$\text{left}(f^*) = \sum \text{ind}(f^*,g^*,\sigma)\,G(S(g^*)), \tag{5.2.30.1}$$

where the sum is taken over all left movers (f^*, g^*, σ) associated with f^* such that g^* has the form

$$g^* = \big(g_1(1)\ldots g_{t(1)}(1)\bar{f}_{r+1}\,|\,\cdots\,|\,g_1(n)\ldots g_{t(n)}(n)\bar{f}_{r+n}\big)^*,$$

$g_\ell(m) \in (\bar{f}_1,\ldots,\bar{f}_r)$ for all ℓ, m; the element $G(S)$ is defined in (4.7.9).

5.2.31. Example. Let $r = 2$, $n = 2$. Then $\text{left}((f_1 f_3 | f_2 f_4)^*) = G(\bar{f}_1 \bar{f}_3 \bar{f}_2 \bar{f}_4)$ $+ q^{-(\Lambda_1,\alpha_2)/2}G(\bar{f}_1 \bar{f}_2 \bar{f}_3 \bar{f}_4) + q^{-(\Lambda_1-\alpha_1,\alpha_2)/2}G(\bar{f}_2 \bar{f}_1 \bar{f}_3 \bar{f}_4)$, cf. (5.2.14).

The Lemma is a direct corollary of definitions.

5.3. Construction of a Monomorphism $\varphi : C.(^+U_q\mathfrak{n}_-;M;{}^+\Delta)_\lambda \mapsto C.(^+U_q\bar{\mathfrak{n}}'_-;\overline{M};2;{}^+\Delta)_{\bar{\lambda}}$

Below we define linear maps

$$\text{twist} : C_j(^+U_q\mathfrak{n}_-;M;{}^+\Delta)_\lambda \longrightarrow C_j(^+U_q\mathfrak{n}_-;M;{}^+\Delta)_\lambda, \tag{5.3.1}$$

$$\text{inj} : C_j(^+U_q\mathfrak{n}_-;M;{}^+\Delta)_\lambda \longrightarrow ((^+U_q\bar{\mathfrak{n}}'_-)^{\otimes j} \otimes \overline{M})_{\bar{\lambda}}, \tag{5.3.2}$$

$$\text{left} : ((^+U_q\bar{\mathfrak{n}}'_-)^{\otimes j} \otimes \overline{M})_{\bar{\lambda}} \longrightarrow ((^+U_q\bar{\mathfrak{n}}'_-)^{\otimes j} \otimes \overline{M})_{\bar{\lambda}} \tag{5.3.3}$$

for all $j \geq 0$, $\lambda = (\ell_1,\ldots,\ell_r) \in \mathbf{N}^r$, $\bar{\lambda} = (\ell_1,\ldots,\ell_r,1,\ldots,1) \in \mathbf{N}^{r+n}$, and set

$$\varphi = \text{left} \circ \text{inj} \circ \text{twist} : C_j(^+U_q\mathfrak{n}_-;M;{}^+\Delta)_\lambda \\ \longrightarrow C_j(^+U_q\bar{\mathfrak{n}}'_-;\overline{M};2;{}^+\Delta)_{\bar{\lambda}}\,. \tag{5.3.4}$$

Note that $(({}^+U_q\bar{\mathfrak{n}}'_-)^{\otimes j} \otimes \overline{M}) \subset C_j({}^+U_q\bar{\mathfrak{n}}'_-;\overline{M};2;{}^+\Delta)$.

Define the map twist by the formula

$$\text{twist}: f \mapsto q^{C(f)} f, \qquad (5.3.5)$$

where

$$\begin{aligned}f = f_1(1)\ldots f_{s(1)}(1) \,|\, \ldots \,|\, f_1(j)\ldots f_{s(j)}(j) \,|\, f_1(j+1)\ldots \\ f_{s(j+1)}(j+1)v_1 \,|\, \ldots \,|\, f_1(j+n)\ldots f_{s(j+n)}(j+n)v_n\end{aligned} \qquad (5.3.6)$$

is a monomial in $C_j({}^+U_q\mathfrak{n}_-;M;{}^+\Delta)_\lambda$, $f_\ell(m) \in \{f_1,\ldots,f_r\}$ for all ℓ, m; $C(f) = C(f^*)$, where f^* is given by (5.2.19), and the number $C(f^*)$ is defined by (5.2.21).

Denote by $a(p)$, $p = 1,\ldots,j+n$, the tensor factors of f, $f = a(1)|\ldots|a(j+n)$.

(5.3.7) The number $C(f)$ depends only on the weights of the elements $a(1) \,|\, \ldots \,|\, a(j)$, $a(j+1),\ldots,a(j+n)$.

Now we define the map inj of (5.3.2).

Define maps

$$\begin{aligned}\text{pr}&: {}^+U_q\mathfrak{n}_- \to {}^+U_q\bar{\mathfrak{n}}'_-, \\ \text{pr}&: M_s \to \overline{M}_s, \\ \text{pr}&: M_s^* \to \overline{M}_s^* \qquad \text{for } s = 1,\ldots,n,\end{aligned} \qquad (5.3.8)$$

by the formulas $f_i \mapsto \bar{f}_i$, $i = 1,\ldots,r$, $v_s \mapsto \bar{f}_{r+s}$, $(f_{j_1}\ldots f_{j_u}v_s)^* \mapsto (\bar{f}_{j_1}\ldots \bar{f}_{j_u}\bar{f}_{r+s})^*$.

Define maps

$$\text{hook}: M_s \mapsto \overline{M}_s, \qquad s = 1,\ldots,n, \qquad (5.3.9)$$

as follows.

Let $f_J = f_{j_1}\ldots f_{j_n}v_s$ be a monomial in M_s, $j_\ell \in \{1,\ldots,r\}$ for all ℓ. Let $\bar{f}_J = \text{pr}\, f_J = \bar{f}_{j_1}\ldots\bar{f}_{j_n}\bar{f}_{r+s}$ be the corresponding monomial in $\overline{M}_s$. Let $\pi: (1,\ldots,n) \to (1,\ldots,n)$ be any (k,n,n)-hook, where k is arbitrary. Set

$$\pi\bar{f}_J = \bar{f}_{j_{\pi(1)}}\bar{f}_{j_{\pi(2)}}\ldots\bar{f}_{j_{\pi(k)}}\bar{f}_{r+s}\bar{f}_{j_{\pi(k+1)}}\ldots\bar{f}_{j_{\pi(n)}}, \qquad (5.3.10)$$

$$\text{ind}\,(\pi,\bar{f}_J) = (-1)^{n-k}q^{A+B}, \qquad (5.3.11)$$

$$A = (\Lambda_s, \alpha_{j_{\pi(1)}} + \cdots + \alpha_{j_{\pi(k)}} - \alpha_{j_{\pi(k+1)}} - \cdots - \alpha_{j_{\pi(n)}})/4,$$

$$B = \sum_{\ell<m} \pm b_{j_\ell,j_m}/4,$$

where the sign of b_{j_ℓ, j_m} is positive if $\pi(\ell) > \pi(m)$ and is negative otherwise.

Set
$$\text{hook} : f_J \mapsto \sum_{k=0}^{n} \sum_{\pi} \text{ind}\,(\pi, \bar{f}_J)\,\pi \bar{f}_J, \qquad (5.3.12)$$

where the second sum is taken over all (k, n, n)-hooks. Set hook: $v_s \mapsto \bar{f}_{r+s}$.
This formula defines the maps (5.3.9).

Set
$$\begin{aligned}\text{inj} = (\text{pr})^{\otimes j} \otimes (\text{hook})^{\otimes n} : (^+U_q\mathfrak{n}_-)^{\otimes j} \otimes M_1 \otimes \cdots \otimes M_n \\ \longrightarrow (^+U_q\bar{\mathfrak{n}}'_-)^{\otimes j} \otimes \overline{M}_1 \otimes \cdots \otimes \overline{M}_n\,.\end{aligned} \qquad (5.3.13)$$

5.3.14. Theorem. *The following diagram is commutative*

$$\begin{array}{ccc} M_s & \xrightarrow{\text{hook}} & \overline{M}_s \\ {\scriptstyle S}\downarrow & & \downarrow{\scriptstyle S} \\ M_s^* & \xrightarrow{\text{pr}} & \overline{M}_s^*\,. \end{array}$$

5.3.15. Example. Let $f_J = f_1 v_1$. Then $\text{pr} \circ S(f_J) = \text{pr}\,((\Lambda_1, \alpha_1)_q(f_1 v_1)^*) = (\Lambda_1, \alpha_1)_q(\bar{f}_1 \bar{f}_{r+1})^*$, $S \circ \text{hook}\,(f_J) = S(q^{(\Lambda_1, \alpha_1)/4}\bar{f}_1\bar{f}_{r+1} - q^{-(\Lambda_1,\alpha_1)/4}\bar{f}_{r+1}\bar{f}_1) = q^{(\Lambda_1,\alpha_1)/4}\big(q^{(\Lambda_1,\alpha_1)/4}(\bar{f}_1\bar{f}_{r+1})^* + q^{-(\Lambda_1,\alpha_1)/4}\bar{f}_{r+1}\bar{f}_1)^*\big) - q^{-(\Lambda_1,\alpha_1)/4}\big(q^{-(\Lambda_1,\alpha_1)/4}(\bar{f}_1\bar{f}_{r+1})^* + q^{(\Lambda_1,\alpha_1)/4}(\bar{f}_{r+1}\bar{f}_1)^*\big) = (\Lambda_1, \alpha_1)_q(\bar{f}_1\bar{f}_{r+1})^*$.

Proof. Let $f_J = f_{j_1} \ldots f_{j_n} v_s \in M_s$. Then
$$\text{pr} \circ S : f_J \mapsto \sum_{f_I} S(f_I, f_J)\,(\bar{f}_{i_1} \ldots \bar{f}_{i_n} \bar{f}_{r+s})^*, \qquad (5.3.16)$$

where $f_I = f_{i_1} \ldots f_{i_n} v_s$ is a monomial in M_s. The sum is taken over all pairwise different monomials f_I having the same weight as f_J. By definition, the number $S(f_I, f_J)$ has the property
$$e_{i_n} \ldots e_{i_1} f_{j_1} \ldots f_{j_n} v_s = S(f_I, f_J) v_s\,. \qquad (5.3.17)$$

The other side of the diagram gives
$$S \circ \text{hook} : f_J \mapsto \sum_{\pi} \sum_{\bar{f}_L} \text{ind}\,(\pi, \bar{f}_J)\,S(\pi \bar{f}_J, \bar{f}_L)\,(\bar{f}_L)^*, \qquad (5.3.18)$$

where $\bar{f}_L$ is a monomial in $\overline{M}_s$ of the form $\bar{f}_{\ell_1}\ldots\bar{f}_{\ell_j}\bar{f}_{r+s}\bar{f}_{\ell_{j+1}}\ldots\bar{f}_{\ell_n}$; the first sum is taken over all hooks; the second sum is taken over all pairwise different monomials in $\overline{M}_s$ having the same weight as $\bar{f}_J$.

To prove the theorem, it is sufficient to check that:

(5.3.19) For any $\bar{f}_L$ such that $j < n$, we have

$$\sum_\pi \mathrm{ind}\,(\pi, \bar{f}_J)\, S(\pi\bar{f}_J, \bar{f}_L) = 0\,.$$

(5.3.20) For any $\bar{f}_L = \bar{f}_{\ell_1}\ldots \bar{f}_{\ell_n}\bar{f}_{r+s}$, we have

$$\sum_\pi \mathrm{ind}\,(\pi, \bar{f}_J)\, S(\pi\bar{f}_J, \bar{f}_L) = S(f_J, f_L)\,,$$

where $f_L = f_{\ell_1}\ldots f_{\ell_n}v_s$.

These equalities are direct corollaries of Lemma (5.1.23) and coincide in the main point with equality (5.1.19) of Theorem (5.1.18). We check (5.3.19) and (5.3.20) in the case where $f_{j_1},\ldots,f_{j_n}$ are pairwise distinct. In this case, $\bar{f}_{\ell_1},\ldots,\bar{f}_{\ell_n}$ are pairwise distinct. The proof of (5.3.20) coincides with the proof of (5.1.20), see (5.1). We check the equality (5.3.19). The left-hand side of (5.3.19) is an alternating sum of 2^n terms. We divide these terms into pairs in such a way that each pair equals zero. The terms are numerated by all (k,n,n)-hooks $\pi = (\pi(1),\ldots,\pi(k)|\pi(k+1),\ldots,\pi(n))$ for all $k = 0, 1, \ldots, n$.

Let $\bar{f}_L = \bar{f}_{\ell_1}\ldots\bar{f}_{\ell_j}\bar{f}_{r+s}\bar{f}_{\ell_{j+1}}\ldots\bar{f}_{\ell_n}$, $j < n$. Let $\sigma \in \Sigma_{n+1}$ send $(j_1,\ldots,j_n, r+s)$ to $(\ell_1,\ldots,\ell_j,r+s,\ell_{j+1},\ldots,\ell_n)$: $\ell_{\sigma(i)} = j_i, i = 1,\ldots,n, \sigma(n+1) = j+1$. Set $a = \max\{i|\sigma(i) > j, i \leq n\}$.

Say that two hooks $\pi_1 = (\pi_1(1),\ldots,\pi_1(k)|\pi_1(k+1),\ldots,\pi_1(n))$, $\pi_2 = (\pi_2(1),\ldots,\pi_2(k+1)|\pi_2(k+2),\ldots,\pi_2(n))$ are adjacent with respect to $(\bar{f}_J,\bar{f}_L)$ if the unordered set $\{\pi_1(1),\ldots,\pi_1(k),a\}$ coincides with the unordered set $\{\pi_2(1),\ldots,\pi_2(k+1)\}$.

5.3.21. Lemma. *If π_1, π_2 are hooks adjacent to $(\bar{f}_J, \bar{f}_L)$, then*

$$\mathrm{ind}\,(\pi_1, \bar{f}_J)\, S(\pi_1\bar{f}_J, \bar{f}_L) + \mathrm{ind}\,(\pi_2, \bar{f}_J)\, S(\pi_2\bar{f}_J, \bar{f}_L) = 0\,.$$

This lemma implies (5.3.19).

Proof of the lemma. According to Lemma (5.1.23),

$$\mathrm{ind}\,(\pi_1, \bar{f}_J)\, S(\pi_1\bar{f}_J, \bar{f}_L) = (-1)^{n-k} q^{\sum (\Lambda, \pm\alpha_{j_\ell})/2 + \sum \pm b_{j_\ell, j_m}/2}\,. \qquad (5.3.22)$$

Here the first sum is taken over all ℓ such that $\sigma(\ell) < j+1$. The sign of α_{j_ℓ} is positive if $\ell \in \{\pi_1(1), \ldots, \pi_1(k)\}$ and the sign is negative otherwise. The second sum is taken over all $\ell < m$ such that $\sigma(\ell) < \sigma(m)$. The sign of $b_{j_\ell j_m}$ is negative if $\ell \in \{\pi_1(1), \ldots, \pi_1(k)\}$, the sign is positive otherwise.

$$\text{ind}\,(\pi_2, \bar{f}_J)\, S(\pi_2 \bar{f}_J, \bar{f}_L) = (-1)^{n-k+1} q^{\sum (\Lambda, \pm \alpha_{j_\ell})/2 + \sum \pm b_{j_\ell, j_m}/2}, \quad (5.3.23)$$

where the sums and the signs are defined analogously to (5.3.22).

By definition of the number a, the first sums of (5.3.22) and (5.3.23) do not contain the expression $(\Lambda, \pm \alpha_{j_a})/2$, and the second sums of (5.3.22) and (5.3.23) do not contain the expression $b_{j_a j_m}$, $a < m$. Moreover, the expressions b_{j_ℓ, j_a}, $\ell < a$, come in the first sums of (5.3.22) and (5.3.23) with the same sign. These remarks prove the lemma.

Now we define the map left of (5.3.3).

(5.3.24) Let
$$f = f_1(1) \ldots f_{s(1)}(1) \,|\, \ldots \,|\, f_1(j) \ldots f_{s(j)}(j) \,|\, f_1(j+1) \ldots$$
$$f_{s(j+1)}(j+1) \bar{f}_{r+1} f_{s(j+1)+1}(j+1) \ldots f_{t(j+1)}(j+1) \,|\, \ldots \,|\, f_1(j+n) \ldots$$
$$f_{s(j+n)}(j+n) \bar{f}_{r+n} f_{s(j+n)+1} \ldots f_{t(j+n)}(j+n),$$
$$g = f_1(1) \ldots f_{s(1)}(1) \,|\, \ldots \,|\, f_1(j) \ldots f_{s(j)}(j) \,|\, g_1(j+1) \ldots$$
$$g_{u(j+1)}(j+1) \bar{f}_{r+1} g_{u(j+1)+1}(j+1) \ldots g_{v(j+1)}(j+1) \,|\, \ldots \,|\, g_1(j+n) \ldots$$
$$g_{u(j+n)}(j+n) \bar{f}_{r+n} g_{u(j+n)+1}(j+n) \ldots g_{v(j+n)}(j+n)$$

be two monomials in $((U_q \bar{\mathfrak{n}}'_-)^{\otimes j} \otimes \overline{M})_{\bar\lambda}$ having the same first j factors, here $f_\ell(m), g_\ell(m) \in \{\bar{f}_1, \ldots, \bar{f}_r\}$ for all ℓ, m.

Let $t = t(j+1) + \cdots + t(j+n) + n = v(j+1) + \cdots + v(j+n) + n$. Let $\sigma \in \Sigma_t$ be any permutation sending the ordered sequence

$$f_1(j+1), \ldots, f_{s(j+1)}(j+1), \bar{f}_{r+1}, f_{s(j+1)+1}(j+1), \ldots,$$
$$f_{t(j+1)}(j+1), f_1(j+2), \ldots, f_1(j+n), \ldots, f_{s(j+n)}(j+n), \quad (5.3.25)$$
$$\bar{f}_{r+n}, f_{s(j+n)+1}(j+n), \ldots, f_{t(j+n)}(j+n)$$

to the ordered sequence

$$g_1(j+1), \ldots, g_{u(j+1)}(j+1), \bar{f}_{r+1}, g_{u(j+1)+1}(j+1), \ldots,$$
$$g_{v(j+1)}(j+1), g_1(j+2), \ldots, g_1(j+n), \ldots, g_{u(j+n)}(j+n), \bar{f}_{r+n}, \quad (5.3.26)$$
$$g_{u(j+n)+1}(j+n), \ldots, g_{v(j+n)}(j+n).$$

A triple (f, g, σ) is called a *left mover* associated with f if the following conditions are fulfilled:

(5.3.27) Let $\sigma f_{\ell_1}(m_1) = g_{\ell_2}(m_2)$. Then $m_2 \leq m_1$. If $m_1 = m_2$, then $\ell_2 < u(m_2)$ iff $\ell_1 < s(m_1)$. If $m_2 < m_1$, then $\ell_2 > u(m_2)$.

(5.3.28) If $\sigma f_{\ell_1}(m_1) = g_{\ell_2}(m_2)$, $\sigma f_{\ell_3}(m_1) = g_{\ell_4}(m_2)$ for some $m_1, m_2 \in \{j+1, \ldots, n\}$, $1 \leq \ell_1 < \ell_3 \leq t(m_1)$, $\ell_2, \ell_4 \in \{1, \ldots, v(m_2)\}$, then $\ell_2 < \ell_4$.

(5.3.29) If $\sigma f_{\ell_1}(m_1) = g_{\ell_2}(m_2)$, $\sigma f_{\ell_3}(m_3) = g_{\ell_4}(m_2)$ for some $j + 1 \leq m_1 < m_3 \leq j + n$ and $\ell_1, \ell_2, m_2, \ell_3, \ell_4$, then $\ell_2 < \ell_4$.

The monomial g will be called the *resulting monomial* of the left mover.

Define the *index* of a left mover as the number

$$\text{ind}(f, g, \sigma) = q^{A_1 + A_2 + A_3} \tag{5.3.30}$$

where $\{A_i\}$ are defined below.

$$A_1 = \sum_{m=1}^{n} \left(\sum (\Lambda_m, \pm \alpha_{f_a(j+m)})/4 + \sum \pm (\alpha_{f_a(j+m)}, \alpha_{f_b(j+m)})/4 \right). \tag{5.3.31}$$

Here $\alpha_{f_a(j+m)}$ denotes α_p if $f_a(j+m) = \bar{f}_p$. The second sum is taken over all $a \in \{1, \ldots, t(j+m)\}$ such that $\sigma f_a(j+m) = g_{a_1}(j+m_1)$, where $m_1 < m$. The sign of α_{f_a} is positive if $a \leq s(j+m)$, and the sign is negative otherwise. The third sum is taken over all $a < b$ such that $\sigma f_a(j+m) = g_{a_1}(j+m_1)$, $\sigma f_b(j+m) = g_{b_1}(j+m_2)$, where $m_1 \neq m_2$. The sign of $(\alpha_{f_a}, \alpha_{f_b})$ is positive if $m_2 < m_1$, the sign is negative otherwise.

$$A_2 = \sum_{m=1}^{n-1} \left(-\Lambda_m, \sum \alpha_{f_a(j+\ell)}/4 + \sum \alpha_{f_a(j+\ell)}/2 \right). \tag{5.3.32}$$

Here the second sum is taken over all $\ell > m$ and a such that $\sigma f_a(j+\ell) = g_{a_1}(j+m)$ for some a_1. The third sum is taken over all $\ell > m$ and a such that $\sigma f_a(j+\ell) = g_{a_1}(j_{\ell_1})$, where $\ell_1 < m$.

$$A_3 = \sum (\alpha_{f_a(j+\ell)}, \alpha_{f_b(j+m)})/4 + \sum (\alpha_{f_a(j+\ell)}, \alpha_{f_b(j+m)})/2, \tag{5.3.33}$$

where the first sum is taken over all $\ell < m$ and a, b such that $\sigma f_a(j+\ell) = g_{a_1}(j+\ell_1)$, $\sigma f_b(j+m) = g_{b_1}(j+\ell_1)$ for some a_1, b_1, ℓ_1. The second sum is taken over all $\ell < m$ and a, b such that $\sigma f_a(j+\ell) = g_{a_1}(j+\ell_1)$, $\sigma f_b(j+m) = g_{b_1}(j+m_1)$, where $\ell_1 > m_1$.

Let f be a monomial of the form (5.3.24). Set

$$\text{left } f = \sum \text{ind}\,(f, g, \sigma)\, g, \tag{5.3.34}$$

where the sum is taken over all left movers (f, g, σ) associated with f. The expression *left f* will be called the *left smash* of f. Formula (5.3.34) defines linear map (5.3.3). Formula (5.3.4) defines the linear map φ.

5.3.35. Examples. Let $r = 3$, $n = 2$, and $\lambda = (1, 1, 1, 1, 1)$. Then

$$\text{left}\,(\bar{f}_1|\bar{f}_2\bar{f}_3|\bar{f}_4\bar{f}_5) = (\bar{f}_1|\bar{f}_2\bar{f}_4|\bar{f}_3\bar{f}_5) + q^{(-\Lambda_1+\alpha_2+\Lambda_2,\alpha_3)/4}(\bar{f}_1|\bar{f}_2\bar{f}_4\bar{f}_3|\bar{f}_5).$$

Let $r = 2$, $n = 2$, $\lambda = (2, 1, 1, 1)$. Then

$$\begin{aligned}\text{left}\,(\bar{f}_1|\bar{f}_3|\bar{f}_1\bar{f}_2\bar{f}_4) &= (\bar{f}_1|\bar{f}_3|\bar{f}_1\bar{f}_2\bar{f}_4) + q^{(\alpha_1,-\Lambda_1+\Lambda_2-\alpha_2)/4}(\bar{f}_1|\bar{f}_3\bar{f}_1|\bar{f}_2\bar{f}_4) \\ &+ q^{(\alpha_2,-\Lambda_1+\Lambda_2+\alpha_1)/4}(\bar{f}_1|\bar{f}_3\bar{f}_2|\bar{f}_1\bar{f}_4) \\ &+ q^{(-\Lambda_1+\Lambda_2,\alpha_1+\alpha_2)/4}(\bar{f}_1|\bar{f}_3\bar{f}_1\bar{f}_2|\bar{f}_4).\end{aligned}$$

The triple (f, f, id), where id is the identity permutation, is a left mover. Its index equals 1.

5.3.36. Lemma. *The following diagram is commutative*

$$\begin{array}{ccc} ((^+U_q\bar{\mathfrak{n}}'_-)^{\otimes j} \otimes \overline{M})_{\bar\lambda} & \xrightarrow{\text{left}} & ((^+U_q\bar{\mathfrak{n}}'_-)^{\otimes j} \otimes \overline{M})_{\bar\lambda} \\ {\scriptstyle S}\downarrow & & \downarrow{\scriptstyle S} \\ ((^+U_q\bar{\mathfrak{n}}'_-)^{\otimes j} \otimes \overline{M})^*_{\bar\lambda} & \xrightarrow{\text{left}} & ((^+U_q\bar{\mathfrak{n}}'_-)^{\otimes j} \otimes \overline{M})^*_{\bar\lambda} \end{array}$$

where the operators left *are defined by (5.3.35) and (5.2.17).*

5.3.37. Examples. Let $r = 1$ and $n = 2$. Then

$$\begin{aligned}S \circ \text{left}\,(\bar{f}_1|\bar{f}_2|\bar{f}_1\bar{f}_3) &= S((\bar{f}_1|\bar{f}_2|\bar{f}_1\bar{f}_3) + q^{(-\Lambda_1+\Lambda_2,\alpha_1)/4}(\bar{f}_1 \mid \bar{f}_2\bar{f}_1|\bar{f}_3)) \\ &= q^{(\Lambda_2,\alpha_1)/4}\cdot(\bar{f}_1|\bar{f}_2|\bar{f}_1\bar{f}_3)^* + q^{-(\Lambda_2,\alpha_1)/4}(\bar{f}_1|\bar{f}_2|\bar{f}_3\bar{f}_1)^* \\ &+ q^{-(\Lambda_1+\Lambda_2,\alpha_1)/4}(q^{(\Lambda_1,\alpha_1)/4}(\bar{f}_1|\bar{f}_2\bar{f}_1|\bar{f}_3)^* \\ &+ q^{-(\Lambda_1,\alpha_1)/4}(\bar{f}_1|\bar{f}_1\bar{f}_2|\bar{f}_3)^*),\end{aligned}$$

$$\text{left} \circ S(\bar{f}_1|\bar{f}_2|\bar{f}_1\bar{f}_3) = \text{left}\big(q^{(\alpha_1,\Lambda_2)/4}(\bar{f}_1|\bar{f}_2|\bar{f}_1\bar{f}_3)^* + q^{-(\alpha_1,\Lambda_2)/4}(\bar{f}_1|\bar{f}_2|\bar{f}_3\bar{f}_1)^*\big)$$

$$= q^{(\alpha_1,\Lambda_2)/4}\big((\bar{f}_1|\bar{f}_2|\bar{f}_1\bar{f}_3)^* + (\bar{f}_1|\bar{f}_2\bar{f}_1|\bar{f}_3)^* + q^{-(\Lambda_1,\alpha_1)/2}(\bar{f}_1|\bar{f}_1\bar{f}_2|\bar{f}_3)^*\big)$$

$$+ q^{-(\alpha_1,\Lambda_2)/4}(\bar{f}_1|\bar{f}_2|\bar{f}_3\bar{f}_1)^*.$$

Proof of the lemma. For any monomial f of the form (5.3.24),

$$S \circ \text{left}(f) = \sum_{(f,h,\sigma)} \sum_g S(g,h)\,\text{ind}(f,h,\sigma)\,g^*, \qquad (5.3.38)$$

where the first sum is taken over all left movers associated with f, and the second sum is taken over all monomials.

$$\text{left} \circ S(f) = \sum_h \sum_{(h^*,g^*,\sigma)} \text{ind}(h^*,g^*,\sigma)\,S(h,f)\,g^*, \qquad (5.3.39)$$

where the first sum is taken over all monomials, the second sum is taken over all left movers associated with h^*.

We check that (5.3.38) and (5.3.39) are equal for the case in which $f_1(j+1), \ldots, f_{t(j+1)}, f_1(j+2), \ldots, f_1(j+n), \ldots, f_{t(j+n)}(j+n)$ are pairwise different. The general case is similar.

It is easy to see that each g^* appearing in (5.3.38) uniquely determines a left mover (f,h,σ) associated with f such that $S(h,g) \neq 0$. Analogously, each monomial g^* appearing in (5.3.39) uniquely determines a monomial h^* and a left mover (h^*,g^*,σ) associated with h^* such that $S(h,f) \neq 0$.

Fix a monomial g^* appearing in the right-hand side of (5.3.38) and (5.3.39). Fix any pair of elements of the monomial f. Such a pair has the form $(f_{\ell_1}(m_1), f_{\ell_2}(m_2))$ or $(\bar{f}_{r+m}, f_{\ell_1}(m_1))$. We check that the impact of such a pair in the corresponding terms $S(h,g)\,\text{ind}(f,h,\sigma)$, $\text{ind}(h^*,g^*,\sigma)\,S(h,f)$ of (5.3.38) and (5.3.39) is the same. The impact depends on the initial position of the elements of the pair inside f and on the final position of the elements of the pair inside g^*. It is necessary to check all possibilities.

For example, let $m_1 = m_2$, $\ell_1 < \ell_2$. Let g^* have the form (5.2.6). Let $f_{\ell_1}(m_1) = g_{\ell_3}(m_3)$, $f_{\ell_2}(m_1) = g_{\ell_4}(m_4)$, where $m_3 > m_4$. Then the impact of the pair $(f_{\ell_1}(m_1), f_{\ell_2}(m_2))$ in (5.3.38) is $q^{(\alpha_{f_{\ell_1}}(m_1), \alpha_{f_{\ell_2}}(m_2))/4}$. It is a part of the corresponding term $\text{ind}(f,h,\sigma)$. The impact of the same pair in (5.3.39) is the same and it is a part of the corresponding term $S(h,f)$.

The cases of other possible positions of pairs of elements are checked analogously. Therefore, the diagram (5.3.36) is commutative.

5.3.40. Lemma. *The following diagram is commutative:*

$$\begin{array}{ccc} C_j(^+U_q\mathfrak{n}_-;M;{^+}\Delta)_\lambda & \xrightarrow{\varphi} & C_j(^+U_q\bar{\mathfrak{n}}'_-;\overline{M};2;{^+}\Delta)_{\bar\lambda} \\ {\scriptstyle S}\downarrow & & \downarrow{\scriptstyle S} \\ C_j^*(^+U_q\mathfrak{n}_-;M;\mu)_\lambda & \xrightarrow{\varphi} & C_j^*(^+U_q\bar{\mathfrak{n}}'_-;\overline{M};2;\mu)_{\bar\lambda}. \end{array}$$

The lemma is a direct corollary of Lemmas (5.3.14) and (5.3.36).

5.3.41. Theorem. *The map φ defines a monomorphism of complexes*

$$\varphi : C.(^+U_q\mathfrak{n}_-;M;{^+}\Delta)_\lambda \longrightarrow C.(^+U_q\bar{\mathfrak{n}}'_-;\overline{M};2;{^+}\Delta)_{\bar\lambda}.$$

Proof. Obviously, φ has no kernel, cf. (5.2.25). It is sufficient to check that φ commutes with differentials.

Consider the following diagram.

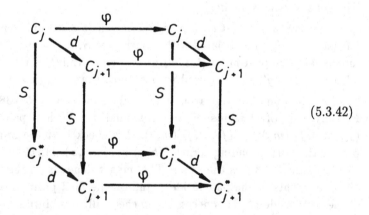

(5.3.42)

According to (4.5.5), (5.1.26), (5.1.25) and (5.3.40), all lateral faces and the bottom face of this cubical diagram are commutative. For general values of all parameters, the maps S are isomorphisms. These remarks imply that the upper face of the cubical diagram is commutative for general values of parameters. Therefore, it is commutative for all values of parameters.

5.3.43. Corollary. *The following diagram is commutative:*

$$
\begin{array}{ccc}
C.(^+U_q\mathfrak{n}_-; M; {}^+\Delta)_\lambda & \xrightarrow{\varphi} & C.(^+U_q\bar{\mathfrak{n}}'_-; \overline{M}; 2; {}^+\Delta)_{\bar\lambda} \\
{\scriptstyle S}\downarrow & & \downarrow{\scriptstyle S} \\
C.^*(^+U_q\mathfrak{n}_-; M; \mu)_\lambda & \xrightarrow{\varphi} & C.^*(^+U_q\bar{\mathfrak{n}}'_-; \overline{M}; 2; \mu)_{\bar\lambda}.
\end{array}
$$

In the next sections, we will prove that the monomorphisms φ are quasi-isomorphisms.

5.4. Theorem

The monomorphism $\varphi : C.^*(^+U_q\mathfrak{n}_-; M; \mu)_\lambda \to C.^*(^+U_q\bar{\mathfrak{n}}'_-; \overline{M}; 2; \mu)_{\bar\lambda}$ is a quasiisomorphism, that is, φ induces an isomorphism of all homology groups.

Proof. According to (4.6.8) and (4.7.7), we have $H_j(C.^*(^+U_q\bar{\mathfrak{n}}'_-; \overline{M}; \mu)_{\bar\lambda}) = 0$ for $j \neq 0$. We prove that $H_j(C.^*(^+U_q\mathfrak{n}_-; M; \mu)_\lambda) = 0$ for $j > 0$ and φ induces an isomorphism of the homology groups in degree 0. This proves the theorem.

5.4.1. Lemma. $H_j(C.^*(^+U_q\mathfrak{n}_-; M; \mu)_\lambda) = 0$ *for $j > 0$.*

Proof. $C_j^* = ((^+U_q\mathfrak{n}_-)^{\otimes j} \otimes M_1 \otimes \cdots \otimes M_n)_\lambda^*$. Let $j = 1$, $x \in C_1^*$, $dx = 0$. We find $y \in C_0^*$ so that $dy = x$. Namely, let $x = \sum a_{f^*} f^*$, where the sum is taken over all monomials $f^* \in C_1^*(^+U_q\mathfrak{n}_-; M; \mu)_\lambda$ of the form

$$f^* = (f_1(1)\ldots f_{s(1)}(1)|f_1(2)\ldots f_{s(2)}(2)v_1|\ldots|f_1(n+1)\ldots f_{s(n+1)}(n+1)v_n)^*$$

and a_{f^*} is a coefficient. Rewrite this sum as $x = \sum' a_{f^*} f^* + \sum'' a_{f^*} f^*$, where the first sum is taken over all monomials f^* such that $s(1) = 1$ and the second sum consists of all of the other monomials.

For each monomial

$$f_0^* = (f_1(1)|f_1(2)\ldots f_{s(2)}(2)v_1|\ldots|f_1(n+1)\ldots f_{s(n+1)}(n+1)v_n)^*$$

with $s(1) = 1$, set

$$f_0^{*\prime} = (f_1(1)f_1(2)\ldots f_{s(2)}(2)v_1|\ldots|f_1(n+1)\ldots f_{s(n+1)}(n+1)v_n)^*.$$

We have $d(f_0^{*\prime}) = \sum b_{f^*} f^*$, where $b_{f_0^*} = -q^{A(f_0^*)}$,

$$A(f_0^*) = \left(\alpha_{f_1(1)}, \sum_{m=3}^{n+1}\left(\Lambda_{m-1} - \sum_{\ell=1}^{s(m)} \alpha_{f_\ell(m)}\right)\right) \Big/ 4.$$

Set
$$y_1 = {\sum}' a_{f^*} q^{-A(f^*)} f^{*\prime},$$
where the sum is taken over all monomials with $s(1) = 1$. Set $x_1 = x + dy_1$.

Applying the same procedure to x_1, we produce x_2, and so on. We have constructed the infinite sequence $x_1, x_2, \ldots, x_m, \ldots$.

It is easy to see that this sequence is stabilized: $x_{m+1} = x_m$ for $m > |\lambda| = \ell_1 + \cdots + \ell_r$. Set $x_0 = x_m$ for $m > |\lambda|$. We prove that $x_0 = 0$.

By construction, $x_0 = \sum c_{f^*} f^*$, where $c_{f^*} = 0$ if a monomial f^* has $s(1) = 1$, and $dx_0 = 0$. We prove that $c_{f^*} = 0$ if a monomial f^* has $s(1) = 2$. In fact, if $c_{f^*} \neq 0$ for some f^* with $s(1) = 2$, then dx_0 contains an element

$$(f_1(1)|f_2(1)|f_1(2)\ldots f_{s(2)}v_1|\ldots|f_1(n+1)\ldots f_{s(n+1)}(n+1)v_n)^*$$

with a nonzero coefficient equal to c_{f^*}. This contradicts to the equality $dx_0 = 0$. Analogously, x_0 does not contain the elements f^* with $s(1) = 3$, and so on.

Thus, we have proved that the initial x is a boundary.

Let $j > 1$, $x \in C_j^*$, $dx = 0$. Let $x = \sum a_{f^*} f^*$, where the sum is taken over all monomials of the form

$$f^* = (f_1(1)\ldots f_{s(1)}(1)|\ldots|f_1(j)\ldots f_{s(j)}(j)|f_1(j+1)\ldots f_{s(j+1)}(j+1)v_1$$
$$|\ldots|f_1(j+n)\ldots f_{s(j+n)}(j+n)v_n)^*.$$

Rewrite this sum as $x = {\sum}' a_{f^*} f^* + {\sum}'' a_{f^*} f^*$, where the first sum is taken over all monomials f^* with $s(1) = 1$.

For each monomial f^* with $s(1) = 1$, set

$$f^{*\prime} = (f_1(1)f_1(2)\ldots f_{s(2)}(2)|f_1(3)\ldots f_{s(3)}(3)$$
$$|\ldots|f_1(j+n)\ldots f_{s(j+n)}(j+n)v_n)^*.$$

Set $y_1 = {\sum}'(-1)^{j+1} a_{f^*} f^{*\prime}$, $x_1 = x + dy_1$. Applying the same procedure to x_1, we produce $x_2, x_3, \ldots$. Analogously to the case $j = 1$, we prove that $x_m = 0$ for $m \gg 1$. The lemma is proved.

5.4.2. Lemma. φ induces an isomorphism

$$H_0(C^*_\cdot({}^+U_q\mathfrak{n}_-; M; \mu)_\lambda) \simeq H_0(C^*_\cdot({}^+U_q\bar{\mathfrak{n}}'_-; \overline{M}; 2; \mu)_{\bar\lambda}).$$

Proof. Since φ is a monomorphism of complexes, it is sufficient to check that φ induces an epimorphism of the homology groups in degree 0. A basis of $H_0(C^{\cdot *}(^+U_q\bar{\mathfrak{n}}_-; \overline{M}; 2; \mu)_{\bar{\lambda}})$ is described in (4.7.10). It is generated by the elements $G(S)$, where S runs through the admissible sequences

$$S = \bar{f}_1(1), \ldots, \bar{f}_{s(1)}(1), \bar{f}_{r+1}, \bar{f}_1(2), \ldots, \bar{f}_1(n), \ldots, \bar{f}_{s(n)}(n), \bar{f}_{r+n} \qquad (5.4.3)$$

of weight $\bar{\lambda}$ starting with $\bar{f}_{r+1}$ (that is, $s(1) = 0$), see (4.7.8) and (4.7.9). Therefore, to prove the lemma, it is sufficient to check that for any admissible sequence S of the form (5.4.3), the element $G(S)$ can be represented as a linear combination,

$$G(S) = \sum a_{f^*} \cdot \text{left} \, (\text{inj} \, (f^*)), \qquad (5.4.4)$$

where f^* runs through all monomials in $(M_1 \otimes \cdots \otimes M_n)_\lambda$ of the form

$$f^* = (f_1(1) \ldots f_{t(1)}(1) v_1 | \ldots | f_1(n) \ldots f_{t(1)} v_n)^*, \qquad (5.4.5)$$

where inj is defined in (5.2.20), $left$ is defined in (5.2.17).

Let $J = \{\bar{f}^*\}$ be any set of monomials in $(\overline{M}_1 \otimes \cdots \otimes \overline{M}_n)_{\bar{\lambda}}$ of the form

$$\bar{f}^* = (\bar{f}_1(1) \ldots \bar{f}_{t(1)}(1) \bar{f}_{r+1} | \ldots | \bar{f}_1(n) \ldots \bar{f}_{t(1)} \bar{f}_{r+n})^*, \qquad (5.4.6)$$

where $\bar{f}_\ell(m) \in \{\bar{f}_1, \ldots, \bar{f}_r\}$ for all ℓ, m. A monomial $\bar{f}^* \in J$ is called an *extreme element* of J if there is no $g^* \in J$ such that there exists a left mover (g^*, f^*, σ) associated with g^*.

According to (5.2.28), the subset of all extreme elements is nonempty.

Let

$$B^1 = \sum b_S^1 G(S)$$

be any sum, where S runs through a subset I_1 of the set of all admissible sequences of the form (5.4.3) of weight $\bar{\lambda}$, $b_S^1 \neq 0$. Consider the set of monomials

$$J_1 = \{\bar{f}(S)^* | S \in I_1\},$$

where $\bar{f}(S)^*$ is the right monomial associated with S, see (4.7.9). J_1 is a set of monomials in $(\overline{M}_1 \otimes \cdots \otimes \overline{M}_n)_{\bar{\lambda}}$. Let

$$K_1 = \{\text{inj}^{-1}(\bar{f}(S)^*) | S \in I_1\},$$

where inj is defined in (5.2.20). K_1 is a set of monomials in $(M_1 \otimes \cdots \otimes M_n)_\lambda$. Set

$$B^2 = B^1 - \sum_{f^* \in K_1} b^1_{S(\text{inj}(f^*))} \, \text{left} \, (\text{inj} \, (f^*)),$$

where $S(\text{inj}(f^*))$ is the admissible sequence associated with a monomial $\text{inj}(f^*)$, see (5.2.29). According to (5.2.30), each term $\text{left}(\text{inj}(f^*))$ is represented as a linear combination

$$\text{left}(\text{inj}(f^*)) = \sum \text{ind}(\text{inj}(f^*), g^*, \sigma) G(S(g^*)).$$

Therefore

$$B^2 = \sum b_S^2 G(S),$$

where S runs through some set I_2 of admissible sequences of weight $\bar{\lambda}$, $b_S^2 \neq 0$. The procedure $B^1 \to B^2$ kills the extreme elements of B^1.

Applying the same procedure to B^2, we can produce the set of J_2 of extreme monomials in $(\overline{M}_1 \otimes \cdots \otimes \overline{M}_n)_{\bar{\lambda}}$, the set K_2 of the corresponding monomials in $(M_1 \otimes \cdots \otimes M_n)_\lambda$, the sum B^3, and so on.

We prove that $B^m = 0$ for $m \gg 1$. In fact, according to (5.2.28), $J_\ell \cap J_m = \emptyset$ for all $\ell \neq m$. The number of monomials in $(\overline{M}_1 \otimes \cdots \otimes \overline{M}_n)_{\bar{\lambda}}$ is finite. Therefore $B^m = 0$ for $m \gg 1$.

We have proved that any linear combination B^1 of the elements $G(S)$ can be represented as a linear combination of the elements $\text{left}(\text{inj}(f^*))$. Applying this to an element $G(S)$, we get (5.4.4). Lemma (5.4.2) is proved.

5.4.7. Example. Let $n = 2$.

$$G(\bar{f}_{r+1}\bar{f}_2\bar{f}_1\bar{f}_{r+2}) = \text{left}(\text{inj}((v_1|f_2f_1v_2)^*)) - q^{-(\alpha_2,\Lambda_1)/2}\text{left}(\text{inj}((f_2v_1|f_1v_2)^*))$$

$$+ q^{-(\alpha_2,\Lambda_1-\alpha_1)/2-(\alpha_1,\Lambda_2)/2}\text{left}(\text{inj}((f_1f_2v_1|v_2)^*))$$

$$= q^{(\Lambda_2,\alpha_1+\alpha_2)/4-(\alpha_1,\alpha_2)/4}\varphi((v_1|f_2f_1v_2)^*)$$

$$- q^{-(\alpha_2,\Lambda_1)/4+(\alpha_1,\Lambda_2)/4}\varphi((f_2v_1|f_1v_2)^*)$$

$$+ q^{-(\alpha_2,\Lambda_1)/4+(\alpha_1,\Lambda_1)/4+(\alpha_1,\alpha_2)/4-(\alpha_1,\Lambda_2)/2}\varphi((f_1f_2v_1|v_2)^*).$$

5.4.8. Remark. Lemma (5.4.2) is a special case of the geometric Theorem (10.1) in [**V4**].

5.5. Theorem

The monomorphism $\varphi : C.(^+U_q\mathfrak{n}_-; M; ^+\Delta)_\lambda \longrightarrow C.(^+U_q\bar{\mathfrak{n}}'_-; \overline{M}; 2; ^+\Delta)_{\bar{\lambda}}$ *is a quasiisomorphism.*

To prove Theorem (5.5) we prove

5.5.1. Theorem. *Let* $c \in C_j(^+U_q\bar{\mathfrak{n}}'_-; \overline{M}; 2; ^+\Delta)_{\bar{\lambda}}$ *and* $dc \in ((^+U_q\mathfrak{n}'_-)^{\otimes(j+1)} \otimes$

$\overline{M})_{\bar{\lambda}} \subset C_{j+1}(^+U_q\bar{\mathfrak{n}}'_-;2;{}^+\Delta)_{\bar{\lambda}}$. Then there exists $e \in C_{j-1}(^+U_q\bar{\mathfrak{n}}'_-;\overline{M};2;{}^+\Delta)_{\bar{\lambda}}$ such that $c + de$ belongs to the image of φ.

Remark. Such an element e will be constructed effectively.

Theorem (5.5.1) is proved in Subsecs. 5.6–5.9.

(5.5.2) Theorem (5.5) follows from (5.5.1). In fact, let $c \in C_j(^+U_q\bar{\mathfrak{n}}_-;\overline{M}; 2;{}^+\Delta)$, $dc = 0$. By (5.5.1) there exists e such that $c + de \in \text{Image}\,\varphi$. Therefore φ induces an epimorphism of the homology groups. Let $dc \in \text{Image}\,\varphi$. By (5.5.1), there exists e such that $c + de \in \text{Image}\,\varphi$. Therefore φ induces a monomorphism of the homology groups.

5.6. Filtration in $C.(^+U_q\bar{\mathfrak{n}}'_-;\overline{M};{}^+\Delta)$

By definition (4.3.2),

$$C_j(^+U_q\bar{\mathfrak{n}}'_-;\overline{M};{}^+\Delta) = \oplus X, \qquad (5.6.1)$$

where $X = X_1 \otimes \cdots \otimes X_{j+n}$ is a j-admissible tensor product of $^+U_q\bar{\mathfrak{n}}'_-, \overline{M}_1, \ldots, \overline{M}_n$, and the sum is taken over all pairwise different j-admissible tensor products.

Define a *module filtration*

$$0 \subset F_j^1 \subset F_j^2 \subset \cdots \subset F_j^{n+1} = C_j(^+U_q\bar{\mathfrak{n}}'_-;\overline{M};{}^+\Delta),$$
$$F_j^k = \oplus X, \qquad (5.6.2)$$

where $X = X_1 \otimes \cdots \otimes X_{j+n}$ is a j-admissible tensor product; the sum is taken over all j-admissible tensor products such that $X_{j+p} = \overline{M}_p$, $p = k, k+1, \ldots, n$.

5.6.3. Example. Let $n = 2$, $j = 1$. The module filtration has the form $F_1^1 = {}^+U_q\bar{\mathfrak{n}}'_- \otimes \overline{M}_1 \otimes \overline{M}_2 \subset F_1^2 = {}^+U_q\bar{\mathfrak{n}}'_- \otimes \overline{M}_1 \otimes \overline{M}_2 \oplus \overline{M}_1 \otimes {}^+U_q\bar{\mathfrak{n}}'_- \otimes \overline{M}_2 \subset F_1^3 = C_1$.

Define
$$\text{hook}_k : M_k \otimes \cdots \otimes M_n \to \overline{M}_k \otimes \cdots \otimes \overline{M}_n \qquad (5.6.4)$$

as the linear map sending $a_k|\ldots|a_n$ to $\text{hook}\,(a_k)\,|\,\cdots\,|\,\text{hook}\,(a_n)$, where $\text{hook}\,(a)$ is defined by (5.3.12).

Define a linear map
$$\text{left}_k : \overline{M}_k \otimes \cdots \otimes \overline{M}_n \to \overline{M}_k \otimes \cdots \otimes \overline{M}_n \qquad (5.6.5)$$

by formula (5.3.35) applied to monomials of $\overline{M}_k \otimes \cdots \otimes \overline{M}_n$.

(5.6.6) An element $c \in C_j(^+U_q\bar{\mathfrak{n}}'_-;\overline{M};2;{}^+\Delta)$ will be called an element *admissible up to level k* if c may be represented as a sum $\Sigma_I c_I$, where each c_I has the form $a_1|\ldots|a_{j+k-1}|a_{j+k}$ and $a_1,\ldots,a_{j+k-1}$ are elements of $^+U_q\bar{\mathfrak{n}}'_-, \overline{M}_1,\ldots, \overline{M}_{k-1}$; $a_{j+k} \in \mathrm{Image}(\mathrm{left}_k \circ \mathrm{hook}_k)$.

Evidently,

(5.6.7) c is admissible up to level 1 iff c belongs to the image of φ.

(5.6.8) Any element c is admissible up to level $n+1$.

(5.6.9) If c is admissible up to level k, then $c \in F_j^k$.

(5.6.10) If c is admissible up to level k, then dc is admissible up to level k.

The last property is a corollary of (5.3.41).

5.6.11. Theorem. Let $c \in C_j(^+U_q\bar{\mathfrak{n}}'_-;\overline{M};2;{}^+\Delta)_{\bar{\lambda}}$ be admissible up to level k and $dc \in F_{j+1}^{k-1}$. Then there exists $e \in C_j(^+U_q\bar{\mathfrak{n}}'_-;\overline{M};2;{}^+\Delta)_{\bar{\lambda}}$ such that $c + de \in F_j^{k-1}$.

5.6.12. Theorem. Let $c \in F_j^k$ and $dc \in F_{j+1}^k$. Then c is admissible up to level k.

Evidently, theorems (5.6.11) and (5.6.12) imply Theorem (5.5.1). Theorem (5.6.11) is proved in (5.9). Theorem (5.6.12) is proved in (5.8).

5.6.13. Remark. The element e in Theorem (5.6.11) is constructed effectively.

5.7. Degree

Let
$$\bar{f} = a_1 \otimes \cdots \otimes a_{j+k} \qquad (5.7.1)$$
be a monomial in F_j^k, where each of $a_1,\ldots,a_{j+k-1}$ is an element of one of $^+U_q\bar{\mathfrak{n}}'_-, \overline{M}_1,\ldots,\overline{M}_{k-1}$, and $a_{j+k} = \bar{f}(j+k)|\ldots|\bar{f}(j+n) \in \overline{M}_k \otimes \cdots \otimes \overline{M}_n$,
$$\bar{f}(j+m) = \bar{f}_1(j+m)\ldots\bar{f}_{s(j+m)}(j+m)\bar{f}_{r+m}\bar{f}_{s(j+m)+1}(j+m)$$
$$\ldots \bar{f}_{t(j+m)}(j+m) \in \overline{M}_m,$$
where $\bar{f}_\ell(j+m) \in \{\bar{f}_1,\ldots,\bar{f}_r\}$ for all ℓ, m. The vector $(t(j+k), s(j+k),\ldots, t(j+n), s(j+n)) \in \mathbb{N}^{2(n-k+1)}$ will be called the *k-degree* of $\bar{f}$ and the *degree* of a_{j+k}.

Say that $\bar{f}$ is a *left monomial up to level k* and a_{j+k} is a *left monomial* if $t(j+k) = s(j+k), \ldots, t(j+n) = s(j+n)$.

Example. $a_{j+k} = \bar{f}_{r+k}| \cdots | \bar{f}_{r+n}$ is a left monomial, deg $a_{j+k} = (0, 0, \ldots, 0, 0)$.

Let $d_i = (t_i(k), s_i(k), \ldots, t_i(n), s_i(n))$, $i = 1, 2$, be two vectors in $\mathbb{N}^{2(n-k+1)}$. Say that d_1 is less than d_2 if

(5.7.2) There is a number m, $k \leq m \leq n$, such that $t_1(j+\ell) = t_2(j+\ell)$, $s_1(j+\ell) = s_2(j+\ell)$, $\ell = 1, \ldots, m-1$, and $t_1(j+m) < t_2(j+m)$ or $t_1(j+m) = t_2(j+m)$, $s_1(j+m) < s_2(j+m)$.

Let $\bar{f}$ be any left monomial up to level k of the form (5.7.1). Set

$$b_{j+k} = f(j+k) | \ldots | f(j+n), \tag{5.7.3}$$

where $f(j+m) = f_1(j+m) \ldots f_{t(j+m)}(j+m)v_m$; $f_\ell(m) = f_a$ if $\bar{f}_\ell(m) = \bar{f}_a$ for all ℓ, m.

Example. $f(j+m) = f_1 f_1 v_m$ if $\bar{f}(j+m) = \bar{f}_1 \bar{f}_1 \bar{f}_{r+m}$.

Set

$$C = -\sum_{m=k}^{n} \Big(\Lambda_m, \sum_{\ell=1}^{t(j+m)} \alpha_{f_\ell(j+m)}\Big)/4 + \sum_{m=k}^{n} \sum_{\ell<p} (\alpha_{f_\ell(j+m)}, \alpha_{f_p(j+m)})/4,$$

$$a_{j+k}^{\text{new}} = q^C \cdot \text{left}_k \circ \text{hook}_k(b_{j+k}). \tag{5.7.4}$$

5.7.5. Lemma. *The element $a_{j+k} - a_{j+k}^{\text{new}}$ is a linear combination of nonleft monomials in $\overline{M}_k \otimes \cdots \otimes \overline{M}_n$.*

The lemma is a corollary of formulas (5.3.12), (5.3.11), (5.3.35) and (5.3.27).

5.8. Proof of Theorem 5.6.12

The element c is represented as a linear combination of monomials such that each of the monomials has the form (5.7.1):

$$c = \sum_\alpha x_\alpha \bar{f}_\alpha,$$

where $\bar{f}_\alpha$ is a monomial and x_α is a coefficient. Rewrite this sum as

$$c = {\sum}' x_\alpha \bar{f}_\alpha + {\sum}'' x_\alpha \bar{f}_\alpha,$$

where the first sum is taken over all left monomials up to level k and the second sum is taken over all of the other monomials.

Let $\bar{f}_\alpha = a_{1,\alpha}|\ldots|a_{j+k,\alpha}$ be any left monomial up to level k. By formula (5.7.4), define $a^{new}_{j+k,\alpha}$ and set $\bar{f}^{new}_\alpha = a_{1,\alpha}|\ldots|a_{j+k-1}|a^{new}_{j+k,\alpha}$,

$$e = {\sum}' x_\alpha(\bar{f}_\alpha - \bar{f}^{new}_\alpha) + {\sum}' x_\alpha \bar{f}_\alpha.$$

To prove the theorem, it is sufficient to check that $e = 0$. By Lemma (5.7.5), e is a linear combination of nonleft monomials of the form (5.7.1). According to the assumptions of Theorem (5.6.12) and to property (5.6.10), $de \in F^k_{j+1}$.

5.8.1. Lemma. *Let $e \in F^k_j$ be a linear combination of nonleft monomials of the form (5.7.1):*

$$e = \sum y_\alpha \bar{f}_\alpha,$$

and $de \in F^k_{j+1}$. Then $e = 0$.

The lemma implies Theorem (5.6.12).

The lemma is proved by induction on k-degree of monomials $\bar{f}_\alpha$. By the assumptions, the sum

$$e = \sum y_\alpha \bar{f}_\alpha \tag{5.8.2}$$

does not contain monomials left up to level k. This is the base of the induction.

Assume that the sum (5.8.2) does not contain monomials $\bar{f}_\alpha$ of k-degree greater than $d = (t(j+k), s(j+k), \ldots, t(j+n), s(j+n))$. We prove that (5.8.2) does not contain monomials of k-degree d.

If $t(j+\ell) = s(j+\ell)$, $\ell = k, \ldots, n$, then a monomial of k-degree d is a left monomial. Therefore it does not appear in (5.8.2).

Let $t(j+\ell) > s(j+\ell)$ for some $\ell \in \{k, \ldots, n\}$, let $\bar{f}_\beta$ be a monomial of degree d, $\bar{f}_\beta = a_1|\ldots|a_{j+k}$, where $a_1, \ldots, a_{j+k}$ are described in (5.7.1). The differential $d\bar{f}_\beta$ is a linear combination of monomials in $C_{j+1}({}^+U_q\bar{n}'_-; \overline{M}; 2; {}^+\Delta)$, and one of these monomials is

$$g = a_1|\ldots|a_{j+k-1}|\bar{f}(j+k)|\ldots|\bar{f}(j+\ell-1)|\bar{f}_{s(j+\ell)+1}(j+\ell)$$
$$\ldots \bar{f}_{t(j+\ell)}(j+\ell)|\bar{f}_1(j+\ell)\ldots \bar{f}_{s(j+\ell)}(j+\ell) \tag{5.8.3}$$
$$\bar{f}_{r+\ell}|\bar{f}(j+\ell+1)|\ldots|\bar{f}(j+n).$$

The monomial g also appears only in the differential of the monomials $\bar{f}$ of the form

$$\bar{f} = a_1|\ldots|a_{j+k-1}|\bar{f}(j+k)|\ldots|f(j+\ell-2)|g_1\ldots g_a \bar{f}_{r+\ell-1}g_{a+1}\ldots g_b$$
$$|\bar{f}_1(j+\ell)\ldots \bar{f}_{s(j+\ell)}\bar{f}_{r+\ell}|\bar{f}(j+\ell+1)|\ldots|\bar{f}(j+n),$$

where $a \geq s(j+\ell-1)$, $b > t(j+\ell-1)$, $g_m \in \{\bar{f}_1, \ldots, \bar{f}_r\}$ for all m, and of the form

$$\bar{f} = a_1 | \ldots | a_{j+k-1} | \bar{f}(j+k) | \cdots | \bar{f}(j+\ell-1) | g_1 \ldots g_a \bar{f}_{r+\ell} g_{a+1}$$
$$\cdots g_{t(j+\ell)} | \bar{f}(j+\ell+1) | \cdots | \bar{f}(j+n),$$

where $a > s(j+\ell)$ and $g_m \in \{\bar{f}_1, \ldots, \bar{f}_r\}$ for all m.

Each of these monomials has k degree greater than d, and, therefore, is absent in the sum (5.8.2). This means that g is not canceled in de unless $y_\beta = 0$. By assumptions we have $de \in F_{j+1}^k$. Therefore g cannot appear in de, and hence $y_\beta = 0$.

5.9. Proof of Theorem 5.6.11

By the assumption, c can be written as

$$c = \sum x_\alpha f_\alpha, \tag{5.9.1}$$

where the right-hand side is a linear combination of elements f_α of the form

$$f_\alpha = a_1 | a_2 | a_3 | a_4, \tag{5.9.2}$$

where $a_4 \in \overline{M}_k \otimes \cdots \otimes \overline{M}_n$ and $a_4 \in \text{Image}(\text{left}_k \circ \text{hook}_k)$, $a_3 \in (^+U_q\bar{\mathfrak{n}}'_-)^{\otimes \ell}$ for some $\ell \in \{0, 1, \ldots, j\}$,

$$a_2 = \bar{f}_1(j+k-\ell-1) \ldots \bar{f}_{s(j+k-\ell-1)}(j+k-\ell-1)$$
$$\bar{f}_{r+k-1} \bar{f}_{s(j+k-\ell+1)+1}(j+k-\ell-1)$$
$$\ldots \bar{f}_{t(j+k-\ell-1)}(j+k-\ell-1) \in \overline{M}_{k-1}$$

and $\bar{f}_\ell(j+k-\ell-1) \in \{\bar{f}_1, \ldots, \bar{f}_r\}$ for all ℓ; $a_1 \in C_{j-\ell}(^+U_q\bar{\mathfrak{n}}'_-; (\overline{M}_1, \ldots, \overline{M}_{k-2}); 2; {}^+\Delta)$.

In particular, if $\ell = 0$, then $f_\alpha = a_1 | a_2 | a_4 | \in F_j^{k-1}$.

We describe some characteristics of an element f_α. Let $\lambda(f_\alpha), \lambda(a_j) \in \mathbf{N}^{r+n}$ be the weights of the corresponding elements, $\lambda(f_\alpha) = \lambda(a_1) + \cdots + \lambda(a_4)$. For any $\lambda = (\ell_1, \ldots, \ell_{r+n})$, set $|\lambda| = \ell_1 + \cdots + \ell_{r+n}$. In particular, $|\lambda(a_2)| = t(j+k-\ell-1)+1$. The vector

$$v(f_\alpha) = (|\lambda(a_2) + \lambda(a_3)|, |\lambda(a_3)|) \in \mathbf{N}^2$$

will be called the *vector of an element* f_α of form (5.9.2). If $v(f_\alpha) = (x_1, x_2)$, then $x_1 \geq x_2 \geq 1$. The equality $x_1 = x_2$ means $\ell = 0$, $f_\alpha \in F_j^{k-1}$.

An element f_α will be called a *left element at level* $\overline{M}_{k-1}$ if $t(j+k-\ell-1) = s(j+k-\ell-1)$ and $\ell > 0$.

Let $v_i = (x_1^i, x_2^i)$, $i = 1, 2$, be two vectors. Say that v^1 is less than v^2 if $x_1^1 < x_1^2$ or $x_2^1 < x_2^2$.

To prove Theorem (5.6.11), it is sufficient to prove the following two lemmas.

5.9.3. Lemma. *There exists $e \in F_{j-1}^k$ admissible up to level k such that the element $c + de$ can be represented as a linear combination*

$$c + de = \sum y_\alpha f_\alpha,$$

where f_α are elements of form (5.9.2) nonleft at level $\overline{M}_{k-1}$.

5.9.4. Lemma. *If the element $c + de$ is a linear combination of elements of form (5.9.2) nonleft at level $\overline{M}_{k-1}$, then $c + de \in F_j^{k-1}$.*

Proof of Lemma 5.9.3. Let $f = a_1|a_2|a_3|a_4$ be any left monomial at level $\overline{M}_{k-1}$ having form (5.9.2). Let

$$a_3 = \bar{f}(j+k-\ell)|\ldots|\bar{f}(j+k-1) \in (^+U_q\bar{\mathfrak{n}}'_-)^{\otimes \ell},$$

where $\bar{f}(m) = \bar{f}_1(m)\ldots\bar{f}_{t(m)}(m)$, $m = j+k-\ell, \ldots, j+k-1$, $\bar{f}_\ell(m) \in \{\bar{f}_1, \ldots, \bar{f}_r\}$ for all ℓ, m.

Set

$$b_3 = \bar{f}(j+k-\ell+1)|\cdots|\bar{f}(j+k-1) \in (U_q\bar{\mathfrak{n}}'_-)^{\otimes \ell}.$$

$$\begin{aligned}b_2 = &\bar{f}_1(j+k-\ell-1)\ldots\bar{f}_{t(j+k-\ell-1)}(j+k-\ell-1)\bar{f}_{r+k-1}\\ &\bar{f}_1(j+k-\ell)\ldots\bar{f}_{t(j+k-\ell)}(j+k-\ell) \in \overline{M}_{k-1}.\end{aligned} \quad (5.9.5)$$

$$g = a_1|b_2|b_3|a_4.$$

We have

$$\begin{aligned}dg = &\pm da_1|b_2|b_3|a_4 \pm a_1|db_2|b_3|a_4 \pm a_1|b_2|db_3|a_4 \\ &\pm a_1|b_2|b_3|da_4,\end{aligned} \quad (5.9.6)$$

where the differential is defined in (4.4.3.1) and (4.5.1).

$$db_2 \in \overline{M}_{k-1} \otimes {}^+U_q\bar{\mathfrak{n}}'_- \oplus {}^+U_q\bar{\mathfrak{n}}'_- \otimes \overline{M}_{k-1}, \quad (5.9.7)$$

where db_2 is a linear combination of monomials, and one of these monomials is equal to $a_2|\bar{f}(j+k-\ell)$. Let z be a number such that $a_2|\bar{f}(j+k-\ell) + zdb_2$ does not contain the monomial $a_2|\bar{f}(j+k-\ell)$. Set

$$f^{\text{new}} = f + zdg. \tag{5.9.8}$$

It is easy to set that

(5.9.9) f^{new} can be represented as a linear combination of elements of form (5.9.2):

$$f^{\text{new}} = \sum y_\beta f_\beta.$$

Moreover, if some element f_β is a left element at level $\overline{M}_{k-1}$, then its vector $v(f_\beta) = (x_{1,\beta}, x_{2,\beta})$ is less than the vector $v(f) = (x_1, x_2)$ of the initial element.

In fact, the first, third, and fourth terms in (5.9.6) are nonleft elements by our construction. If a left element f_β comes from the first term of (5.9.7), then $x_{1,\beta} = x_1$, $x_{2,\beta} < x_2$. If a left element f_β comes from the second term of (5.9.7), then $x_{1,\beta} < x_1$, $x_{2,\beta} \leq x_2$.

We have described a procedure transforming any element f which is left at level $\overline{M}_{k-1}$ into a sum $f^{\text{new}} = \sum y_\beta f_\beta$ whose left elements at level $\overline{M}_{k-1}$ have the weight vector less than $v(f)$.

Now consider the element c of Theorem (5.6.11). Rewrite (5.9.1) as

$$c = {\sum}' x_\alpha f_\alpha + {\sum}'' x_\alpha f_\alpha,$$

where the first sum is taken over all monomials f_α left at level $\overline{M}_{k-1}$ and the second sum is taken over all of the other monomials.

Applying the procedure described above, we produce a new element

$$c_1 = {\sum}' x_\alpha f_\alpha^{\text{new}} + {\sum}'' x_\alpha f_\alpha.$$

Applying the same procedure to c_1 we produce c_2 and so on. According to (5.9.9), after a finite number of steps we will get an element c_m which is a linear combination of elements of form (5.9.2) nonleft at level $\overline{M}_{k-1}$. This proves Lemma (5.9.3).

Proof of Lemma 5.9.4. Let

$$c + de = \sum y_\alpha f_\alpha \tag{5.9.10}$$

be a linear combination of elements of form (5.9.2) nonleft at level $\overline{M}_{k-1}$. To each f_α assign number $d(f_\alpha) = t(j+k-\ell-1) - s(j+k-\ell-1)$.

We prove that each f_α of (5.9.10) belongs to F_j^{k-1}; in other words, it has $\ell = 0$. This is done by induction on the number $d(f_\alpha)$. Assume that (5.9.10) does not contain the elements f_α with $\ell > 0$ and $d(f_\alpha) < d$. We prove that (5.9.10) does not contain the elements f_α with $\ell > 0$ and $d(f_\alpha) = d$.

Assume that for some α, we have $y_\alpha \neq 0$ and f_α has $\ell > 0$ and $d(f_\alpha) = d$. Let $f_\alpha = a_1|a_2|a_3|a_4$, where $a_1, \ldots, a_4$ are described in (5.9.2). Then, $df_\alpha = \pm da_1|a_2|a_3|a_4 \pm a_1|da_2|a_3|a_4 \pm \cdots \pm a_1|a_2|a_3|da_4$; da_2 is a linear combination of monomials, and one of these monomials is the monomial

$$b_2 = \bar{f}_1(j+k-\ell-1)\ldots \bar{f}_{s(j+k-\ell-1)}(j+k-\ell-1)\bar{f}_{r+k-1}|$$
$$\bar{f}_{s(j+k-\ell-1)+1}(j+k-\ell-1)\ldots \bar{f}_{t(j+k-\ell-1)}(j+k-\ell-1)$$
$$\in \overline{M}_{k-1} \otimes U_q\bar{\mathfrak{n}}'_-.$$

Set $g_\alpha = a_1|b_2|a_3|a_4$. It is easy to see that g_α is not canceled in $d\left(\sum y_\alpha f_\alpha\right)$ unless $y_\alpha = 0$. This contradicts to the fact that $dc \in F_{j+1}^{k-1}$. Theorem (5.5) is proved.

5.10. Remark

In this book, our quantum groups $U_q\mathfrak{g}$, $U_q\bar{\mathfrak{n}}_-$, $U_q\bar{\mathfrak{n}}'_-$, ... depend on the parameter $q = \exp(2\pi i/\varkappa)$, where $\varkappa \in \mathbb{C}$ is a fixed number. We can consider $\varkappa$ as a variable and define the same objects $M_1, \ldots, M_n, \overline{M}_1, \ldots, \overline{M}_n$. In this new setting, Theorems (5.4) and (5.5) and their proofs do not change. In fact, all maps defined in constructions of Sec. 5 are polynomial functions in variables $q^{\pm(\alpha_i,\alpha_j)/4}$ and $q^{\pm(\Lambda_i,\alpha_j)/4}$, see, for example, formulas (5.1.17), (5.2.25), (5.3.4), (5.3.12), (5.3.34), (5.7.4), and (5.9.8).

5.11. Geometric Interpretation of Theorems 5.3.34, 5.4, and 5.5

Combining diagrams (4.7.5) and (5.3.43), we receive the following commutative diagram

$$\begin{array}{ccccc}
C.(^+U_q\mathfrak{n}_-; M; {}^+\Delta)_\lambda & \xrightarrow{\varphi} & C.(^+U_q\bar{\mathfrak{n}}'_-; \overline{M}; 2; {}^+\Delta)_{\bar{\lambda}} & \xrightarrow{p} & \mathcal{X}_{\ell-}.(\mathcal{C}_{n,\ell,\lambda})^- \\
s\downarrow & & s\downarrow & & s\downarrow \quad (5.11.1) \\
C.^*(^+U_q\mathfrak{n}_-; M; \mu)_\lambda & \xrightarrow{\varphi} & C.^*(^+U_q\bar{\mathfrak{n}}'_-; \overline{M}; 2; \mu)_{\bar{\lambda}} & \xrightarrow{p} & \mathcal{Q}_{\ell-}.(\mathcal{C}_{n,\ell,\lambda})^-.
\end{array}$$

We review the notations. The complexes of the left column are defined in (5.1.1) and (5.1.24), $\lambda = (\ell_1, \ldots, \ell_r) \in \mathbb{N}^r$. The left S is the contravariant form, see (5.1.26).

The complexes of the middle column are define in (4.5) and (5.2), see (5.2.3). Here $\bar{\lambda} = (\ell_1, \ldots, \ell_r, 1, \ldots, 1) \in \mathbb{N}^{r+n}$. The middle S is the contravariant form, see (4.5.4) and (4.5.5).

The homomorphisms φ are defined in (5.2) and (5.3), see (5.2.26), (5.3.40) and (5.3.41). The homomorphisms φ are monomorphisms of complexes and are quasiisomorphisms, see (5.4) and (5.5).

The complexes of the right column are defined in (4.7). Namely, consider a discriminantal configuration $\mathcal{C}_{n,\ell}(z)$. Here $z = (z_1, \ldots, z_n) \in \mathbb{R}^n$ is an arbitrary point such that $z_1 < \cdots < z_n$, $\ell = \ell_1 + \cdots + \ell_r$. The configuration $\mathcal{C}_{n,\ell}(z)$ is defined in (3.1). It is a configuration of hyperplanes in $\mathbb{R}^\ell$. Let $x_1(1), \ldots, x_{\ell_1}(1), \ldots, x_1(r), \ldots, x_{\ell_r}(r)$ be coordinates in $\mathbb{R}^\ell$. Define the weights of the configuration as in (4.7.1), (5.2.1) and (5.2.2):

$$\begin{aligned} a(x_i(j), x_u(v)) &= (\alpha_j, \alpha_v)/\varkappa, \\ a(x_i(j), z_u) &= (\alpha_j, -\Lambda_u)/\varkappa \end{aligned} \tag{5.11.2}$$

for all i, j, u, v. Here $\alpha_1, \ldots, \alpha_r$ are simple roots of $U_q\mathfrak{n}_-$, $\Lambda_1, \ldots, \Lambda_n$ are highest weights of Verma modules $M_1, \ldots, M_n$, $\varkappa$ is the parameter of the quantum group, and $q = \exp\left(\frac{2\pi i}{\varkappa}\right)$. Let $\Sigma' = \Sigma_{\ell_1} \times \cdots \times \Sigma_{\ell_r}$ be the group of permutations of coordinates $\{x_i(j)\}$ in $\mathbb{R}^\ell$ preserving the index j. Σ' preserves the weights of $\mathcal{C}_{n,\ell}$. In this situation, the configuration $\mathcal{C}_{n,\ell}(z)$ was denoted by $\mathcal{C}_{n,\ell,\lambda}(z)$, see (4.7) and (3.16). Now $\mathcal{X}.(\mathcal{C}_{n,\ell,\lambda})^-$, $Q.(\mathcal{C}_{n,\ell,\lambda})^-$ are the abstract complexes of $\mathcal{C}_{n,\ell,\lambda}$ introduced in (3.16).

The isomorphisms p in (5.11.1) are defined in (4.7).

The complexes of the right column of (5.11.1) compute homology groups of the pair $\mathbb{C}^\ell$, $\mathcal{C}_{n,\ell,\lambda}(z)$ as described in (2.9). Therefore, the complexes of the left column of (5.11.1) compute homology groups of the pair $\mathbb{C}^\ell$, $\mathcal{C}_{n,\ell,\lambda}(z)$.

Namely, let

$$U_{n,\ell} = \left(\bigcup_{H \in \mathcal{C}_{n,\ell,\lambda}(z)} H_{\mathbb{C}}\right) \subset \mathbb{C}^\ell.$$

Let $\mathcal{S}^*(a)$ be the local system on $\mathbb{C}^\ell - U_{n,\ell}$ defined by the weights a of (5.11.2). Let

$$\begin{aligned} w\pi &: \mathcal{X}.(\mathcal{C}_{n,\ell,\lambda}(z))^- \to C.(X, \mathcal{S}^*(a))^- \subset C.(\mathbb{C}^\ell - U_{n,\ell}, \mathcal{S}^*(a)), \\ w\pi &: Q.(\mathcal{C}_{n,\ell,\lambda}(z))^- \to C.(Q'/Q'', \mathcal{S}^*(a))^- \subset C_!.(\mathbb{C}^\ell - U_{n,\ell}, \mathcal{S}^*(a)) \end{aligned}$$

be a realization of the abstract complexes, see (3.16.7) and (2.9.7). The realization and the diagram (5.11.1) induce canonical isomorphisms,

$$\begin{aligned} H_j(C.(^+U_q\mathfrak{n}_-; M; {}^+\Delta)_\lambda) &\longrightarrow H_{\ell-j}(\mathbb{C}^\ell - U_{n,\ell}, \mathcal{S}^*(a))^-, \\ H_j(C.^*(^+U_q\bar{\mathfrak{n}}_-; M; \mu)_\lambda) &\longrightarrow H_{\ell-j}(\mathbb{C}^\ell, U_{n,\ell}, \mathcal{S}^*(a))^-, \end{aligned} \qquad (5.11.3)$$

for all j. Here the superscript '$-$' denotes the subspace of all elements x such that $\sigma x = (-1)^{|\sigma|} x$ for all $\sigma \in \Sigma'$.

The following diagram is commutative

$$\begin{array}{ccc} H_j(C.(^+U_q\mathfrak{n}_-; M; {}^+\Delta)_\lambda) & \xrightarrow{S} & H_j(C.^*(^+U_q\mathfrak{n}_-; M; 2; \mu)_\lambda) \\ {\scriptstyle f}\downarrow & & {\scriptstyle f}\downarrow \\ H_{\ell-j}(\mathbb{C}^\ell - U_{n,\ell}, \mathcal{S}^*(a))^- & \xrightarrow{S} & H_{\ell-j}(\mathbb{C}^\ell, U_{n,\ell}, \mathcal{S}^*(a))^-, \end{array} \qquad (5.11.4)$$

where the upper S is induced by the contravariant form and the lower S is the canonical homomorphism.

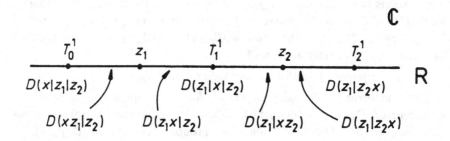

Fig. 5.1

5.11.5. *Example*. Let $n = 2$, $\lambda = (1, 0, \ldots, 0) \in \mathbb{N}^r$, cf. (5.1.2) and (5.1.25). The configuration $\mathcal{C}_{n,\ell,\lambda}(z)$ is a configuration of two points $z_1 < z_2$ in $\mathbb{R}$. The group Σ' is trivial.

The weights of z_1 and z_2 are $-(\alpha_1, \Lambda_1)/\varkappa$ and $-(\alpha_2, \Lambda_2)/\varkappa$, respectively. The cells of complexes $C.(Q'/Q'', \mathcal{S}^*(a))^- = C.(Q'/Q'', \mathcal{S}^*(a))$ and

$C.(X, S^*(a))^- = C.(X, S^*(a))$ are shown in Figs. 5.1 and 5.2, respectively.

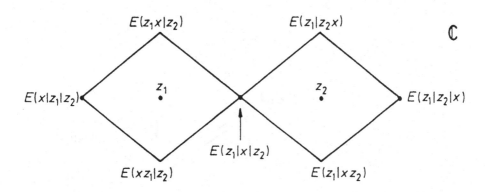

Fig. 5.2

The image of elements of $C.^*(^+U_q\mathbf{n}_-; M; \mu)_\lambda$ under the map $w\pi p\varphi$ is shown in Fig. 5.3.

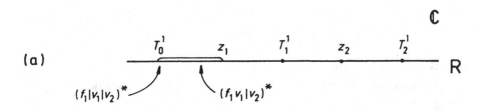

Fig. 5.3

The image of elements of $C.(^+U_q\mathbf{n}_-; M; ^+\Delta)_\lambda$ under the map $w\pi p\varphi$ is shown in Fig. 5.4.

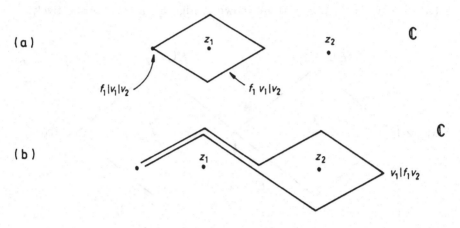

Fig. 5.4

5.11.6. Example. Let $n = 2$, $\lambda = (1, 1, 0, \ldots, 0) \in \mathbb{N}^r$. The corresponding configuration $\mathcal{C}_{n,\ell,\lambda}(z)$ is a configuration of five lines in $\mathbb{R}^r$ shown in Fig. 3.1. The weights of the configuration are given by the formulas:

$$a(x_1, x_2) = (\alpha_1, \alpha_2)/\varkappa,$$
$$a(x_i, z_j) = (\alpha_i, -\Lambda_j)/\varkappa \qquad \text{for } i, j = 1, 2.$$

In this example the group Σ' is trivial.

The image of some elements of $C^{\cdot*}(^+U_q\mathfrak{n}_-; M; \mu)_\lambda$ is schematically shown in Fig. 5.5.

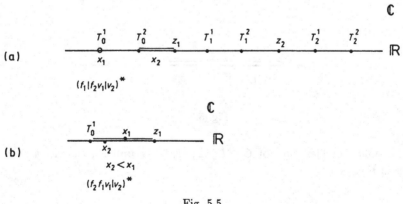

Fig. 5.5

The image of some elements of $C.(^+U_q\mathfrak{n}_-; M; {}^+\Delta)_\lambda$ is shown schematically in Fig. 5.6.

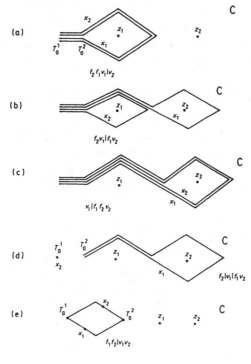

Fig. 5.6

For example, the element $f_2 f_1 v_1 | v_2 \in C_0(^+U_q\mathfrak{n}_-; M; {}^+\Delta)_\lambda$ is represented by a two-chain in $\mathbb{C}^2$. This chain consists of points (x_1, x_2) such that x_1 and x_2 independently run through the curves in $\mathbb{C}$ shown in Fig. 5.6a. Namely, x_1 runs around z_1, and x_2 runs around the curve traced by x_1. More precisely, $(f_2 f_1 v_1 | v_2)$ is represented by the chain

$$w\pi p\,\varphi(f_2 f_1 v_1 | v_2) = q^{(b_{0\,1}+b_{0\,2}+b_{1\,2})/4}\big(q^{(-b_{0\,1}-b_{0\,2}-b_{1\,2})/4} E(x_2 x_1 z_1 | z_2)$$
$$- q^{(b_{0\,1}-b_{0\,2}-b_{1\,2})/4} E(x_2 z_1 x_1 | z_2)$$
$$- q^{(b_{0\,2}-b_{0\,1}+b_{1\,2})/4} E(x_1 z_1 x_2 | z_2)$$
$$+ q^{(b_{0\,1}+b_{0\,2}+b_{1\,2})/4} E(z_1 x_1 x_2 | z_2)\big)$$

of the complex $C.(X, \mathcal{S}^*(a))$ of the configuration $\mathcal{C}_{n,\ell,\lambda}(z)$. Here $b_{0,i} = (-\Lambda_1, \alpha_i)$, $b_{1\,2} = (\alpha_1, \alpha_2)$, see (5.3.5) and (5.13.12).

We have $w\pi p\varphi(f_1 f_2 | v_1 | v_2) = q^{b/4} E(x_1 x_2 | z_1 | z_2)$, where $b = (\Lambda_1 + \Lambda_2, -\alpha_1 - \alpha_2) + (\alpha_1, \alpha_2)$, see Fig. 5.6e. In this figure, x_2 runs over x_1 and $x_1 + x_2 = T_0^1 + T_0^2$.

5.11.7. Example. Let n be an arbitrary natural number and $\lambda = (\ell, 0, \ldots, 0) \in \mathbb{N}^r$. The corresponding configuration is a discriminantal configuration $C_{n,\ell,\lambda}(z)$ in $\mathbb{R}^\ell$. The weights of the configuration are given by the formula:

$$a(x_i, x_j) = (\alpha_1, \alpha_1)/\varkappa,$$
$$a(x_i, z_j) = (\alpha_1, -\Lambda_j)/\varkappa$$

for all i, j. In this case, the group Σ' coincides with the group of all permutations of the coordinates in $\mathbb{R}^\ell$. The superscript in the right-hand side of (5.11.3) in this case denotes the subspace of all skewsymmetric elements with respect to the group of all permutations of coordinates.

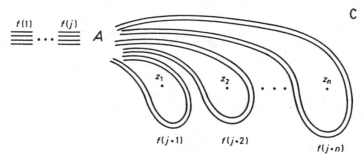

Fig. 5.7

The pictures of Fig. 5.6 have a natural generalization. Let n be an arbitrary natural number and $\lambda = (1, \ldots, 1) \in \mathbb{N}^r$. Any monomial f of form (5.3.6) is represented in $C.(X, \mathcal{S}^*(a))$ by a chain shown in Fig. 5.7. Some details of Fig. 5.7 are shown in Fig. 5.8.

The left picture of Fig. 5.8 shows that a monomial $f(p) = f_1(p) \ldots f_{s(p)}(p)$ is represented by points $x_1(p), \ldots, x_{s(p)}(p)$ running one over another, subject to one linear relation. The right picture shows that a monomial $f(j+p)v_p = f_1(j+p) \ldots f_{s(j+p)}(j+p)v_p$ is represented by points $x_1(j+p), \ldots, x_{s(j+p)}(j+p)$ running around z_p as shown in Fig. 5.8. These points run independently as long as they are out of the region A in Fig. 5.7, but they are subject to some restrictions if some of them are close to region A. A precise description of the geometric image of a monomial f is given by constructions of Sec. 5.3 and by diagram (5.11.1).

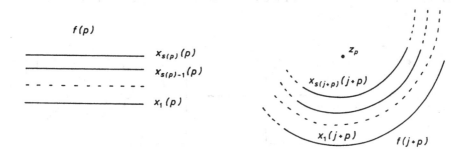

Fig. 5.8

The multiplication $U_q\mathfrak{n}_- \otimes M_1 \otimes \cdots \otimes M_n \to M_1 \otimes \cdots \otimes M_n$ has a simple interpretation in terms of Fig. 5.7. For a monomial f shown in Fig. 5.7, a monomial $f' = f(1)|\ldots|f(j)\otimes f_i(f(j+1)v_1|\ldots|f_{(j+n)}v_n)$ is shown in Fig. 5.9. The image of f' has an additional loop, the point x_i runs around all previous loops.

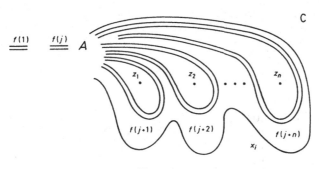

Fig. 5.9

For any $h \in U_q\mathfrak{h}$, we have $f(1)|\ldots|f(j)\otimes h(f(j+1)v_1|\ldots|f(j+n)v_n) = $ const f. Thus, the geometric image of this element is the same as that of f.

The multiplication $e_i \otimes f \mapsto f(1)|\ldots|f(j)\otimes e_i(f(j+1)v_1|\ldots|f(j+n)v_n)$ for any i is geometrically interpreted as the deletion of the corresponding loop from the picture shown in Fig. 5.7. The action of an element e_i is a part of the boundary operator in $C.(U_q \mathfrak{n}_-; M; {}^+\Delta)$, see (5.1.6).

Any isotopy of $\mathbb{C}$ induces an isotopy of $\mathbb{C}^r$ preserving all diagonal hyperplanes. Fix an isotopy changing the real line $\mathbb{R} \subset \mathbb{C}$ as shown in Fig. 5.10. This isotopy induces an isotopy of Figures 5.7 and 5.9, see Fig. 5.11. For details of these pictures, see Fig. 5.12.

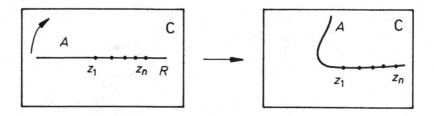

Fig. 5.10

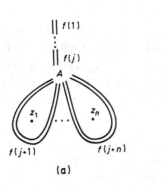
(a)

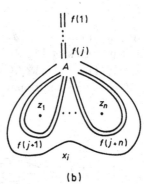
(b)

Fig. 5.11

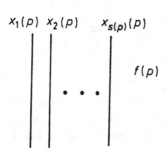

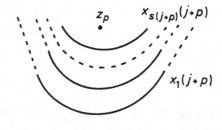

Fig. 5.12

Consider the monomial f represented by the chain shown in Fig. 5.7. The map $w\pi p\varphi$ fixes a certain section of $\mathcal{S}^*(a)$ over the chain. The isotopy in Fig. 5.10 induces analytic continuation of this distinguished section. Thus, the chains shown in Fig. 5.11 have distinguished sections over them. The section over the left chain (the right chain, resp.) in Fig. 5.11 will be denoted by $s_f (s_{f'}$, resp.). Each of the sections s_f, $s_{f'}$ has the following form: it is a branch of the corresponding multivalued function $\prod (t_i - t_j)^{a_{i,j}} \prod (t_i - z_j)^{a_{i,j}+r}$ multiplied by a suitable complex number. The sections s_f, $s_{f'}$ have the following simple definitions.

(5.11.8) The sections s_f and $s_{f'}$ are the sections that tend to the section

$$s_0 = \prod |t_i - t_j|^{a_{i,j}} \prod |t_i - z_j|^{a_{i,j}+r}$$

if the points $x_1, \ldots, x_r$ tend to the special position as explained below. More precisely, $s_f/s_0 \to 1$ and $s_{f'}/s_0 \to 1$.

(5.11.9) Special position of points $x_1, \ldots, x_r$ for the monomial f.

$$\arg (x_a(p) - x_b(p)) \to 0 \qquad \text{for } p \leq j \text{ and } b < a,$$
$$\arg (x_a(p) - x_b(s)) \to \pi/2 \qquad \text{for } p \leq j \text{ and } p < s,$$
$$\arg (x_a(p) - z_s) \to \pi/2 \qquad \text{for } p \leq j,$$
$$\arg (x_a(p) - x_b(s)) \to 0 \qquad \text{for } j < s < p,$$
$$\arg (x_a(p) - z_s) \to 0 \qquad \text{for } s + j < p,$$
$$\arg (x_a(p) - z_s) \to \pi \qquad \text{for } j < p < s + j,$$
$$\arg (x_a(p) - z_{p-j}) \to -\pi/2 \qquad \text{for } j < p,$$

see Fig. 5.13a.

(5.11.10) Special positions of points $x_1, \ldots, x_r$ for the monomial f'.

Points $x_1, \ldots, x_r$ satisfy the conditions (5.11.9) and in addition we have

$$\arg (x_i - x_a(p)) \to -\pi/2 \qquad \text{for all } a, p,$$
$$\arg (x_i - z_p) \to -\pi/2 \qquad \text{for all } p,$$

see Fig. 5.13b.

(5.11.11) Thus, for any monomial f we have constructed a chain shown in Fig. 5.11 with a coefficient in $\mathcal{S}^*(a)$ as defined in (5.11.8) and (5.11.9). This construction defines a monomorphism

$$C.(^+U_q\mathfrak{n}_-; M; {}^+\Delta)_\lambda \longrightarrow C.(\mathbb{C}^r - U_{n,r}(z), \mathcal{S}^*(a)),$$

which is a quasiisomorphism.

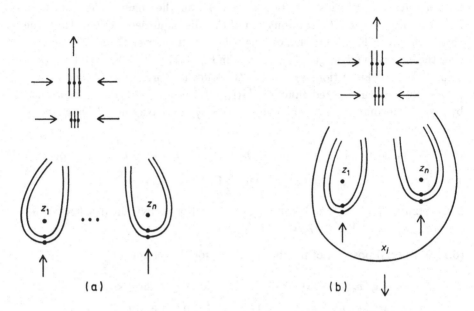

Fig. 5.13

5.11.12. Remark. We came to definitions (5.11.8)–(5.11.10) starting with the definition of the cells $\{E(f,a)\}$ in Sec. 3, constructions of (5.3), and the isotopy in Fig. 5.10. In reality, the direction was the opposite. I knew that the homology of a discriminantal configuration can be computed in terms of cells $\{E(f,a)\}$. I had the chains in Fig. 5.11 and definitions (5.11.8)–(5.11.10). I knew that the homological complex of such chains coincides with $C.(^+U_q\mathfrak{n}_-; M; {}^+\Delta)_\lambda$. I wanted to prove that this complex computes the homology of a discriminant configuration. So it was necessary to turn the tails of the chains in Fig. 5.11 to the left by the isotopy inverse to the isotopy shown in Fig. 5.10, and then to compare the chains in Figs. 5.7 and 5.9 with the cells $\{E(f,a)\}$. Thus, the awful algebraic formulas of Sec. 5.3 were invented.

The construction of homomorphism (5.11.11) has a further generalization. Fix a fork F with n tines in $\mathbb{C}$ as shown in Fig. 5.14. Call the points $z_1, z_2, \ldots, z_n$ the boundary of the fork. The fork in Fig. 5.15 is the standard fork.

Geometric Interpretation of Theorems 5.3.34, 5.4, and 5.5   137

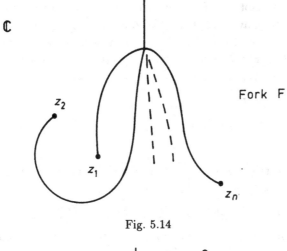

Fig. 5.14

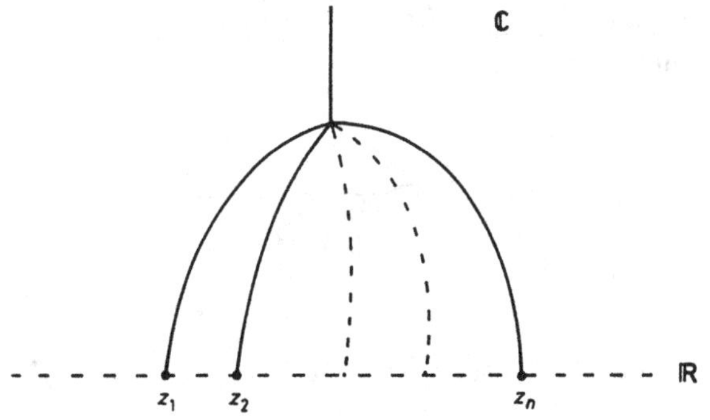

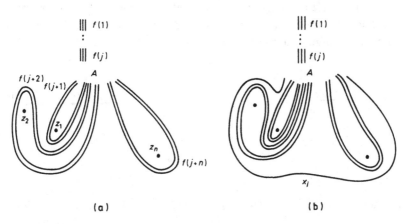

Fig. 5.15

Fig. 5.16

There is an isotopy of $\mathbb{C}$ sending the standard fork to a given one. Such an isotopy induces an isotopy of the pictures of Fig. 5.11 to the pictures shown in Fig. 5.16. One may think that the loops go along the tines of the fork. Analytic continuation of the sections s_f, $s_{f'}$ determines a certain section $s_{f,F}$ over the left cell in Fig. 5.16 and a certain section $s_{f',F}$ over the right cell.

(5.11.13) *This construction defines a monomorphism*

$$\psi_F : C.(^+U_q\mathfrak{n}_-; M; {^+\Delta})_\lambda \longrightarrow C.(\mathbb{C}^r - U_{n,r}(z), \mathcal{S}^*(a))$$

for any fork F with boundary $z_1, \ldots, z_n$. This monomorphism is a quasiisomorphism.

A similar construction may be done for the complex $C.^*(^+U_q\mathfrak{n}_-; M; \mu)_\lambda$. As above, the maps φ, p, $w\pi$ define a monomorphism

$$w\,\pi\,p\,\varphi : C.^*(^+U_q\mathfrak{n}_-; M; \mu)_\lambda \longrightarrow C_!.(\mathbb{C}^r - U_{n,r}(z), \mathcal{S}^*(a)).$$

Let f^* be any monomial of form (5.2.19). Then f^* may be represented by Fig. 5.17. Some details of Fig. 5.17 are shown in Fig. 5.18. The left picture of Fig. 5.18 shows that

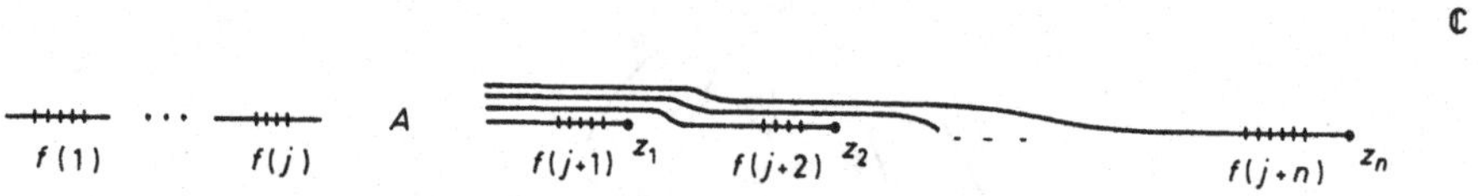

$$f^* = (f(1)| \ldots |f(j)|f(j+1)v_1| \ldots |f(j+n)v_n)^*$$

Fig. 5.17

a factor $f(p) = f_1(p) \ldots f_{s(p)}(p)$ is represented by moving points $x_1(p), \ldots, x_{s(p)}(p)$. These points run preserving the order $x_1(p) < \cdots < x_{s(p)}(p)$ subject to one linear relation. The right

Fig. 5.18

picture shows that a factor $f(j+p)v_p = f_1(j+p) \ldots f_{s(j+p)}(j+p)v_p$ is represented by moving points $x_1(j+p), \ldots, x_{s(j+p)}(j+p)$. These points run preserving the order $x_1(j+p) < \cdots < x_{s(j+p)}(j+p)$. These points run from z_p to the region A over the points $z_1, \ldots, z_{p-1}$. They run independently as long as they are out of the region A in Fig. 5.17. But they are subject to some restrictions if some of them are close to the region A. The precise description of the geometric image of the monomial f^* is given by constructions in Sec. 5.2 and by diagram (5.11.1). An isotopy of $\mathbb{C}$ in Fig. 5.10 induces an isotopy of the chain in Fig. 5.17 to the chain shown in Fig. 5.19.

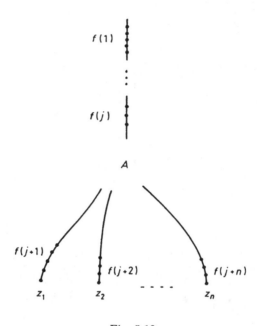

Fig. 5.19

The monomial f^* is represented by the chain shown in Fig. 5.17. The map $w\pi p\varphi$ fixes a certain section of $S^*(a)$ over the chain shown in Fig. 5.17. The isotopy in Fig. 5.10 induces analytic continuation of this distinguished section. Thus, the chain in Fig. 5.19 has a distinguished section over itself. Denote it by s_{f^*}. The section s_{f^*} has the following simple definition.

(5.11.14) s_{f^*} is the section such that $s_{f^*}/s_0 \to 1$, if the points $x_1, \ldots, x_r$, tend to the special positions as explained below.

(5.11.15) Special positions of points $x_1, \ldots, x_r$ for the monomial f^*.

$\arg(x_a(p) - x_b(s)) \longrightarrow \pi/2 \quad$ if $p = s$ and $a < b$, or if $p < s$ and $p \leq j$,
$\arg(x_a(p) - x_b(s)) \longrightarrow 0 \quad$ if $j < s < p$,
$\arg(x_a(p) - z_s) \longrightarrow 0 \quad$ for $s + j < p$,
$\arg(x_a(p) - z_s) \longrightarrow \pi \quad$ for $j < p < s + j$,
$\arg(x_a(p) - z_{p-j}) \longrightarrow \pi/2 \quad$ for $j < p$,

see Fig. 5.20.

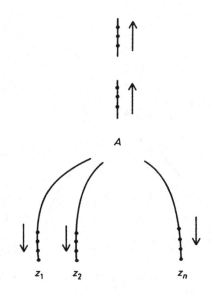

Fig. 5.20

(5.11.16) For any monomial f^* we have constructed a chain shown in Fig. 5.19 with a coefficient in $S^*(a)$ as defined in (5.11.14) and (5.11.15). This construction defines a monomorphism

$$C_\cdot^*(^+U_q\mathfrak{n}_-; M; \mu)_\lambda \longrightarrow C_{!\cdot}(\mathbb{C}^r - U_{n,r}(z), S^*(a)),$$

this monomorphism is a quasiisomorphism.

For any fork F with n tines in $\mathbb{C}$ as in Fig. 5.14, there is an isotopy of $\mathbb{C}$

sending the standard fork to the fork F. Such an isotopy induces an isotopy of the chain shown in Fig. 5.19 to a chain shown in Fig. 5.21. One may think that points $x_1(j+p),\ldots,x_{s(j+p)}(j+p)$ run along the pth tine of the fork as long as none of them is in the region A in Fig. 5.21. Analytic continuation of the section s_{f^*} determines a certain section $s_{f^*,F}$ over the chain shown in Fig. 5.21.

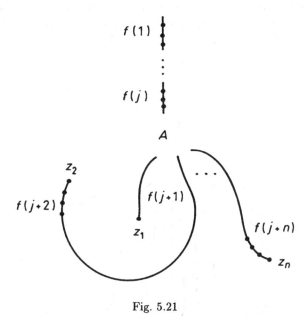

Fig. 5.21

(5.11.17) *This construction defines a monomorphism*

$$\psi_F : C.^*(^+U_q\mathfrak{n}_-;M;\mu)_\lambda \longrightarrow C_!.(\mathbb{C}^r - U_{n,r}(z), \mathcal{S}^*(a)),$$

for any fork F with boundary $z_1,\ldots,z_n$. This monomorphism is an isomorphism.

For any $h \in U_q\mathfrak{h}$, we have $(f(1)|\ldots|f(j))^* \otimes h(f(j+1)v_1|\ldots|f(j+n)v_n)^* = \text{const } f^*$. Thus, the geometric image of this element is the same as that of f^*, see Fig. 5.19.

The multiplication $e_i \otimes f^* \mapsto (f(1)|\ldots|f(j))^* \otimes e_i(f(j+1)v_1|\ldots|f(j+n)v_n)^*$ for any i is geometrically interpreted as follows. Denote the result of the multiplication by F_i^*. Then $F_i^* = 0$ if $f_i \notin \{f_1(j+1),\ldots,f_1(j+n)\}$. Let $f_i = f_1(j+p)$ for some p. Then $F_i^* = \text{const}\,(f(1)|\ldots|f(j)|f(j+1)v_1|\ldots|f(j+p-$

$1)v_{p-1}|f_2(j+p)\ldots f_{s(j+p)}(j+p)v_p|(f(j+p+1)v_{p+1}|\ldots |f(j+n)v_n)^*$. Thus, the multiplication by e_i is interpreted as the deletion of the moving point x_i, if this point is the highest point in one of the tines. The action of an element e_i is a part of the boundary operator in $C.^*(^+U_q\mathfrak{n}_-;M;\mu)_\lambda$, see (5.1.3). The cell $G_i^* = (f(1)|\ldots |f(j)|f_i|f(j+1)v_1|\ldots |f_2(j+p)\ldots f_{s(j+p)}(j+p)v_p|\ldots |f(j+n)v_n)^*$ is a part of the boundary of the cell f^*. In the geometric image of G_i^*, the point x_i is the first point over the region A, and x_i does not move. The above *const* may be computed by comparing the sections distinguished by rules (5.11.14) and (5.11.15) for monomials f^* and G_i^*.

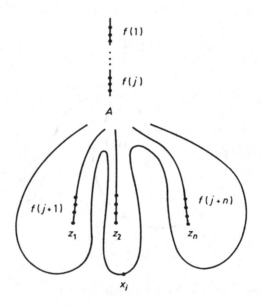

Fig. 5.22

The multiplication $f_i \otimes f^* \mapsto (f(1)|\ldots |f(j))^* \otimes f_i(f(j+1)v_1|\ldots |f(j+n)v_n)^*$ for any i is geometrically interpreted as follows (the multiplication is with respect to the contragradient structure). Denote the result by F_i^*. F_i^* is a linear combination of monomials in $((U_q\mathfrak{n}_-)^{\otimes j} \otimes M)^*$. The images of these monomials are given by the following construction. Draw a loop around the tines in Fig. 5.19, see Fig. 5.22. Let x_i move along this loop. Contract the new loop to the tines of the fork. Now x_i moves along each tine and may be in any position on any tine. Thus, we get a union of cells of the form shown in Fig. 5.19 with one additional point x_i. The definition of the distinguished section over F_i^* is similar to that given by (5.11.8) and (5.11.10), and Fig. 5.13b. Comparison of this section with the sections defined by (5.11.14) and (5.11.15) gives the algebraic formula for the multiplication by f_i.

There is a natural contraction of the picture 5.16a to the fork, see Fig. 5.23.

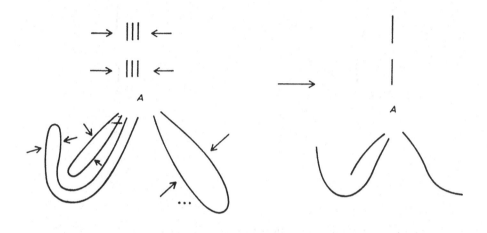

Fig. 5.23

The contractions send the chain shown in Fig. 5.16a to a linear combination of the chains of the form shown in Fig. 5.21. After an identification of these chains with monomials, the contraction of the chains with coefficients in $S^*(a)$ coincides with the contravariant form $S : C.(^+U_q\mathfrak{n}_-; M; ^+\Delta)_\lambda \longrightarrow C.^*(^+U_q\mathfrak{n}_-; M; \mu)_\lambda$, see diagram (5.11.1).

For example, the monomial $f_1 v_1$ is represented by the loop shown in Fig. 5.24a and the section s over the loop which tends to $|x_1 - z_1|^a$ when x_1 tends to z_1 from below, here $a = b/\varkappa = (\alpha_1, -\Lambda_1)/\varkappa$. The monomial $(f_1 v_1)^*$ is represented by the curve shown in Fig. 5.24b and the section s' over the curve which tends to $|x_1 - z_1|^a$ when x_1 tends to z_1 from above. The contraction of the loop to the curve gives the curve twice, first with coefficient $q^{b/2}$ and second with coefficient $-q^{-b/2}$. Thus, we get the formula for the contravariant form on a Verma module:

$$S : f_1 v_1 \mapsto (q^{(\Lambda_1, \alpha_1)/2} - q^{-(\Lambda_1, \alpha_1)/2})(f_1 v_1)^*.$$

Above, we have considered the case $\lambda = (1,\ldots,1) \in \mathbf{N}^r$. The case of an arbitrary weight $\lambda = (\ell_1,\ldots,\ell_r) \in \mathbf{N}^r$ is similar. The standard symmetrization procedures reduce this case to the case where $\lambda = (1,\ldots,1)$, see (4.7) and (4.8).

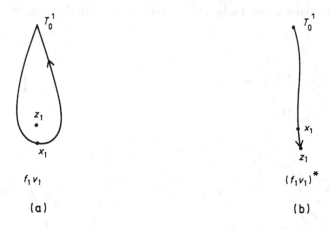

Fig. 5.24

For example, the monomial $f^2 v_1$ is the sum of two chains shown in Fig. 5.25.

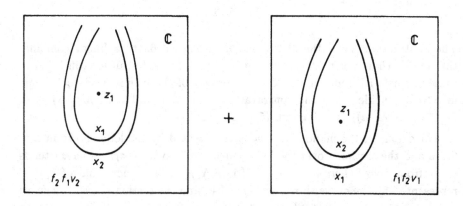

Fig. 5.25 $f^2 v_1 = f_1 f_2 v_1 + f_2 f_1 v_1$

6

Bundle Properties of a Discriminantal Configuration

In Sec. 8, we will describe the monodromy in homology groups of a discriminantal configuration $\mathcal{C}_{n,k}(z)$ under deformations of the parameters z. This description is based on the bundle properties of the discriminantal configuration given in this section.

6.1. Subordinated Monomials

Set $x = (x_1, \ldots, x_k)$, $k \in \mathbb{N}$, $x' = (x_1, \ldots x_l)$, $x'' = (x_{l+1}, \ldots, x_k)$ for some $0 < l < k$, $z = (z_1, \ldots, z_n)$.

Let $f \in \text{Adm}(x''; z)$ be an admissible monomial for symbols x'', z, see Subsec. 3.8. Let $F \in \text{Adm}(x; z)$ be an admissible monomial for symbols x, z. Say that F is subordinated to f, $f \prec F$, if F and f have the same number of factors and f can be obtained from F by deleting all symbols of x'.

Example. The monomials $x_1 z_1 x_2 | x_3 | z_2$, $z_1 x_2 | x_3 x_1 | z_2$, are subordinated to $z_1 | x_3 | z_2$.

6.2. Leaves

Let $z_1 < z_2 < \cdots < z_n$ be real numbers. Consider the discriminantal configuration $\mathcal{C}_{n,k}(z)$ in $\mathbb{R}^k$. Fix its equipment, real numbers $\{T_j^i\}$, $j = 0, 1, \ldots, n$, $i = 1, \ldots, k$, with property (3.3.1) and consider the cell complex

Q'/Q'' constructed for $C_{n,k}(z)$, $\{T_j^i\}$ in Sec. 3. This complex consists of the cells $D(F)$ for $F \in \text{Adm}\,(x;z)$. Each cell is a convex polytope. The complex and its cells depend on the configuration and the numbers $\{T_j^i\}$: $Q'/Q'' = Q'/Q''(C_{n,k}(z);\{T_j^i\})$, $D(F) = D(F;C_{n,k}(z);\{T_j^i\})$.

Let
$$Q = \bigcup_{F \in \text{Adm}\,(x;z)} D(F) \subset \mathbb{R}^k \qquad (6.2.1)$$
$$\pi : \mathbb{R}^k \to \mathbb{R}^l, (x_1, \ldots, x_k) \mapsto (x_1, \ldots, x_l), \quad Y = \pi(Q).$$

For any admissible monomial $f \in \text{Adm}\,(x'',z)$, introduce a *leaf* of f as the union
$$L(f) = \bigcup_{F, f \prec F} D(F). \qquad (6.2.2)$$

6.2.3. Example. A configuration $C_{2,2}(z)$ and the cells $\{D(F)\}$ are shown in Fig. 3.2. For $l = 1$ the leaves are shown in Fig. 6.1.

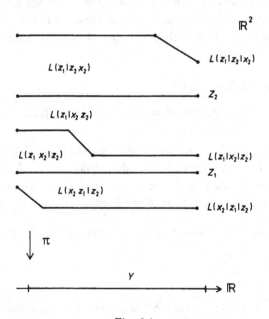

Fig. 6.1

Let $y \in Y$. For any $f \in \text{Adm}\,(x'',z)$, set $D_y(f) = L(f) \cap \pi^{-1}(y)$.

6.2.4. Theorem. *There exist continuous functions* $T_j^i : Y \to \mathbb{R}, i = 1, \ldots, k -$

$l, j = 0, 1, \ldots, n$, such that for any $y \in Y$, we have

(i) $T_0^1(y) < T_0^2(y) < \cdots < T_0^{k-l}(y) < z_1 < T_1^1(y) < \cdots < T_1^{k-l}(y) < z_2 < \cdots < z_n < T_n^1(y) < \cdots < T_n^{k-l}(y)$,

(ii) $D_y(f) = D(f; \mathcal{C}_{n,k-l}(z); \{T_j^i(y)\})$ for all $f \in \mathrm{Adm}\,(x''; z)$.

The theorem is proved in (6.4).

Property (ii) means that the sets $\{D_y(f)\}$ for all $f \in \mathrm{Adm}\,(x''; z)$ are convex polytopes and they form the cell complex $Q'/Q''(\mathcal{C}_{n,k-l}(z), \{T_j^i(y)\})$ of the configuration $\mathcal{C}_{n,k-l}(z)$ in the fiber over y.

6.2.5. Corollary. *The map*

$$\pi|_{\pi^{-1}(Y)} : \left(\{L(f)\}_{f \in \mathrm{Adm}\,(x''; z)}, Q\right) \to Y$$

is a trivial bundle canonically trivialized.

The corollary follows from (2.2.8).

6.3. Properties of Leaves

6.3.1. Lemma. *The set Q is a convex polytope in $\mathbb{R}^k$. It is the basic polytope of the second type generated by numbers $T_0^1 < \cdots < T_0^k < T_n^1 < \cdots < T_n^k$, see (3.4). The set $Y = \pi(Q)$ is the convex polytope in $\mathbb{R}^l$ given by the system of inequalities parameterized by the subsets $I = \{i_1, \ldots, i_p\} \subset \{1, \ldots, l\}$. The inequality*

$$T_0^1 + \cdots + T_0^p \leq \sum_{i \in I} x_i \leq T_n^k + T_n^{k-1} + \cdots + T_n^{k-p+1}$$

corresponds to a subset I.

Proof. The boundary of Q consists of all polytopes $D(f)$, such that $F \in \mathrm{Adm}\,(x; z)$ has $k+1$ factors and its single algebra factor is the first factor or the last. If the algebra factor $x_{i_1} \ldots x_{i_m}$ is the first factor, then $D(F)$ lies in the hyperplane $x_{i_1} + \cdots + x_{i_m} = T_0^1 + \cdots + T_0^m$. If the algebra factor is the last factor, then $D(F)$ lies in the hyperplane $T_n^{k-m+1} + \cdots + T_n^k$. This reason easily implies the lemma.

(6.3.2) Each leaf $L(f)$ is a union of convex polytopes of the same codimension c in $\mathbb{R}^k$, where c is the number of algebra factors in f, see (6.2.2).

The maximal possible codimension of a leaf is $k - l$. A leaf of codimension $k - l$ will be called *thin*. For example, there are three thin leaves in Fig. 6.1.

The projection p restricted to each polytope constituting a thin leaf is a one-to-one map.

6.3.3. Theorem. *For each thin leaf $L(f)$, the projection $\pi|_{L(f)} : L(f) \to Y$ is a homeomorphism.*

Proof of the Theorem.

6.3.4. Lemma. *For a thin leaf $L(f)$ the map $p|_{L(f)} : L(f) \to Y$ is a covering away from a union of polytopes in Y of codimension not less than 2.*

6.3.5. Corollary. $p(L(f)) = Y$ *(as $L(f)$ is closed).*

Proof of the Lemma. f has the form

$$f = u_1|u_2|\ldots|u_{i(1)}|z_1|u_{i(1)+1}|\ldots|u_{i(2)}|z_2|\ldots \\ |u_{i(n)}|z_n|u_{i(n)+1}\ldots|u_{k-l}|, \quad (6.3.6)$$

where $u_1, \ldots, u_{k-l}$ is a permutation of $x'' = (x_{l+1}, \ldots, x_k)$. Let $f \prec F$. Then F has the form

$$F = U_1|\ldots|U_{i(1)}|Z_1|U_{i(1)+1}|\ldots|U_{i(n)}|Z_n|\cdots|U_{k-l}, \quad (6.3.7)$$

where

$$U_j = y_1(u_j)\ldots y_{a(u_j)}(u_j)u_j y_{a(u_j)+1}(u_j)\ldots y_{b(u_j)}(u_j), \quad j = 1, \ldots, k-l,$$
$$Z_j = y_1(z_j)\ldots y_{a(z_j)}(z_j)z_j y_{a(z_j)+1}(z_j)\ldots y_{b(z_j)}(z_j), \quad j = 1, \ldots, n,$$

and

$$\{y_\alpha(\beta)|\beta \in \{u_1,\ldots,u_{k-l},z_1,\ldots,z_n\}, \alpha \in \{1,\ldots,b(\beta)\}\} = x' \\ = \{x_1,\ldots,x_l\}. \quad (6.3.8)$$

The boundary of $D(F)$ is a union of polytopes of codimension $k - l + 1$. It is easy to see that if $\Delta \subset \partial D(F)$ is one of these polytopes, and $\pi(\Delta) \not\subset \partial Y$, then there is a unique polytope $D(F') \subset L(f)$, $f \prec F'$, different from F, such that its boundary also contains Δ.

Hence the leaf $L(f)$ has the form $D(F) \cup D(F')$ in a neighborhood of Δ and is a covering over a neighborhood of $\pi(\Delta)$. This reason proves the lemma.

Example. Let $n = 2, k = 2, l = 1$, see Fig. 6.1. Let $f = z_1|x_2|z_2$, $F = z_1|x_1x_2|z_2$. The boundary $\partial D(F)$ consists of the two points a, b, where

$a = D(z_1|x_1|x_2|z_2)$, b is singled out in $D(F)$ by the equality $x_1 = x_2$. Then $a \in \partial D(z_1 x_1|x_2|z_2) \subset L(f)$, $b \in D(z_1|x_2 x_1|z_2) \subset L(f)$, and $\pi : D(z_1 x_1|x_2|z_2) \cup D(z_1)|x_1 x_2|z_2) \cup D(z_1|x_2 x_1|z_2) \to \mathbb{R}$ forms a covering over a neighborhood of $\pi(D(F))$.

Let f, $f' \in \mathrm{Adm}\,(x''; z)$ be monomials having $k - l$ algebra factors each. Let F, $F' \in \mathrm{Adm}\,(x; z)$, $f \prec F$, $f' \prec F'$. Let f have the form (6.3.6), F the form (6.3.7). Let f' have the form

$$f' = u'_1|u'_2|\ldots|u'_{i'(1)}|z_1|u'_{i'(1)+1}|\ldots|u'_{i'(2)}|z_2| \qquad (6.3.9)$$
$$\ldots |z_n|u'_{i'(n)+1}\ldots|u'_{k-l}|.$$

Let F' have the form

$$F' = U'_1|\ldots|U'_{i'(1)}|Z'_1|U'_{i'(1)}|Z'_n|\ldots|U'_{k-l}, \qquad (6.3.10)$$

where

$$U'_j = y'_1(u'_j)\ldots y'_{a'(u'_j)}(u'_j) u'_j y'_{a'(u'_j)+1}(u'_j)\ldots y'_{b'(u'_j)}(u'_j), \quad j = 1, \ldots, k-l,$$
$$Z'_j = y'_1(z_j)\ldots y'_{a'(z_j)}(z_j) z_j y'_{a'(z_j)+1}(z_j)\ldots y'_{b'(z_j)}(z_j), \quad j = 1, \ldots, n.$$

Assume that some numbers $s \in \{1, \ldots, k-l\}$, $j \in \{0, 1, \ldots, n\}$ have the property: The symbol u_s stands in between z_j, z_{j+1} in f and the symbol u'_s stands in between z_j, z_{j+1} in f'.

Assume that interiors of $\pi(D(F))$ and $\pi(D(F'))$ have a nonempty intersection.

6.3.11. Lemma. *Under these assumptions, we have* $a'(u'_s) = a(u_s)$, $b'(u'_s) = b(u_s)$; $y'_i(u'_s) = y_i(u_s)$ *for* $i = 1, \ldots, b(u_s)$.

6.3.12. Corollary. *Under these assumptions, if* $f = f'$, *then* $F = F'$. *Hence* $p|_{L(f)} : L(f) \to Y$ *is a one-sheeted covering over a generic point of* Y.

Lemma (6.3.4) and Corollary (6.3.12) imply Theorem (6.3.3).

Proof of the Lemma. Let $y = (x_1, \ldots, x_l) \in p(D(F))$. Then the sequence of coordinates of the point y written in the increasing order coincides with the sequence $y_1(u_1), \ldots, y_{b(u_1)}(u_1), y_1(u_2), \ldots, y_1(u_{i(1)}), \ldots, y_{b(u_{i(1)})}(u_{i(1)})$, $y_1(z_1), \ldots, y_{b(z_1)}(z_1), \ldots, y_{b(u_{k-l})}(u_{k-l})$, cf. (6.3.8). Now the lemma easily follows from inequalities (3.9.8)–(3.9.10).

According to Lemma (6.3.11), the numbers $a(u_s)$, $b(u_s)$ and the symbols $y_1(u_1),\ldots,y_{b(u_s)}(u_s)$ are functions of s, j, and a point $y \in p(D(F)) \subset Y$. Namely, $a(u_s) = a(s,j,y)$, $b(u_s) = b(s,j,y)$, and there exists an inclusion $\delta = \delta_{s,j,y} : \{1, 2, \cdots, b(s,j,y)\} \to \{1, \cdots, l\}$ such that $y_p(u_s) = x_{\delta(p)}$ for $p = 1, 2, \cdots, b(s, j, y)$.

These remarks imply that the number

$$\sigma = b(u_1) + \cdots + b(u_{s-1}) + b(z_1) + \cdots + b(z_j)$$

is also a function of s, j, y.

Then the u_s coordinate of the point $A = \pi^{-1}(y) \cap D(F)$ is also a function of s, j, y:

$$u_s|_A = T_j^s(y) := \sum_{m=1}^{b(s,j,y)+1} T_j^{\sigma+m} - \sum_{m=1}^{b(s,j,y)} y_m(u_s)|_y, \qquad (6.3.13)$$

see (3.9.3).

The functions $T_j^s : Y \to \mathbb{R}$, $j = 0, \ldots, n, s = 1, \ldots, k - l$, are continuous by Theorem (6.3.3). Obviously, they satisfy inequalities (6.2.4.i).

We need the following generalization of Lemma (6.3.11).

A monomial f in (6.3.6) is the tensor product of $n + k - l$ symbols $u_1, \ldots, u_{i(1)}, z_1, \ldots$. For any $1 \le \alpha \le \beta \le n + k - l$, the set of the symbols with numbers $\alpha, \alpha + 1, \ldots, \beta$ will be called the $[\alpha, \beta]$ segment of f. For example, $\{u_2, u_3, \ldots, u_{i(1)}, z_1\}$ is the $[2, i(1) + 1]$ segment of f. Notation: Seg $(f, [\alpha, \beta])$.

A monomial F in (6.3.7) is the tensor product of $n + k - l$ monomials in x, z. The set of all symbols of x, z entering the monomials with numbers $\alpha, \alpha + 1, \ldots, \beta$ will be called the $[\alpha, \beta]$-segment of F and denoted by Seg $(F, [\alpha, \beta])$. For example:

$$\{y_1^1(u_1), \ldots, y_{a(u_1)}(u_1), u_1, y_{a(u_1)+1}(u_1), \ldots, y_{b(u_1)}(u_1)\} = \text{Seg}(F, [1, 1]).$$

Let $f, f' \in \text{Adm}(x''; z)$ be monomials having $k - l$ algebra factors each. Let $F, F' \in \text{Adm}(x; z)$, $f \prec F$, $f' \prec F'$. Let f, f', F, F' have form (6.3.6), (6.3.9), (6.3.7) and (6.3.10), respectively.

Assume that some numbers $\alpha, \beta, 1 \le \alpha \le \beta \le n + k = l$, have the following properties:

$$\sharp \text{Seg}(f, [1, \alpha - 1]) \cap x'' = \sharp \text{Seg}(f', [1, \alpha - 1]) \cap x'',$$
$$\sharp \text{Seg}(f, [\beta + 1, n + k - l]) \cap x'' = \sharp \text{Seg}(f', [\beta + 1, n + k - l]) \cap x'' \qquad (6.3.14)$$

where # means the number of elements.

Assume that the interiors of $\pi(D(F))$ and $\pi(D(F'))$ have a nonempty intersection.

6.3.15. Lemma. *Under these assumptions, we have*

$$\operatorname{Seg}(F, [\alpha, \beta]) \cap x' = \operatorname{Seg}(F', [\alpha, \beta]) \cap x'.$$

The lemma easily follows from inequalities (3.9.8)–(3.9.10).

6.4. Proof of Theorem 6.2.4

In (6.3.13), we have introduced continuous functions $T_j^s : Y \to \mathbb{R}$, $j = 0, \ldots, n$, $s = 1, \ldots, k - l$, satisfying (6.2.4.i). We will prove (6.2.4.ii) for these functions.

For any $y \in Y$, we have two collections of sets in the fiber $p^{-1}(y)$:

$$\{D_y(f)\}, \{D(f; C_{n, k-l}(z); \{T_j^s(y)\})\}$$

for all $f \in \operatorname{Adm}(x''; z)$.

By definition of numbers $\{T_j^s(y)\}$, we have

$$D_y(f) = D(f; C_{n,k-l}(z); \{T_j^s(y)\}) \qquad (6.4.1)$$

for all monomials f having $k - l$ algebra factors. To prove equality (6.4.1) for all $f \in \operatorname{Adm}(x''; z)$ we will prove two lemmas.

6.4.2. Lemma.
$$D_y(f) \supset D(f; C_{n,k-l}(z); \{T_j^s(y)\})$$

for all $f \in \operatorname{Adm}(x'', z)$.

6.4.3. Lemma.

$$\bigcup_{f \in \operatorname{Adm}(x''; z)} D_y(f) \subset \bigcup_{f \in \operatorname{Adm}(x''; z)} D(f; C_{n,k-l}(z); \{T_j^s(y)\}).$$

These lemmas imply Theorem (6.2.4) since open leaves have no pairwise intersections.

Proof of Lemma (6.4.2). The lemma easily follows from Lemma (6.3.15). To simplify notations, we will prove the lemma for an example. We will use

152 Bundle Properties of a Discriminantal Configuration

notations of Subsecs. 3.2–3.6.

Example. Let $n = 2, k = 4, l = 2, f = x_3z_1|x_4|z_2$, and $y \in Y \subset \mathbb{R}^2$.

By definition,

$$D(f;\mathcal{C}_{2,2}(z);\{T_j^s(y)\}) = \Delta(\tilde{F}) \cap \{x \in \mathbb{R}^4 | x_3 \leq z_1 \leq x_4 \leq z_2\}, \quad (6.4.4)$$

where $\Delta(\tilde{F})$ is the polytope corresponding to the facet $\tilde{F}$ of $\mathcal{C}_{2,2}(z) \subset p^{-1}(y)$ given by admissible ordered decomposition $\{x_3, z_1\} < \{x_4\} < \{z_2\}$, see Subsec. 3.2.

(6.4.5) $\Delta(\tilde{F})$ is the convex hull of points $\left(T_0^1(y), T_1^2(y)\right), \left(T_1^1(y), T_1^2(y)\right)$.

$$\begin{aligned}(T_0^1(y), T_1^2(y)) &= D(x_3|z_1|x_4|z_4; \mathcal{C}_{2,2}(y);\{T_j^i(y)\}), \\ (T_1^1(y), T_1^2(y)) &= D(z_1|x_3|x_4|z_4; \mathcal{C}_{2,2}(y);\{T_j^i(y)\}).\end{aligned} \quad (6.4.6)$$

Let $F^1, F^2 \in \text{Adm}(x_1, x_2, x_3, x_4; z_1, z_2)$ be monomials such that $x_3|z_1|x_4|z_4 \prec F^1$, $z_1|x_3|x_4|z_4 \prec F^2$ and

$$\begin{aligned}D(F^1) \cap p^{-1}(y) &= D(x_3|z_1|x_4|z_4; \mathcal{C}_{2,2}(z);\{T_j^i(y)\}), \\ D(F^2) \cap p^{-1}(y) &= D(z_1|x_3|x_4|z_4; \mathcal{C}_{2,2}(z);\{T_j^i(y)\}).\end{aligned} \quad (6.4.7)$$

Let $F^1 = X_3^1|Z_1^1|X_4^1|Z_4^1$, $F^2 = Z_1^2|X_3^2|Z_4^2|X_4^2$, where $\{X_j^s, Z_j^s\}$ are monomials in x, z. By Lemma 6.3.15, we have $\text{Seg}(F^1, [1,2]) = \text{Seg}(F^2, [1,2])$, $\text{Seg}(F^1, [3,3]) = \text{Seg}(F^2, [3,3])$, $\text{Seg}(F^1, [4,4]) = \text{Seg}(F^2, [4,4])$.

Let S be the admissible ordered decomposition

$$\text{Seg}(F^1, [1,2]) < \text{Seg}(F^1, [3,3]) < \text{Seg}(F^1, [4,4])$$

of the set $\{x_1, x_2, x_3, x_4, z_1, z_2\}$. Let $Fa(S)$ be the facet of $\mathcal{C}_{2,4}(z) \subset \mathbb{R}^4$ corresponding to S. Let $\Delta(Fa(S))$ be the convex polytope constructed in (3.5) for the facet $Fa(S)$.

By construction,

$$\Delta(Fa(S)) \cap \{x \in \mathbb{R}^4 | x_3 \leq z_1 \leq x_4 \leq z_2\} \subset L(x_3z_1|x_4|z_2). \quad (6.4.8)$$

$$D(F^1), D(F^2) \text{ belong to the boundary of } \Delta(Fa(S)). \quad (6.4.9)$$

Properties (6.4.9), (6.4.5)–(6.4.7) imply inclusion $\Delta(\tilde{F}) \subset \Delta(Fa(S))$. Moreover, properties (6.4.4) and (6.4.8) imply inclusion $D(f; \mathcal{C}_{2,2}(z); \{T_j^s(y)\}) \subset D_y(f)$. Lemma (6.4.2) is proved for this example. The general case is similar.

Proof of Lemma 6.4.3 easily follows from the first statement of Lemma (6.3.1). In fact, Q and

$$Q_y = \bigcup_{f \in \text{Adm}(x'';z)} D(f; \mathcal{C}_{n,k-l}(z); \{T_j^s(y)\})$$

are convex polytopes. Each of them is the convex hull of its vertices.

Consider the union of all thin leaves and its intersection with $p^{-1}(y)$. Then by construction, Q_y is the convex hull of this set. On the other side, it is easy to see that the vertices of the convex polytope $Q \cap p^{-1}(y)$ belong to the union of thin leaves. Hence $Q_y \supset Q \cap p^{-1}(y)$. The lemma and Theorem 6.2.4 are proved.

7

R-Matrix for the Two-sided Hochschild Complexes

In Sec. 4, we have identified the complexes computing cohomology groups of a discriminantal configuration $\mathcal{C}_{n,k}(z)$ with the suitable two-sided Hochschild complexes. In Sec. 8, we will describe the monodromy in homology groups of the discriminantal configuration under deformations of z. The description will be given in terms of the above identification. Namely, an elementary monodromy transformation will be identified with a suitable linear operator acting on the Hochschild complexes. These linear operators are defined in this section.

7.1. Bistructures on $(^+U_q\mathfrak{n}_-)^{\otimes n}, (^+U_q\mathfrak{n}_-)^{*\otimes n}$

Define a right $U_q\mathfrak{n}_-$-module structure on $(^+U_q\mathfrak{n}_-)^{\otimes n}$,

$$\mu_r : (^+U_q\mathfrak{n}_-)^{\otimes n} \otimes (U_q\mathfrak{n}_-) \longrightarrow (^+U_q\mathfrak{n}_-)^{\otimes n}, \qquad (7.1.1)$$

as follows. Let $a_1|\ldots|a_n \in (^+U_q\mathfrak{n}_-)_{\lambda_1} \otimes \cdots \otimes (^+U_q\mathfrak{n}_-)_{\lambda_n}$, $\lambda_i = (l_{1,i}, \ldots, l_{r,i}) \in \mathbb{N}^r$. Set

$$\mu_r : a_1|\ldots|a_n|f_j \mapsto \sum_{i=1}^n q^{A_i} a_1|\ldots|a_{i-1}|a_i f_j|a_{i+1}|\ldots|a_n, \qquad (7.1.2)$$

where

$$A_i = -\sum_{k=1}^{i-1}\Big(\alpha_j, \sum_{m=1}^r l_{m,k}\alpha_m\Big)\Big/4 + \sum_{k=i+1}^n \Big(\alpha_j, \sum_{m=1}^r l_{m,k}\alpha_m\Big)\Big/4.$$

It is easy to see that this formula defines a $U_q\mathfrak{n}_-$-module structure on $({}^+U_q\mathfrak{n}_-)^{\otimes n}$.

Define a left $U_q\mathfrak{n}_-$-module structure on $({}^+U_q\mathfrak{n}_-)^{\otimes n}$,

$$\mu_l : (U_q\mathfrak{n}_-) \otimes ({}^+U_q\mathfrak{n}_-)^{\otimes n} \longrightarrow ({}^+U_q\mathfrak{n}_-)^{\otimes n}, \qquad (7.1.3)$$

by the formula

$$\mu_l : f_j|a_1|\ldots|a_n \longrightarrow \sum_{i=1}^{n} q^{-A_i} a_1|\ldots|a_{i-1}|f_j a_i|a_{i+1}|\ldots|a_n. \qquad (7.1.4)$$

It is easy to see that the left and right $U_q\mathfrak{n}_-$-module structures on $({}^+U_q\mathfrak{n}_-)^{\otimes n}$ form a two-sided $U_q\mathfrak{n}_-$-module structure on $({}^+U_q\mathfrak{n}_-)^{\otimes n}$ (or $U_q\mathfrak{n}_-$-bimodule structure).

Describe μ_r, μ_l in terms of $\overline{\Delta}$.

Let $\overline{\Delta} : U_q\mathfrak{n}_- \to U_q\mathfrak{n}_- \otimes U_q\mathfrak{n}_-$ be the comultiplication defined in (4.3.3). Let $\overline{\Delta}^{(n)} : U_q\mathfrak{n}_- \to (U_q\mathfrak{n}_-)^{\otimes n}$ be the map defined by iteration of $\overline{\Delta}$. Define the map

$$m : (U_q\mathfrak{n}_-)^{\otimes n} \otimes (U_q\mathfrak{n}_-)^{\otimes n} \longrightarrow (U_q\mathfrak{n}_-)^{\otimes n} \qquad (7.1.5)$$

as follows.

Let $a_1|\ldots|a_n, b_1|\ldots|b_n \in (U_q\mathfrak{n}_-)^{\otimes n}$, $a_j \in (U_q\mathfrak{n}_-)_{\lambda_j}$, $b_j \in (U_q\mathfrak{n}_-)_{\nu_j}$; $\lambda_j = (\lambda_{j,1},\ldots,\lambda_{j,r})$, $\nu_j = (\nu_{j,1},\ldots,\nu_{j,r}) \in \mathbf{N}^r$. Set

$$m : a_1|\ldots|a_n|b_1|\ldots|b_n \mapsto q^{A(\lambda,\nu)} a_1 b_1|\ldots|a_n b_n,$$

$$A(\lambda,\nu) = \frac{1}{4}\sum_{j=1}^{n}\Big(-\sum_{i=1}^{j-1}\sum_{k=1}^{r}\lambda_{i,k}\alpha_k + \sum_{i=j+1}^{n}\sum_{k=1}^{r}\lambda_{i,k}\alpha_k, \sum_{k=1}^{r}\nu_{j,k}\alpha_k\Big), \qquad (7.1.6)$$

where $\alpha_1,\ldots,\alpha_r$ are simple roots of the quantum group.

7.1.7. Lemma. *The map μ_r coincides with the composition*

$$({}^+U_q\mathfrak{n}_-)^{\otimes n} \otimes (U_q\mathfrak{n}_-) \xrightarrow{id\otimes\overline{\Delta}^{(n)}} ({}^+U_q\mathfrak{n}_-)^{\otimes n} \otimes (U_q\mathfrak{n}_-)^{\otimes n} \xrightarrow{m} ({}^+U_q\mathfrak{n}_-)^{\otimes n}.$$

The map μ_l coincides with the composition

$$(U_q\mathfrak{n}_-) \otimes ({}^+U_q\mathfrak{n}_-)^{\otimes n} \xrightarrow{\overline{\Delta}^{(n)}\otimes id} (U_q\mathfrak{n}_-)^{\otimes n} \otimes ({}^+U_q\mathfrak{n}_-)^{\otimes n} \xrightarrow{m} ({}^+U_q\mathfrak{n}_-)^{\otimes n}.$$

The proof follows easily from definitions.

7.1.8. Example.
$$\mu_r : (f_3|f_4) \otimes (f_1 f_2) \mapsto q^{(b_{14}+b_{24})/4}(f_3 f_1 f_2 | f_4) + q^{(b_{14}-b_{12}-b_{23})/4}(f_3 f_1 | f_4 f_2)$$
$$+ q^{(b_{24}+b_{12}-b_{13})/4}(f_3 f_2 | f_4 f_1)$$
$$+ q^{(-b_{13}-b_{12})/4}(f_3 | f_4 f_1 f_2),$$

cf. (5.3.35).

Define a right $(U_q \mathfrak{n}_-)^*$-module structure on $(^+U_q\mathfrak{n}_-)^{*\otimes n}$,
$$\Delta_r^* : (^+U_q\mathfrak{n}_-)^{*\otimes n} \otimes (U_q\mathfrak{n}_-)^* \longrightarrow (^+U_q\mathfrak{n}_-)^{*\otimes n}, \qquad (7.1.9)$$
as follows.

Let $\mu^* : (U_q\mathfrak{n}_-)^* \to (U_q\mathfrak{n}_-)^* \otimes (U_q\mathfrak{n}_-)^*$ be the map dual to the multiplication, $\mu^{*(n)} : (U_q\mathfrak{n}_-)^* \to (U_q\mathfrak{n}_-)^{*\otimes n}$ be the map defined by iteration of μ^*. Define
$$D : (U_q\mathfrak{n}_-)^{*\otimes n} \otimes (U_q\mathfrak{n}_-)^{*\otimes n} \longrightarrow (U_q\mathfrak{n}_-)^{*\otimes n} \qquad (7.1.10)$$
as follows.

For $a_1|\ldots|a_n, b_1|\ldots|b_n \in (U_q\mathfrak{n}_-)^{*\otimes n}$, $a_j \in (U_q\mathfrak{n}_-)^*_{\lambda_j}$, $b_j \in (U_q\mathfrak{n}_-)^*_{\nu_j}$; $\lambda_j, \nu_j \in \mathbf{N}^r$, set
$$D : a_1|\ldots|a_n|b_1|\ldots|b_n \longrightarrow q^{A(\lambda,\nu)}\overline{\Delta}^*(a_1|b_1)|\ldots|\overline{\Delta}^*(a_n|b_n),$$
where $\overline{\Delta}^*$ is the map dual to $\overline{\Delta}$, $A(\lambda,\nu)$ is defined in (7.1.6).

Define Δ_r^* as the composition
$$(^+U_q\mathfrak{n}_-)^{*\otimes n} \otimes (U_q\mathfrak{n}_-)^* \xrightarrow{D \circ (id \otimes \mu^{*(n)})} (^+U_q\mathfrak{n}_-)^{*\otimes n}. \qquad (7.1.11)$$

Define a left $(U_q\mathfrak{n}_-)^*$-module structure on $(^+U_q\mathfrak{n}_-)^{*\otimes n}$ as the composition
$$\Delta_l^* : (U_q\mathfrak{n}_-)^* \otimes (^+U_q\mathfrak{n}_-)^{*\otimes n} \xrightarrow{D \circ (\mu^{*(n)} \otimes id)} (^+U_q\mathfrak{n}_-)^{*\otimes n}. \qquad (7.1.12)$$

It is easy to see that the left and right multiplications commute.

7.1.13. Lemma. *The following diagrams are commutative.*

$$\begin{array}{ccc} U_q\mathfrak{n}_- \otimes (^+U_q\mathfrak{n}_-)^{\otimes n} & \xrightarrow{\mu_l} & (^+U_q\mathfrak{n}_-)^{\otimes n} \\ S \downarrow & & \downarrow S \\ (U_q\mathfrak{n}_-)^* \otimes (^+U_q\mathfrak{n}_-)^{*\otimes n} & \xrightarrow{\Delta_l^*} & (U_q\mathfrak{n}_-)^{*\otimes n} \end{array}$$

$$\begin{array}{ccc}
(^+U_q\mathfrak{n}_-)^{\otimes n} \otimes U_q\mathfrak{n}_- & \xrightarrow{\mu_r} & (^+U_q\mathfrak{n}_-)^{\otimes n} \\
s \downarrow & & \downarrow s \\
(^+U_q\mathfrak{n}_-)^{*\otimes n} \otimes (U_q\mathfrak{n}_-)^* & \xrightarrow{\Delta_r^*} & (^+U_q\mathfrak{n}_-)^{*\otimes n}
\end{array}$$

The lemma is a corollary of (4.3.6) and (7.1.7).

The lemma implies that Δ_l^*, Δ_r^* define module structures and, therefore, form a $(U_q\mathfrak{n}_-)^*$-bimodule structure on $(^+U_q\mathfrak{n}_-)^{*\otimes n}$.

7.1.14. Example.

$$\Delta_r^* : (f_1|f_2)^* \otimes (f_3)^* \longrightarrow q^{(-b_{13}-b_{23})/4}(f_1|f_2f_3)^* + q^{(b_{23}-b_{13})/4}(f_1|f_3f_2)^*$$
$$+ q^{(b_{23}-b_{13})/4}(f_1f_3|f_2) + q^{(b_{13}+b_{23})/4}(f_3f_1|f_2)^*.$$

cf. (5.2.31).

(7.1.15) It is easy to see that μ_l, μ_r, Δ_r^*, Δ_l^* preserve the weight decomposition.

7.2. $U_q\mathfrak{n}_-$-bimodule Structure on $(U_q\mathfrak{n}_-)^*$

Define the left and right multiplications

$$\begin{aligned} m_l : U_q\mathfrak{n}_- \otimes (U_q\mathfrak{n}_-)^* &\longrightarrow (U_q\mathfrak{n}_-)^*, \\ m_r : (U_q\mathfrak{n}_-)^* \otimes U_q\mathfrak{n}_- &\longrightarrow (U_q\mathfrak{n}_-)^* \end{aligned} \qquad (7.2.1)$$

by the formulas

$$\begin{aligned} m_l : f_I | \phi(\cdot) &\mapsto \phi(f_{\tilde{I}} \cdot), \\ m_r : \phi(\cdot) | f_I &\mapsto \phi(\cdot f_{\tilde{I}}), \end{aligned} \qquad (7.2.2)$$

where $f_I = f_{i_1}\ldots f_{i_p} \in U_q\mathfrak{n}_-$, $f_{\tilde{I}} = f_{i_p}\ldots f_{i_1}$, $\phi \in (U_q\mathfrak{n}_-)^*$.

7.2.3. Example. Let $f_I = f_{i_1}\ldots f_{i_k}$, $(f_J)^* = (f_{j_1}\ldots f_{j_l})^*$, then $m_l : f_I|(f_J)^* \mapsto (f_{j_{k+1}}\ldots f_{j_l})^*$ if $l > k$, $i_p = j_{k-p}$, $p = 1,\ldots,k$, and $m_l : f_I|(f_J)^* \mapsto 0$, otherwise.

It is easy to see that m_l and m_r commute and define a $U_q\mathfrak{n}_-$-bimodule structure on $(U_q\mathfrak{n}_-)^*$.

7.3. $(U_q\mathfrak{n}_-)^*$-bimodule Structure on $U_q\mathfrak{n}_-$

Define a map

$$m_l : (U_q\mathfrak{n}_-)^* \otimes U_q\mathfrak{n}_- \longrightarrow U_q\mathfrak{n}_- \qquad (7.3.1)$$

as follows.

Let $f_I \in U_q\mathfrak{n}_-$ be a monomial. Let $\overline{\Delta}(f_I) = \sum_{f_J} f_J \otimes a_{I,J}$, where the sum is taken over all monomials $f_J = f_{j_1} \ldots f_{j_p} \in U_q\mathfrak{n}_-$, $a_{I,J}$ are suitable elements of $U_q\mathfrak{n}_-$. Set
$$m_l : (f_J)^* \otimes f_I \mapsto a_{I,\bar{J}},$$
where $\bar{J} = (j_p, \ldots, j_1)$.

Example. $m : (f_1 f_2)^* \otimes f_2 f_3 f_1 \mapsto q^{(b_{13}-b_{23})/4} f_3$.

7.3.2. Lemma. m_l is a left $(U_q\mathfrak{n}_-)^*$-module structure.

Proof. $(\mathrm{id} \otimes \overline{\Delta}) \circ \overline{\Delta}(f_I) = \sum_{J,K} f_J \otimes f_K \otimes m(f_K \otimes m(f_J^* \otimes f_I))$.

$$(\mathrm{id} \otimes \overline{\Delta}) \circ \overline{\Delta}(f_I) = (\overline{\Delta} \otimes \mathrm{id}) \circ \overline{\Delta}(f_I) = \sum_{J,L} f_J \otimes m(f_J^* \otimes f_L) \otimes m(f_L^* \otimes f_I).$$

The lemma follows from a comparison of the two sums.

Define a map
$$m_r : U_q\mathfrak{n}_- \otimes (U_q\mathfrak{n}_-)^* \longrightarrow U_q\mathfrak{n}_- \qquad (7.3.3)$$

as follows.

Let $f_I \in U_q\mathfrak{n}_-$ be a monomial. Let $\overline{\Delta} f_I = \sum_{f_J} a_{I,J} \otimes f_J$, where the sum is taken over all monomials $f_J \in U_q\mathfrak{n}_-$, $a_{I,J}$ are suitable elements of $U_q\mathfrak{n}_-$. Set

$$m_r : f_I \otimes (f_J)^* \mapsto a_{I,\bar{J}}.$$

It is easy to see that m_r is a right $(U_q\mathfrak{n}_-)^*$-module structure.

Moreover, m_l, m_r commute and, therefore, define a bimodule structure.

7.4. R-Matrix

Let $U_q\bar{\mathfrak{n}}'_-, \overline{M}_1, \ldots, \overline{M}_n$ be the objects introduced in (4.5). Let $s_1, s_2 \in \{1, \ldots, n\}$, $t \in \mathbb{N}$,
$$X = \overline{M}_{s_1} \otimes (^+U_q\bar{\mathfrak{n}}'_-)^{\otimes t} \otimes \overline{M}_{s_2}. \qquad (7.4.1)$$

Define a linear map
$$R : X \longrightarrow X$$

as follows.

Let $x = x_1 | \ldots | x_{t+2} \in X_\lambda$, $\lambda = (l_1, \ldots, l_k, 0, \ldots, 1_{k+s_1}, 0, \ldots, 0, 1_{k+s_2}, 0, \ldots, 0) \in \mathbb{N}^{k+n}$. Set $\alpha(x) = \sum l_j \bar{\alpha}_j + \bar{\alpha}_{k+s_1} + \bar{\alpha}_{k+s_2} \in \bar{\mathfrak{h}}^*$.

For any monomial $f_I = \bar{f}_{i_1} \ldots \bar{f}_{i_l}$, set $\alpha(f_I) = \bar{\alpha}_{i_1} + \cdots + \bar{\alpha}_{i_l} \in \bar{\mathfrak{h}}^*$.
For any $\alpha = \bar{\alpha}_{i_1} + \cdots + \bar{\alpha}_{i_l}$, set $|\alpha| = l$,

$$d(\alpha) = q^{(b_{i_1 i_1} + \cdots + b_{i_l i_l})/4}, \quad d(0) = 1. \tag{7.4.2}$$

We have $d(\alpha(f_I))d(\alpha(f_J)) = d(\alpha(f_I f_J))$.
Set

$$c(\alpha) = q^{\sum_{l<m} b_{i_l i_m}/2}. \tag{7.4.3}$$

Define a map R by

$$\begin{aligned} Rx &= (-1)^{|\alpha(x)|+t(t+1)/2} c(\alpha(x)) \sum_{f_I} \sum_{f_J} q^{-(\alpha(f_I f_J), \alpha(x))/4} \\ &\quad \times (-1)^{|\alpha(f_I f_J)|} q^{(\alpha(f_I), \alpha(f_J))/4} \\ &\quad \times d(\alpha(f_I f_J)) f_J((f_I^* x_1) | x_2 | \ldots | x_{t+1} | (x_{t+2} f_J^*)) f_I, \end{aligned} \tag{7.4.4}$$

where the first and second sums are taken over all monomials of $U_q \bar{\mathfrak{n}}'_-$. $f_I^* x_1$ is the multiplication of x_1 from the left by f_I^* defined in (7.3). $x_{t+2} f_J^*$ is the multiplication of x_{t+2} from the right by f_J^*. $f_J y f_I$, $y \in X$, is the multiplication of y from the left by f_J and from the right by f_I defined in (7.1). The multiplications from the right and left commute.

The map R will be called the R-*matrix*.

R preserves the weight decomposition: $Rx \in X_\lambda$.

The sum (7.4.4) has a finite number of nonzero terms for each x.

7.4.5. Example. Let $n = 2, s_1 = 1, s_2 = 2, t = 0, x = \bar{f}_1 \bar{f}_{k+1} | \bar{f}_{k+2}$. Then

$$\begin{aligned} Rx &= -q^{(b_{1,k+1}+b_{1,k+2}+b_{k+1,k+2})/2}(\bar{f}_1 \bar{f}_{k+1} | \bar{f}_{k+2} \\ &\quad - q^{-b_{1,k+1}/2} \bar{f}_{k+1} \bar{f}_1 | \bar{f}_{k+2} \\ &\quad - q^{-3b_{1,k+1}/4 - b_{1,k+2}/4} \bar{f}_{k+1} | \bar{f}_{k+2} \bar{f}_1). \end{aligned}$$

Define a map

$$R : X^* \longrightarrow X^*, \tag{7.4.6}$$

where X^* is the space dual to X given by (7.4.1).

Let $x = x_1|\ldots|x_{t+2} \in X_\lambda$ be a monomial, let $x^* = x_1^*|\ldots|x_{t+2}^* \in X_\lambda^*$ be the corresponding δ-function. Set

$$Rx^* = (-1)^{|\alpha(x)|+t(t+1)/2}c(\alpha(x))\sum_{f_I}\sum_{f_J}q^{-(\alpha(f_If_J),\alpha(x))/4}$$

$$\times (-1)^{|\alpha(f_If_J)|}q^{(\alpha(f_I),\alpha(f_J))/4}d(\alpha(f_If_J)) \tag{7.4.7}$$

$$\times f_J^*((f_Ix_1^*)\,|x_2^*|\ldots|x_{t+1}^*|(x_{t+2}^*f_J))f_I^*,$$

where each of the sums is over all monomials in $U_q\bar{\mathfrak{n}}'_-$. $f_Ix_1^*$ is the multiplication from the left by f_I, $x_{t+2}^*f_J$ is the multiplication from the right by f_J. $f_J^*yf_I^*$, $y \in X^*$, is the multiplication from the left and right. See (7.1).

R will be called the R-*matrix*.

R preserves the weight decomposition.

7.4.8. Example. Let $x^* = x_1^*|\ldots|x_{t+2}^*$, $x_1 = \bar{f}_{k+s_1}\bar{f}_{a_1}\ldots\bar{f}_{a_n}$, $x_{t+2} = \bar{f}_{b_1}\ldots\bar{f}_{b_v}\bar{f}_{k+s_2}$. Then $Rx^* = (-1)^{|\alpha(x)|+t(t+1)/2}c(\alpha(x))x^*$, since $f_Ix_1^* = x_{t+2}^*f_I = 0$ if $f_I \ne 1$.

7.4.9. Example. Let $n = 2, s_1 = 1, s_2 = 2, t = 0, x = \bar{f}_1\bar{f}_{k+1}|\bar{f}_{k+2}$. Then

$$Rx^* = -q^{(b_{1,k+1}+b_{1,k+2}+b_{k+1,k+2})/2}(-q^{-b_{1,k+1}/2}(\bar{f}_{k+1}\bar{f}_1|\bar{f}_{k+2})^*$$

$$-q^{-b_{1,k+1}/2}(\bar{f}_{k+1}|\bar{f}_1\bar{f}_{k+2})^* - q^{-(b_{1,k+1}+b_{1,k+2})/2}(\bar{f}_{k+1}|\bar{f}_{k+2}\bar{f}_1)^*),$$

cf. (7.4.5) and (5.2.27).

Let X be given by (7.4.1) and let

$$S : X \longrightarrow X^* \tag{7.4.10}$$

be the contravariant form defined as the tensor product of the contravariant forms on factors.

7.4.11. Lemma. $SR = RS$

Proof. Let $x = x_1|\ldots|x_{t+2} \in X_\lambda$, $x_1 \in (\overline{M}_{s_1})_{\lambda_1}, x_{t+2} \in (\overline{M}_{s_2})_{\lambda_{t+2}}$, $x_j \in (^+U_q\bar{\mathfrak{n}}'_-)_{\lambda_j}$. Let $F_\nu = \{f_I\}$ be the set of all monomials in $(U_q\bar{\mathfrak{n}}'_-)_\nu$. It is easy to see that to prove the lemma it is sufficient to check the following three equalities.

$$S\bigg(\bigg(\sum_{f_I \in F_{\nu_1}} f_I \otimes f_I^* x_1\bigg) \otimes x_2 \cdots \otimes x_{t+1} \otimes \bigg(\sum_{f_J \in F_{\nu_2}} x_{t+2}f_J^* \otimes f_J\bigg)\bigg)$$

$$= \bigg(\sum_{f_I \in F_{\nu_1}} f_I^* \otimes f_I Sx_1\bigg) \otimes Sx_2 \cdots \otimes Sx_{t+1} \otimes \bigg(\sum_{f_J \in F_{\nu_2}} (Sx_2)f_J \otimes f_J^*\bigg) \tag{7.4.12}$$

for all ν_1, ν_2. Here $f_{\bar{I}} = \bar{f}_{i_p} \ldots \bar{f}_{i_1}$ if $f_I = \bar{f}_{i_1} \ldots \bar{f}_{i_p}$.

$$S(f_I, f_J) = S(f_{\bar{I}}, f_{\bar{J}}) \tag{7.4.13}$$

for all monomials $f_I, f_J \in U_q \bar{\mathfrak{n}}_-$.

$$S(abc) = (Sa)(Sb)(Sc), \tag{7.4.14}$$

where $a, c \in U_q \bar{\mathfrak{n}}'_-, b \in X$, ab means $\mu_l(a \otimes b)$, $(Sa)(Sb)$ means $\Delta_l^*(Sa \otimes Sb)$ and so on.

Equality (7.4.12) is a corollary of (4.3.6). Equality (7.4.13) is obvious. Equality (7.4.14) is a corollary of (7.1.13). The lemma is proved.

Cf. (7.4.5), (7.4.8), and (7.4.9) as an example for Lemma (7.4.11).

Define the involution $p : U_q \bar{\mathfrak{n}}_- \to U_q \bar{\mathfrak{n}}_-$ by the formula $p : f_I \to f_{\bar{I}}$. Define the involution $P : (U_q \bar{\mathfrak{n}}_-)^{\otimes (t+2)} \to (U_q \bar{\mathfrak{n}}_-)^{\otimes (t+2)}$, $P : (U_q \bar{\mathfrak{n}}_-)^{*\otimes (t+2)} \to (U_q \bar{\mathfrak{n}}_-)^{*\otimes (t+2)}$ by the formulas

$$\begin{aligned} P &: x_1|\ldots|x_{t+2} \longrightarrow p(x_{t+2})|\ldots|p(x_1), \\ P &: x_1^*|\ldots|x_{t+2}^* \longrightarrow p(x_{t+2})^*|\ldots|p(x_1)^* . \end{aligned} \tag{7.4.15}$$

The restriction of P to X given by (7.4.1) induces maps

$$\begin{aligned} P &: X \longrightarrow X', & P &: X' \longrightarrow X, \\ P &: X^* \longrightarrow X'^*, & P &: X'^* \longrightarrow X^*, \end{aligned} \tag{7.4.16}$$

where $X' = \overline{M}_{s_2} \otimes (^+U_q\bar{\mathfrak{n}}'_-)^{\otimes t} \otimes \overline{M}_{s_1}$, $P^2 = \mathrm{id}$.

The maps P commute with the contravariant form: $PS = SP$.

Define the map $R' : X \to X$ by the formula

$$\begin{aligned} R'x = (-1)^{|\alpha(x)|+t(t+1)/2} c(\alpha(x))^{-1} \sum_{f_I} \sum_{f_J} q^{(\alpha(f_I f_J), \alpha(x))/4} \\ \times (-1)^{|\alpha(f_I f_J)|} q^{-(\alpha(f_I), \alpha(f_J))/4} d(\alpha(f_I f_J))^{-1} \\ \times f_I((f_I^* x_1) | x_2 | \ldots | x_{t+1} | (x_{t+2} f_J^*)) f_J . \end{aligned} \tag{7.4.17}$$

Define the map $R' : X^* \to X^*$ by the formula

$$\begin{aligned} R'x^* = (-1)^{|\alpha(x)|+t(t+1)/2} c(\alpha(x))^{-1} \sum_{f_I} \sum_{f_J} q^{(\alpha(f_I f_J), \alpha(x))/4} \\ \times (-1)^{|\alpha(f_I f_J)|} q^{-(\alpha(f_I), \alpha(f_J))/4} d(\alpha(f_I f_J))^{-1} \\ \times f_I^*((f_I x_1^*) | x_2^* | \ldots | x_{t+1}^* | (x_{t+2}^* f_J)) f_J^* . \end{aligned} \tag{7.4.18}$$

In these formulas, the sums are over all monomials of $U_q\bar{\mathfrak{n}}'_-$.

7.4.19. Lemma. *The map $PR'P$ is inverse to the map R, namely, $PR'PR$ = id on X and on X^*.*

The lemma is proved by easy direct verification. Moreover, it is sufficient to check the equality $PR'PR =$ id for one of X and X^* in virtue of (7.4.11) and nondegeneracy of S for general values of $(\bar{\alpha}_i, \bar{\alpha}_j)$.

7.4.20. Example. Let $x^* = (\bar{f}_1\bar{f}_{k+1}|\bar{f}_{k+2})$, see (7.4.9). Then

$$PRx^* = q^{(b_{1,k+1}+b_{1,k+2}+b_{k+1,k+2})/2}\big(q^{-b_{1,k+1}/2}(\bar{f}_{k+2}|\bar{f}_1\bar{f}_{k+1})^*$$
$$+ q^{-b_{1,k+1}/2}(\bar{f}_{k+2}\bar{f}_1|\bar{f}_{k+1})^*$$
$$+ q^{-(b_{1,k+1}+b_{1,k+2})/2}(\bar{f}_1\bar{f}_{k+2}|\bar{f}_{k+1})^*\big)$$
$$\xrightarrow{R'} -q^{-b_{1,k+1}/2}(\bar{f}_{k+2}|\bar{f}_1\bar{f}_{k+1})^* - q^{b_{1,k+1}/2}(f_{k+2}f_1|f_{k+1})^*$$
$$- q^{-(b_{1,k+1}+b_{1,k+2})/2}\big((\bar{f}_1\bar{f}_{k+2}|\bar{f}_{k+1})^*$$
$$- q^{(b_{1,k+1}+b_{1,k+2})/4}\big(q^{(-b_{1,k+1}-b_{1,k+2})/4}(\bar{f}_1\bar{f}_{k+2}|\bar{f}_{k+1})^*$$
$$+ q^{(-b_{1,k+1}+b_{1,k+2})/4}(\bar{f}_{k+2}\bar{f}_1|\bar{f}_{k+1})^*$$
$$+ q^{(-b_{1,k+1}+b_{1,k+2})/4}(\bar{f}_{k+2}|\bar{f}_1\bar{f}_{k+1})^*$$
$$+ q^{(b_{1,k+1}+b_{1,k+2})/4}(\bar{f}_{k+2}|\bar{f}_{k+1}\bar{f}_1)^*\big)\big)$$
$$= (\bar{f}_{k+2}|\bar{f}_{k+1}\bar{f}_1)^*.$$

Transformations R^p.

Fix $p \in \{1,\ldots,n-1\}, \delta = (\delta(1),\ldots,\delta(n)) \in \Sigma_n$, set

$$\delta^p = \big(\delta(1),\ldots,\delta(p-1),\delta(p+1),\delta(p),\delta(p+2),\ldots,\delta(n)\big) \in \Sigma_n.$$

Set $A = {}^+U_q\bar{\mathfrak{n}}_-$,

$$Y_\delta = A^{\otimes s_0} \otimes \overline{M}_{\delta(1)} \otimes A^{\otimes s_1} \otimes \overline{M}_{\delta(2)} \otimes \ldots \overline{M}_{\delta(n)} \otimes A^{\otimes s_n},$$
$$Y_{\delta^p} = A^{\otimes s_0} \otimes \overline{M}_{\delta(1)} \otimes \cdots \otimes \overline{M}_{\delta(p-1)} \otimes A^{\otimes s_{p-1}} \otimes \overline{M}_{\delta(p+1)} \otimes \quad (7.4.21)$$
$$A^{s_p} \otimes \overline{M}_{\delta(p)} \otimes A^{\otimes s_{p+1}} \otimes \overline{M}_{\delta(p+2)} \otimes \ldots \overline{M}_{\delta(n)} \otimes A^{\otimes s_n}$$

for some $s_0,\ldots,s_n \in \mathbb{N}$.

Let
$$R^p : Y_\delta \longrightarrow Y_{\delta^p} \quad (7.4.22)$$

be the linear map acting as PR on
$$\overline{M}_{\delta(p)} \otimes A^{\otimes s_p} \otimes \overline{M}_{\delta(p+1)}$$
and as the identity on all of the other factors, where P, R are defined by (7.4.4) and (7.4.15).

Consider the complex $C.(^+U_q\bar{\mathfrak{n}}'_-; \overline{M}_\delta; 2; {}^+\Delta)$ defined in (4.5). Define linear maps
$$R^p : C_j(^+U_q\bar{\mathfrak{n}}'_-; \overline{M}_\delta; 2; {}^+\Delta) \longrightarrow C_j(^+U_q\bar{\mathfrak{n}}'_-; \overline{M}_{\delta^p}; 2; {}^+\Delta) \qquad (7.4.23)$$
for all j as follows.

$C_j(^+U_q\bar{\mathfrak{n}}'_-; \overline{M}_\delta; 2; {}^+\Delta)$ is the direct sum of all tensor products j-admissible for the sequence $\overline{M}_{\delta(1)}, \ldots, \overline{M}_{\delta(n)}$, see (4.4.2.1). Each of the j-admissible tensor products has the form of Y_δ in (7.4.21). Analogously, $C_j(^+U_q\bar{\mathfrak{n}}'_-; \overline{M}_{\delta^p}; 2; {}^+\Delta)$ is the direct sum of the tensor products of the form of Y_{δ^p} in (7.4.21). Let R^p in (7.4.23) act on each Y_δ as in (7.4.22).

Let $Y_\delta^*, Y_{\delta^p}^*$ be the spaces dual to Y_δ, Y_{δ^p}.

Let
$$R^p : Y_\delta^* \longrightarrow Y_{\delta^p}^* \qquad (7.4.24)$$
be the linear map acting as PR on $\overline{M}^*_{\delta(p)} \otimes A^{*\otimes s_p} \otimes \overline{M}_{\delta(p+1)}$ and as the identity on all of the other factors, where P, R are defined by (7.4.7) and (7.4.13).

Consider the complex $C^{\cdot*}(^+U_q\bar{\mathfrak{n}}'_-; \overline{M}_\delta; 2; \mu)$ defined in (4.5). Define linear maps
$$R^p : C_j^*(^+U_q\bar{\mathfrak{n}}'_-; \overline{M}_\delta; 2; \mu) \longrightarrow C_j^*(^+U_q\bar{\mathfrak{n}}'_-; \overline{M}_\delta; 2; \mu) \qquad (7.4.25)$$
for all j as follows. $C_j^*(^+U_q\bar{\mathfrak{n}}'_-; \overline{M}_\delta; 2; \mu)$ is the direct sum of the spaces of the form Y_δ^* with $s_0 + \cdots + s_n = j$. Let R^p in (7.4.25) act on each Y_δ^* as in (7.4.24).

7.4.26. Lemma. *The transformations R^p preserve the weight decomposition and commute with the contravariant form.*

The first statement is obvious, the second statement follows from (7.4.11).

7.4.27. Lemma. *Linear maps R^p in (7.4.23), (7.4.25) define isomorphisms of complexes*
$$R^p : C.(^+U_q\bar{\mathfrak{n}}'_-; \overline{M}_\delta; 2; {}^+\Delta)_\lambda \longrightarrow C.(^+U_q\bar{\mathfrak{n}}'_-; \overline{M}_{\delta^p}; 2; {}^+\Delta)_\lambda$$
$$R^p : C^{\cdot*}(^+U_q\bar{\mathfrak{n}}'_-; \overline{M}_\delta; 2; \mu)_\lambda \longrightarrow C^{\cdot*}(^+U_q\bar{\mathfrak{n}}'_-; \overline{M}_{\delta^p}; 2; \mu)_\lambda$$

for all weights λ.

To prove the lemma, it is sufficient to check that the maps R^p commute with the differentials of the complexes. It is sufficient to check that $R^p d = dR^p$ for one of the two isomorphisms of the lemma by virtue of $SR^p = R^p S$. The equality $R^p d = dR^p$ for each of the isomorphisms is proved by direct verification.

7.4.28. Remark. The equality $R^p d = dR^p$ will follow also from Theorem (8.3.4) formulated in Sec. 8. The proof of Theorem (8.3.4) does not use Lemma (7.4.27) and is purely topological.

7.5. Symmetrization and the R-Matrix

Let $U_q\bar{\mathfrak{g}}$, $^+U_q\bar{\mathfrak{n}}'_-$, $\overline{M}_1, \ldots, \overline{M}_n$ be the objects introduced in (4.5). Let $U_q\tilde{\mathfrak{g}}$ be a quantum group adjacent to $U_q\bar{\mathfrak{g}}$ at weight $\lambda = (l_1, \ldots, l_k, 1, \ldots, 1) \in \mathbb{N}^{k+n}$. $U_q\tilde{\mathfrak{g}}$ is generated by elements 1, $e_i(j)$, $f_i(j)$ for $j = 1, \ldots, k$, $i = 1, \ldots, l_j$, by elements e_j, f_j for $j = k+1, \ldots, k+n$ and by a space $\mathfrak{h}$, see (4.8) and (4.9).

Let $U_q\tilde{\mathfrak{n}}'_- \subset U_q\tilde{\mathfrak{g}}$ (resp. $^+U_q\tilde{\mathfrak{n}}'_-$) be the subalgebra generated by 1, $f_i(j)$ (resp. $f_i(j)$) for all i, j, $\widetilde{M}_i \subset U_p\tilde{\mathfrak{g}}$ be the two-sided submodule over $U_q\tilde{\mathfrak{n}}'_-$ generated by f_{k+i}, $i = 1, \ldots, n$.

Set $\widetilde{M}_\delta = (\widetilde{M}_{\delta(1)}, \ldots, \widetilde{M}_{\delta(n)})$ for any $\delta \in \Sigma_n$. Set $\tilde{\lambda} = (1, \ldots, 1) \in \mathbb{N}^{n+l}$, where $l = l_1 + \cdots + l_k$.

The group $\Sigma' = \Sigma_{l_1} \times \cdots \times \Sigma_{l_k}$ acts on $C.(^+U_q\tilde{\mathfrak{n}}'_-; \overline{M}_\delta; 2; {}^+\Delta)_{\tilde{\lambda}}$, $C^*(^+U_q\tilde{\mathfrak{n}}'_-; \overline{M}_\delta; 2; \mu)_{\tilde{\lambda}}$ permuting $f_1(j), \ldots, f_{l_j}(j)$ for $j = 1, \ldots, k$, see (4.8). Denote by $^+$ the invariant part of this action.

In (4.8) isomorphisms

$$\pi : C.(^+U_q\bar{\mathfrak{n}}'_-; \overline{M}_\delta; 2; {}^+\Delta)_\lambda \longrightarrow C.(^+U_q\tilde{\mathfrak{n}}'_-; \widetilde{M}_\delta; 2; {}^+\Delta)_{\tilde{\lambda}}^+$$
$$\pi : C^*(^+U_q\bar{\mathfrak{n}}'_-; \overline{M}_\delta; 2; \mu)_\lambda \longrightarrow C^*(^+U_q\tilde{\mathfrak{n}}'_-; \widetilde{M}_\delta; 2; \mu)_{\tilde{\lambda}}^+ ,$$

(7.5.1)

were defined. See (4.8.8) and (4.9.5).

7.5.2. Theorem. *For any $\delta \in \Sigma_n$ and any $p = 1, \ldots, n-1$, the following diagrams are commutative:*

$$\begin{array}{ccc}
C.(^+U_q\bar{\mathfrak{n}}'_-; \overline{M}_\delta; 2; {}^+\Delta)_\lambda & \xrightarrow{R^p} & C.(^+U_q\bar{\mathfrak{n}}'_-; \overline{M}_{\delta^p}; 2; {}^+\Delta)_\lambda \\
\pi \downarrow & & \downarrow \pi \\
C.(^+U_q\tilde{\mathfrak{n}}'_-; \widetilde{M}_\delta; 2; {}^+\Delta)_{\tilde{\lambda}} & \xrightarrow{R^p} & C.(^+U_q\tilde{\mathfrak{n}}'_-; \widetilde{M}_{\delta^p}; 2; {}^+\Delta)_{\tilde{\lambda}}
\end{array}$$

$$C^{\cdot *}(^+U_q\bar{\mathfrak{n}}'_-; \overline{M}_\delta; 2; \mu)_\lambda \xrightarrow{R^p} C^{\cdot *}(^+U_q\bar{\mathfrak{n}}'_-; \overline{M}_{\delta^p}; 2; \mu)_\lambda$$
$$\pi \downarrow \qquad\qquad\qquad \downarrow \pi$$
$$C^{\cdot *}(^+U_q\bar{\mathfrak{n}}'_-; \widetilde{M}_\delta; 2; \mu)_{\bar{\lambda}} \xrightarrow{R^p} C^{\cdot *}(^+U_q\bar{\mathfrak{n}}'_-; \widetilde{M}_{\delta^p}; 2; \mu)_{\bar{\lambda}}.$$

The theorem is proved by easy direct verification.

7.5.3. Example. Let $n = 2$ and $x = f_{k+1}f_1^2 | f_{k+2}$. Then

$$R^1 x = P(c(\alpha(x))[x - q^{-(\alpha(f_1),\alpha(x))/4}d(\alpha(f_1)) \cdot (f_1^*(f_{k+1}f_1^2) | f_{k+2})f_1$$
$$+ q^{-(\alpha(f_1^2),\alpha(x))/4}d(\alpha(f_1^2)) \cdot ((f_1^2)^*(f_{k+1}f_1^2) | f_{k+2})f_1^2]$$
$$= Bf_1^2 f_{k+2} | f_{k+1},$$

where $B = q^{(b_{1,k+1} + b_{1,k+2} + b_{k+1,k+2})/2}$, see (7.4.4).

It is easy to see that $\pi x = x_1 + x_2$, where $x_1 = f_{k+1}f_1(1)f_2(1) | f_{k+2}$, $x_2 = f_{k+1}f_2(1)f_1(1) | f_{k+2}$, and $R^1 x_1 = Bf_1(1)f_2(1)f_{k+2} | f_{k+1}$, $R^1 x_2 = Bf_2(1)f_1(1) f_{k+2} | f_{k+1}$. Hence $\pi R^1 x = R^1 \pi x$.

7.6. R-Matrix for Verma Modules (cf. [SV4])

Let $U_q\mathfrak{g}$ be a quantum group of type (4.1.a)–(4.1.d), $M_1, \ldots, M_n$, $M'_1, \ldots, M'_{n'}$ Verma modules over $U_q\mathfrak{g}$ with highest weights $\Lambda_1, \ldots, \Lambda_n$, $\Lambda'_1, \ldots, \Lambda'_{n'} \in \mathfrak{h}^*$. Set $M = M_1 \otimes \cdots \otimes M_n$, $M' = M'_1 \otimes \cdots \otimes M'_{n'}$.

Consider the expression

$$R = \sum_{\lambda \in \mathbb{N}^r} q^{\Omega_0/2 + \frac{1}{4}(h_\lambda \otimes 1 - 1 \otimes h_\lambda) + D(\lambda)} \Omega_\lambda . \qquad (7.6.1)$$

Here $\Omega_0 \in \mathfrak{h} \otimes \mathfrak{h}$ is the element corresponding to the form $(\,,\,)$, see (4.1); for $\lambda = (b_1, \ldots, b_r)$, $h_\lambda := b^{-1}(\sum_{i=1}^r l_i \alpha_i) \in \mathfrak{h}$; $D(\lambda) \in \mathbb{C}$ is a constant defined as follows: represent $\alpha(\lambda) = \sum l_i \alpha_i$ as a sum of simple roots $\alpha(\lambda) = \alpha_{i_1} + \cdots + \alpha_{i_N}$. Then

$$D(\lambda) := -\sum_{p \leq q} b_{i_p i_q}/4.$$

Ω_λ is defined by

$$\Omega_\lambda := \sum g_J \otimes f_J, \qquad (7.6.2)$$

where the sum is taken over all monomials $f_J = f_{j_1} \cdots f_{j_N} \in (U_q\mathfrak{n}_-)_\lambda$, g_J is defined by (5.1.6) and (5.1.10).

R defines an operator

$$R : M \otimes M' \longrightarrow M \otimes M'. \tag{7.6.3}$$

7.6.4. Example. (i) For $U_q sl_2$ generated by e, f, h subject to the relations $[e, f] = q^{h/2} - q^{-h/2}$, $[h, e] = 2e$, $[h, f] = -2f$, the R-matrix operator (7.6.3) has the form

$$R = q^{h \otimes h/4} \sum_{k \geq 0} q^{-k(k+1)/4} (1)_q^k / (k)_q! \, q^{kh/4} e^k \otimes q^{-kh/4} f^k.$$

(ii) Suppose $M = M_1, M' = M_1'$ and $\lambda = (\delta_{i1}, \ldots, \delta_{ir})$ for some i, $1 \leq i \leq r$. $(M \otimes M')_\lambda$ is generated by $f_i v \otimes v'$, $v \otimes f_i v'$. The part of R acting nontrivially on this subspace is

$$q^{\Omega_0/2} + q^{\Omega_0/2 + \frac{1}{4}(h_i \otimes 1 - 1 \otimes h_i)} q^{-b_{ii}/4} e_i \otimes f_i.$$

Hence

$$R(v \otimes f_i v') = q^{(\Lambda_1, \Lambda_1' - \alpha_i)/2} v \otimes f_i v',$$
$$R(f_i v \otimes v') = (\Lambda_1, \alpha_i)_q q^{(\Lambda_1, \Lambda_1')/2 - (\alpha_i, \Lambda_1 + \Lambda_1')/4} v \otimes fv'$$
$$+ q^{(\Lambda_1 - \alpha_i, \Lambda_1')/2} f_i v \otimes v'.$$

For any $J = (j_1, \ldots, j_N)$ such that $j_i \in \{1, \ldots, r\}$ for all i, define a linear operator

$$u_J : (M')^* \longrightarrow (M')^* \tag{7.6.5}$$

by $\phi(\cdot) \to \phi(g_J \cdot)$ for all $\phi \in (M')^*$, here $\bar{J} = (j_N, \cdots, j_1)$.

Consider the expression

$$R = \sum_{\lambda \in \mathbb{N}^r} q^{\Omega_0/2 + \frac{1}{4}(h_\lambda \otimes 1 - 1 \otimes h_\lambda) + D(\lambda)} \Phi_\lambda. \tag{7.6.6}$$

Here $\Phi_\lambda := \sum e_J \otimes u_J$ and the sum is taken over all monomials $e_J = e_{j_1} \ldots e_{j_N} \in (U_q n_+)_\lambda$. This R defines a linear operator

$$R : M^* \otimes M'^* \longrightarrow M^* \otimes M'^*.$$

7.6.7. Example. (i) For $U_q sl_2$, the R-matrix operator (7.6.6) has the form (7.6.4.i). (ii) Suppose $M = M_1, M' = M_1'$, and $\lambda = (\delta_{i1}, \ldots, \delta_{ir})$ for some *i*. Denote by $(fv \otimes v')^*$, $(v \otimes fv')^*$ the basis of $(M \otimes M')^*_\lambda$ dual to $fv \otimes v'$, $v \otimes fv'$. The part of R acting nontrivially on $(M \otimes M')^*_\lambda$ is the same as in (7.6.4). Hence

$$R((v \otimes f_i v')^*) = q^{(\Lambda_1, \Lambda'_1 - \alpha_i)/2}(v \otimes f_i v')^*,$$
$$R((f_i v \otimes v')^*) = (\Lambda'_1, \alpha_i)_q q^{(\Lambda_1, \Lambda'_1)/2 - (\alpha_i, \Lambda_1 + \Lambda'_1)/4}(v \otimes f_i v')^*$$
$$+ q^{(\Lambda_1 - \alpha_i, \Lambda'_1)/2}(f_i v \otimes v')^*.$$

7.6.8. Theorem. *The following diagram is commutative*

$$\begin{array}{ccc} M \otimes M' & \xrightarrow{R} & M \otimes M' \\ S \downarrow & & \downarrow S \\ M^* \otimes M'^* & \xrightarrow{R} & M^* \otimes M'^* \end{array}$$

where S is the contravariant form defined in (4.2).

Proof. The theorem easily follows from (5.1.6) and (5.1.19).

Consider the complexes $C.(^+U_q \mathfrak{n}_-; M_\delta; ^+\Delta)$, $C.^*(^+U_q \mathfrak{n}_-; M_\delta; \mu)$ defined in (5.1). Here $\delta \in \Sigma_n$ and $M_\delta = M_{\delta(1)} \otimes \cdots \otimes M_{\delta(n)}$. We have $C_j(^+U_q\mathfrak{n}_-; M_\delta; ^+\Delta) = (^+U_q\mathfrak{n}_-)^{\otimes j} \otimes M_\delta$, $C_j^* = \left((^+U_q\mathfrak{n}_-)^{\otimes j} \otimes M_\delta\right)^*$. For any $p = 1, \ldots, n-1$, define linear operators

$$R^p : (^+U_q\mathfrak{n}_-)^{\otimes j} \otimes M_\delta \longrightarrow (^+U_q\mathfrak{n}_-)^{\otimes j} \otimes M_{\delta^p},$$
$$R^p : \left((^+U_q\mathfrak{n}_-)^{\otimes j} \otimes M_\delta\right)^* \longrightarrow \left((^+U_q\mathfrak{n}_-)^{\otimes j} \otimes M_{\delta^p}\right)^*, \quad (7.6.9)$$

acting as PR on the pth and $(p+1)$th factors of M_δ and as the identity operator on all other factors. Here P is the transposition of the pth and $(p+1)$th factors. By Theorem (7.6.8), these operators commute with the contravariant form S, see (5.1.26).

In the next section, we will show that these linear operators commute with the differentials of the complexes.

7.7. Connection of the R-Matrices for Two- and One-sided Hochschild Complexes

In Sec. 5, we have constructed monomorphisms

$$\phi : C.(^+U_q\mathfrak{n}_-; M_\delta; ^+\Delta)_\lambda \longrightarrow C.(^+U_q\bar{\mathfrak{n}}'_-; \overline{M}_\delta; 2; ^+\Delta)_{\bar\lambda},$$
$$\phi : C.^*(^+U_q\mathfrak{n}_-; M_\delta; \mu)_\lambda \longrightarrow C.^*(^+U_q\bar{\mathfrak{n}}'_-; \overline{M}_\delta; 2; \mu)_{\bar\lambda}, \quad (7.7.1)$$

see (5.2.3).

7.7.2. Theorem. *These monomorphisms commute with the action of the operators R^p defined in (7.4) and (7.6), that is, for any p, we have $\phi R^p = R^p \phi$.*

Proof. It suffices to check the commutativity for one of the monomorphisms (7.7.1), as ϕ and R^p commute with the contravariant form S and S is an isomorphism for general values of parameters of $U_q \mathfrak{g}$, M.

For the second monomorphism of (7.7.1), the commutativity is proved by easy direct verification.

7.7.3. Example. For $x^* = (f_1 v_1 | v_2)^* \in C_0^*(U_q \mathfrak{n}_-; M_1 \otimes M_2; \mu)$, check that $\phi R^1 x^* = R^1 \phi x^*$. By (5.2.27) and (7.4.9), we have

$$R^1 \phi x^* = q^{-(\alpha_1, \Lambda_1)/4} R^1 (\bar{f}_1 \bar{f}_2 | \bar{f}_3)^*$$
$$= q^{-(\alpha_1, \Lambda_1)/4 - (\alpha_1, \Lambda_2)/2 + (\Lambda_1, \Lambda_2)/2} ((\bar{f}_3 | \bar{f}_1 \bar{f}_2)^*$$
$$+ (\bar{f}_3 \bar{f}_1 | \bar{f}_2)^* + q^{(\alpha_1, \Lambda_2)/2} (\bar{f}_1 \bar{f}_3 | \bar{f}_2)^*).$$

By (5.2.27), we have

$$\phi R^1 x^* = \phi \big((q^{(\Lambda_2, \alpha_1)/2} - q^{-(\Lambda_2, \alpha_1)/2}) q^{(\Lambda_1, \Lambda_2)/2 - (\alpha_1, \Lambda_1 + \Lambda_2)/4} (f_1 v_2 | v_1)^*$$
$$+ q^{(\Lambda_1 - \alpha_1, \Lambda_2)/2} (v_2 | f_1 v_1)^* \big)$$
$$= (q^{(\Lambda_2, \alpha_1)/2} - q^{-(\Lambda_2, \alpha_1)/2}) q^{(\Lambda_1, \Lambda_2)/2 - (\alpha_1, \Lambda_1 + \Lambda_2)/4} q^{-(\alpha_1, \Lambda_2)/4} (\bar{f}_1 \bar{f}_3 | \bar{f}_2)^*$$
$$+ q^{(\Lambda_1 - \alpha_1, \Lambda_2)/2} q^{-(\alpha_1, \Lambda_1)/4} \big[(\bar{f}_3 | \bar{f}_1 \bar{f}_2)^* + (\bar{f}_3 \bar{f}_1 | \bar{f}_2)^*$$
$$+ q^{-(\alpha_1, \Lambda_2)/2} (\bar{f}_1 \bar{f}_3 | \bar{f}_2)^* \big].$$

Obviously, this is the same as $R^1 \phi x$.

7.7.4. Theorem. *For any p, the linear operators R^p defined in (7.6.9) define isomorphisms of complexes*

$$R^p : C.(^+U_q \mathfrak{n}_-; M_\delta; {}^+\Delta) \longrightarrow C.(^+U_q \mathfrak{n}_-; M_{\delta^p}; {}^+\Delta),$$
$$R^p : C^*.(^+U_q \mathfrak{n}_-; M_\delta; \mu) \longrightarrow C.(^+U_q \mathfrak{n}_-; M_{\delta^p}; \mu).$$

Proof. R^p are morphisms of complexes by (7.7.2). R^p are invertible as the R-matrices of the right-hand side of (7.7.1) are invertible, see (7.4.19).

Remark. The commutativity of R^p and the differentials of complexes (7.7.4) can be verified directly without a reference to (7.7.2).

8

Monodromy

In this section, we describe the monodromy in homology groups of a discriminantal configuration $\mathcal{C}_{n,k}(z_1, \ldots, z_n)$ under deformations of the parameters $z_1, \ldots, z_n$. It turns out that the elementary monodromy transformations are described in terms of the R-matrix operator of the corresponding quantum group.

8.1. Gauss–Manin Connection

Let $\mathcal{C}_l$ be the configuration of all diagonal hyperplanes $t_i = t_j$, $1 \leq i < j \leq l$, in $\mathbb{C}^l$. Set

$$U_l = \bigcup_{H \in \mathcal{C}_l} H \subset \mathbb{C}^l. \tag{8.1.1}$$

Let $p_{n,k} : \mathbb{C}^{n+k} \to \mathbb{C}^n$, $(t_1, \ldots, t_{n+k}) \to (t_{k+1}, \ldots, t_{k+n})$ be the projection on the last n coordinates. Let $P_{n,k}(z) = p_{n,k}^{-1}(z)$ be its fiber over $z \in \mathbb{C}^n$.

The configuration $\mathcal{C}_{n+l}$ cuts a configuration of hyperplanes in each fiber $P_{n,k}(z)$. Denote it by $\mathcal{C}(z)$.

Set

$$U_{n,k}(z) = \bigcup_{H \in \mathcal{C}(z)} H \subset P_{n,k}(z). \tag{8.1.2}$$

A point of the base $\mathbb{C}^n$ is called discriminantal if the configuration $\mathcal{C}(z)$ in the fiber over it is degenerate. The discriminantal points form the discriminant, the configuration $\mathcal{C}_n$ of all diagonal hyperplanes in $\mathbb{C}^n$. Over the complement of the discriminant, fibers with distinguished configurations form a locally trivial bundle:

$$p_{n,k} : \left(p_{n,k}^{-1}(\mathbb{C}^n \backslash U_n), U_{n+k} \bigcap p_{n,k}^{-1}(\mathbb{C}^n \backslash U_n) \right) \longrightarrow \mathbb{C}^n \backslash U_n. \qquad (8.1.3)$$

Let the configuration $\mathcal{C}_{n+k}$ be weighted, let a_{ij} be the weight of the hyperplane $t_i = t_j$, $A = \{a_{ij}\}$, $\mathcal{S}^*(A)$ be the local system of coefficients on $\mathbb{C}^{n+k} - U_{n+k}$ associated with the weights A, see (2.5). Let $\mathcal{S}^*(A, z)$ be the restriction of $\mathcal{S}^*(A)$ to $P_{n,k}(z)$.

We will be interested in the chain complexes

$$\begin{aligned} C_{\cdot}(z) &:= C_{\cdot}(P_{n,k}(z) - U_{n,k}(z), \mathcal{S}^*(A, z)), \\ C_{!\cdot}(z) &:= C_{!\cdot}(P_{n,k}(z) - U_{n,k}(z), \mathcal{S}^*(A, z)), \end{aligned} \qquad (8.1.4)$$

see (2.3). The first complex calculates $H_{\cdot}(P_{n,k}(z) - U_{n,k}(z), \mathcal{S}^*(A, z))$, the second calculates $H_{\cdot}(P_{n,k}(z), U_{n,k}(z), \mathcal{S}^*(A, z))$.

For any $l \in \mathbb{N}$, there are two holomorphic vector bundles over the complement of the discriminant. The first bundle h_l has a fiber $H_l(P_{n,k}(z) - U_{n,k}(z), \mathcal{S}^*(A, z))$ over a point z. The second bundle $h_{!l}$ has a fiber $H_l(P_{n,k}(z), U_{n,k}(z), \mathcal{S}^*(A, z))$. The natural inclusion $C_{\cdot}(z) \hookrightarrow C_{!\cdot}(z)$ induces a morphism of vector bundles

$$S : h_l \longrightarrow h_{!l}. \qquad (8.1.5)$$

The bundles h_l, $h_{!l}$ will be called the homology bundles associated with the projection $p_{n,k}$ and the weighted configuration $\mathcal{C}_{n+k}$.

The homology bundles have a canonical holomorphic integrable connection called the Gauss–Manin connection. The morphism (8.1.5) commutes with the Gauss–Manin connection.

We remind the construction of the connection. Let $\gamma : [0, 1] \to \mathbb{C}^n \backslash U_n$ be any curve in the base,

$$\gamma^* p : (X, Y) \longrightarrow [0, 1], \qquad (8.1.6)$$

the bundle induced on the curves by the bundle (8.1.3), where $(X(t), Y(t))$ is its fiber over $t \in [0, 1]$. Let $tr : (X(0) \times [0, 1], Y(0) \times [0, 1]) \to (X, Y)$ be any trivialization of the bundle $\gamma^* p$, and $\mathcal{S}$ be the local system on $X \backslash Y$ induced by $\mathcal{S}^*(A)$.

Let $f : \triangle \to X(0) \backslash Y(0)$ be any singular l-cell, s any section of $f^* \mathcal{S}$ over $\triangle$. Then $F : \triangle \times [0, 1] \to X \backslash Y$, $(a, t) \to tr(f(a), t)$, is an $(l+1)$-cell in $X \backslash Y$. Denote by $\mathcal{S}$ the section of $F^* \mathcal{S}$ over $\triangle \times [0, 1]$ such that $\mathcal{S}|_{\triangle \times 0} = s$. Then

$F|_{\triangle \times 1} : \triangle \times 1 \to X(1) \backslash Y(1)$ is a singular l-cell, and $S|_{\triangle \times 1}$ is a section of $(F|_{\triangle \times 1})^* \mathcal{S}$ over it.

This construction defines isomorphisms of complexes

$$\begin{aligned} T_\gamma &: C.(\gamma(0)) \longrightarrow C.(\gamma(1)), \\ T_\gamma &: C_!.(\gamma(0)) \longrightarrow C_!.(\gamma(1)). \end{aligned} \qquad (8.1.7)$$

These isomorphisms define isomorphisms of homology groups of the complexes. The isomorphisms of homology groups do not depend on the choice of the trivialization tr and define the Gauss–Manin connection.

8.2. Chain Complexes over Real Points of the Base

Let $z^0 = (z_1^0, \ldots, z_n^0) \in \mathbb{R}^n \subset \mathbb{C}^n$, and

$$z_{\delta(1)}^0 < z_{\delta(2)}^0 < \cdots < z_{\delta(n)}^0 \qquad (8.2.1)$$

for some permutation $\delta \in \Sigma_n$. Set $y = (y_1, \ldots, y_n)$, where $y_j = z_{\delta(j)}^0$.

The fiber over the point z^0 of the bundle (8.1.3) is the affine space $P_{n,k}(z^0)$ and the configuration $\mathcal{C}(z^0)$ in it.

(8.2.2) The space $P_{n,k}(z^0)$ and the configuration $\mathcal{C}(z^0)$ are identified naturally with $\mathbb{C}^k$ and the discriminantal configuration $\mathcal{C}_{n,k}(y)$ considered in (3.1), if we identify the coordinates $t_1, \ldots, t_k$ in $P_{n,k}(z^0)$ with the standard coordinates $x_1, \ldots, x_k$ in $\mathbb{C}^k$, resp.

The weights A of the configuration $\mathcal{C}_{n+k}$ define the weights of its section $\mathcal{C}(z^0)$: for any hyperplane, $H \in \mathcal{C}_{n+k}$, define the weight of $H \cap p_{n,k}^{-1}(z^0)$ as the weight of H. Denote the set of the weights of $\mathcal{C}_{n,k}(y)$ by a. Let $\mathcal{S}^*(a)$ be the local system defined on the complement of $\mathcal{C}_{n,k}(y)$ in $\mathbb{C}^k$ and associated with the weights a, see (2.5).

We construct an isomorphism

$$i : \mathcal{S}^*(a) \simeq \mathcal{S}^*(A, z^0) \qquad (8.2.3)$$

as follows. Any section s of $\mathcal{S}^*(a)$ has the form $s = \text{const } bl_a$, where const $\in \mathbb{C}$, bl_a is a branch of the multivalued function

$$l_a = \prod_{1 \leq i < j \leq k} (t_i - t_j)^{a_{ij}} \prod_{i=1}^{n} \prod_{j=1}^{k} (z_i^0 - t_j)^{a_{k+i,j}},$$

see (2.5.6). Any section S of $\mathcal{S}^*(A, z^0)$ has the form $S = \text{const } bl_A$, where const $\in \mathbb{C}$, bl_A is a branch of the multivalued function

$$l_A = \prod_{1 \leq i < j \leq n+k} (t_i - t_j)^{a_{ij}}.$$

Set

$$i : s \mapsto s \prod_{1 \leq i < j \leq n} |z_i^0 - z_j^0|^{a_{k+i,k+j}}. \qquad (8.2.4)$$

Recall that $t_{k+i} = z_i^0$, $i = 1, \ldots, n$, on $P_{n,k}(z^0)$.

The chain complexes $C.(X, \mathcal{S}^*(a))$, $C.(Q'/Q'', \mathcal{S}^*(a))$ of a discriminantal configuration and the homorphism $S : C.(X, \mathcal{S}^*(a)) \to C.(Q'/Q'', \mathcal{S}^*(a))$ have been constructed in (2.5), (2.7), and (3.12).

The identification of $(P_{n,k}(z^0), \mathcal{C}(z^0), \mathcal{S}^*(A, z))$ with $(\mathbb{C}^k, \mathcal{C}_{n,k}(y), \mathcal{S}^*(a))$ identifies $C.(X, \mathcal{S}^*(a))$ with a subcomplex of $C.(z^0)$ and identifies $C.(Q'/Q'', \mathcal{S}^*(a))$ with a subcomplex of $C_!.(z^0)$. We denote these subcomplexes by

$$\begin{aligned} C.(X, \mathcal{S}^*(A), z^0) &\hookrightarrow C.(z^0), \\ C.(Q'/Q'', \mathcal{S}^*(A), z^0) &\hookrightarrow C_!.(z^0). \end{aligned} \qquad (8.2.5)$$

The homomorphism of the projection on the real part induces a homomorphism

$$S : C.(X, \mathcal{S}^*(A), z^0) \longrightarrow C.(Q'/Q'', \mathcal{S}^*(A), z^0). \qquad (8.2.6)$$

The complexes $C.(X, \mathcal{S}^*(a))$, $C.(Q'/Q'', \mathcal{S}^*(a))$ of the configuration $\mathcal{C}_{n,k}(y)$ have bases formed by the cells $E(f, a)$, $D(f, a)$, resp., where f runs through the set of all monomials admissible to $(x_1, \ldots, x_k; y_1, \ldots, y_n)$, $f \in \text{Adm } (x_1, \ldots, x_k; y_1, \ldots, y_n)$, see (3.12). The corresponding cells of the complexes $C.(X, \mathcal{S}^*(A), z^0)$, $C.(Q'/Q'', \mathcal{S}^*(A), z^0)$ will be denoted by $E(f, A, z^0)$, $D(f, A, z^0)$, resp.

According to (2.2.5), (2.3.3) and (2.3.2), inclusions (8.2.5) are quasiisomorphisms.

The abstract complexes $\mathcal{X}.(\mathcal{C}_{n,k}(y))$, $\mathcal{Q}.(\mathcal{C}_{n,k}(y))$ of the weighted discriminantal configuration $\mathcal{C}_{n,k}(y)$ and the homomorphism

$$S : \mathcal{X}.(\mathcal{C}_{n,k}(y)) \longrightarrow \mathcal{Q}.(\mathcal{C}_{n,k}(y))$$

have been described in (3.14). These objects are determined by the numbers n, k and the weights of $\mathcal{C}_{n,k}(y)$ and do not depend on y.

Let
$$w : \mathcal{X}.(\mathcal{C}_{n,k}(y)) \longrightarrow C.(X, \mathcal{S}^*(a)),$$
$$w : \mathcal{Q}.(\mathcal{C}_{n,k}(y)) \longrightarrow C.(Q'/Q'', \mathcal{S}^*(a))$$

be a realization of $\mathcal{X}.$, $\mathcal{Q}.$, see (2.9). The realization and the identifications (8.2.2), (8.2.3) induce isomorphisms:

$$w : \mathcal{X}.(\mathcal{C}_{n,k}(y)) \longrightarrow C.(X, \mathcal{S}^*(A), z^0),$$
$$w : \mathcal{Q}.(\mathcal{C}_{n,k}(y)) \longrightarrow C.(Q'/Q'', \mathcal{S}^*(A), z^0).$$

These isomorphism will be called the twisted realization of the abstract complexes $\mathcal{X}.(\mathcal{C}_{n,k}(y))$, $\mathcal{Q}.(\mathcal{C}_{n,k}(y))$.

A twisted realization induces canonical isomorphisms

$$w : H.(\mathcal{X}.(\mathcal{C}_{n,k}(y))) \longrightarrow H.(P_{n,k}(z^0) - U_{n,k}(z^0), \mathcal{S}^*(A, z^0)),$$
$$w : H.(\mathcal{Q}.(\mathcal{C}_{n,k}(y))) \longrightarrow H.(P_{n,k}(z^0), U_{n,k}(z^0), \mathcal{S}^*(A, z^0)).$$
(8.2.7)

These isomorphisms do not depend on the choice of a twisted realization.

8.2.8. Lemma. *These canonical isomorphism are horizontal with respect to the Gauss–Manin connection.*

Namely, let $\gamma : [0,1] \to \mathbb{R}^n \setminus \mathbb{R} \cap U_n$ be a curve, and let

$$T_\gamma : H.(P_{n,k}(\gamma(0)) - U_{n,k}(\gamma(0)), \mathcal{S}^*(A, \gamma(0))) \longrightarrow$$
$$H.(P_{n,k}(\gamma(1)) - U_{n,k}(\gamma(1)), \mathcal{S}^*(A, \gamma(1)))$$

be the isomorphism of the Gauss–Manin connection, then the following diagram is commutative:

$$\begin{array}{ccc} H.(\mathcal{X}.(\mathcal{C}_{n,k}(\gamma(0)))) & \xrightarrow{\omega} & H.(P_{n,k}(\gamma(0)) - U_{n,k}(\gamma(0)), \mathcal{S}^*(A, \gamma(0))) \\ \sim \downarrow & & \downarrow T_\gamma \\ H.(\mathcal{X}.(\mathcal{C}_{n,k}(\gamma(1)))) & \xrightarrow{\omega} & H.(P_{n,k}(\gamma(1)) - U_{n,k}(\gamma(1)), \mathcal{S}^*(A, \gamma(1))), \end{array}$$
(8.2.9)

where the left isomorphism is the canonical isomorphism, see (2.9.9) and (3.14.4).

A similar statement holds for the second isomorphism of (8.2.7).

The lemma follows easily from definitions.

Let $U_q\bar{\mathfrak{g}}$ be a quantum group of the form (4.1) such that $U_q\bar{\mathfrak{g}}$ is generated by 1, $\bar{e}_i$, $\bar{f}_i$, $i = 1,\ldots,n+k$, and the space $\bar{\mathfrak{h}}$, and let $\bar{\alpha}_1,\ldots,\bar{\alpha}_{n+k}$ be its simple roots. Let $U_q\bar{\mathfrak{n}}_- \subset U_q\bar{\mathfrak{g}}$ (resp. $^+U_q\bar{\mathfrak{n}}'_-$) be the subalgebra generated by 1, $\bar{f}_1,\ldots,\bar{f}_k$, (resp. $\bar{f}_1,\ldots,\bar{f}_k$), let $\overline{M}_i \subset U_q\bar{\mathfrak{g}}$ be the two-sided submodule over $U_q\bar{\mathfrak{n}}'_-$ generated by $\bar{f}_{k+i}$, $i = 1,\ldots,n$, see (4.5). Set $b_{ij} = (\bar{\alpha}_i, \bar{\alpha}_j)$, $1 \leq i < j \leq n+k$.

Define the weights of $C_{n,k}$ by the formula

$$a_{ij} = b_{ij}/\kappa, \quad 1 \leq i < j \leq n+k, \qquad (8.2.10)$$

where κ, $q = \exp(2\pi i/\kappa)$, is the parameter of the quantum group.

In this case, the complexes $\mathcal{X}.(C_{n,k}(y))$, $Q.(C_{n,k}(y))$ are interpreted in terms of the quantum group $U_q\bar{\mathfrak{g}}$, see (4.6). Namely, let $\delta \in \Sigma_n$ be the permutation defined in (8.2.1), then the following diagram is commutative:

$$\begin{array}{ccc} C.(^+U_q\bar{\mathfrak{n}}'_-;\overline{M}_\delta;2;^+\Delta)_\lambda & \xrightarrow{S} & C.^*(^+U_q\bar{\mathfrak{n}}'_-;\overline{M}_\delta;2;\mu)_\lambda \\ {\scriptstyle p}\downarrow & & \downarrow{\scriptstyle p} \\ \mathcal{X}_{k-}.(C_{n,k}(y)) & \xrightarrow{S} & Q_{k-}.(C_{n,k}(y)), \end{array} \qquad (8.2.11)$$

where $\lambda = (1,\ldots,1) \in \mathbb{N}^{n+k}$, $\overline{M}_\delta = (\overline{M}_{\delta(1)},\ldots,\overline{M}_{\delta(n)})$, p are the isomorphisms defined in (4.6), see (4.6.6).

(8.2.12) The objects $U_q\bar{\mathfrak{g}}$, $^+U_q\bar{\mathfrak{n}}'_-$, $\overline{M}_1$, ..., $\overline{M}_n$, $C.(^+U_q\bar{\mathfrak{n}}'_-;\overline{M}_\delta;2;^+\Delta)_\lambda$, $C.^*(^+U_q\bar{\mathfrak{n}}'_-;\overline{M}_\delta;2;\mu)_\lambda$ will be called associated with the pair $(p_{n,k}, C_{n,k})$. The isomorphisms

$$\begin{aligned} wp : C.(^+U_q\bar{\mathfrak{n}}'_-;\overline{M}_\delta;2;^+\Delta)_\lambda &\longrightarrow C_{k-}.(X, \mathcal{S}^*(A), z), \\ wp : C.^*(^+U_q\bar{\mathfrak{n}}'_-;\overline{M}_\delta;2;\mu)_\lambda &\longrightarrow C_{k-}.(Q'/Q'', \mathcal{S}^*(A), z) \end{aligned} \qquad (8.2.13)$$

will be called the twisted realization of the left complexes.

8.3. Parallel Translations Along Special Curves

Fix $z^0 = (z_1^0,\ldots,z_n^0) \in \mathbb{R}^n$. Then $z_{\delta(1)}^0 < \cdots < z_{\delta(n)}^0$ for some $\delta \in \Sigma_n$. Let $z^u \in \mathbb{R}^n$ be the point obtained from z^0 by the transposition of its $\delta(u)$-th and $\delta(u+1)$-th coordinates, $u = 1,\ldots,n-1$.

For example, for $z^0 = (3,1,2)$, one has $z^1 = (3,2,1)$, $z^2 = (2,1,3)$.

Fix a curve

$$\gamma^u : [0,1] \longrightarrow \mathbb{C}^n \backslash U_n \qquad (8.3.1)$$

such that

(a) $\gamma^u(0) = z^0$, $\gamma^u(1) = z^u$,
(b) The coordinates $z_{\delta(1)}(\gamma(t)), \ldots, z_{\delta(u-1)}(\gamma(t))$, $z_{\delta(u+2)}(\gamma(t)), \ldots, z_{\delta(n)}(\gamma(t))$ do not depend on t,
(c) The complex numbers $z_{\delta(u)}(\gamma^u(t))$, $z_{\delta(u+1)}(\gamma^u(t))$ change their positions as shown in Fig. 8.1.

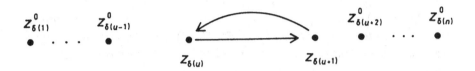

Fig. 8.1

We will describe parallel translations with respect to the Gauss–Manin connection along the curves γ^u, $u = 1, \ldots, n-1$. It is easy to see that knowing these translations, one can determine the whole monodromy representation of the Gauss–Manin connection.

Let $^+U_q\bar{\mathfrak{n}}'_-, \overline{M}_1, \ldots, \overline{M}_n$ be the objects associated with $(p_{n,k}, \mathcal{C}_{n+k})$, see (8.2.16). Set $\overline{M}_\delta = (\overline{M}_{\delta(1)}, \ldots, \overline{M}_{\delta(n)})$, $\overline{M}_{\delta^u} = (\overline{M}_{\delta(1)}, \ldots, \overline{M}_{\delta(u-1)}, \overline{M}_{\delta(u+1)}, \overline{M}_{\delta(u)}, \overline{M}_{\delta(u+2)}, \ldots, \overline{M}_{\delta(n)})$.

Fix real numbers T_j^i, $j = 0, \ldots, n$, $i = 1, \ldots, k$, such that $T_0^1 < \cdots < T_0^k < z_{\delta(1)}^0 < T_1^1 < \cdots < T_1^k < z_{\delta(2)}^0 < \cdots < T_{n-1}^k < z_{\delta(n)}^0 < T_n^1 < \cdots < T_n^k$. $\{T_j^i\}$ is an equipment for $\mathcal{C}_{n,k}(z^0)$.

Let

$$
\begin{aligned}
w &: C_\cdot(^+U_q\bar{\mathfrak{n}}'_-; \overline{M}_\delta; 2; {}^+\Delta)_\lambda \longrightarrow C_{k-\cdot}(X, \mathcal{S}^*(A), z^0), \\
w &: C_\cdot^*(^+U_q\bar{\mathfrak{n}}'_-; \overline{M}_\delta; 2; \mu)_\lambda \longrightarrow C_{k-\cdot}(Q'/Q'', \mathcal{S}^*(A), z^0)
\end{aligned}
\tag{8.3.2}
$$

be the twisted realization defined by the equipment $\{T_j^i\}$ for $\mathcal{C}_{n,k}(z^0)$. Let

$$
\begin{aligned}
w &: C_\cdot(^+U_q\bar{\mathfrak{n}}'_-; \overline{M}_{\delta^u}; 2; {}^+\Delta)_\lambda \longrightarrow C_{k-\cdot}(X, \mathcal{S}^*(A), z^u), \\
w &: C_\cdot^*(^+U_q\bar{\mathfrak{n}}'_-; \overline{M}_{\delta^u}; 2; \mu)_\lambda \longrightarrow C_{k-\cdot}(Q'/Q'', \mathcal{S}^*(A), z^u)
\end{aligned}
\tag{8.3.3}
$$

be the twisted realization defined by the equipment $\{T_j^i\}$ for $\mathcal{C}_{n,k}(z^u)$. In these formulas, $\lambda = (1, \ldots, 1) \in \mathbb{N}^{n+k}$, see (8.2).

178 *Monodromy*

Let $X(z^0)$ be the union of all cells of $C.(X, \mathcal{S}^*(a), z^0)$, $Q(z^0)$ the union of all cells of $C.(Q'/Q'', \mathcal{S}^*(a), z^0)$.

8.3.4. Theorem. *There exist continuous homotopies*

$$T_t^u : X(z^0) \longrightarrow P_{n,k}(\gamma^u(t)) \backslash U_{n,k}(\gamma^u(t)),$$

$$T_{!t}^u : (Q(z^0) \cap U_{n,k}(z^0), Q(z^0)) \longrightarrow (U_{n,k}(\gamma^u(t)), P_{n,k}(\gamma^u(t))),$$

for $t \in [0, 1]$ such that

(i) T_1^u *sends any cell of $C.(X, \mathcal{S}^*(A), z^0)$ onto a union of suitable cells of $C.(X, \mathcal{S}^*(A), z^u)$. $T_{!1}^u$ sends any cell of $C.(Q'/Q'', \mathcal{S}^*(A), z^0)$ onto a union of suitable cells of $C.(Q'/Q'', \mathcal{S}^*(A), z^u)$. Hence, the construction of the Gauss-Manin connection applied to T_t^u, $T_{!1}^u$ gives complex homomorphisms*

$$T^u := T_1^u : C.(X, \mathcal{S}^*(A), z^0) \longrightarrow C.(X, \mathcal{S}^*(A), z^u),$$

$$T^u := T_{!1}^u : C.(Q'/Q'', \mathcal{S}^*(A), z^0) \longrightarrow C.(Q'/Q'', \mathcal{S}^*(A), z^u).$$

(ii) $T^u w = w R^u$, *where R^u is defined in (7.4.23) and (7.4.25).*

8.3.5. Corollary. *The following diagrams are commutative.*

$$
\begin{array}{ccc}
H.C.(^+U_q \bar{\mathfrak{n}}'_-; \overline{M}_\delta; 2; ^+\Delta)_\lambda & \xrightarrow{R^u} & H.C.(^+U_q \bar{\mathfrak{n}}'_-; \overline{M}_{\delta^u}; 2; ^+\Delta)_\lambda \\
{\scriptstyle w} \downarrow & & \downarrow {\scriptstyle w} \\
H_{k-.}(P_{n,k}(z^0) \backslash U_{n,k}(z^0), \mathcal{S}^*(A, z^0)) & \xrightarrow{T^u} & H_{k-.}(P_{n,k}(z^u) \backslash U_{n,k}(z^u), \mathcal{S}^*(A, z^u)),
\end{array}
$$

$$
\begin{array}{ccc}
H.C.^*(^+U_q \bar{\mathfrak{n}}'_-; \overline{M}_\delta; 2; \mu)_\lambda & \xrightarrow{R^u} & H.C.^*(^+U_q \bar{\mathfrak{n}}'_-; \overline{M}_{\delta^u}; 2; \mu)_\lambda \\
{\scriptstyle w} \downarrow & & \downarrow {\scriptstyle w} \\
H_{k-.}(P_{n,k}(z^0), U_{n,k}(z^0), \mathcal{S}^*(A, z^0)) & \xrightarrow{T^u} & H_{k-.}(P_{n,k}(z^u), U_{n,k}(z^u), \mathcal{S}^*(A, z^u)),
\end{array}
$$

where T^u is the isomorphism of parallel translation along γ^u and $\lambda = (1, \ldots, 1) \in \mathbb{N}^{n+k}$.

Theorem (8.3.4) is proved in (8.8) and (8.12).

8.4. Isotopy of the Real Line

In Subsecs. (8.4)–(8.11), we will construct $T_{!t}^u$ and prove Theorem (8.3.4) for $T_{!t}^u$.

To simplify notations, we will write T_t instead of T^u_{lt}. We will assume that $z^0 = (z_1^0, \ldots, z_n^0)$ is such that $z_1^0 < \cdots < z_n^0$. Hence $\delta = (1, \ldots, n)$, $\delta^u = (1, \ldots, u-1, u+1, u, u+2, \ldots, n)$. We fix u.

In this section, we construct an auxiliary isotopy of the real line. Namely, define a continuous isotopy $h_t : \mathbb{R} \to \mathbb{C}$ for $t \in [0,1]$ as shown in Fig. 8.2.

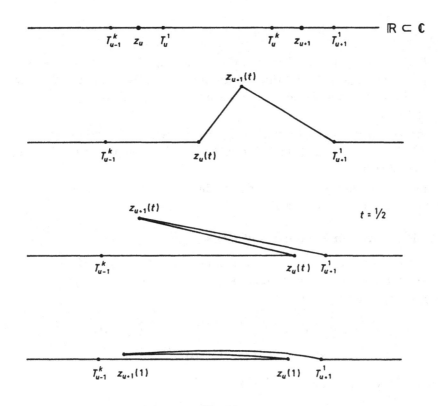

Fig. 8.2

Here $z_j(t) := z_j(\gamma^u(t))$, $j = u, u+1$, and we set

$$h_t : z_u^0 \mapsto z_u(t), \qquad z_{u+1}^0 \mapsto z_{u+1}(t). \tag{8.4.1}$$

h_t to be the identity on $(-\infty, T^k_{u-1}]$ and $[T^1_{u+1}, \infty)$. (8.4.2)

h_t to be linear on $[T^k_{u-1}, z_u^0]$, $[z_u^0, z_{u+1}^0]$, $[z_{u+1}^0, T^1_{u+1}]$. (8.4.3)

Set $A_t := h_t(\mathbb{R}) \subset \mathbb{C}$.

(8.4.4) We will assume that $\operatorname{Im} z_{u+1}(t) > 0$, $\operatorname{Im} z_u(t) = 0$, for $t \in (0,1)$.

(8.4.5) Define an isotopy $h_t^0 : [z_u^0, z_{u+1}^0] \to \mathbb{C}$ for $t \in [0,1]$ by the formula $h_t^0 = h_t \circ (h_1|_{[z_u^0, z_{u+1}^0]})^{-1}$.

Recall that by construction the map $h_t|_{[z_u^0, z_{u+1}^0]} : [z_u^0, z_{u+1}^0] \to \mathbb{C}$ is a homomorphism onto its image and $h_1([z_u^0, z_{u+1}^0]) = [z_u^0, z_{u+1}^0]$.

We have

(i) $h_1^0 = \text{id}$,

(ii) $h_0^0(x) = -x$ for all $x \in [z_u^0, z_{u+1}^0]$.

Each fiber $P_{n,k}$ is identified with $\mathbb{C}^k$ by the map $(t_{n+1}, \ldots, t_{n+k}) \mapsto (x_1 = t_{n+1}, \ldots, x_k = t_{n+k})$. Under this identification, $Q(z^0)$ is a subset of $\mathbb{R}^k \subset \mathbb{C}^k$. We will construct the desired deformation T_t such that $T_t(Q(z^0)) \subset (A_t)^k \subset \mathbb{C}^k$, $t \in [0, 1]$.

We will construct the deformation T_t on each cell of the complex $C.(Q'/Q'', S^*(A), z^0)$ separately, and these deformations will be compatible on the intersections of cells.

In the next three sections, we give auxiliary statements that are necessary to construct the deformation T_t.

8.5. Factorization Properties of Cells

Let $f \in \text{Adm}(x; z)$, and let f have form (3.8.1). Then $f = f(1)|\ldots|f(p)$, where $\{f(i)\}$ are its tensor factors, see (3.8). Assume that $z_u \in f(a)$, $z_{u+1} \in f(b)$, where $1 \leq a < b \leq p$. Let

$$f(a) = g_1(a) \ldots g_{\alpha(a)}(a) z_u g_{\alpha(a)+1}(a) \ldots g_{\beta(a)}(a),$$
$$f(b) = g_1(b) \ldots g_{\alpha(b)}(b) z_{u+1} g_{\alpha(b)+1}(b) \ldots g_{\beta(b)}(b),$$

where $g_i(j) \in \{x_1, \ldots, x_k\}$ for all i, j. Introduce

$$f^l = f(1)|\ldots|f(a-1), \quad f^r = f(b+1)|\ldots|f(p),$$
$$f^m = z_u g_{\alpha(a)+1}(a) \ldots g_{\beta(a)}(a)|f(a+1)|\ldots|f(b-1)|g_1(b) \ldots g_{\alpha(b)}(b) z_{u+1}$$
$$f^{lm} = g_1(a) \ldots g_{\alpha(a)}(a), \quad f^{mr} = g_{\alpha(b)+1}(b) \ldots g_{\beta(b)}(b),$$

then

$$f = f^l | f^{lm} f^m f^{mr} | f^r. \tag{8.5.1}$$

Introduce x^l (resp. x^r, x^m, x^{lm}, x^{mr}) as the subset of $x = \{x_1, \ldots, x_k\}$ consisting of all x_j such that x_j belongs to f^l (resp. f^r, f^m, f^{lm}, f^{mr}). Set $k^l = \#x^l$, $k^r = \#x^r$ and so on.

Consider x^l, x^r, x^m, x^{lm}, x^{mr} as coordinates in $\mathbb{R}^{k^l}$, $\mathbb{R}^{k^r}$, $\mathbb{R}^{k^m}$, $\mathbb{R}^{k^{lm}}$, $\mathbb{R}^{k^{mr}}$, resp.

Let $\{T_j^i\}$ be the equipment of $\mathcal{C}_{n,k}(z^0)$ fixed in Theorem (8.3.4).

Let $\{T_j^i(l)\}$ be the equipment of the configuration $\mathcal{C}_{u-1,k^l}(z_1^0,\ldots,z_{n-1}^0)$ in $\mathbb{R}^{k^l}$ such that

$$T_j^i(l) = T_j^i \text{ for } j = 0,\ldots,n-1 \text{ and } i = 1,\ldots,k^l. \tag{8.5.2}$$

Let $T_{u-1}^1(lm) < \cdots < T_{u-1}^{k^{lm}}(lm) < z_u < T_u^1(lm) < \cdots < T_u^{k^{lm}}(lm)$ be any equipment of the configuration $\mathcal{C}_{1,k^{lm}}(z_u^0)$ in $\mathbb{R}^{k^{lm}}$ such that

$$T_{u-1}^i(lm) = T_{u-1}^{k^l+i} \text{ for all } i. \tag{8.5.3}$$

Let $T_{u-1}^1(m) < \cdots < T_{u-1}^{k^m}(m) < z_u^0 < T_u^1(m) < \cdots < T_u^{k^m}(m) < z_{u+1}^0 < T_{u+1}^1(m) < \cdots < T_{u+1}^{k^m}(m)$ be any equipment of the configuration $\mathcal{C}_{2,k^m}(z_u^0, z_{u+1}^0)$ in $\mathbb{R}^{k^m}$ such that

$$T_u^i(m) = T_u^{k^l+k^{lm}+i} \text{ for all } i. \tag{8.5.4}$$

Let $T_u^1(mr) < \cdots < T_u^{k^{mr}}(mr) < z_{u+1}^0 < T_{u+1}^1(mr) < \cdots < T_u^{k^{mr}}(mr)$ be any equipment of the configuration $\mathcal{C}_{1,k^{mr}}(z_{u+1}^0)$ such that

$$T_{u+1}^i(mr) = T_{u+1}^{k^l+k^{lm}+k^m+i} \text{ for all } i. \tag{8.5.5}$$

Let $T_{u+1}^1(r) < \cdots < T_{u+1}^{k^r} < z_{u+2}^0 < \cdots < z_n^0 < T_n^1(r) < \cdots < T_n^{k^r}(r)$ be the equipment of the configuration $\mathcal{C}_{n-u-1}(z_{u+2}^0,\ldots,z_n^0)$ such that

$$T_j^i(r) = T_j^{k^l+k^{lm}+k^m+k^{mr}+i} \text{ for all } i,j. \tag{8.5.6}$$

Let

$$T_{u-1}^1(M) < \cdots < T_{u-1}^{k^{lm}+k^m+k^{mr}}(M) < z_u^0 < T_u^1(M) < \cdots < T_u^{k^{lm}+k^m+k^{mr}}(M)$$

$$< z_{u+1}^0 < T_{u+1}^1(M) < \cdots < T_{u+1}^{k^{lm}+k^m+k^{mr}}(M)$$

be the equipment of the configuration $\mathcal{C}_{2,k^{lm}+k^m+k^{mr}}(z_u^0, z_{u+1}^0)$ such that

$$T_j^i(M) = T_j^{k^l+i} \text{ for all } i,j. \tag{8.5.7}$$

Let $F^{lm} = f^{lm}z_u$, $F^{mr} = z_{u+1}f^{mr}$, $F^m = f^{lm}f^m f^{mr}$. Then $f^l \in$ Adm$(x^l; z_1, \ldots, z_{u-1})$, $F^{lm} \in$ Adm$(x^{lm}; z_u)$, $f^m \in$ Adm$(x^m; z_u, z_{u+1})$, $F^m \in$ Adm$(x^{lm} \cup x^m \cup x^{mr}; z_u, z_{u+1})$, $F^{mr} \in$ Adm$(x^{mr}; z_{u+1})$, $f^r \in$ Adm$(x^r; z_{u+2}, \ldots, z_n)$.

Let $D(f; \{T_j^i\})$ (resp. $D(f^l; T_j^i(l))$, $D(F^{lm}; \{T_j^i(lm)\})$, $D(f^m; \{T_j^i(m)\})$, $D(F^{mr}; \{T_j^i(mr)\})$, $D(f^r; \{T_j^i(r)\})$, $D(F^m; \{T_j^i(M)\})$) be the cell corresponding to f (resp. f^l, F^{lm}, f^m, F^{mr}, f^r, F^m) of the complex Q'/Q'' of the configuration $C_{n,k}(z^0)$ (resp. $C_{u-1,k^l}(z_1^0, \ldots, z_{u-1}^0)$, $C_{1,k^{lm}}(z_u^0)$, $C_{2,k^m}(z_u^0, z_{u+1}^0)$, $C_{1,k^{mr}}(z_{u+1}^0)$, $C_{n-u-1,k^r}(z_{u+2}^0, \ldots, z_n^0)$, $C_{2,k^{lm}+k^m+k^{mr}}(z_u^0, z_{u+1}^0)$)) equipped with $\{T_j^i\}$ (resp. $\{T_j^i(l)\}$, $\{T_j^i(lm)\}$, $\{T_j^i(m)\}$, $\{T_j^i(mr)\}$, $\{T_j^i(r)\}$, $\{T_j^i(M)\}$). These cells are convex polytopes in $\mathbb{R}^k$, $\mathbb{R}^{k^l}$, $\mathbb{R}^{k^{lm}}$, $\mathbb{R}^{k^m}$, $\mathbb{R}^{k^{mr}}$, $\mathbb{R}^{k^r}$, $\mathbb{R}^{k^{lm}+k^m+k^{mr}}$, respectively. It is easy to see that these polytopes do not depend on the choice of equipment with properties (8.5.2)–(8.5.7).

8.5.8. Lemma. *The direct products* $\mathbb{R}^k = \mathbb{R}^{k^l} \times \mathbb{R}^{k^{lm}} \times \mathbb{R}^{k^m} \times \mathbb{R}^{k^{mr}} \times \mathbb{R}^{k^r}$, $\mathbb{R}^{k^{lm}+k^m+k^{mr}} = \mathbb{R}^{k^{lm}} \times \mathbb{R}^{k^m} \times \mathbb{R}^{k^r}$ *induce the equalities*

$$D(f, \{T_j^i\}) = D(f^l; \{T_j^i(l)\}) \times D(F^{lm}; \{T_j^i(lm)\}) \times D(f^m; \{T_j^i(m)\})$$
$$\times D(F^{mr}; \{T_j^i(mr)\}) \times D(f^r; \{T_j^i(r)\}),$$
$$D(F^m; \{T_j^i(M)\}) = D(F^{lm}; \{T_j^i(lm)\}) \times D(f^m; \{T_j^i(m)\})$$
$$\times D(F^{mr}; \{T_j^i(mr)\}).$$

The lemma is a corollary of (3.9).

8.5.9. Remark. According to (3.9),

$$D(F^{lm}; \{T_j^i(lm)\}) = \{x^{lm} \in \mathbb{R}^{k^{lm}} | g_1(a) \leq g_2(a) \leq \cdots \leq g_{\alpha(a)}(a) \leq z_u^0$$
$$\text{and} \sum_{i=1}^p g_i(a) \geq \sum_{i=1}^p T_{u-1}^i(lm) \text{ for } p = 1, \ldots, \alpha(a)\},$$

where, as before, $x^{lm} = \{g_1(a), \ldots, g_{\alpha(a)}(a)\}$, $k^{lm} = \alpha(a)$.

$$D(F^{mr}; \{T_j^i(mr)\}) = \{x^{mr} \in \mathbb{R}^{k^{mr}} | z_{u+1}^0 \leq g_{\alpha(b)+1}(b) \leq g_{\alpha(b)+2}(b)$$
$$\leq \cdots \leq g_{\beta(b)}(b)$$
$$\text{and} \sum_{i=1}^p g_{\beta(b)-i+1}(b) \leq \sum_{i=1}^p T_{u+1}^{k^{mr}-i+1}(mr) \text{ for } p = 1, \ldots, \beta(b) - \alpha(b)\},$$

where, as before, $x^{mr} = \{g_{\alpha(b)+1}(b), \ldots, g_{\beta(b)}(b)\}$, $k^{mr} = \beta(b) - \alpha(b)$.

8.6. Involution

The involution $\phi : \mathbb{R}^{k^m} \to \mathbb{R}^{k^m}$, $x_j \mapsto (z_u^0 + z_{u+1}^0) - x_j$ for $x_j \in x^{k^m}$, preserves $C_{2,k^m}(z_u^0, z_{u+1}^0)$. Introduce a new equipment $\{T_j^i(m)'\}$ for this configuration. Set $T_j^i(m)' = T_j^i(m)$ for $j = u-1, u+1$ and all i; set $T_u^i(m)' = (z_u^0 + z_{u+1}^0) - T_u^{k^m - i + 1}(u)$ for all i.

Set

$$\bar{f}^m = z_u g_{\alpha(b)}(b) g_{\alpha(b)-1}(b) \ldots g_1(b) |\overline{f(b-1)}| \overline{f(b-2)}| \ldots |\overline{f(a+1)}|$$
$$\times g_{\beta(a)}(a) g_{\beta(a)-1}(a) \ldots g_{\alpha(a)+1}(a) z_{u+1},$$

where $\overline{f(i)} = f_{q(i)}(i) f_{q(i)-1}(i) \ldots f_1(i)$, for $i = a+1, \ldots, b-1$, see notations in (3.8). By definition, $\bar{f}^m \in \mathrm{Adm}\,(x^m; z_u, z_{u+1})$.

8.6.1. Lemma. $\phi(D(f^m; \{T_j^i(m)\})) = D(\bar{f}^m; \{T_j^i(m)'\})$.

The lemma is a corollary of (3.9).

8.7. Bundle Property

Consider the configuration $C_{2,k^{lm}+k^m+k^{mr}}(z_u^0, z_{u+1}^0)$ in $\mathbb{R}^{k^{lm}+k^m+k^{mr}}$ equipped with $\{T_j^i(m)\}$ and the natural projection $\pi : \mathbb{R}^{k^{lm}+k^m+k^{mr}} \to \mathbb{R}^{k^{lm}+k^{mr}}$, $\{x^{lm} \cup x^m \cup x^{mr}\} \mapsto \{x^{lm} \cup x^{mr}\}$. We apply Theorem (6.2.4) to this pair.

The theorem states that there exists a continuous family of equipment of configuration $C_{2,k^m}(z_u^0, z_{u+1}^0)$,

$$T_{u-1}^1(y) < \cdots < T_{u-1}^{k^m}(y) < z_u^0 < T_u^1(y) < \cdots < T_u^{k^m}(y) < z_{u+1}^0 < T_{u+1}^1(y)$$
$$< \cdots < T_{u+1}^{k^m}(y),$$

with property (6.2.4.ii).

We apply the theorem to the monomial $\bar{f}^m \in \mathrm{Adm}\,(x^m; z_u, z_{u+1})$. Namely, let

$$\Delta^l = \{x^{lm} \in \mathbb{R}^{k^{lm}} \mid g_1(a) \leq g_2(a) \leq \cdots \leq g_{\alpha(a)}(a) \leq z_{u+1}^0$$
$$\text{and } \sum_{i=1}^p g_i(a) \geq \sum_{i=1}^p T_{u-1}^i(M) \text{ for } p = 1, \ldots, \alpha(a)\},$$

$$\Delta^r = \{x^{mr} \in \mathbb{R}^{k^{mr}} | z_u^0 \leq g_{\alpha(b)+1}(b) \leq g_{\alpha(b)+2}(b) \cdots \leq g_{\beta(b)}(b)$$

$$\text{and } \sum_{i=1}^{p} g_{\beta(b)-i+1}(b) \leq \sum_{i=1}^{p} T_{u+1}^{k^{lm}+k^m+k^{rm}-i+1}(M) \qquad (8.7.1)$$

$$\text{for } p = 1, \ldots, \beta(b) - \alpha(b)\}.$$

Note that

$$\sum_{i=1}^{p} T_{u-1}^{i}(M) = \sum_{i=1}^{p} T_{u-1}^{i}(lm) \text{ for } p = 1, \ldots, \alpha(a),$$

$$\sum_{i=1}^{p} T_{u+1}^{k^{lm}+k^m+k^{rm}-i+1}(M) = \sum_{i=1}^{p} T_{u+1}^{k^{mr}-i+1}(mr) \qquad (8.7.2)$$

$$\text{for } p = 1, \ldots, \beta(b) - \alpha(b),$$

see (8.5.3), (8.5.5) and (8.5.7). Cf. (8.5.9).

Now let $L(\bar{f}^m) \subset \mathbb{R}^{k^{lm}+k^m+k^{mr}}$ be the leaf of $\bar{f}^m$, see (6.2). By Theorem (6.2.4), for any $y \in \Delta^l \times \Delta^r \in \mathbb{R}^{k^{lm}+k^{mr}}$, we have

$$\pi^{-1}(y) \cap L(\bar{f}^m) = D(\bar{f}^m; \{T_j^i(y)\}). \qquad (8.7.3)$$

8.7.4. Lemma.

$$\pi^{-1}(\Delta^l \times \Delta^r) \cap L(\bar{f}^m) = \cup D(F; \{T_j^i(M)\})$$

where $D(F; \{T_j^i(M)\})$ is a cell of the complex Q'/Q'' of $C_{2,k^{lm}+k^m+k^{mr}}(z_u^0, z_{u+1}^0)$ equipped with $\{T_j^i(M)\}$, the sum is over all $F \in \text{Adm}(x^{lm} \cup x^m \cup x^{mr}; z_u, z_{u+1})$ such that

(i) F is subordinated to $\bar{f}^m$,
(ii) the order of the symbols of x^{lm} in F is the same as their order in f^{lm},
(iii) the order of the symbols of x^{mr} in F is the same as their order in f^{mr},
(iv) the symbols of x^{lm} are on the left of z_{u+1} in F, and the symbols of x^{mr} are on the right of z_u in F.

Proof. It suffices to check that vertices of any polytope of the sum are projected into $\Delta^l \times \Delta^r$ by π, and this is obvious.

8.7.5. Remark. Set $D_{l,r} := D(F^{lm}; \{T_j^i(lm)\}) \times D(F^{mr}; \{T_j^i(mr)\})$. We have $D_{l,r} \subset \Delta^l \times \Delta^r$, see (8.7.2). It is easy to see that $\pi^{-1}(D_{l,r}) \cap L(\bar{f}^m) =$

$D(f^{lm}\bar{f}^m f^{mr}; \{T^i_j(M)\})$, and, consequently, $T^i_u(y) = T^{k^{lm}+i}_u(M) = T^i_u(m) = T^{k^l+k^m+i}_u$ for $y \in \Delta_{l,r}$ and $i = 1, \ldots, k^m$.

Let

$$h^b(y) : D(\bar{f}^m; \{T^i_j(m)'\}) \longrightarrow D(\bar{f}^m; \{T^i_j(y)\}) \subset \mathbb{R}^{k^m} \qquad (8.7.6)$$

be the canonical piecewise linear homeomorphism, where $y \in \Delta^l \times \Delta^r$.

For $t \in [0,1]$ and $y \in \Delta^l \times \Delta^r$, define a continuous family of maps

$$h^b(y,t) : D(\bar{f}^m; \{T^i_j(m)'\}) \longrightarrow \mathbb{R}^{k^m} \qquad (8.7.7)$$

by the formula $h^b(y,t) = t\, h^b(y) + (1-t)\,\mathrm{id}$. For fixed y and t, the map $h^b(y,t)$ is a homeomorphism of $D(\bar{f}^m; \{T^i_j(m)'\})$ onto its image.

8.8. Construction of T_t

Let f be any monomial of $\mathrm{Adm}\,(x;z)$ as in Secs. 8.5–8.7. Define the deformation

$$T_t : D(f; \{T^i_j\}) \longrightarrow \mathbb{C}^k \text{ for } t \in [0,1] \qquad (8.8.1)$$

as follows. Let $y = (y_1, \ldots, y_k) \in D(f; \{T^i_j\})$. Set $T_t(y) = (T_{1,t}(y), \ldots, T_{k,t}(y)) \in \mathbb{C}^k$. Set

$$\begin{aligned} T_{j,t}(y) &= y_j && \text{if } x_j \in x^l \cup x^r, \\ T_{j,t}(y) &= h_t(y_j) && \text{if } x_j \in x^{lm} \cup x^{mr}. \end{aligned} \qquad (8.8.2)$$

Now define all other coordinates $\{T_{j,t}(y)\}$, where j runs through all indices such that $x_j \in x^m$.

Set $y^m = \{y_j | x_j \in x^m\}$, $y^{l,r} = \{y_j | x_j \in x^{lm} \cup x^{mr}\}$. y^m is a point of $\mathbb{C}^{k^m}$, $y^{l,r}$ is a point of $\mathbb{C}^{k^{lm}+k^{mr}}$.

For $t \in [0,1]$, define a map $H^m_t : [z^0_u, z^0_{u+1}]^{k^m} \to \mathbb{C}^{k^m}$ by $\{x_j\} \mapsto \{h^0_t(x_j)\}$, where h^0_t is defined in (8.4.5).

For $t \in [0,1]$, define a map $H^{l,r} : \mathbb{R}^{k^{lm}+k^{mr}} \to \mathbb{R}^{k^{lm}+k^{mr}}$ by $\{x_j\} \mapsto \{h_1(x_j)\}$, where $h_{t=1}$ is defined in (8.4.1)–(8.4.3).

Set

$$T_t(y)^m = H^m_t \circ h^b(H^{l,r}(y^{l,r}), t) \circ \phi(y^m)\,. \qquad (8.8.3)$$

Here ϕ is defined in (8.6), $h^b(y,t)$ is defined in (8.7.7).

If y runs through $D(f;\{T_j^i\})$, then $y^{l,r}$ runs through $D(F^{lm};\{T_j^i(lm)\}) \times D(F^{mr};\{T_j^i(mr)\})$, see (8.5.8). $H^{l,r}(y^{l,r})$ runs through $\Delta^l \times \Delta^r$, see (8.4.1), (8.4.3), (8.5.9), and (8.7.1). Hence $h^b(H^{l,r}(y^{l,r}),t)$ is well defined.

Now all coordinates $T_{j,t}(y)$ are defined, the deformation T_t is defined.

8.8.4. Lemma. $T_0(y)^m = y^m$.

Proof. $H_0^m = \phi$, $h^b(H_1^{l,r}(y^{l,r}),0) = \text{id}$. $T_0(y)^m = \phi \circ \phi(y^m) = y^m$.

8.8.5. Lemma. $T_1(D(f;\{T_j^i\})) = \cup D(F;\{T_j^i\})$, where the sum is taken over all $F = F(1)|\ldots|F(p) \in \text{Adm}(x,z)$ such that

(i) F is subordinate to $\bar{f}^m$,
(ii) $F(1)|\ldots|F(a-1) = f^l$, $F(b+1)|\ldots|F(p) = f^r$,
(iii) the order of symbols of x^{lm} in F is the same as their order in f^{lm},
(iv) the order of symbols of x^{mr} in F is the same as their order in f^{mr},
(v) the symbols of x^{lm} are on the left of z_{u+1} in F, and the symbols of x^{mr} are on the right of z_u in F.

Remark. According to (ii), symbols of x^{lm} and x^{mr} are sitting in $F(a)|\ldots|F(b)$.

Proof. $T_1(y)^m = h^b(H^{l,r}(y^{l,r})) \circ \phi(y^m)$. The lemma follows from (8.6.1), (8.7.4), and (8.5.8).

8.9. Lemma.

The deformation T_t defined on each cell separately is compatible on intersections of cells.

Proof. It suffices to check the following two cases.

Case 1. Let $f, f' \in \text{Adm}(x_1,\ldots,x_k;z_1,\ldots,z_n)$, and let f' be adjacent to f, $f' < f$, see (3.8). In this case, $D(f',\{T_j^i\}) \subset \partial D(f,\{T_j^i\})$. The lemma states that T_t, constructed on $D(f,\{T_j^i\})$ and then restricted to $D(f',\{T_j^i\})$, coincides with T_t constructed on $D(f',\{T_j^i\})$.

Case 2. Let $f, f' \in \text{Adm}(x_1,\ldots,x_k;z_1,\ldots,z_n)$, and let f' be obtained from f by a transposition of two neighboring symbols in one of its tensor factors. For example: $f = z_1 f_1 | z_2 f_2 f_3$, $f' = f_1 z_1 | z_2 f_2 f_3$ or $f' = z_1 f_1 | z_2 f_3 f_2$. The lemma states that T_t constructed separately on $D(f,\{T_j^i\})$, $D(f',\{T_j^i\})$ coincide on their intersection.

Proof of Case 1. f' is obtained from f by an insertion of one vertical line. The line may be inserted in one of f^l, f^{lm}, f^m, f^{mr}, f^r.

(a) Insertion in f^l or f^r. The lemma is obvious.

(b) Insertion in f^m. The lemma easily follows from Theorem (6.2.4).

(c) Insertion in f^{lm}. (Insertion in f^{mr} is similar.)

Let $f = f^l | f^{lm} f^m f^{mr} | f^r$ as in (8.5). Let f' have the form $f' = f^l | f_1^{lm} | f_2^{lm} f^m f^{mr} | f^r$, where $f_1^{lm} = g_1(a) \ldots g_\gamma(a)$, $f_2^{lm} = g_{\gamma+1}(a) \ldots g_{\alpha(a)}(a)$ for some γ, $1 < \gamma \le \alpha(a)$. Thus $f^{lm} = f_1^{lm} f_2^{lm}$. Set $x_1^{lm} = \{g_1(a), \ldots, g_\gamma(a)\}$, $x_2^{lm} = \{g_{\gamma+1}(a), \ldots, g_\alpha(a)\}$. $k_1^{lm} = \gamma$, $k_2^{lm} = \alpha(a) - \gamma$.

By (3.9), we have

$$D(f', \{T_j^i\}) = D(f^l, \{T_j^i(l)\}) \times \Delta_1^{lm} \times \Delta_2^{lm} \times D(f^m; \{T_j^i(m)\}) \times D(F^{mr}; \{T_j^i(mr)\}) \times D(f^r; \{T_j^i(r)\}),$$

where we use the notations of (8.5) and

$$\Delta_1^{lm} = \Big\{ x_1^{lm} \in \mathbb{R}^{k_1^{lm}} \mid \sum_{i=1}^{\gamma} g_i(a) = \sum_{i=1}^{\gamma} T_{u-1}^{k^l+i}, g_1(a) \le g_2(a) \le \cdots \le g_\gamma(a),$$

$$\text{and } \sum_{i=1}^{p} g_i(a) \ge \sum_{i=1}^{p} T_{u-1}^{k^l+i} \text{ for } p = 1, \ldots, \gamma - 1 \Big\},$$

$$\Delta_2^{lm} = \Big\{ x_2^{lm} \in \mathbb{R}^{k_2^{lm}} \mid g_\gamma(a) \le g_{\gamma+1}(a) \le \cdots \le \gamma_{\beta(a)}(a) \le z_u^0$$

$$\text{and } \sum_{i=1}^{p} g_{\gamma+i}(a) \ge \sum_{i=1}^{p} T_{u-1}^{k^l+k_1^{lm}+i} \text{ for } p = 1, \ldots, \alpha(a) - \gamma \Big\}.$$

We have $\Delta_1^{lm} \times \Delta_2^{lm} \subset D(F^{lm}; \{T_j^i(lm)\})$, see (8.5).

Constructing T_t on $D(f, \{T_j^i(y)\})$, we apply Theorem (6.2.4) to $\pi: R^{k^{lm}+k^m+k^{mr}} \to \mathbb{R}^{k^{lm}+k^{mr}}$. Constructing T_t on $D(f', \{T_j^i\})$, we apply Theorem (6.2.4) to $\pi': \mathbb{R}^{k_2^{lm}+k^m+k^{mr}} \to R^{k_2^{lm}+k^{mr}}$.

For the first case, the theorem states the existence of functions $T_j^i(x^{lm} \cup x^{mr}) = T_j^i(x_1^{lm} \cup x_2^{lm} \cup x^{mr})$ with property (6.2.4.ii). For the second case, the theorem states the existence of functions $\bar{T}_j^i(x_2^{lm} \cup x^{mr})$ with property (6.2.4.ii).

It is easy to see from the construction in Theorem (6.2.4) that $T_u^i(x_1^{lm} \cup x_2^{lm} \cup x^{mr}) = \bar{T}_u^i(x_2^{lm} \cup x^{mr})$ for $x_1^{lm} \in \Delta_1^{lm}$ and all i. This equality implies the required property of the deformation T_t, see (8.8.3), (8.7.7) and (8.7.6).

Proof of Case 2. If f' is obtained from f by transposition of two neighboring symbols of $\{x_1,\ldots,x_k\}$, the required property of T_t follows from Theorem (6.2.4).

If f' is obtained from f by transposition of some neighboring symbols x_j, z_i and $i \neq u, u+1$, then the required property of T_t is obvious. If i equals u or $u+1$, then the verification of the required property of T_t is similar to Case 1.c above.

8.10. Example. Let $k = 2$, $n = 2$. Consider $T^1_{!t} = T_t$. Cells $D(f) = D(f, \{T_j^i\})$ for some $f \in \text{Adm}(x_1, x_2; z_1, z_2)$ are shown in Fig. 8.3. By (8.85), $T_1(D(f)) = \cup D(F)$, where F runs through $z_1|x_1x_2z_2$, $z_1x_1|x_2z_2$, $x_1z_1|x_2z_2$, $z_1x_1x_2|z_2$, $x_1z_1x_2|z_2$, $x_1x_2z_1|z_2$ for $f = x_1x_2z_1|z_2$; F runs through $z_1|x_2x_1z_2$, $z_1|x_1x_2z_2$, $z_1x_1|x_2z_2$, $x_1z_1|x_2z_2$ for $f = x_1z_1x_2|z_2$, F runs through $z_1x_2|x_1z_2$, $z_1x_2x_1|z_2$, $z_1x_1x_2|z_2$, $x_1z_1x_2|z_2$, $z_1|x_2x_1z_2$, $z_1|x_1x_2z_2$, $z_1x_1|x_2z_2$, $x_1z_1|x_2z_2$, $z_1|x_1z_2x_2$, $z_1x_1|z_2x_2$, $x_1z_1|z_2x_2$ for $f = x_1z_1|x_2z_2$; F runs through $x_1|z_1|x_2z_3$ for $f = x_1|z_1x_2|z_2$; F runs through $z_1|x_2|x_1z_2$, $z_1|x_2x_1|z_2$, $z_1|x_1x_2|z_2$, $z_1x_1|x_2|z_2$, $x_1z_1|x_2|z_2$ for $f = x_1z_1|x_2|z_2$, see Fig. 8.3.

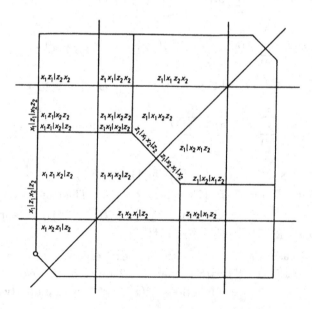

Fig. 8.3

8.11. Computation of the Action of $T_1 : C.(Q'/Q'', S^*(A), z^0) \to$ $C.(Q'/Q'', S^*(A), z^0)$ for the Isotropy T_t Constructed in (8.8)

The cells of these complexes are $\{D(f, A, z^0)\}$, $\{D(f, A, z^u)\}$, resp. Here, $f \in \text{Adm }(x, z)$. Each cell is a convex polytope $D(f, \{T_j^i\})$ in $\mathbb{R}^k$ with a section of the corresponding local system over this polytope.

To compute the action of T_1, we start with a distinguished section over $D(f, \{T_j^i\})$ of the local system $S^*(A, z^0)$, then horizontally continue it to a section s of $S^*(A)$ over $\bigcup_{t \in [0,1]} T_t(D(f, \{T_j^i\}))$ and then compare the restriction $s|_{T_1(D(f; \{T_j^i\}))}$ with the distinguished section of $S^*(A, z^u)$ over $T_1(D(f; \{T_j^i\}))$.

These procedure and Lemma (8.8.5) lead to formula

$$T_1(D(f; A, z^0)) = \sum_F C_{f,F} D(f, A, z^u) \qquad (8.11.1)$$

where the sum is taken over all $F \in \text{Adm }(x; z)$ satisfying (8.8.5.i)–(8.8.5.v), and $C_{f,F}$ are some complex numbers.

8.11.2. Lemma. $C_{f,F} = (-1)^{k^m + (b-a)(b-a-1)/2} q^{A_u + A_{u+1} + B + b_{k+u,k+u+1}/2}$.

Here $\{b_{ij}/\kappa\}$ are weights of $C_{n+k} \subset \mathbb{C}^{n+k}$, see (8.2.10),

$$A_{u+1} = \sum b_{j,k+u+1}/2,$$

where the sum is taken over all j such that $x_j \in x^{lm} \cup x^m$ and x_j stands on the right of z_u in F;

$$A_u = \sum b_{j,k+u}/2,$$

where the sum is taken over all j such that $x_j \in x^m \cup x^{mr}$ and x_j stands on the left of z_{u+1} in F;

$$B = \sum b_{ij}/2,$$

where the sum is taken over all pairs $(x_i, x_j) \subset x^{lm} \cup x^m \cup x^{mr}$ such that x_i stands on the left of x_j in f and stands on right of x_j in F.

Proof. The isotopy h_t shown in Fig. 8.2 reverses the orientation of the coordinates on $D(f)$ that run inside $[z_u^0, z_{u+1}^0]$. Comparison of the distinguished orientations of $D(f)$ and $D(F)$ defined in (3.10) gives the sign $(-1)^{k^m + \frac{(b-a)(b-a-1)}{2}}$ in the formula.

$D(f, A, z^0)$ is the polytope $D(f, \{T_j^i\})$ with the distinguished section over it. The section is defined in (2.5.7), (2.5.8), (8.2.3) and (8.2.4). It is determined by the choice of arguments of all functions $t_i - t_j$, $1 \leq i < j \leq k+n$, on $D(f, \{T_j^i\})$. Under the deformation T_t for $t \in [0,1]$, some $\arg(t_i - t_j)$ increase by π. Each such a change gives a factor $q^{b_{ij}/2}$ in $C_{f,F}$. The lemma is proved.

8.11.3. Example. Let $k = 1$, $n = 2$. Then

$$T_1(D(x_1 z_1 | z_2)) = q^{b_{2,3}/2}\big(D(x_1 z_1 | z_2) + q^{b_{13}/2} D(z_1 x_1 | z_2) \\ + q^{b_{13}/2} D(z_1 | x_1 z_2)\big),$$

cf. (7.4.9), (7.4.20).

Comparison of (8.11.1), (7.4.7) and (7.4.22) proves (8.3.4.ii) for $T^u = T^u_{!1} = T_1$. Theorem (8.3.4) is proved for $T^u_{!t}$.

8.12. Proof of Theorem 8.3.4 for T^u_t

Proof of Theorem (8.3.4) for T^u_t is analogous to the proof of Theorem (8.3.4) for $T^u_{1t} = T_t$ in Secs. 8.4–8.11. Cells $E(f)$ of the complex X are deformations of unions of suitable cells of the complex Q'/Q'', see (2.1.7), (2.2.1), and (2.3.4). Cells $E(f)$ lie in a neighborhood of $\mathbb{R}^k$ in $\mathbb{C}^k$. The construction of the deformation T^u_t repeats, with minor changes, the construction of the deformation $T^u_{!t}$ in (8.4) and (8.8). As the previous construction, it is based on Theorem (6.2.4). We omit details.

Example. Let $k = 2$, $n = 2$, $u = 1$. Let $z^0 = (z^0_1, z^0_2) \in \mathbb{R}^2$, $z^0_1 < z^0_2$. $C_2(X, A, z^0)$ is generated by $E(f; A, z^0)$, where f runs through $\text{Adm}(x_1, x_2; z_1, z_2)$. These cells can be drawn on the complex line $\mathbb{C}$, see examples in Fig. 8.4. For example, $E(z_1 x_1 | z_2 x_2)$ is the direct product of two curves: x_1 runs over z^0_1, x_2 runs over z^0_2. Explicit description is given in Sec. 3.

Now consider the deformation $\gamma^1(t) = (z_1(t), z_2(t))$ as shown in Fig. 8.2. Consider as an example the cell $E(z_1 x_1 | z_2 x_2)$. The deformation of the pair $\mathbb{R} \subset \mathbb{C}$ shown in Fig. 8.2 induces a deformation of $E(z_1 x_1 | z_2 x_2)$ shown in Fig. 8.5. It is easy to see that this cell can be naturally deformed onto the union of the last three cells shown in Fig. 8.4. Comparing the distinguished sections over these cells, distinguished orientations and changes of arguments of $x_j - z_i$, $x_j - x_i$ under the deformation, we get

$$T_1(E(z_1 x_1 | z_2 x_2; A, z^0)) = q^{(b_{12} + b_{13} + b_{14} + 2b_{23} + 2b_{24} + 2b_{34})/4}(-E(x_1 x_2 z_1 | z_2; A, z^1)$$
$$+ q^{-b_{24}/2} E(x_1 z_1 x_2 | z_2; A, z^1) + q^{-(b_{12} + b_{23} + 3b_{24})/4} E(x_1 z_1 | z_2 x_2; A, z^1)) \quad (8.12.1)$$

as predicted by (7.4.4) and (7.4.22).

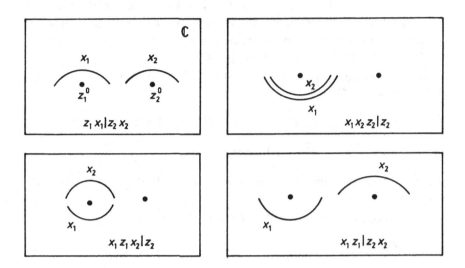

Fig. 8.4

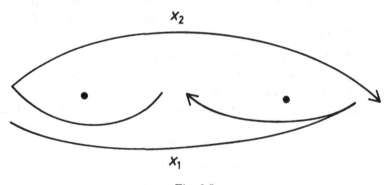

Fig. 8.5

8.13. Geometric Interpretation of the R-Matrix Operators Acting on the Two-sided Hochschild Complexes Constructed in Sec. 4

Consider the Hochschild complexes

$$C.(^{+}U_q\bar{\mathfrak{n}}'_{-};\overline{M}_\delta;2;{}^{+}\Delta)_\lambda\,,\qquad C^{*}(^{+}U_q\bar{\mathfrak{n}}'_{-};\overline{M}_\delta;2;\mu)_\lambda \qquad (8.13.1)$$

constructed in Sec. 4.

Fix $\lambda = (l_1, \ldots, l_k, 1, \ldots, 1) \in \mathbb{N}^{n+k}$. The R-matrix operators defined in Sec. 7 act on these complexes. In this section, we give a geometric interpretation of this action. This interpretation generalizes the interpretation for $\lambda = (1, \ldots, 1) \in \mathbb{N}^{n+k}$ given by Theorem (8.3.4).

Let $U_q\tilde{\mathfrak{g}}$ be a quantum group adjacent to $U_q\bar{\mathfrak{g}}$ at weight $\lambda = (l_1, \ldots, l_k, 1, \ldots, 1) \in \mathbb{N}^{n+k}$, see (7.5). Let

$$\pi : C.(^+U_q\tilde{\mathfrak{n}}'_-; \overline{M}_\delta; 2; {}^+\Delta)_\lambda \longrightarrow C.(^+U_q\tilde{\mathfrak{n}}'_-; \widetilde{M}_\delta; 2; {}^+\Delta)^+_{\tilde\lambda},$$
$$\pi : C^*.(^+U_q\tilde{\mathfrak{n}}'_-; \overline{M}_\delta; 2; \mu)_\lambda \longrightarrow C^*.(^+U_q\tilde{\mathfrak{n}}'_-; \widetilde{M}_\delta; 2; \mu)^+_{\tilde\lambda} \tag{8.13.2}$$

be the isomorphisms constructed in (7.5) and (4.8). Here $\tilde\lambda = (1, \ldots, 1) \in \mathbb{N}^{l+n}$ and $l = l_1 + \cdots + l_k$.

The right-hand sides of (8.13.2) are subcomplexes of

$$C.(^+U_q\tilde{\mathfrak{n}}'_-; \widetilde{M}_\delta; 2; {}^+\Delta)_{\tilde\lambda}, \qquad C^*.(^+U_q\tilde{\mathfrak{n}}'_-; \widetilde{M}_\delta; 2; \mu)_{\tilde\lambda}, \tag{8.13.3}$$

resp. These complexes and the action of the operators R^p, $p = 1, \ldots, n - 1$, are geometrically interpreted by Theorem (8.3.4). The image of π in (8.13.3) is geometrically interpreted by Theorem (4.7.5). Thus the action of the operators R^p on (8.13.1) is the action of the monodromy operators T^p of Theorem (8.13.3) on the anti-invariant part of the chain complexes $C.(X, \mathcal{S}^*(A), z^0)$, $C.(Q'/Q'', \mathcal{S}^*(A), z^0)$.

More precisely, let $\mathbb{C}^{n+l}$ have coordinates $t_1(1), \ldots, t_{l_1}(1), t_1(2), \ldots, t_{l_2}(2), \ldots, t_1(k), \ldots, t_{l_k}(k)$, $t_{k+1}, \ldots, t_{k+n}$. Let $p_{n,l} : \mathbb{C}^{n+l} \to \mathbb{C}^n$, $(\{t_i(j)\}, t_{k+1}, \ldots, t_{k+n}) \mapsto (t_{k+1}, \ldots, t_{k+n})$ be the projection.

Introduce the weights A for the configuration $\mathcal{C}_{n+l}$ in $\mathbb{C}^{n+l}$ of all diagonal hyperplanes:

$$a(t_i, t_j) = b_{ij}/\kappa,$$
$$a(t_m(i), t_j) = b_{ij}/\kappa, \tag{8.13.4}$$
$$a(t_m(i), t_s(j)) = b_{ij}/\kappa$$

for all i, j, m, s. Here κ is the parameter of our quantum groups, $q = \exp(2\pi i/\kappa)$. The numbers b_{ij} are scalar products of the simple roots of $U_q\bar{\mathfrak{g}}$, see (8.2.10).

Let $\mathcal{S}^*(A)$ be the local system of coefficients on $\mathbb{C}^{n+l} - U_{n+l}$ associated with A. The group $\Sigma' = \Sigma_{l_1} \times \Sigma_{l_2} \times \cdots \times \Sigma_{l_k}$ naturally acts on $\mathbb{C}^{n+l}$ permuting the coordinates $\{t_j(i)\}$ in such a way that the index i is preserved. The action of Σ' preserves fibers of the projection $p_{n,l}$ and the local system $\mathcal{S}^*(A)$.

Let $z^0 = (z_1^0, \ldots, z_n^0) \in \mathbb{R}^n \subset \mathbb{C}^n$, and $z_{\delta(1)}^0 < z_{\delta(2)}^0 < \cdots < z_{\delta(n)}^0$ for some $\delta \in \Sigma_n$.

Consider twisted realizations

$$\omega : C.(^+U_q\tilde{\mathfrak{n}}'_-; \widetilde{M}_\delta; 2; {}^+\Delta)_{\tilde{\lambda}} \xrightarrow{\sim} C_{l-.}(X, \mathcal{S}^*(A), z^0),$$
$$\omega : C^*(^+U_q\tilde{\mathfrak{n}}'_-; \widetilde{M}_\delta; 2; \mu)_{\tilde{\lambda}} \xrightarrow{\sim} C_{l-.}(Q'/Q'', \mathcal{S}^*(A), z^0),$$
(8.13.5)

defined in (8.2). The group Σ' acts on both sides of (8.13.5), see (4.7) and (4.8). The isomorphisms (8.13.5) restricted to the invariant parts of the left-hand sides define the isomorphisms

$$\omega : C.(^+U_q\tilde{\mathfrak{n}}'_-; \widetilde{M}_\delta; 2; {}^+\Delta)_{\tilde{\lambda}}^+ \xrightarrow{\sim} C_{l-.}(X, \mathcal{S}^*(A), z^0)^-,$$
$$\omega : C^*(^+U_q\tilde{\mathfrak{n}}'_-; \widetilde{M}_\delta; 2; \mu)_{\tilde{\lambda}}^+ \xrightarrow{\sim} C_{l-.}(Q'/Q'', \mathcal{S}^*(A), z^0)^-,$$
(8.13.6)

where the superscript $^-$ denotes the skew-invariant part of the Σ' action, see (4.7)–(4.9).

Applying Theorem (8.3.4) to the projection $p_{n,l}$ the local system $\mathcal{S}^*(A)$, the point z^0, we get:

8.13.7. Theorem. *The homomorphisms of Theorem (8.3.4),*

$$T^u : C.(X, \mathcal{S}^*(A), z^0) \longrightarrow C.(X, \mathcal{S}^*(A), z^u),$$
$$T^u : C.(Q'/Q'', \mathcal{S}^*(A), z^0) \longrightarrow C.(Q'/Q'', \mathcal{S}^*(A), z^u),$$

for $u = 1, \ldots, n-1$, preserve the Σ'-skew-invariant part. Moreover, the following diagrams are commutative.

$$\begin{array}{ccc}
C.(^+U_q\tilde{\mathfrak{n}}'_-; \overline{M}_\delta; 2; {}^+\Delta)_\lambda & \xrightarrow{R^u} & C.(^+U_q\tilde{\mathfrak{n}}'_-; \overline{M}_{\delta^p}; 2; {}^+\Delta)_\lambda \\
{\scriptstyle \omega\pi}\downarrow & & \downarrow{\scriptstyle \omega\pi} \\
C_{l-.}(X, \mathcal{S}^*(A), z^0)^- & \xrightarrow{T^u} & C_{l-.}(X, \mathcal{S}^*(A), z^u)^-
\end{array}$$

$$\begin{array}{ccc}
C^*(^+U_q\tilde{\mathfrak{n}}'_-; \overline{M}_\delta; 2; \mu)_\lambda & \xrightarrow{R^u} & C.(^+U_q\tilde{\mathfrak{n}}'_-; \overline{M}_{\delta^p}; 2; \mu)_\lambda \\
{\scriptstyle \omega\pi}\downarrow & & \downarrow{\scriptstyle \omega\pi} \\
C_{l-.}(Q'/Q'', \mathcal{S}^*(A), z^0)^- & \xrightarrow{T^u} & C_{l-.}(Q'/Q'', \mathcal{S}^*(A), z^u)^-.
\end{array}$$

The theorem is a corollary of Theorems (8.3.4) and (7.5.2).

8.13.8. Remark. Recall that $S\omega\pi = \omega\pi S$, where the right S is the contravariant form and the left S is the homomorphism of the projection on the real part, see (4.7.5).

8.14. Geometric Interpretation of the R-Matrix Operators on the Complexes $C.(^+U_q\mathfrak{n}_-; M_\delta; ^+\Delta)_{\lambda'}$, $C^*(^+U_q\mathfrak{n}_-; M_\delta; \mu)_{\lambda'}$ Constructed in Sec. 5

Let $U_q\mathfrak{g}$ be a quantum of type (4.1.a)–(4.1.d); let $M_1, \ldots, M_n$ be the Verma modules over $U_q\mathfrak{g}$ with highest weights $\Lambda_1, \ldots, \Lambda_n \in \mathfrak{h}^*$. Set $M_\delta = M_{\delta(1)} \otimes \cdots \otimes M_{\delta(n)}$. Consider the complexes $C.(^+U_q\mathfrak{n}_-; M_\delta; ^+\Delta)_{\lambda'}$, $C^*(^+U_q\mathfrak{n}_-; M_\delta; \mu)_{\lambda'}$, constructed in (5.1), and the R-matrix operators R^u for $u = 1, \ldots, n-1$ acting on these complexes, see (7.6). Here $\lambda' = (l_1, \ldots, l_k) \in \mathbb{N}^k$.

Let $\lambda = (l_1, \ldots, l_k, 1, \ldots, 1) \in \mathbb{N}^{n+k}$. In Sec. 5, we have constructed monomorphisms

$$\begin{aligned}\phi &: C.(^+U_q\mathfrak{n}_-; M_\delta; ^+\Delta)_{\lambda'} \longrightarrow C.(^+U_q\bar{\mathfrak{n}}'_-; \overline{M}_\delta; 2; ^+\Delta)_\lambda, \\ \phi &: C^*(^+U_q\mathfrak{n}_-; M_\delta; \mu)_{\lambda'} \longrightarrow C^*(^+U_q\bar{\mathfrak{n}}'_-; \overline{M}_\delta; 2; \mu)_\lambda,\end{aligned} \qquad (8.14.1)$$

see (5.2.26) and (5.3.41). These monomorphisms are quasiisomorphisms and commute with the contravariant form, see (5.3.40), (5.4), and (5.5). These monomorphisms commute with the action of the R-matrix operator by (7.7.2).

The isomorphisms $\omega\pi$ of (8.13) identify the right-hand side with the Σ' skew-invariant part of the corresponding chain complexes. Thus we have monomorphisms

$$\begin{aligned}\omega\pi\phi &: C.(^+U_q\mathfrak{n}_-; M_\delta; ^+\Delta)_{\lambda'} \longrightarrow C_{l-.}(X, S^*(A), z^0)^- \\ \omega\pi\phi &: C^*(^+U_q\mathfrak{n}_-; M_\delta; \mu)_{\lambda'} \longrightarrow C_{l-.}(Q'/Q'', S^*(A), z^0)^- \\ \omega\pi\phi &: C.(^+U_q\mathfrak{n}_-; M_{\delta^u}; ^+\Delta)_{\lambda'} \longrightarrow C_{l-.}(X, S^*(A), z^u)^- \\ \omega\pi\phi &: C^*(^+U_q\mathfrak{n}_-; M_{\delta^u}; \mu)_{\lambda'} \longrightarrow C_{l-.}(Q'/Q'', S^*(A), z^u)^-\end{aligned} \qquad (8.14.2)$$

for all $u = 1, \ldots, n-1$. Here we use the notations of (8.13).

(8.14.3) These monomorphisms are quasiisomorphisms by (5.4) and (5.5).

(8.14.4) These monomorphisms send the R-matrix action to the monodromy action:

$$\omega\pi\phi R^u = T^u \omega\pi\phi$$

for all u, see (8.13.7) and (7.7.2).

(8.14.5) These monomorphisms send the contravariant form to the homomorphism of the projection on real part:

$$\omega\pi\phi S = S\omega\pi\phi,$$

see (4.7.5) and (5.3.40).

8.14.6. Corollary. *Monomorphisms $\omega\pi\phi$ induce the isomorphisms*

$$H.C.(^+U_q\mathfrak{n}_-; M_\delta; {}^+\Delta)_{\lambda'} \simeq H_{l-\cdot}(P_{n,l}(z^0)\backslash U_{n,l}(z^0), \mathcal{S}^*(A, z^0))^-,$$
$$H.C.^*(^+U_q\mathfrak{n}_-; M_\delta; \mu)_{\lambda'} \simeq H_{l-\cdot}(P_{n,l}(z^0), U_{n,l}(z^0), \mathcal{S}^*(A, z^0))^-,$$
$$H.C.(^+U_q\mathfrak{n}_-; M_{\delta^u}; {}^+\Delta)_{\lambda'} \simeq H_{l-\cdot}(P_{n,l}(z^u)\backslash U_{n,l}(z^u), \mathcal{S}^*(A, z^u))^-,$$
$$H.C.^*(^+U_q\mathfrak{n}_-; M_{\delta^u}; \mu)_{\lambda'} \simeq H_{l-\cdot}(P_{n,l}(z^u), U_{n,l}(z^u), \mathcal{S}^*(A, z^u))^-$$

for all $u = 1, \ldots, n-1$. Here the superscript $^-$ denotes the subspace of all elements x such that $\sigma x = (-1)^{|\sigma|}x$ for all $\sigma \in \Sigma'$, cf. (5.11.3) and (5.11.4). These isomorphisms send the R-matrix action to the monodromy action ($\omega\pi\phi R^u = T^u \omega\pi\phi$) and the contravariant form to the canonical homomorphism ($\omega\pi\phi S = S\omega\pi\phi$), see (5.11.4).

8.14.7. Example. $U_q sl_2$ is generated by three elements e, f, h subject to the relations $[h, e] = 2e$, $[h, f] = -2f$, $[e, f] = q^{h/2} - q^{-h/2}$. Fix $\lambda' = l \in \mathbb{N}$. Fix Verma modules $M_1, \ldots, M_n$ with highest weights $\Lambda_1, \ldots, \Lambda_n$.

Consider the projection $p_{n,l} : \mathbb{C}^{n+l} \to \mathbb{C}^n$, $(t_1, \ldots, t_{n+l}) \to (t_{l+1}, \ldots, t_{l+n})$. Define the weights of the configuration $C_{n+l} \subset \mathbb{C}^{n+l}$ as

$$a(t_i, t_j) = (-\alpha_1, -\alpha_1)/\kappa = 2/\kappa \text{ for } 1 \leq i < j \leq l,$$
$$a(t_i, t_{l+j}) = (-\alpha_1, \Lambda_j)/\kappa \text{ for } 1 \leq i \leq l \text{ and } 1 \leq j \leq n,$$
$$a(t_{l+i}, t_{l+j}) = (\Lambda_i, \Lambda_j)/\kappa \text{ for } 1 \leq i < j \leq n.$$

Let $\Sigma' = \Sigma_l$ be the group of all permutations of the first coordinates.

Let $z^0 \in \mathbb{R}^n \in \mathbb{C}^n$ be a point such that $z^0_{\delta(1)} < z^0_{\delta(2)} < \cdots < z^0_{\delta(n)}$ for some $\delta \in \Sigma_n$. Then we have canonical monomorphisms (8.14.2) with properties (8.14.3)–(8.14.6).

9

R-matrix Operator as the Canonical Element, Quantum Doubles

9.1. Quantum Double [D1, R2, Le]

Let A be a Hopf algebra with multiplication $\mu : A \otimes A \to A$, comultiplication $\Delta : A \to A \otimes A$, unit $i : \mathbb{C} \to A$, counit $\epsilon : A \to \mathbb{C}$ and antipode $a : A \to A$.

Fix a linear basis $\{e_i\}$ of A. Then

$$\mu : e_i \otimes e_j \longmapsto \mu_{ij}^k e_k,$$
$$\Delta : e_i \longmapsto \Delta_i^{jk} e_j \otimes e_k,$$
$$a : e_i = a_i^j e_j.$$

A^*, dual to A, is a Hopf algebra with the basis $\{e^i\}$ dual to $\{e_i\} : \langle e^i, e_j \rangle = \delta_j^i$. Multiplication, comultiplication, unit, counit, and antipode are given by

$$\Delta^* : A^* \otimes A^* \to A^*, \quad \mu^* : A^* \to A^* \otimes A^*, \quad \epsilon^* : \mathbb{C} \to A^*, \quad i^* : A^* \to \mathbb{C},$$

$a^* : A^* \to A^*$, respectively. In particular, $\langle \Delta^*(e^u \otimes e^v), e_k \rangle = \langle e^u \otimes e^v, \Delta e_k \rangle = \langle e^u \otimes e^v, \Delta_k^{ij} e_i \otimes e_j \rangle = \Delta_k^{uv}$, $\langle \mu^* e^u, e_i \otimes e_j \rangle = \mu_{ij}^u$ and so on.

The map $a' := a^{-1}$ inverse to a is called the skew antipode. $a' : A \to A$ is the antipode for the algebra A with the opposite comultiplication and the same multiplication. a' is also the antipode for the algebra A with the opposite multiplication and the same comultiplication.

Denote by A^0 the algebra A^* with the opposite comultiplication.

A quasitriangular Hopf algebra is a pair (A, R) consisting of a Hopf algebra A and an invertible element $R \in A \otimes A$ such that:

$$R\Delta(x)R^{-1} = \Delta'(x) \qquad \text{for all } x \in A, \tag{9.1.1}$$

where Δ' is the opposite comultiplication,

$$\begin{aligned}(\Delta \otimes id)(R) &= R_{13}R_{23}, \\ (id \otimes \Delta)(R) &= R_{13}R_{12},\end{aligned} \tag{9.1.2}$$

where, if $R = \sum a_i \otimes b_i$, then $R_{13} = \sum a_i \otimes 1 \otimes b_i, \cdots$

For any Hopf algebra A, there exists a unique quasitriangular Hopf algebra $(\mathcal{D}(A), R)$ such that:

(9.1.3) $\mathcal{D}(A)$ contains A and A^0 as subalgebras;

(9.1.4) R is the image of the canonical element of $A \otimes A^0$ by the embedding $A \otimes A^0 \to \mathcal{D}(A) \otimes \mathcal{D}(A)$;

(9.1.5) the linear map: $A \otimes A^0 \to \mathcal{D}(A)$ is bijective, $a \otimes b \to ab$.

So, as a linear space, $\mathcal{D}(A)$ can be identified with $A \otimes A^0$ and its algebra and coalgebra structure will be completely determined as soon as one knows how to compute a product $e^t e_s$ as a sum of products $e_j e^\ell$. In fact, one can give an intrinsic formula for the product $A^0 \otimes A \to \mathcal{D}(A)$:

$$e^t e_s = \Delta_s^{njk} \mu_{k\ell p}^t a_n'^p e_j e^\ell. \tag{9.1.6}$$

Here $\mu_{k\ell p}^t$, Δ_s^{njk}, $a_n'^p$ are the matrices of the multiplication $A \otimes A \otimes A \to A$, the comultiplication $A \to A \otimes A \otimes A$, the skew antipode $A \to A$. Namely, to get (9.1.6), one takes

$$e_s \xrightarrow{\Delta} \Delta_s^{njk} e_n \otimes e_j \otimes e_k \xrightarrow{a' \otimes id \otimes id} \Delta_s^{njk} a_n'^p e_p \otimes e_j \otimes e_k$$
$$\longmapsto \Delta_s^{njk} a_n'^p e_k \otimes e_j \otimes e_p,$$
$$e^t \xrightarrow{\mu^*} \mu_{k\ell p}^t e^k \otimes e^\ell \otimes e^p,$$

and then pairs the first and the third factors.

This formula follows from the equalities $\Delta'(e_s)R = R\Delta(e_s)$, $\Delta'(e^s)R = R\Delta(e^s)$, where $R = \sum e_i \otimes e^i$. The formula (9.1.6) implies (9.1.2).

9.2. Quantum Doubles $\mathcal{D}((U_q\mathfrak{b}_-)')$ and $\mathcal{D}(U_q\mathfrak{b}_+)$

Let $U_q\mathfrak{g}$ be a quantum group of type (4.1.a)–(4.1.d). Let $h_1, \ldots, h_r, h_{r+1}, \ldots, h_k$ be a basis of $\mathfrak{h}$ such that $h_i = b^{-1}(\alpha_i)$ for $i \leq r$.

For any $\lambda = (\ell_1,\ldots,\ell_r) \in \mathbf{N}^r$, set $h_\lambda = \ell_1 h_1 + \cdots + \ell_r h_r$. For any $p = (p_1,\ldots,p_k) \in \mathbf{N}^k$, set $h^p = h_1^{p_1}\ldots h_k^{p_k}$. Then $\{h^p q^{-h_\lambda/4} f_J\}$ is a linear basis of $U_q\mathfrak{b}_-$, $\{h^p q^{h_\lambda/4} e_J\}$ is a linear basis of $U_q\mathfrak{b}_+$. Here $p \in \mathbf{N}^k$, $\lambda \in \mathbf{N}^r$, f_J runs through all monomials in $(U_q\mathfrak{n}_-)_\lambda$, e_J runs through all monomials in $(U_q\mathfrak{n}_+)_\lambda$, see (4.1). Denote by $\{(h^p q^{-h_\lambda/4} f_J)^*\}$ (resp. $\{(h^p q^{h_\lambda/4} e_J)^*\}$) the dual basis of $(U_q\mathfrak{b}_-)^*$ (resp. $(U_q\mathfrak{b}_+)^*$).

Denote by $(U_q\mathfrak{b}_-)'$ the Hopf algebra $(U_q\mathfrak{b}_-)^*$ with the opposite multiplication. Its multiplication, comultiplication and antipode are given by the maps $(P\Delta)^* : (U_q\mathfrak{b}_-)^* \otimes (U_q\mathfrak{b}_-)^* \to (U_q\mathfrak{b}_-)^*$, $\mu^* : (U_q\mathfrak{b}_-)^* \to (U_q\mathfrak{b}_-)^* \otimes (U_q\mathfrak{b}_-)^*$, $(A^{-1})^* : (U_q\mathfrak{b}_-)^* \to (U_q\mathfrak{b}_-)^*$, respectively, where P is the transposition of factors and A is the antipode of $U_q\mathfrak{b}_-$, see (4.1.7).

We have $((U_q\mathfrak{b}_-)')^0 = U_q\mathfrak{b}_-$, where $U_q\mathfrak{b}_-$ has the standard multiplication, comultiplication and antipode.

Consider the quantum double $\mathcal{D}((U_q\mathfrak{b}_-)')$ of $(U_q\mathfrak{b}_-)'$. Consider the quantum double $\mathcal{D}(U_q\mathfrak{b}_+)$ of $U_q\mathfrak{b}_+$. We will discuss multiplication, comultiplication and commutation relations in these doubles.

Let S be the contravariant form on $U_q\mathfrak{n}_-$ and on $U_q\mathfrak{n}_+$ defined in (4.2). Let (S_{IJ}^{-1}) be the matrix inverse to the matrix $(S_{IJ}) = (S(f_I,f_J)) = (S(e_I,e_J))$, if it exists. Here f_I, f_J (resp. e_I, e_J) run through all monomials in $U_q\mathfrak{n}_-$ (resp. $U_q\mathfrak{n}_+$).

For any $f_I \in (U_q\mathfrak{b}_-)_\lambda$, $e_I \in (U_q\mathfrak{b}_+)_\lambda$ and $h^p \in U_q\mathfrak{h}$, set

$$\begin{aligned}(h^p q^{-h_\lambda/4} F_I)^* &:= \sum_J S_{IJ}(h^p q^{-h_\lambda/4} f_J)^*, \\ (h^p q^{h_\lambda/4} E_I)^* &:= \sum_J S_{IJ}(h^p q^{h_\lambda/4} e_J)^*.\end{aligned} \qquad (9.2.1)$$

If S is nondegenerate, then the elements (9.2.1) form the basis of $U_q\mathfrak{b}_-$ (resp. $U_q\mathfrak{b}_+$) dual to the basis $\{h^p q^{-h_\lambda/4} F_I\}$ (resp. $\{h^p q^{h_\lambda/4} E_I\}$), where

$$F_I = \sum_j S_{JI}^{-1} f_J, \qquad E_I = \sum_j S_{JI}^{-1} e_J. \qquad (9.2.2)$$

For these elements, we have $S(F_I,f_J) = S(E_I,e_J) = \delta_{I,J}$. Note that $E_I = g_{\bar I}$, where $g_{\bar I}$ is defined in (5.1.6).

9.2.3. Example.

$$F_{12} = (q^{-b/4} f_1 f_2 - q^{b/4} f_2 f_1)/(q^{-b/2} - q^{b/2})$$

and
$$F_{21} = (-q^{b/4} f_1 f_2 + q^{-b/4} f_2 f_1)/(q^{-b/2} - q^{b/2}),$$
where $b = b_{12}$.

$$(q^{-(h_1+h_2)/4} F_{12})^* = (q^{-(h_1+h_2)/4} f_1 f_2)^* q^{-b/4} + (q^{-(h_1+h_2)/4} f_2 f_1)^* q^{b/4}$$

and

$$(q^{-(h_1+h_2)/4} F_{21})^* = (q^{-(h_1+h_2)/4} f_1 f_2)^* q^{b/4} + (q^{-(h_1+h_2)/4} f_2 f_1)^* q^{-b/4}$$

are elements of form (9.2.1). They are linearly dependent if $q^{-b/2} - q^{b/2} = 0$.

9.2.4. Lemma.

$$\Delta(h^p q^{-(h_{i_1}+\cdots+h_{i_N})/4} F_{i_1\ldots i_N})$$
$$= \sum_{j=0}^{N} \Delta(h^p)(q^{-(h_{i_{j+1}}+\cdots+h_{i_N})/2} \otimes 1) \cdot (F_{i_1\ldots i_j} \otimes F_{i_{j+1}\ldots i_N}),$$
$$\Delta(h^p q^{(h_{i_1}+\cdots+h_{i_N})/4} E_{i_1\ldots i_N})$$
$$= \sum_{j=0}^{N} \Delta(h^p)(1 \otimes q^{(h_{i_{j+1}}+\cdots+h_{j_N})/2}) \cdot (E_{i_{j+1}\ldots i_N} \otimes E_{i_1\ldots i_j}).$$

The proof is analogous to (5.1.8).

Hence in $(U_q \mathfrak{b}_-)'$ and $(U_q \mathfrak{b}_+)^0$, we have

$$(P\Delta)^* : (h^\ell q^{-h_\lambda/4} F_I)^* \otimes (h^m q^{-h_\nu/4} F_J)^* \mapsto \sum_p A_p^{\ell,m} (h^p q^{-(h_\lambda+h_\nu)/4} F_{JI})^*$$
$$\Delta^* : (h^\ell q^{h_\lambda/4} E_I)^* \otimes (h^m q^{h_\nu/4} E_J)^* \mapsto \sum_p B_p^{\ell,m} (h^p q^{(h_\lambda+h_\nu)/4} E_{JI})^*. \tag{9.2.5}$$

Here $I = (i_1,\ldots,i_N)$, $J = (j_1,\ldots,j_L)$, $JI = (j_1,\ldots,j_L,i_1,\ldots,i_N)$, $F_I \in (U_q \mathfrak{n}_-)_\lambda$, $F_J \in (U_q \mathfrak{n}_-)_\nu$, $h^\ell = h_1^{\ell_1},\ldots,h_k^{\ell_k}$, $h^m = h_1^{m_1},\ldots,h_k^{m_k}$, $h^p = h_1^{p_1},\ldots,h_k^{p_k}$ and $\ell_j \le p_j \le \ell_j + m_j$ for all j.

$$A_p^{\ell,m} = \langle \Delta(h^p) \cdot (1 \otimes q^{-(h_{i_1}+\cdots+h_{i_N})/2}), (h^\ell)^* \otimes (h^m)^* \rangle,$$
$$B_p^{\ell,m} = \langle \Delta(h^p) \cdot (1 \otimes q^{(h_{i_1}+\cdots+h_{i_N})/2}), (h^\ell)^* \otimes (h^m)^* \rangle. \tag{9.2.6}$$

Moreover, let $\varphi := 2\pi i/\kappa$ and $q = \exp\varphi$. Let $h_{i_1} + \cdots + h_{i_N} = s_1 h_1 + \cdots + s_k h_k$. (Note that $s_{r+1} = \ldots s_k = 0$.) Then

$$A_p^{\ell,m} = \prod_{j=1}^{k} \frac{p_j!}{\ell_j!(p_j - \ell_j)!(m_j + \ell_j - p_j)!} \left(-\frac{\varphi}{2} s_j\right)^{(m_j+\ell_j-p_j)},$$
$$B_p^{\ell,m} = \prod_{j=1}^{k} \frac{p_j!}{\ell_j!(p_j - \ell_j)!(m_j + \ell_j - p_j)!} \left(\frac{\varphi}{2} s_j\right)^{(m_j+\ell_j-p_j)}.$$
(9.2.7)

9.2.8. Corollary. *Elements $\{(h^p q^{-h_\lambda/4} F_I)^*\}$ (resp. $(h^p q^{h_\lambda/4} E_I)^*$) form a subalgebra of $(U_q \mathfrak{b}_-)'$ (resp. $(U_q \mathfrak{b}_+)^0$). This subalgebra is generated by elements $\{(h_i)^*, (q^{-h_j/4} f_j)^*\}$ (resp. $\{(h_i)^*, (q^{h_j/4} e_j)^*\}$), where $i = 1, \ldots, k$, and $j = 1, \ldots, r$.*

Denote this subalgebra by $\overline{(U_q \mathfrak{b}_-)}\,'$ (resp. by $\overline{(U_q \mathfrak{b}_+)}\,^0$).

9.2.9. Remarks. 1. For a quantum group $U_q \mathfrak{g}$ of type (4.1.a)–(4.1.d) with matrix $B = (b_{ij}) = ((\alpha_i, \alpha_j))$ and parameter q in general position, we have $\overline{(U_q \mathfrak{b}_-)}\,' = (U_q \mathfrak{b}_-)'$ and $\overline{(U_q \mathfrak{b}_+)}\,^0 = (U_q \mathfrak{b}_+)^0$, see (4.6.10) and (3.12.9).

2. According to (9.2.5),

$$(q^{-h_\lambda/4} F_I)^* (q^{-h_\nu/4} F_J)^* = (q^{-(h_\lambda+h_\nu)/4} F_{JI})^*,$$
$$(q^{h_\lambda/4} E_I)^* (q^{h_\nu/4} E_J)^* = (q^{(h_\lambda+h_\nu)/4} E_{JI})^*$$

in $(U_q \mathfrak{b}_-)'$ and $(U_q \mathfrak{b}_+)^0$, respectively. If $\sum a_I f_I \in \ker S$ on $(U_q \mathfrak{n}_-)_\lambda$ (resp. $\sum a_I e_I \in \ker S$ on $(U_q \mathfrak{n}_+)_\lambda$), then $\sum a_I (q^{-h_\lambda/4} F_I)^* = 0$ (resp. $\sum a_I (q^{h_\nu/4} E_J)^* = 0$), cf. (4.2.10).

The comultiplication in $(U_q \mathfrak{b}_-)'$ and $(U_q \mathfrak{b}_+)^0$ is given by the following formulas:

$$\mu^*((h^p)^*) = (\mu P)^*((h^p)^*) = \sum_{\substack{\ell_j + m_j = p_j \\ j=1,\ldots,k}} \left[\prod \frac{p_j!}{\ell_j! m_j!}\right] (h^\ell)^* \otimes (h^m)^*,$$

$$\mu^*((q^{-h_j/4} f_j)^*) = 1^* \otimes (q^{-h_j/4} f_j)^*$$
$$+ (q^{-h_j/4} f_j)^* \otimes \sum_p \langle h_1, \alpha_j\rangle^{p_1} \ldots \langle h_k, \alpha_j\rangle^{p_k} (h^p)^*,$$
(9.2.10)
$$(\mu P)^*((q^{h_j/4} e_j)^*) = (q^{h_j/4} e_j)^* \otimes 1^*$$
$$+ \sum_p \langle -h_1, \alpha_j\rangle^{p_1} \ldots \langle -h_k, \alpha_j\rangle^{p_k} (h^p)^* \otimes (q^{h_j/4} e_j)^*.$$

The multiplication in $\mathcal{D}((U_q\mathfrak{b}_-)')$ and $\mathcal{D}(U_q\mathfrak{b}_+)$ is given by the following lemma.

9.2.11. Lemma.

1. $h_i(h_j)^* = (h_j)^* h_i$

2. $(q^{-h_j/4}f_j)(h_i)^* = (h_i)^*(q^{-h_j/4}f_j) + \delta_{ij}\dfrac{\varphi}{2}q^{-h_j/4}f_j$

3. $h_i(q^{-h_j/4}f_j)^* = (q^{-h_j/4}f_j)^* h_i + \langle h_i, \alpha_j\rangle(q^{-h_j/4}f_j)^*$,

4. $(q^{-h_j/4}f_i)(q^{-h_j/4}f_j)^* = (q^{-h_j/4}f_j)^*(q^{-h_i/4}f_i) + \delta_{ij}q^{-h_i/2}$
   $-\delta_{ij}\sum_p\langle h_1, \alpha_j\rangle^{p_1}\ldots\langle h_k, \alpha_j\rangle^{p_k}(h^p)^*$,

5. $(h_i)^*(q^{h_j/4}e_j) = (q^{h_j/4}e_j)(h_i)^* + \delta_{ij}\dfrac{\varphi}{2}(q^{h_j/4}e_j)$,

6. $(q^{h_j/4}e_j)^* h_i = h_i(q^{h_j/4}e_j)^* + \langle h_i, \alpha_j\rangle(q^{h_j/4}e_j)^*$,

7. $(q^{h_i/4}e_i)^*(q^{h_j/4}e_j)^* = (q^{h_j/4}e_j)(q^{h_i/4}e_i)^* - \delta_{ij}q^{h_j/2}$
   $+\delta_{ij}\sum_p\langle -h_1, \alpha_j\rangle^{p_1}\ldots\langle -h_k, \alpha_j\rangle^{p_k}(h^p)^*$.

Proof. The lemma is proved by direct verification. For example, we check (9.2.11.2).

$$(h_i)^* \xmapsto{\Delta} (h_i)^* \otimes 1^* \otimes 1^* + 1^* \otimes (h_i)^* \otimes 1^* + 1^* \otimes 1^* \otimes (h_i)^*$$
$$\xmapsto{(A^{-1})\otimes \text{id}\otimes \text{id}} -(h_i)^* \otimes 1^* \otimes 1^* \qquad (9.2.12)$$
$$+ 1^* \otimes (h_i)^* \otimes 1^* + 1^* \otimes 1^* \otimes (h_i)^*.$$

$$(q^{-h_j/4}f_j) \xmapsto{\Delta} q^{-h_j/4}f_j \otimes 1 \otimes 1 + q^{-h_j/2} \otimes q^{-h_j/4}f_j \otimes 1 \qquad (9.2.13)$$
$$+ q^{-h_j/2} \otimes q^{-h_j/2} \otimes q^{-h_j/4}f_j.$$

By (9.1.6),

$$(q^{-h_j/4}f_j)(h_i)^* = -\langle (h_i)^*, q^{-h_j/2}\rangle q^{-h_j/4}f_j + (h_i)^*(q^{-h_j/4}f_j)$$
$$= \delta_{ij}\dfrac{\varphi}{2}(q^{-h_j/4}f_j) + (h_i)^*(q^{-h_j/4}f_j).$$

9.3. The Action of the Quantum Doubles on Verma Modules and Their Duals

Let $M_1, \ldots, M_n$ be Verma modules over $U_q\mathfrak{g}$ with highest weights $\Lambda_1, \ldots,$

$\Lambda_n \in \mathfrak{h}^*$, respectively. Set $M = M_1 \otimes \cdots \otimes M_n$. Define maps

$$\mathcal{D}((U_q\mathfrak{b}_-)') \otimes M \longrightarrow M,$$
$$\mathcal{D}(U_q\mathfrak{b}_+) \otimes M^* \longrightarrow M^*.$$

$\mathcal{D}((U_q\mathfrak{b}_-)') \cong (U_q\mathfrak{b}_-)' \otimes U_q\mathfrak{b}_-$, $\mathcal{D}(U_q\mathfrak{b}_+) = U_q\mathfrak{b}_+ \otimes (U_q\mathfrak{b}_+)^0$ as vector spaces. Let $U_q\mathfrak{b}_- \otimes M \to M$, $U_q\mathfrak{b}_+ \otimes M^* \to M^*$ be the standard actions, see (4.1). Define $(U_q\mathfrak{b}_-)' \otimes M \to M$ and $(U_q\mathfrak{b}_+)^0 \otimes M^* \to M^*$ by the formulas

$$(h^p q^{-h_\lambda/4} f_{i_1} f_{i_2} \ldots f_{i_N})^* \otimes x \mapsto q^{D(h_\lambda)} \left[\prod_{j=1}^k \frac{Q(h_j^*)^{p_j}}{p_j!} \right] q^{h_\lambda/4} g_{i_1 \ldots i_N}(x),$$

$$(h^p q^{h_\lambda/4} e_{i_1} \ldots e_{i_N})^* \otimes x \mapsto q^{D(h_\lambda)} \left[\prod_{j=1}^k \frac{Q(h_j^*)^{p_j}}{p_j!} \right] q^{-h_\lambda/4} u_{i_1 \ldots i_N}(x).$$

(9.3.1)

Here $h^p = h_1^{p_1} \ldots h_k^{p_k}$; $f_{i_1} \ldots f_{i_N} \in (U_q\mathfrak{b}_-)_\lambda$; $e_{i_1} \ldots e_{i_N} \in (U_q\mathfrak{b}_+)_\lambda$; $D(h_\lambda) = -\sum_{a \leq b} b_{i_a i_b}/4$, see (6.2); the operator $g_I : M \to M$ is defined in (5.1); the operator $u_I : M^* \to M^*$ is defined in (7.6.5); $Q : \mathfrak{h}^* \to \mathfrak{h}$ is defined as $\frac{\varphi}{2}\Omega_0$, where $\Omega_0 : \mathfrak{h}^* \to \mathfrak{h}$ is the bilinear form corresponding to $U_q\mathfrak{g}$, see (4.1.b) and (7.6.1).

Formulas (9.3.1) imply

$$(h^p q^{-h_\lambda/4} F_{i_1 \ldots i_N})^* \longmapsto q^{D(h_\lambda)} \left[\prod_{j=1}^k \frac{Q(h_j^*)^{p_j}}{p_j!} \right] q^{h_\lambda/4} e_{i_N} \ldots e_{i_1},$$

$$(h^p q^{h_\lambda/4} E_{i_1 \ldots i_N})^* \longmapsto q^{D(h_\lambda)} \left[\prod_{j=1}^k \frac{Q(h_j^*)^{p_j}}{p_j!} \right] q^{-h_\lambda/4} f_{i_N} \ldots f_{i_1}.$$

(9.3.2)

Formulas (9.3.3) follow from the definitions of g_I and u_I.

9.3.3. Theorem. *Formulas (9.3.1) well define homomorphisms*

(1) $\quad \mathcal{D}((U_q\mathfrak{b}_-)') \to \mathrm{End}\,(M)$,

(2) $\quad \mathcal{D}(U_q\mathfrak{b}_+) \to \mathrm{End}\,(M^*)$.

With respect to these homomorphisms, the R-matrix operators $R : M \otimes M' \to M \otimes M'$ and $R : M^* \otimes M'^* \to M^* \otimes M'^*$ defined in (7.6.1) and (7.6.6) are the images of the canonical element.

9.3.4. Corollary. *R-matrix operators (7.6.1) and (7.6.6) have quasitriangular properties (9.1.1) and (9.1.2).*

Remark. A geometrical proof of the quasitriangular properties of the R-matrix operators (7.6.1) and (7.6.6) is provided by Theorem (8.3.4). Theorem (8.3.4) identifies the R-matrix operators with monodromy transformations in the corresponding chain complexes. The equality $(\Delta \otimes \text{id})(R) = R_{13}R_{23}$ states that the monodromy transformation of the chain complexes corresponding to the rotation of the point z_3 over the pair of points z_1 and z_2 is the composition of rotations of z_3 over z_2 and then z_3 over z_1, see Fig. 9.1, cf. Figs. 8.2, 5.17(a) and 5.22.

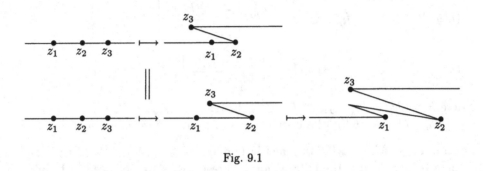

Fig. 9.1

The equality $(id \otimes \Delta)(R) = R_{13}R_{12}$ is shown in Fig. 9.2.

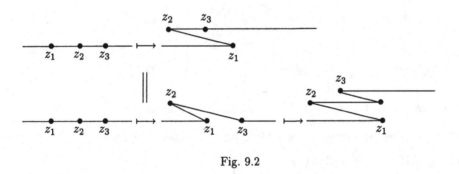

Fig. 9.2

The equality (9.1.1) corresponds to the fact that the monodromy commutes with the boundary operator.

Proof of Theorem 9.3.3. We will prove that (9.3.1) well defines homomorphisms (1) and (2) for general values of the matrix (b_{ij}) and the parameter q. Formulas (9.3.2) analytically depend on $(b_{ij}), q$. Hence they will define homomorphisms for all $(b_{ij}), q$.

For a general value of $(b_{ij}), q$, the algebras $(U_q\mathfrak{b}_-)'$ and $(U_q\mathfrak{b}_+)^0$ are generated by elements $\{(h_i)^*, (q^{-h_j/4}f_j)^*\}$ and $\{(h_i)^*, (q^{h_j/4}e_j)^*\}$, resp. We will use this remark. Fix a general value of $(b_{ij}), q$.

First, we prove that formulas (9.3.2) define homomorphisms $(U_q\mathfrak{b}_-)' \to \text{End}(M)$. In fact,

$$(h^\ell q^{-h_\lambda/4} F_{i_1\ldots i_N})^* (h^m q^{-h_\nu/4} F_{j_1\ldots j_L})^* \longmapsto$$

$$q^{D(h_\lambda)} \left[\prod_{j=1}^k \frac{Q(h_j^*)^{\ell_j}}{\ell_j!}\right] q^{h_\lambda/4} e_{i_N} \ldots e_{i_1} q^{D(h_\nu)} \left[\prod_{j=1}^k r\frac{Q(h_j^*)^{m_j}}{m_j!}\right] q^{h_\nu/4} e_{j_L} \ldots e_{j_1}$$

$$= q^{D(h_\lambda)+D(h_\nu)-\sum_{a=1}^N \sum_{b=1}^L b_{i_a j_b}/4} \left[\prod_{j=1}^k \frac{Q(h_j^*)^{\ell_j}}{\ell_j!}\right] \cdot$$

$$\left[\prod_{j=1}^k \frac{\left[Q(h_j^*) - \left\langle\sum_{a=1}^N \alpha_{i_a}, Q(h_j^*)\right\rangle\right]^{m_j}}{m_j!}\right] q^{(h_\lambda+h_\nu)/4} e_{i_N} \ldots e_{i_1} e_{j_L} \ldots e_{j_1} \quad (9.3.5)$$

$$= q^{D(h_\lambda+h_\nu)} \left[\prod_{j=1}^k \frac{Q(h_j^*)^{\ell_j}}{\ell_j!}\right] \cdot \left[\prod_{j=1}^k \frac{(Q(h_j^*) - s_j\frac{\varphi}{2})^{m_j}}{\ell_j!}\right] q^{(h_\lambda+h_\nu)/4} e_{\bar{I}} e_{\bar{J}}$$

$$= q^{D(h_\lambda+h_\nu)} \left[\prod_{j=1}^k \frac{Q(h_j^*)^{\ell_j}}{\ell_j!}\right] \cdot \left[\prod_{j=1}^k \sum_{a_j=0}^{m_j} \frac{Q(h_j^*)^{a_j}}{a_j!} \frac{[-s_j\frac{\varphi}{2}]^{m_j-a_j}}{(m_j-a_j)!}\right]$$

$$\cdot q^{(h_\lambda+h_\nu)/4} e_{\bar{I}} e_{\bar{J}}.$$

Here $s_1, \ldots, s_r$ are coefficients in the decomposition $\alpha_{i_1} + \cdots + \alpha_{i_N} = s_1\alpha_1 + \cdots + s_r\alpha_r$ and $s_{r+1} = \cdots = s_k = 0$. In (9.3.5), we use the equality $\langle \alpha_{i_a}, Q(h_j^*)\rangle = \frac{\varphi}{2}\langle \alpha_{i_a}, \Omega_0(h_j^*)\rangle = \frac{\varphi}{2}\delta_{i_a,j}$. It is easy to see that the right-hand side of (9.3.5) is the image of the right-hand side of the first formula in (9.2.5).

The fact that formulas (9.3.2) define a homomorphism $(U_q\mathfrak{b}_+)^0 \to \text{End}(M^*)$ is proved analogously.

Now we will check that the map $(U_q\mathfrak{b}_-)' \otimes U_q\mathfrak{b}_- \to \text{End}(M)$ preserves the commutation relations (9.2.11) of the quantum double $D((U_q\mathfrak{b}_-)')$. For example, check (9.2.11.4). Its left-hand side goes to

$$\left[q^{-h_i/4} f_i\right]\left[q^{-b_{jj}/4} q^{h_j/4} e_j\right] = \left[q^{-b_{jj}/4} q^{h_j/4} e_j\right]\left[q^{-h_i/4} f_i\right]$$
$$- \delta_{ij}\left[q^{h_i/2} - q^{-h_i/2}\right]. \quad (9.3.6)$$

The right-hand side of (9.2.11.4) goes to

$$\left[q^{-b_{jj}/4}q^{h_j/4}e_j\right]\left[q^{-h_i/4}f_i\right] + \delta_{ij}q^{-h_i/2} - \delta_{ij}\sum_{p_1\ldots p_k}\frac{\langle h_1,\alpha_j\rangle^{p_1}}{p_1!} \quad (9.3.7)$$

$$\ldots\frac{\langle h_k,\alpha_j\rangle^{p_k}}{p_k!}\cdot\frac{Q(h_1^*)^{p_1}}{p_1!}\ldots\frac{Q(h_k^*)^{p_k}}{p_k!}.$$

It is easy to see that (9.3.6) equals (9.3.7).

The fact that the map $U_q\mathfrak{b}_+ \otimes (U_q\mathfrak{b}_+)^0 \to \operatorname{End}(M')$ defines a homomorphism of $D(U_q\mathfrak{b}_+)$ is proved analogously.

The last statement of Theorem (9.3.3) is a direct corollary of formulas (9.3.1). In fact, the formulas (9.3.1) were guessed in an attempt to find a $D((U_q\mathfrak{b}_-)')$ structure on M such that R is its canonical element.

9.3.8. Remarks.

1. The comultiplication of operators g_I and u_I is defined by Lemma (5.18). It is easy to check that the homomorphisms $D((U_q\mathfrak{b}_-)') \to \operatorname{End}(M)$ and $D(U_q\mathfrak{b}_+) \to \operatorname{End}(M^*)$ are compatible with the comultiplication.

2. Let $\sum a_I f_I := \sum a_{i_1\ldots i_N}f_{i_1}\ldots f_{i_N} \in \ker(S:(U_q\mathfrak{n}_-)_\lambda \to (U_q\mathfrak{n}_-)^*_\lambda)$. Then the map $M \to M$, $x \mapsto \sum a_I e_{\bar{I}}x$, is the zero map.

In fact, the zero element $0 = \sum a_I(q^{-h_\lambda/4}F_I)^*$ acts on M as $q^{D(h_\lambda)+h_\lambda/4}\sum a_I e_{\bar{I}}$.

Corollary. $(\sum a_I f_I)M(\Lambda) \subset \ker(S:M(\Lambda) \to M(\Lambda)^*)$, where S is the contravariant form and $M(\Lambda)$ is a Verma module with an arbitrary highest weight Λ.

Corollary. $\sum a_I f_I$ acts as zero in any irreducible representation with highest weight.

Analogously, let $\sum a_I e_I \in \ker(S:(U_q\mathfrak{b}_+)_\lambda \to (U_q\mathfrak{b}_+)^*_\lambda)$. Then the map $M^* \to M^*$, $x \mapsto \sum a_I f_{\bar{I}}x$, is the zero map.

3. Consider the complex $C.(^+U_q\mathfrak{n}_-;M_\sigma;^+\Delta)_\lambda$ described in (5.1.24). Then

$$H_0(C.(^+U_q\mathfrak{n}_-;M_\sigma;^+\Delta)_\lambda) = \operatorname{Vac}(M_\sigma)_\lambda := \{m \in M_\sigma \mid g_J m = 0 \text{ for all } J\}.$$

The double $D((U_q\mathfrak{b}_-)')$ acts on M_σ as explained above. Consider the subalgebra $\mathfrak{N} \subset (U_q\mathfrak{b}_-)'$ generated by all elements of the form $(q^{-h_\lambda/4}f_{i_1}\ldots f_{i_N})^*$ for

$N > 0$. Here, the notations are the same as in (9.3.1). It is easy to see that the product of two elements of this type is a linear combination of elements of this type, cf. (9.2.5). We have

$$H_0(C.(^+U_q\mathfrak{n}_-; M_\sigma; {}^+\Delta)_\lambda) = \{m \in M_\lambda \mid \mathfrak{N}m = 0\}.$$

If $S : U_q\mathfrak{n}_- \to (U_q\mathfrak{n}_-)^*$ is nondegenerate, then

$$H_0(C.(^+U_q\mathfrak{n}_-; M_\sigma; {}^+\Delta)_\lambda) = \{m \in M_\lambda \mid e_1 m = \cdots = e_r m = 0\}.$$

9.4. The Quotient Complex $C.(^+U_q\mathfrak{n}_-; M_\sigma; {}^+\Delta)/\ker S$

Consider the morphism $S : C.(^+U_q\mathfrak{n}_-; M_\sigma; {}^+\Delta) \to C.^*(^+U_q\mathfrak{n}_-; M_\sigma; \mu)$ described in (5.1.26). In this section, we will describe $C.(^+U_q\mathfrak{n}_-; M_\sigma; {}^+\Delta)/\ker S$.

Let $S : U_q\mathfrak{n}_\pm \to (U_q\mathfrak{n}_\pm)^*$ be the contravariant form. Set $\overline{U_q\mathfrak{n}_\pm} = U_q\mathfrak{n}_\pm/\ker S$ and $^+\overline{U_q\mathfrak{n}_\pm} = {}^+U_q\mathfrak{n}_\pm/\ker S$. The contravariant form induces the nondegenerate symmetric bilinear form

$$\bar{S} : \overline{U_q\mathfrak{n}_\pm} \longrightarrow (\overline{U_q\mathfrak{n}_\pm})^*. \tag{9.4.1}$$

Let $^+\Delta : {}^+U_q\mathfrak{n}_- \to {}^+U_q\mathfrak{n}_- \otimes {}^+U_q\mathfrak{n}_-$ be the coalgebra structure introduced in (4.3.10). By Lemma (4.3.12), this coalgebra structure induces a coalgebra structure

$$^+\Delta : {}^+\overline{U_q\mathfrak{n}_-} \longrightarrow {}^+\overline{U_q\mathfrak{n}_-} \otimes {}^+\overline{U_q\mathfrak{n}_-}. \tag{9.4.2}$$

Consider the anti-isomorphism

$$\tau : U_q\mathfrak{n}_- \longrightarrow U_q\mathfrak{n}_+ \tag{9.4.3}$$

defined by formula $f_{i_1} \ldots f_{i_p} \mapsto e_{i_p} \ldots e_{i_1}$. This map induces an anti-isomorphism

$$\tau : \overline{U_q\mathfrak{n}_-} \longrightarrow \overline{U_q\mathfrak{n}_+}. \tag{9.4.4}$$

Define the element

$$\delta = \sum a_k \otimes b_k \in {}^+\overline{U_q\mathfrak{n}_-} \otimes {}^+\overline{U_q\mathfrak{n}_+} \tag{9.4.5}$$

as the image of the canonical element $\sum a_k \otimes a^k \in {}^+\overline{U_q\mathfrak{n}_-} \otimes {}^+\overline{U_q\mathfrak{n}_-}^*$ under the map

$$^+\overline{U_q\mathfrak{n}_-} \otimes {}^+\overline{U_q\mathfrak{n}_-}^* \xrightarrow{\text{id} \otimes \bar{S}^{-1}} {}^+\overline{U_q\mathfrak{n}_-} \otimes {}^+\overline{U_q\mathfrak{n}_-} \xrightarrow{\text{id} \otimes \tau} {}^+\overline{U_q\mathfrak{n}_-} \otimes {}^+\overline{U_q\mathfrak{n}_+}.$$

Let $M_1, \ldots, M_n$ be Verma modules over $U_q\mathfrak{g}$ with highest weights $\Lambda_1, \ldots, \Lambda_n \in \mathfrak{h}^*$, respectively. Let $\sigma \in \Sigma_n$ and $M_\sigma = M_{\sigma(1)} \otimes \cdots \otimes M_{\sigma(n)}$.

Let $S : M_j \to M_j^*$ be the contravariant form. Then $L_j = M_j/\ker S$ is the irreducible representation with the highest weight Λ_j, $j = 1, \ldots, n$.

The tensor product of contravariant forms defines a symmetric contravariant form S on M_σ such that $S(f_i x, y) = S(x, e_i y)$ and $S(hx, y) = S(x, hy)$ for all $x, y \in M_\sigma$, $h \in \mathfrak{h}$, $i = 1, \ldots, r$. This form induces an isomorphism

$$S : L_\sigma \longrightarrow L_\sigma^*.$$

Let $\overline{U_q\mathfrak{g}}$ be the quotient algebra of $U_q\mathfrak{g}$ by the two-sided ideal generated by $\ker(S : U_q\mathfrak{n}_- \to (U_q\mathfrak{n}_-)^*)$ and $\ker(S : U_q\mathfrak{n}_+ \to (U_q\mathfrak{n}_+)^*)$.

9.4.6. Lemma. *The action*

$$\overline{U_q\mathfrak{g}} \otimes L_\sigma \longrightarrow L_\sigma$$

is well defined.

Proof. Any $x = \sum a_I e_I \in \ker(S : U_q\mathfrak{n}_+ \to (U_q\mathfrak{n}_+)^*)$ acts on M_σ as zero by (9.3.8.2). For any $y = \sum b_I f_I \in \ker(S : U_q\mathfrak{n}_- \to (U_q\mathfrak{n}_-)^*)$, we have $y(M_\sigma) \in \ker(S : M_\sigma \to M_\sigma^*)$ by (9.3.8.2). Hence, y acts as zero on L_σ. It is easy to see that $\ker(S : M_\sigma \to M_\sigma^*)$ is invariant under the action of $U_q\mathfrak{g}$.

Define the map

$$\delta : L_\sigma \longrightarrow \overline{{}^+U_q\mathfrak{n}_-} \otimes L_\sigma, \qquad x \longmapsto \sum a_k \otimes b_k x, \qquad (9.4.7)$$

where $\sum a_k \otimes b_k$ is defined in (9.4.5).

9.4.8. Lemma. $\delta : L_\sigma \to \overline{{}^+U_q\mathfrak{n}_-} \otimes L_\sigma$ *is a comodule structure and the diagram*

$$\begin{array}{ccc} L_\sigma & \xrightarrow{\delta} & \overline{{}^+U_q\mathfrak{n}_-} \otimes L_\sigma \\ \uparrow & & \uparrow \\ M_\sigma & \xrightarrow{{}^+\Delta} & {}^+U_q\mathfrak{n}_- \otimes M_\sigma \end{array}$$

is commutative. Here the vertical maps are the natural projections.

Proof. The maps $S : M_\sigma \to M_\sigma^*$ and $S \otimes S : {}^+U_q\mathfrak{n}_- \otimes M_\sigma \to ({}^+U_q\mathfrak{n}_-)^* \otimes M_\sigma^*$

identify L_σ with Im S and $\overline{{}^+U_q\mathfrak{n}_-} \otimes L_\sigma$ with Im $(S \otimes S)$. The diagram

$$\begin{array}{ccc} M_\sigma^* & \xrightarrow{\mu^*} & ({}^+U_q\mathfrak{n}_-)^* \otimes M_\sigma^* \\ S\uparrow & & \uparrow S\otimes S \\ M_\sigma & \xrightarrow{{}^+\Delta} & ({}^+U_q\mathfrak{n}_-) \otimes M_\sigma \end{array}$$

is commutative and $\mu^* : \varphi(\cdot) \mapsto \sum (f_I)^* \otimes e_I \varphi(\cdot) = \sum (f_I)^* \otimes \varphi(f_I \cdot)$, where f_I runs through all monomials in ${}^+U_q\mathfrak{n}_-$. Choose a new basis in ${}^+U_q\mathfrak{n}_-$, $\{y_\beta, x_\lambda \mid \beta \in B, \lambda \in \Lambda\}$ such that $\{x_\lambda\}$ form a basis of ker S on ${}^+U_q\mathfrak{n}_-$. Let $\{f_\beta^*, x_\lambda^*\}$ be the dual basis of $({}^+U_q\mathfrak{n}_-)^*$. Then by (9.3.8.2), we have

$$\sum f_I^* \otimes S(x, f_I \cdot) = \sum_\beta y_\beta^* \otimes S(x, y_\beta \cdot) + \sum_\lambda x_\lambda^* \otimes S(x, x_\lambda \cdot)$$

$$= \sum_\beta y_{\beta'}^* \otimes S(x, y_\beta \cdot) = \sum_\beta S(S^{-1}(y_\beta^*)) \otimes S(\tau(y_\beta)x, \cdot)$$

$$= S \otimes S \left[\sum_\beta S^{-1}(y_\beta^*) \otimes \tau(y_\beta)x \right].$$

This formula shows that $\mu^* S[x] = (S \otimes S) \delta[x]$ for any $[x] \in L_\sigma$.

9.4.9. Corollary. *The quotient complex $C.({}^+U_q\mathfrak{n}_-; M_\sigma; {}^+\Delta)/\ker S$ is canonically isomorphic to the standard Hochschild complex of the coalgebra $\overline{{}^+U_q\mathfrak{n}_-}$ with coefficients in the comodule L_σ denoted by $C.(\overline{{}^+U_q\mathfrak{n}_-}; L_\sigma; \delta)$, see (4.4.3).*

9.4.10. Remark. $H_0(C.(\overline{{}^+U_q\mathfrak{n}_-}; L_\sigma; \delta)) = \{x \in L_\sigma \mid e_1 x = \cdots = e_r x = 0\}$.

9.4.11. For any $\lambda \in \mathbb{N}^r$, introduce the element $\bar\Omega_\lambda \in (\overline{U_q\mathfrak{n}_+})_\lambda \otimes (\overline{U_q\mathfrak{n}_-})_\lambda$ as the image of the canonical element $\sum a_k \otimes a^k \in (\overline{U_q\mathfrak{n}_-})_\lambda^* \otimes (\overline{U_q\mathfrak{n}_-})_\lambda$ under the map

$$(\overline{U_q\mathfrak{n}_-})_\lambda^* \otimes (\overline{U_q\mathfrak{n}_-})_\lambda \xrightarrow{\bar S^{-1} \otimes id} (\overline{U_q\mathfrak{n}_-})_\lambda \otimes (\overline{U_q\mathfrak{n}_-})_\lambda \xrightarrow{\tau \otimes id} (\overline{U_q\mathfrak{n}_+})_\lambda \otimes (\overline{U_q\mathfrak{n}_-})_\lambda.$$

Consider the R-matrix operators $R : M \otimes M' \to M \otimes M'$ and $R : M^* \otimes M'^* \to M^* \otimes M'^*$, introduced in (7.6.1) and (7.6.6), and the commutative diagram

$$\begin{array}{ccc} M \otimes M' & \xrightarrow{R} & M \otimes M' \\ S\downarrow & & \downarrow S \\ M^* \otimes M'^* & \xrightarrow{R} & M^* \otimes M'^* \end{array} \qquad (9.4.12)$$

9.4.13. Corollary. *The action of the R-matrix operator on $(M \otimes M')/\ker S$ is given by the operator*

$$R = \sum_{\lambda \in \mathbb{N}^r} q^{\Omega_0/2 + \frac{1}{4}(h_\lambda \otimes 1 - 1 \otimes h_\lambda) + D(\lambda)} \bar{\Omega}_\lambda .$$

Here $\bar{\Omega}_\lambda$ is defined in (9.4.11) and all other notations are explained in (7.6.1).

9.5. Quantum Groups Corresponding to Kac–Moody Algebras

Fix the following data (4.1.a)–(4.1.d): a finite-dimensional complex space $\mathfrak{h}$; a nondegenerate symmetric bilinear form (,) on $\mathfrak{h}$; linearly independent covectors $\alpha_1, \ldots, \alpha_r \in \mathfrak{h}^*$; and a complex nonzero number κ. As in (4.1), introduce the notation $b_{ij} = (\alpha_i, \alpha_j)$, $h_i = b^{-1}(\alpha_i)$, $q = \exp(2\pi\sqrt{-1}/\kappa)$ and so on.

Suppose that, for any $i, b_{ii} \neq 0$. Set $a_{ij} = 2b_{ij}/b_{ii}$. Suppose that the condition

$$a_{ij} \in \mathbb{Z}, \qquad a_{ij} \leq 0, \qquad \text{for } i \neq j, \qquad (9.5.1)$$

holds. So, (a_{ij}) is a Generalized Cartan Matrix. Suppose that the data (4.1.a)–(4.1.d) is its realization [**K**, Subsec. 1.1].

Introduce the standard quantum group $U_q\mathfrak{g}^0$ corresponding to the Generalized Cartan Matrix as the $\mathbb{C}$ algebra generated by the elements x_i^+, x_i^-, $i = 1, \ldots, r$, and the space $\mathfrak{h}$, subject to the relations (9.5.2) and (9.5.3):

$$[h, x_i^\pm] = \pm \langle \alpha_i, h \rangle x_i^\pm ,$$

$$[x_i^+, x_j^-] = \delta_{ij} \frac{q^{h_i/2} - q^{-h_i/2}}{q_i^{1/2} - q_i^{-1/2}} , \qquad (9.5.2)$$

$$[h, h'] = 0$$

for all i, j and $h, h' \in \mathfrak{h}$. Here $q_i = q^{b_{ii}/2}$.

$$\sum_{v=0}^{n_{ij}} (-1)^v (x_i^+)^v (x_j^+)(x_i^+)^{n_{ij}-v} \begin{bmatrix} n_{ij} \\ v \end{bmatrix}_{q_i} = 0,$$

$$\sum_{v=0}^{n_{ij}} (-1)^v (x_i^-)^v (x_j^-)(x_i^-)^{n_{ij}-v} \begin{bmatrix} n_{ij} \\ v \end{bmatrix}_{q_i} = 0 \qquad (9.5.3)$$

for all $i \neq j$. Here $n_{ij} = -a_{ij} + 1$ and $(\ )_{q_i}$ is a q-binomial coefficient introduced in (4.1.10).

$U_q\mathfrak{g}^0$ is a Hopf algebra, see [**D1, J, R**].

Let $U_q\mathfrak{g}$ be the quantum group corresponding to the data (4.1.a)–(4.1.d) and introduced in (4.1). Let $U_q\mathfrak{g}^{DJ}$ be the quotient algebra of $U_q\mathfrak{g}$ by the two-sided ideal generated by all Chevalley–Serre elements $R_{ij}(f)$, $R_{ij}(e)$ introduced in (4.1.11).

9.5.4. Lemma. *The map $x_i^+ \mapsto e_i$, $x_i^- \mapsto (q_i^{1/2} - q_i^{-1/2})f_i$, $h \mapsto h$ defines an isomorphism $U_q\mathfrak{g}^0 \simeq U_q\mathfrak{g}^{DJ}$.*

Consider the contravariant form $S : U_q\mathfrak{b}_- \to (U_q\mathfrak{b}_-)^*$ and $S : U_q\mathfrak{b}_+ \to (U_q\mathfrak{b}_+)^*$ introduced in (4.2). The contravariant form induces a well-defined symmetric bilinear form

$$S : U_q\mathfrak{b}_-^{DJ} \longrightarrow (U_q\mathfrak{b}_-^{DJ})^*,$$
$$S : U_q\mathfrak{b}_+^{DJ} \longrightarrow (U_q\mathfrak{b}_+^{DJ})^*, \qquad (9.5.5)$$

see Lemma (4.2.10).

(9.5.6) Assume that the numbers $b_{11}/2, \ldots, b_{rr}/2$ are positive relatively prime integers.

9.5.7. Theorem. *Let κ be not a rational number. Then the contravariant form S is nondegenerate on $U_q\mathfrak{b}_-^{DJ}$ and $U_q\mathfrak{b}_+^{DJ}$.*

Proof. Consider the Kac–Moody algebra $\mathfrak{g}$ corresponding to (a_{ij}). A Verma module for $\mathfrak{g}$ with highest weight Λ is irreducible for general values of Λ. Fix such Λ_0. The Verma module for $U_q\mathfrak{g}^{DJ}$ with highest weight Λ_0 is irreducible for general values of q. Hence $S : U_q\mathfrak{n}_-^{DJ} \to (U_q\mathfrak{n}_-^{DJ})^*$ is nondegenerate for general values of q, see (9.3.8). By (2.6.6) (see also (3.11)), S is not degenerate if q is not a root of unity.

9.5.8. Corollary. *The natural epimorphism*

$$U_q\mathfrak{g}^{DJ} \longrightarrow \overline{U_q\mathfrak{g}}$$

is an isomorphism if κ is not a rational number.

9.5.9. Corollary. *The maps $U_q\mathfrak{b}_+^{DJ} \to (U_q\mathfrak{b}_-^{DJ})'$, $U_q\mathfrak{b}_-^{DJ} \to (U_q\mathfrak{b}_+^{DJ})^0$ defined by formulas*

$$q^{D(h_\lambda)} \left[\prod_{j=1}^k \frac{Q(h_j^*)^{p_j}}{p_j!} \right] q^{h_\lambda/4} e_{i_N} \ldots e_{i_1} \mapsto \left[h^p q^{-h_\lambda/4} F_{i_1 \ldots i_N} \right]^*,$$

$$q^{D(h_\lambda)} \left[\prod_{j=1}^{k} \frac{Q(h_j^*)^{p_j}}{p_j!} \right] q^{-h_\lambda/4} f_{i_N} \cdots f_{i_1} \mapsto \left[h^p q^{h_\lambda/4} E_{i_1 \ldots i_N} \right]^*$$

are isomorphisms if κ is not a rational number. For notation, see (9.3.2).

Assume that $q^\ell = 1$, where $\ell \in \mathbb{N}$, and q is a primitive root. Assume that ℓ is odd and coprime with each $b_{11}/2, \ldots, b_{rr}/2$. For any j,

$$S(f_j^\ell, f_j^\ell) = S(e_j^\ell, e_j^\ell) = (\ell)_{q_j}!/(1)_{q_j}^\ell = 0. \qquad (9.5.10)$$

Let $I_\ell^- \subset U_q \mathfrak{n}_-^{DJ}$ (resp. $I_\ell^+ \subset U_q \mathfrak{n}_+^{DJ}$) be the two-sided ideal generated by $f_1^\ell, \ldots, f_r^\ell$ (resp. $e_1^\ell, \ldots, e_r^\ell$). Let $u_q \mathfrak{g}$ be the quotient algebra of $U_q \mathfrak{g}^{DJ}$ be the two-sided ideal generated by $\{f_1^\ell, \ldots, f_r^\ell, e_1^\ell, \ldots, e_r^\ell\}$. We call $u_q \mathfrak{g}$ the restricted quantum group.

9.5.11. Theorem. *Let (a_{ij}) be a Generalized Cartan Matrix, let (b_{ij}) be positive definite and let q be as above. Then $I_\ell^- = \ker(S : U_q \mathfrak{n}_-^{DJ} \to (U_q \mathfrak{n}_-^{DJ})^*)$ and $I_\ell^+ = \ker(S : U_q \mathfrak{n}_+^{DJ} \to (U_q \mathfrak{n}_+^{DJ})^*)$.*

Proof. By (9.5.10), $I_\ell^- \subset \ker S$. On the other side, if $x \in \ker S$ on $U_q \mathfrak{n}_-^{DJ}$, then x acts as zero in any irreducible representation with a highest weight, see (9.3.8). By [Ku, Prop. 4.2], there exists a $u_q \mathfrak{g}$-irreducible Verma module. This implies that $x \in I_\ell^-$.

Consider the R-matrix operators $R : M \otimes M' \to M \otimes M'$ and $R : M^* \otimes M'^* \to M^* \otimes M'^*$, introduced in (7.6.1) and (7.6.6) and the commutative diagram (9.4.12). The action of the R-matrix operator on $(M \otimes M')/\ker S$ is given by the R-matrix operator (9.4.13). $(M \otimes M')/\ker S$ is the tensor product of irreducible representations of $U_q \mathfrak{g}^{DJ}$ for irrational κ, and $(M \otimes M')/\ker S$ is the tensor product of irreducible representations of the restricted quantum group $u_q \mathfrak{g}$, if q is a primitive root of unity.

Drinfeld introduced the universal R-matrix $R \subset U_q \mathfrak{g}^{DJ} \otimes U_q \mathfrak{g}^{DJ}$ as the image of the R-matrix of the quantum double $D(U_q \mathfrak{b}_+^{DJ})$ under the natural projection $D(U_q \mathfrak{b}_+^{DJ})^{\otimes 2} \to (U_q \mathfrak{g}_+^{DJ})^{\otimes 2}$, see [**D1, R, LS**].

9.5.12. Theorem. *Let (a_{ij}) be a Generalized Cartan Matrix. Then for all q the action of the R-matrix operator (9.4.13) on $(M \otimes M')/\ker S$ coincides with the action of the Drinfeld universal R-matrix of $U_q \mathfrak{g}^{DJ}$.*

Proof. The Drinfeld R-matrix depends analytically on q. The R-matrix operator (9.4.13) depends analytically on q for $\kappa \notin \mathbb{Q}$ and its action for a rational κ_0 may be computed as a limit for $q \to \exp(2\pi i/\kappa_0)$. Therefore, it suffices to prove the theorem for $\kappa \notin \mathbb{Q}$ and q close to 1. The R-matrix operator (9.4.13) has triangular properties (9.1.1) and (9.1.2), see (9.3.4). Its expansion in $1/\kappa$ starts with id. By the Drinfeld uniqueness theorem [**D1**], the R-matrix operator (9.4.13) coincides with the Drinfeld universal R-matrix for small $1/\kappa$, see [**D1**]. The theorem is proved.

9.5.13. Remark. Theorems (9.5.7) and (9.5.12) show that for the special case, where the matrix $(a_{ij}) = (2b_{ij}/b_{ii})$ is a Generalized Cartan Matrix and q is not a root of unity, the quotient complex $C.(^+U_q\mathfrak{n}_-; M_\sigma; {}^+\Delta)/\ker S$ is canonically isomorphic to the standard Hochschild complex of the coalgebra $^+U_q\mathfrak{n}_-^{DJ}$ with coefficients in the comodule L_σ which is the tensor product of irreducible $U_q\mathfrak{g}^{DJ}$-modules with highest weights, see (9.4.9). The induced R-matrix action on this complex is given by Drinfeld's R-matrix

Theorems (9.5.11) and (9.5.12) show that under conditions of Theorem (9.5.11), the quotient complex $C.(^+U_q\mathfrak{n}_-; M_\sigma; {}^+\Delta)/\ker S$ is canonically isomorphic to the standard Hochschild complex of the finite-dimensional coalgebra $^+u_q\mathfrak{n}_-$ with coefficients in the tensor product of finite-dimensional irreducible representations with highest weights. Here $^+u_q\mathfrak{n}_- \subset u_q\mathfrak{g}$ is the nilpotent subalgebra of the restricted quantum group. This quantum group and these representations in particular were studied in [**DK, DKP**]. The induced R-matrix action on the image of S is given by Drinfeld's R-matrix, see also [**R4, R5**].

10

Hypergeometric Integrals

10.1. Orlik–Solomon Algebra and Flags

In this subsection, we formulate results of [OS, SV3]. Let V be an affine real or complex k-dimensional space and let $\mathcal{C}$ be a configuration of hyperplanes in V.

Define abelian groups $\mathcal{A}^p(\mathcal{C}, \mathbb{Z})$, $0 \leq p \leq k$. For $p = 0$, set $\mathcal{A}^p(\mathcal{C}, \mathbb{Z}) = \mathbb{Z}$. For $p \geq 1$, $\mathcal{A}^p(\mathcal{C})$ is generated by p-tuples $(H_1, \ldots, H_p)$, $H_i \in \mathcal{C}$, subject to the following relations:

$$(H_1, \ldots, H_p) = 0 \tag{10.1.1}$$

if $H_1, \ldots, H_p$ are not in the general position,

$$(H_{\sigma(1)}, \ldots, H_{\sigma(p)}) = (-1)^{|\sigma|}(H_1, \ldots, H_p), \tag{10.1.2}$$

for any permutation $\sigma \in \Sigma_p$,

$$\sum_{i=1}^{p+1}(-1)^i(H_1, \ldots, \hat{H}_i, \ldots, H_{p+1}) = 0 \tag{10.1.3}$$

for any $(p+1)$-tuple $(H_1, \ldots, H_{p+1})$ which is not in general position and such that $H_1 \cap \cdots \cap H_{p+1} \neq 0$.

The direct sum $\mathcal{A}^{\cdot}(\mathcal{C}, \mathbb{Z}) = \bigoplus_{p=0}^{k} \mathcal{A}^p(\mathcal{C}, \mathbb{Z})$ is a graded skew commutative algebra with respect to the multiplication

$$(H_1, \ldots, H_p) \cdot (H'_1, \ldots, H'_q) = (H_1, \ldots, H_p, H'_1, \ldots, H'_q).$$

$\mathcal{A}^{\cdot}(\mathcal{C})$ is called the *Orlik–Solomon algebra* of the configuration $\mathcal{C}$.

10.1.4. Theorem. [Bj], see also [SV3]. *All groups $\mathcal{A}^p(\mathcal{C}, \mathbb{Z})$ are free over $\mathbb{Z}$.*

10.1.5. Flags [SV3]. For $0 \leq p \leq k$, denote by $\operatorname{Flag}^p(\mathcal{C})$, the set of all flags,

$$L^0 \supset L^1 \supset \cdots \supset L^p,$$

where L^i is an edge of $\mathcal{C}$ of codimension i. Denote by $\overline{\operatorname{Flag}}^p(\mathcal{C})$ the free abelian group on $\operatorname{Flag}^p(\mathcal{C})$ and by $F\ell^p(\mathcal{C}, \mathbb{Z})$ the quotient of $\overline{\operatorname{Flag}}^p(\mathcal{C})$ by the following relations.

For every i, $0 < i < p$, and a flag with a gap,

$$\hat{F} = (L^0 \supset L^1 \supset \cdots \supset L^{i-1} \supset L^{i+1} \supset \cdots \supset L^p),$$

where L^j is an edge of codimension j, we set

$$\sum_{F \supset \hat{F}} F = 0 \tag{10.1.6}$$

in $F\ell^p(\mathcal{C}, \mathbb{Z})$, where the summation is over all $F = (\tilde{L}^0 \supset \tilde{L}^1 \supset \cdots \supset \tilde{L}^p)$ $\in \operatorname{Flag}^p(\mathcal{C})$ such that $\tilde{L}^j = L^j$ for all $j \neq i$.

10.1.7. Theorem. [SV3]. *All groups $F\ell^p(\mathcal{C}, \mathbb{Z})$ are free over $\mathbb{Z}$.*

$\mathcal{A}^p(\mathcal{C}, \mathbb{Z})$ and $F\ell^p(\mathcal{C}, \mathbb{Z})$ are dual. Namely, define a homomorphism $\varphi^p : \mathcal{A}^p(\mathcal{C}, \mathbb{Z}) \to F\ell^p(\mathcal{C}, \mathbb{Z})^*$.

For $(H_1, \ldots, H_p)$ in the general position, $H_i \in \mathcal{C}$, define a flag

$$F(H_1, \ldots, H_p) = (H_1 \supset H_{12} \supset \cdots \supset H_{12\ldots p}) \in \operatorname{Flag}^p(\mathcal{C}), \tag{10.1.8}$$

where $H_{12\ldots i} = H_1 \cap \cdots \cap H_i$.

For a flag $F \in \operatorname{Flag}^p(\mathcal{C})$, define a functional $\delta_F \in F\ell^p(\mathcal{C}, \mathbb{Z})^*$ as $\delta_F(F') = 1$ if $F' = F$ and $\delta_F(F') = 0$ otherwise.

For $(H_1, \ldots, H_p)$ in a general position, set

$$\varphi^p(H_1, \ldots, H_p) = \sum_{\sigma \in \Sigma_p} (-1)^{|\sigma|} \delta_{F(H_{\sigma(1)}, \ldots, H_{\sigma(p)})}. \tag{10.1.9}$$

10.1.10. Theorem [SV3]. *All maps φ^p are isomorphisms.*

Set $\mathcal{A}^p(\mathcal{C}) = \mathcal{A}^p(\mathcal{C}, \mathbb{Z}) \otimes_{\mathbb{Z}} \mathbb{C}$ and $F\ell^p(\mathcal{C}) = F\ell^p(\mathcal{C}, \mathbb{Z}) \otimes_{\mathbb{Z}} \mathbb{C}$ for all p.

10.2. Framings and Bases

For a configuration $\mathcal{C}$ in an affine k-dimensional space, denote by $\mathcal{C}^p$ the set of all edges of $\mathcal{C}$ of codimension p, where $p = 1, \ldots, k$.

For an edge L of $\mathcal{C}$, set $\mathcal{C}_L = \{H \in \mathcal{C} | H \supset L\}$. Inclusions $\mathcal{C}_L \hookrightarrow \mathcal{C}$ induce a map.

$$\bigoplus_{L \in \mathcal{C}^p} \mathcal{A}^p(\mathcal{C}_L, \mathbb{Z}) \longrightarrow \mathcal{A}^p(\mathcal{C}, \mathbb{Z}). \tag{10.2.1}$$

This map is an isomorphism, see [**OS, SV3**].

We denote by $\mathcal{A}^p(\mathcal{C}, \mathbb{Z})_L$ the image of $\mathcal{A}^p(\mathcal{C}_L, \mathbb{Z})$ in $\mathcal{A}^p(\mathcal{C}, \mathbb{Z})$. So, we have

$$\mathcal{A}^p(\mathcal{C}, \mathbb{Z}) = \bigoplus_{L \in \mathcal{C}^p} \mathcal{A}^p(\mathcal{C}, \mathbb{Z})_L. \tag{10.2.2}$$

For $L^p \in \mathcal{C}^p$ and $H_0 \in \mathcal{C}_{L^p}$, set

$$\mathcal{A}^{p-1}(\mathcal{C}_{L^p}, H_0, \mathbb{Z}) = \bigoplus \mathcal{A}^{p-1}(\mathcal{C}, \mathbb{Z})_L \hookrightarrow \mathcal{A}^{p-1}(\mathcal{C}, \mathbb{Z}), \tag{10.2.3}$$

where the sum is taken over all edges $L \in \mathcal{C}^{p-1}$ such that $L^p \subset L$ and $L \not\subset H_0$.

Consider a map

$$\pi(L^p, H_0) : \mathcal{A}^{p-1}(\mathcal{C}_{L^p}, H_0, \mathbb{Z}) \to \mathcal{A}^p(\mathcal{C}, \mathbb{Z})_{L^p}, \tag{10.2.4}$$

sending $(H_1, \ldots, H_{p-1})$ to $(H_0, H_1, \ldots, H_{p-1})$.

10.2.5. Lemma. cf. [**SV3**, (1.5.2)]. *This map is an isomorphism.*

Proof. The inverse map may be defined as the composition of the map $\mathcal{A}^p(\mathcal{C})_{L^p} \to \mathcal{A}^{p-1}(\mathcal{C})$, which sends a p-tuple in general position $(H_1, \ldots, H_p)$ to $\sum_{i=1}^{p}(-1)^{i-1}(H_1, \ldots, \hat{H}_i, \ldots, H_p)$, and the projection $\mathcal{A}^{p-1}(\mathcal{C}) \to \mathcal{A}^{p-1}(\mathcal{C}_{L^p}, H_0)$ induced by the decomposition (10.2.2).

Inclusions $\mathcal{C}_L \hookrightarrow \mathcal{C}$ induce a map

$$\bigoplus_{L \in \mathcal{C}^p} F\ell^p(\mathcal{C}_L, \mathbb{Z}) \longrightarrow F\ell^p(\mathcal{C}, \mathbb{Z}). \tag{10.2.6}$$

This map is an isomorphism, see [**SV3**]. We denote by $F\ell^p(\mathcal{C}, \mathbb{Z})_L$, the image of $F\ell^p(\mathcal{C}_L, \mathbb{Z})$ in $F\ell^p(\mathcal{C}, \mathbb{Z})$. We have

$$F\ell^p(\mathcal{C}, \mathbb{Z}) = \bigoplus_{L \in \mathcal{C}^p} F\ell^p(\mathcal{C}, \mathbb{Z})_L. \tag{10.2.7}$$

For $L^p \subset \mathcal{C}^p$ and $H_0 \in \mathcal{C}_{L^p}$, set

$$F\ell^{p-1}(\mathcal{C}_{L^p}, H_0, \mathbb{Z}) = \bigoplus F\ell^{p-1}(\mathcal{C}, \mathbb{Z})_L \hookrightarrow F\ell^{p-1}(\mathcal{C}, \mathbb{Z}), \tag{10.2.8}$$

where the sum is taken over all edges $L \in C^{p-1}$ such that $L^p \subset L$ and $L \not\subset H_0$. Consider the map

$$F\ell^{p-1}(C_{L^p}, H_0, \mathbb{Z}) \longrightarrow F\ell^p(C, \mathbb{Z})_{L^p} \qquad (10.2.9)$$

sending $(L^0 \supset \cdots \supset L^{p-1})$ to $(L^0 \supset \cdots \supset L^{p-1} \supset L^p)$.

10.2.10. Lemma. *This map is an isomorphism. This isomorphism multiplied by $(-1)^{p-1}$ is compatible with the isomorphisms (10.2.4) and (10.1.9).*

A framing $\mathcal{O}$ of a configuration C is a choice for every edge L of C of a hyperplane $H(L) \in C$ containing L. For such a framing, define inductively sets $\mathcal{O}_j$ for $j = 0, \ldots, k-1$. Set $\mathcal{O}_0 = C^k$,

$$\mathcal{O}_1 = \{L^k \subset L^{k-1} \mid L^k \in \mathcal{O}_0, L^{k-1} \in C^{k-1} \text{ and } L^{k-1} \not\subset H(L^k)\}$$

$$\mathcal{O}_j = \{L^k \subset L^{k-1} \subset \cdots \subset L^{k-j} \mid (L^k \subset \cdots \subset L^{k-j+1}) \in \mathcal{O}_{j-1} \qquad (10.2.11)$$

and $L^{k-j} \not\subset H(L^p)$ for $p > k - j\}$.

Define the groups $\mathcal{A}^{k-j,j} = \mathcal{A}^{k-j,j}(C, \mathcal{O}, \mathbb{Z})$ for $j = 0, \ldots, k-1$ as follows:

$$\mathcal{A}^{k,0} = \mathcal{A}^k(C, \mathbb{Z}) \qquad (10.2.12)$$

and

$$\mathcal{A}^{k-j,j} = \bigoplus_{(L^k \subset \cdots \subset L^{k-j+1}) \subset \mathcal{O}_{j-1}} \mathcal{A}^{k-j}(C_{L^{k-j+1}}, H(L^{k-j-1}), \mathbb{Z}).$$

We have $\mathcal{A}^{k-j,j} \simeq \mathcal{A}^{k-j+1,j-1}$ for all $j = 0, \ldots, k-2$, see [**SV3**, (1.6)].

Define a set

$$B(\mathcal{O}) = \{(H(L^1), \ldots, H(L^k))\} \subset \mathcal{A}^k(C, \mathbb{Z}) \qquad (10.2.13)$$

whose elements are indexed by all elements of $\mathcal{O}_{k-1}$.

From the above it follows

10.2.14. Theorem, [**Bj**], cf. [**SV3**, (1.6.4)]. $B(\mathcal{O})$ *is a basis of* $\mathcal{A}^k(C, \mathbb{Z})$ *over* $\mathbb{Z}$.

10.2.15. Theorem. $\mathcal{O}_{k-1}$ *is a basis of* $F\ell^k(C, \mathbb{Z})$ *over* $\mathbb{Z}$. *This basis multiplied by* $(-1)^{k(k-1)/2}$ *is the basis dual to* $B(\mathcal{O})$.

10.3. Quasiclassical Contravariant Form of a Configuration, [SV3]

Assume that the configuration C is weighted, that is, to any hyperplane $H \in C$ its weight, a number $a(H)$, is assigned. Define the quasiclassical weight of any edge L of $\mathcal{C}$ as the sum of the weights of all hyperplanes which contain the edge.

Say that a p-tuple $\bar{H} = (H_1, \ldots, H_p)$, $H_i \in \mathcal{C}$, is adjacent to a flag F if there exists $\sigma \in \Sigma_p$ such that $F = F(H_{\sigma(1)}, \ldots, H_{\sigma(p)})$, see (10.1.8). This permutation σ is unique. Denote it by $\sigma(\bar{H}, F)$.

Define a symmetric bilinear form S^p in $F\ell^p(\mathcal{C})$. For $F, F' \in \operatorname{Flag}^p(\mathcal{C})$, set

$$S^p(F, F') = \frac{1}{p!} \sum (-1)^{\sigma(\bar{H},F)\sigma(\bar{H},F')} a(H_1) \ldots a(H_p), \qquad (10.3.1)$$

where the summation is over all $\bar{H} = (H_1, \ldots, H_p)$ which are adjacent to F and F'.

10.3.2. Theorem [SV3]. *Formula (10.3.1) defines a bilinear symmetric form on $F\ell^p(\mathcal{C})$.*

The form S^p is called the *quasiclassical contravariant form* of the configuration $\mathcal{C}$, cf. (2.6).

For the configuration of all diagonal hyperplanes, the quasiclassical contravariant form is interpreted in **[SV3]** as the contravariant form of a suitable Kac–Moody algebra.

The determinant of the matrix S^p in any basis of $F\ell^p(\mathcal{C}, \mathbb{Z})$ is denoted by $\det S^p$. $\det S^p$ is well defined.

10.3.3. Theorem, [SV3], cf. [V4]. *For any p,*

$$\det S^p = \operatorname{const} \prod_L a(L)^{e(L)},$$

where the product is over all edges of $\mathcal{C}$, $a(L)$ is the quasiclassical weight of the edge L, $e(L)$ is a suitable non-negative integer.

See its definition in **[SV3]** and **[V4]**.

10.3.4. Remark. For any weighted configuration of hyperplanes in a real affine space, two bilinear forms are defined: the quasiclassical bilinear form and the quantum bilinear form, see (2.6). The quasiclassical bilinear form is an appropriate limit of the quantum bilinear form, see **[V4]**.

10.3.5. Flag complex. Define a differential $d : F\ell^p \to F\ell^{p+1}$ by

$$d(L^0 \supset L^1 \supset \cdots \supset L^p) = \sum_{L^{p+1}} (L^0 \supset \cdots \supset L^p \supset L^{p+1}),$$

where the sum is taken over all edges L^{p+1} of codimension $p+1$ such that $L^p \supset L^{p+1}$. From (10.1.6), it follows that $d^2 = 0$.

10.3.6. Complex $(\mathcal{A}^{\cdot}, d(a))$. Set

$$\omega = \omega(a) = \sum_{H \in \mathcal{C}} a(H) H, \qquad \omega(a) \in \mathcal{A}^1.$$

Define a differential $d = d(a) : \mathcal{A}^p \to \mathcal{A}^{p+1}$ by the rule

$$dx = \omega(a) \cdot x.$$

It is clear that $d^2 = 0$.

For any p, the quasiclassical bilinear form of $\mathcal{C}$ defines a homomorphism

$$S : F\ell^p \longrightarrow (F\ell^p)^* \simeq \mathcal{A}^p. \qquad (10.3.7)$$

10.3.8. Lemma [SV3]. **S** *defines a homomorphism of complexes* $S = S(a)$: $(F\ell^{\cdot}, d) \to (\mathcal{A}^{\cdot}, d(a))$.

For any edge L, set

$$S(L) = \sum_{\substack{H \in \mathcal{C} \\ L \subset H}} a(H) H, \qquad S(L) \in \mathcal{A}^1. \qquad (10.3.9)$$

It is easy to see that the homomorphism (10.3.7) is defined by the formula

$$S : (L^0 \supset L^1 \supset \cdots \supset L^p) \longmapsto S(L^1) \cdot S(L^2) \ldots S(L^p). \qquad (10.3.10)$$

In other words,

$$S(L^0 \supset \cdots \supset L^p) = \sum a(H_1) \ldots a(H_p)(H_1, \ldots, H_p), \qquad (10.3.11)$$

where the sum is over all p-tuples $(H_1, \ldots, H_p)$ such that $H_i \supset L^i$ for all i.

10.4. Relative Complexes

Let V^{n+k} and B^n be affine spaces of dimension $n+k$ and n, respectively. Let $pr : V^{n+k} \to B^n$ be an affine epimorphism. Let $\mathcal{C}$ be a weighted configuration of hyperplanes in V, and let C' be a configuration of hyperplanes in B. Assume that

(a) The preimage $pr^{-1}(H)$ of any hyperplane $H \in C'$ is a hyperplane of C.

(b) The image $pr(L)$ of any edge L of C is B or is contained in a hyperplane of C'.

10.4.1. Example. Let $pr : \mathbb{C}^{n+k} \to \mathbb{C}^n$ be the projection on the last n coordinates, $C = C_{n+k}$ the configuration of all diagonal hyperplanes in $\mathbb{C}^{n+k}$, $C' = C_n$ the configuration of all diagonal hyperplanes in $\mathbb{C}^n$.

10.4.2. Remark. Let $pr : V \to B$ be an affine epimorphism. Let C be a configuration in V and C' a configuration in B. Then there exists a configuration $\bar{C}$ in V and a configuration $\bar{C}'$ in B such that $C \subset \bar{C}$, $C' \subset \bar{C}'$, and the triple pr, $\bar{C}$, $\bar{C}'$ satisfy properties (a) and (b).

An edge L of C has *relative codimension p* if $pr(L)$ has codimension p in B. An edge will be called *horizontal*, if it has relative codimension 0, and will be called *vertical*, otherwise.

For all p and i, introduce

$$\mathcal{A}_i^p = \bigoplus \mathcal{A}^p(C)_L, \qquad (10.4.3)$$

where the sum is taken over all edges $L \in C^p$ of relative codimension i.

(10.4.4) If $\mathcal{A}_i^p$ is nontrivial, then $\max(0, p-k) \leq i \leq \min(p, n)$.

(10.4.5) The subspace $\mathcal{A}_v^p = \bigoplus_{i>0} \mathcal{A}_i^p \subset \mathcal{A}^p(C)$ will be called *vertical* and the subspace $\mathcal{A}_h^p := \mathcal{A}_0^p \subset \mathcal{A}^p(C)$ will be called *horizontal*.

We have

$$\mathcal{A}^p(C) = \mathcal{A}_h^p \oplus \mathcal{A}_v^p = \bigoplus_{i \geq 0} \mathcal{A}_i^p. \qquad (10.4.6)$$

Let a be the weights of C. Consider the differential $d(a)$ in $\mathcal{A}^{\cdot}(C)$ introduced in (10.3). It is easy to see that

$$d(a)(\mathcal{A}_i^p) \subset \mathcal{A}_i^{p+1} \oplus \mathcal{A}_{i+1}^{p+1}. \qquad (10.4.7)$$

Let

$$d(a) = d_h(a) + d_v(a) \qquad (10.4.8)$$

be the corresponding horizontal and vertical decomposition, that is, $d_h(a) : \mathcal{A}_i^p \to \mathcal{A}_i^{p+1}$ and $d_v(a) : \mathcal{A}_i^p \to \mathcal{A}_{i+1}^{p+1}$ for all p and i. We have

$$d_h^2(a) = 0, \qquad d_v^2(a) = 0,$$
$$d_h(a)d_v(a) + d_v(a)d_h(a) = 0. \qquad (10.4.9)$$

The complex $(\mathcal{A}_h^p(\mathcal{C}), d_h(a))$ will be called the *horizontal complex* of $\mathcal{C}$ and pr.

More generally, let M be any edge $\mathcal{C}'$. For any p, set

$$\mathcal{A}^p(\mathcal{C}, M) = \bigoplus \mathcal{A}^p(\mathcal{C})_L, \qquad (10.4.10)$$

where the sum is taken over all edges $L \in \mathcal{C}^p$ such that $pr(L) = M$. If $\mathcal{A}^p(\mathcal{C}, M)$ is nontrivial, then p is not less than the codimension of M in B. It is easy to see that $d_h(\mathcal{A}^p(\mathcal{C}, M)) \subset \mathcal{A}^{p+1}(\mathcal{C}, M)$. Hence,

$$(\mathcal{A}^{\cdot}(\mathcal{C}, M), d_h(a)) \qquad (10.4.11)$$

form a complex corresponding to an edge $M \subset B$. In particular,

$$(\mathcal{A}_h^{\cdot}(\mathcal{C}), d_h(a)) = (\mathcal{A}^{\cdot}(\mathcal{C}, B), d_h(a)). \qquad (10.4.12)$$

For all p and i, introduce

$$F\ell_i^p = \bigoplus F\ell^p(\mathcal{C})_L, \qquad (10.4.13)$$

where the sum is taken over all edges $L \in \mathcal{C}^p$ of relative codimension i.

(10.4.14) The subspace $F\ell_v^p := \bigoplus_{i>0} F\ell_i^p \subset F\ell^p(\mathcal{C})$ will be called vertical and the subspace $F\ell_h^p := F\ell_0^p \subset F\ell^p(\mathcal{C})$ will be called horizontal.

$$F\ell^p(\mathcal{C}) = F\ell_h^p \oplus F\ell_v^p = \bigoplus_{i \geq 0} F\ell_i^p. \qquad (10.4.15)$$

Isomorphisms (10.1.9), (10.2.1), and (10.2.6) induce isomorphisms

$$(\mathcal{A}_i^p)^* \simeq F\ell_i^p \quad \text{for all } p \text{ and } i. \qquad (10.4.16)$$

Consider the differential d in $F\ell^{\cdot}(\mathcal{C})$ introduced in (10.3). It is easy to see that

$$d(F\ell_i^p) \subset F\ell_i^{p+1} \oplus F\ell_{i+1}^{p+1}. \qquad (10.4.17)$$

Let
$$d = d_h + d_v \qquad (10.4.18)$$
be the corresponding horizontal and vertical components, $d_h : F\ell_i^p \to F\ell_i^{p+1}$ and $d_v : F\ell_i^p \to F\ell_{i+1}^{p+1}$ for all p and i. The complex $(F\ell_h^{\cdot}(\mathcal{C}), d_h)$ will be called the *horizontal flag complex* of $\mathcal{C}$ and pr.

Consider the quasiclassical contravariant form $S : (F\ell^{\cdot}(\mathcal{C}), d) \to (\mathcal{A}^{\cdot}(\mathcal{C}), d(a))$ introduced in (10.3). It is easy to see that $S(F\ell_i^p) \subset \mathcal{A}_i^p$ for all p and i and S commutes with the horizontal and vertical differentials (10.4.8) and (10.4.18). In particular, the restriction of S to the horizontal complex $(F\ell_h^{\cdot}(\mathcal{C}), d_h)$ defines a horizontal quasiclassical contravariant form

$$S_h : (F\ell_h^{\cdot}(\mathcal{C}), d_h) \longrightarrow (\mathcal{A}_h^{\cdot}(\mathcal{C}), d_h(a)). \qquad (10.4.19)$$

S_h is a homomorphism of the complexes.

Let $z \in B$ be a point and $z \notin \bigcup_{H \in \mathcal{C}'} H$. Let $F(z) = pr^{-1}(z)$ be the fiber of pr over z. The configuration $\mathcal{C}$ cut in $F(z)$ a configuration of hyperplanes denoted by $\mathcal{C}(z)$. The hyperplanes of $\mathcal{C}(z)$ are numerated by horizontal hyperplanes of $\mathcal{C}$. Define the weights for $\mathcal{C}(z)$ by the rule: $a(H \cap F(z)) = a(H)$ for any horizontal $H \subset \mathcal{C}$.

Consider the complexes $(F\ell^{\cdot}(\mathcal{C}(z)), d)$ and $(\mathcal{A}^{\cdot}(\mathcal{C}(z)), d(a))$ of the configuration $\mathcal{C}(z)$ and the corresponding quasiclassical contravariant form $S(z) : (F\ell^{\cdot}(\mathcal{C}(z)), d) \to (\mathcal{A}^{\cdot}(\mathcal{C}(z)), d(a))$.

There are natural isomorphisms

$$\begin{aligned}\mu : (F\ell_h^{\cdot}(\mathcal{C}), d_h) &\longrightarrow (F\ell^{\cdot}(\mathcal{C}(z)), d), \\ \nu : (\mathcal{A}_h^{\cdot}(\mathcal{C}), d_h(a)) &\longrightarrow (\mathcal{A}^{\cdot}(\mathcal{C}(z)), d(a)).\end{aligned} \qquad (10.4.20)$$

These isomorphisms are defined by the rule

$$\begin{aligned}\mu : (L^0 \supset \cdots \supset L^p) &\longmapsto (L^0 \cap F(z) \supset \cdots \supset L^p \cap F(z)), \\ \nu : (H_1, \ldots, H_p) &\longmapsto (H_1 \cap F(z), \ldots, H_p \cap F(z)).\end{aligned} \qquad (10.4.21)$$

These isomorphisms identify the horizontal contravariant form S_h with the contravariant form $S(z)$.

10.5. Integrable Connection on the Horizontal Complexes

Under the assumptions of (10.4), we will construct homomorphisms $\Omega(H) : (\mathcal{A}_h^{\cdot}(\mathcal{C}), d_h) \to (\mathcal{A}_h^{\cdot}(\mathcal{C}), d_h(a))$ and $\Omega(H) : (F\ell_h^{\cdot}(\mathcal{C}), d_h) \to (F\ell_h^{\cdot}(\mathcal{C}), d_h)$ for any

$H^1 \in \mathcal{C}'$. These operators will play the role of the Casimir operators for the horizontal complex of a special triple $pr, \mathcal{C}, \mathcal{C}'$ described below in Sec. 12.

Informally speaking, $\Omega(H^1)$ is the part of the differential $d_v(a) : \mathcal{A}_0^p \to \mathcal{A}_1^{p+1}$ corresponding to the vertical hyperplane $pr^{-1}(H^1)$ (after division of the image by $pr^{-1}(H^1)$).

Let $L^p \subset \mathcal{C}^p$ be an edge of codimension p and relative codimension 1. Hence $pr(L^p) \subset \mathcal{C}'$ and $pr^{-1}(pr(L^p))$ is a hyperplane in $\mathcal{C}$. Denote it by $H(L^p)$. Consider the monomorphism

$$\pi(L^p) : \mathcal{A}^p(\mathcal{C})_{L^p} \longrightarrow \mathcal{A}_h^{p-1}(\mathcal{C}) \tag{10.5.1}$$

which is the composition of the isomorphism $\pi(L^p, H(L^p))^{-1} : \mathcal{A}^p(\mathcal{C})_{L^p} \to \mathcal{A}^{p-1}(\mathcal{C}_{L^p}, H(L^p))$ and the inclusion $\mathcal{A}^p(\mathcal{C}_{L^p}, H(L^p)) \subset \mathcal{A}_h^p$.

Let $L^p \in \mathcal{C}^p$ be a horizontal edge and $H^1 \in \mathcal{C}'$, set $L^{p+1} = L^p \cap pr^{-1}(H^1)$ and $C(L^p, H^1) = \{H \in \mathcal{C} | H \cap L^p = L^{p+1}\}$. Then L^{p+1} is an edge of codimension $p+1$ and relative codimension 1.

For every $H \in C(L^p, H^1)$, define a map

$$\Omega(L^p, H^1, H) : \mathcal{A}^p(\mathcal{C})_{L^p} \longrightarrow \mathcal{A}_h^p(\mathcal{C}) \tag{10.5.2}$$

as the composition of the map $H \wedge : \mathcal{A}^p(\mathcal{C})_{L^p} \to \mathcal{A}^{p+1}(\mathcal{C})_{L^{p+1}}$, given by the left multiplication by H, and the monomorphism $\pi(L^{p+1}) : \mathcal{A}^{p+1}(\mathcal{C})_{L^{p+1}} \to \mathcal{A}_h^p(\mathcal{C})$.

Define a map

$$\Omega(L^p, H^1) : \mathcal{A}^p(\mathcal{C})_{L^p} \longrightarrow \mathcal{A}_h^p(\mathcal{C}) \tag{10.5.3}$$

by the formula

$$\Omega(L^p, H^1) = \sum_{H \in C(L^p, H^1)} a(H) \Omega(L^p, H^1, H).$$

Define a map

$$\Omega(H^1) : \mathcal{A}_h^p(\mathcal{C}) \longrightarrow \mathcal{A}_h^p(\mathcal{C}) \tag{10.5.4}$$

as follows. $\mathcal{A}_h^p(\mathcal{C}) = \oplus \mathcal{A}^p(\mathcal{C})_{L^p}$, where L^p runs through all horizontal edges of codimension p. Set $\Omega(H^1)$ to be the map whose restriction to $\mathcal{A}^p(\mathcal{C})_{L^p}$ is $\Omega(L^p, H^1)$.

10.5.5. Lemma. *The map $\Omega(H^1)$ induces an endomorphism of the horizontal complex,*

$$\Omega(H^1) : (\mathcal{A}_h^{\cdot}(\mathcal{C}), d_h(a)) \longrightarrow (\mathcal{A}_h^{\cdot}(\mathcal{C}), d_h(a)).$$

The lemma follows from (10.4.9) and (10.2.5).

More generally, we have the following picture. The projection $pr: V \to B$ induces a monomorphism

$$pr^*: \mathcal{A}^\cdot(\mathcal{C}') \longrightarrow \mathcal{A}^\cdot(\mathcal{C}). \tag{10.5.6}$$

We identify $\mathcal{A}^\cdot(\mathcal{C}')$ with its image $pr^*(\mathcal{A}^\cdot(\mathcal{C}'))$. Isomorphism (10.2.4) induces an isomorphism

$$\pi: \mathcal{A}_i^p(\mathcal{C}) \longrightarrow \mathcal{A}^i(\mathcal{C}') \otimes \mathcal{A}_0^{p-i}(\mathcal{C}) \tag{10.5.7}$$

for all p. The following diagrams are commutative for all p:

$$\begin{array}{ccc}
\mathcal{A}_i^p(\mathcal{C}) & \xrightarrow{d_h(a)} & \mathcal{A}_i^{p+1}(\mathcal{C}) \\
\pi \downarrow & & \searrow \pi \\
\mathcal{A}^i(\mathcal{C}') \otimes \mathcal{A}_0^{p-i}(\mathcal{C}) & \xrightarrow{(-1)^i \otimes d_h(a)} & \mathcal{A}^i(\mathcal{C}') \otimes \mathcal{A}_0^{p-i+1}(\mathcal{C})
\end{array} \tag{10.5.8}$$

$$\begin{array}{ccc}
 & \mathcal{A}_i^p(\mathcal{C}) \xrightarrow{d_v(a)} \mathcal{A}_{i+1}^{p+1}(\mathcal{C}) & \\
\pi \swarrow & & \searrow \pi \\
\mathcal{A}^i(\mathcal{C}') \otimes \mathcal{A}_0^{p-i}(\mathcal{C}) & \xrightarrow{(-1)^i \Omega} & \mathcal{A}^{i+1}(\mathcal{C}') \otimes \mathcal{A}_0^{p-i}(\mathcal{C}),
\end{array} \tag{10.5.9}$$

where

$$\Omega = \sum_{H \in \mathcal{C}'} (H \wedge) \otimes \Omega(H). \tag{10.5.10}$$

Here $H\wedge$ is the left multiplication by H.

Let $\mathcal{A}_h^\cdot(\mathcal{C})$ be the trivial bundle over B with fiber $\mathcal{A}_h^\cdot(\mathcal{C})$. Define a connection $\nabla(a)$ in $\mathcal{A}_h^\cdot(\mathcal{C})$ as follows. Fix an affine equation $\ell_H = 0$ for any $H \in \mathcal{C}'$. For any $x \in \mathcal{A}_h^p(\mathcal{C})$, denote by $\bar{x}$ the constant section of $\mathcal{A}_h^p(\mathcal{C})$ generated by x. Set

$$\nabla(a)\bar{x} = \sum_{H \in \mathcal{C}'} \Omega(H)\bar{x}\, d\ell_H/\ell_H. \tag{10.5.11}$$

The connection has poles of the first order at $\bigcup_{H \in \mathcal{C}'} H$.

The connection $\nabla(a)$ will be called the *hypergeometric connection*.

10.5.12. Theorem. *The hypergeometric connection commutes with the differ-*

ential $d_h(a) : \mathcal{A}_h^{\cdot}(\mathcal{C}) \to \mathcal{A}_h^{\cdot+1}(\mathcal{C})$. *The hypergeometric connection is integrable.*

Proof. The first statement follows from (10.5.5). The integrability condition is $\Omega \wedge \Omega = 0$. It follows from the equality $d_v^2(a) = 0$ and (10.5.9).

Let $L^p \subset \mathcal{C}^p$ be a horizontal edge and $H^1 \in \mathcal{C}'$. Set $L^{p+1} = L^p \cap pr^{-1}(H^1)$. Set
$$F\ell^p(\mathcal{C}_{L^{p+1}}, H^1) = \bigoplus F\ell^p(\mathcal{C})_L \hookrightarrow F\ell^p(\mathcal{C}), \qquad (10.5.13)$$
where the sum is taken over all edges $L \in \mathcal{C}^p$ such that $L^{p+1} \subset L$ and $L \not\subset pr^{-1}(H^1)$.

We define a map
$$\Omega(L^p, H^1) : F\ell^p(\mathcal{C})_{L^p} \longrightarrow F\ell^p(\mathcal{C}_{L^{p+1}}, H(L^{p+1})) \qquad (10.5.14)$$
as follows.

Let $F = (L^1 \supset L^2 \supset \cdots \supset L^p)$ be a flag in $F\ell^p(\mathcal{C})_{L^p}$. Let $H = (H_1, \ldots, H_p)$ be any p-tuple of hyperplanes of $\mathcal{C}$ such that $F(H) = F$, see (10.1.8). Let $H_0 \in \mathcal{C}$ be any hyperplane such that $H_0 \cap H_1 \cap \cdots \cap H_p = L^{p+1}$. For any $i = 1, \ldots, p$, define a flag
$$F(H, H_0, i) = (L^1(H, H_0, i) \supset \cdots \supset L^p(H, H_0, i)), \qquad (10.5.15)$$
where
$$L^j(H, H_0, i) = L^j \quad \text{for } j < i,$$
$$L^j(H, H_0, i) = H_1 \cap \cdots \cap \hat{H}_i \cap \cdots \cap H_{j+1} \quad \text{for } i \leq j < p,$$
$$L^p(H, H_0, i) = H_1 \cap \cdots \cap \hat{H}_i \cap \cdots \cap H_p \cap H_0.$$

Say that $\bar{F} = (\bar{L} \supset \cdots \supset \bar{L}^p)$ is admissible for F if it has the form $\bar{F} = F(H, H_0, i)$ and $\bar{F}$ is horizontal.

For such an admissible $\bar{F} = F(H, H_0, i)$, define the index $\text{ind}\,(F, \bar{F})$ by the formula
$$\text{ind}\,(F, \bar{F}) = (-1)^{p+i+1}(a(L^i) - a(L^{i-1})), \qquad (10.5.16)$$
where $a(L^j)$ is the quasiclassical weight of L^j, see (10.3), and we set $a(L^0) = 0$. Set
$$\text{ind}\,(F, F) = a(L^{p+1}) - a(L^p). \qquad (10.5.17)$$

Define a map $\Omega(L^p, H^1)$ by the formula
$$\Omega(L^p, H^1) : F \mapsto \text{ind}\,(F, F)F + \sum \text{ind}\,(F, \bar{F})\bar{F}, \qquad (10.5.18)$$

where the sum is taken over all flags $\bar{F}$ admissible to F.

10.5.19. Lemma. *The following diagram is commutative:*

$$\begin{array}{ccc} F\ell^p(\mathcal{C})_{L^p} & \xrightarrow{\Omega(L^p,H^1)} & F\ell^p(\mathcal{C}_{L^{p+1}}, H(L^{p+1})) \\ S\downarrow & & \downarrow S \\ \mathcal{A}^p(\mathcal{C})_{L^p} & \xrightarrow{\Omega(L^p,H^1)} & \mathcal{A}^p(\mathcal{C}_{L^{p+1}}, H(L^{p+1})). \end{array}$$

Here S is the quasiclassical contravariant form and the lower $\Omega(L^p, H^1)$ is defined by (10.5.3).

The lemma follows from definitions by direct verification.

Define the map

$$\Omega(H^1) : F\ell_h^p(\mathcal{C}) \longrightarrow F\ell_h^p(\mathcal{C}) \tag{10.5.20}$$

as follows. $F\ell_h^p(\mathcal{C}) = \bigoplus F\ell^p(\mathcal{C})_{L^p}$, where L^p runs through all horizontal edges of codimension p. Set $\Omega(H^1)$ to be the map whose restriction to $F\ell^p(\mathcal{C})_{L^p}$ is $\Omega(L^p, H^1)$.

10.5.21. Lemma. *The map $\Omega(H^1)$ induces an endomorphism of the horizontal complex, $\Omega(H^1) : (F\ell_h^\cdot(\mathcal{C}), d_h) \to (F\ell_h^\cdot(\mathcal{C}), d_h)$.*

Proof. The horizontal quasiclassical contravariant form defines a homomorphism of horizontal complexes (10.4.19). This homomorphism is an isomorphism for general values of weights of hyperplanes. So, for general values of weights, the lemma follows from (10.5.5) and (10.5.19). By continuity, the lemma is valid for arbitrary values.

Let $\boldsymbol{F\ell_h^\cdot(\mathcal{C})}$ be the trivial bundle over B with fiber $F\ell_h^\cdot(\mathcal{C})$. Define a connection $\nabla(a)$ in $\boldsymbol{F\ell_h^\cdot(\mathcal{C})}$ as follows. Set

$$\nabla(a)x = dx + \sum_{H \in \mathcal{C}'} \Omega(H) x d\ell_H / \ell_H \tag{10.5.22}$$

for an arbitrary section x. The connection will be called the hypergeometric connection.

10.5.23. Theorem. *The hypergeometric connection on $\boldsymbol{F\ell_h^\cdot(\mathcal{C})}$ commutes with the differential $d_h : F\ell_h^\cdot(\mathcal{C}) \to F\ell_h^{\cdot+1}(\mathcal{C})$. The hypergeometric connection is integrable. The horizontal contravariant form S_h induces a morphism of bundles with hypergeometric connections,*

$$S_h : \boldsymbol{F\ell_h^\cdot(\mathcal{C})} \to \boldsymbol{\mathcal{A}_h^\cdot(\mathcal{C})}.$$

Proof. S_h is a morphism of bundles with connections by (10.5.19). The integrability of the connection in the first bundle follows from the integrability of the connection in the second bundle, as S_h is an isomorphism for general values of weights.

10.5.24. Corollary. *The image $S_h(Fl_h^\cdot(\mathcal{C})) \subset A_h^\cdot(\mathcal{C})$ of the horizontal contravariant form is invariant under the hypergeometric connection.*

10.6. Hypergeometric Differential Forms

Let $\mathcal{C}$ be a weighted configuration of hyperplanes in a complex k-dimensional space V. Let $a = \{a(H)|H \in \mathcal{C}\}$ be the weights. Fix an affine equation $\ell_H = 0$ for each hyperplane $H \in \mathcal{C}$. Set

$$Y = V \setminus \bigcup_{H \in \mathcal{C}} H.$$

Consider the trivial line bundle $\mathcal{L}(a)$ over Y with an integrable connection

$$d(a) : \mathcal{O} \to \Omega^1 \qquad (10.6.1)$$

given by

$$d + \Omega(a) = d + \sum_{H \in \mathcal{C}} a(H) d\log \ell_H,$$

where d is the de Rham differential, see (2.5). Denote by $\Omega^\cdot(\mathcal{L}(a))$ the complex of Y-sections of the holomorphic de Rham complex of $\mathcal{L}(a)$.

To any $H \in \mathcal{C}$, assign the one-form

$$i(H) = d\log \ell_H \in \Omega^1(\mathcal{L}(a)). \qquad (10.6.2)$$

This construction defines a monomorphism

$$i(a) : (\mathcal{A}^\cdot(\mathcal{C}), d(a)) \longrightarrow (\Omega^\cdot(\mathcal{L}(a)), d(a)) \qquad (10.6.3)$$

see [OS], [SV3]. The image of this homomorphism is called the *complex of hypergeometric differential forms of weight a*. We denote it by $(\mathcal{A}^\cdot(\mathcal{C}, a), d(a))$. The image of the homomorphism

$$i(a)S : (F\ell^\cdot(\mathcal{C}), d) \longrightarrow (\Omega^\cdot(\mathcal{L}(a)), d(a)) \qquad (10.6.4)$$

is called the *complex of flag hypergeometric differential forms of weight a*. We denote it by $(F\ell^\cdot(C,a), d(a))$, see [**V2**], [**SV3**].

10.6.5. Example. Let C be a configuration of three lines in $\mathbb{C}^2$: $t_1 = 0$, $t_2 = 0$, $t_2 - t_1 = 0$, with weights a_{13}, a_{23}, and a_{12}, respectively. Then $F\ell^\cdot(C,a)$ has the following generators:

$F\ell^0 : 1$

$F\ell^1 : \omega_{13} = a_{13} dt_1/t_1, \quad \omega_{23} = a_{23} dt_2/t_2, \quad \omega_{12} = a_{12} d(t_1 - t_2)/(t_1 - t_2)$.

$F\ell^2 : \omega_{13\cdot} = \omega_{13} \wedge (\omega_{23} + \omega_{12}), \quad \omega_{23\cdot} = \omega_{23} \wedge (\omega_{12} + \omega_{13})$,

$\quad\quad \omega_{12\cdot} = \omega_{12} \wedge (\omega_{13} + \omega_{23})$

and one relation

$$\omega_{12\cdot} + \omega_{13\cdot} + \omega_{23\cdot} = 0.$$

The differential $d(a)$ is given by

$$1 \longmapsto \omega_{13} + \omega_{23} + \omega_{12},$$
$$\omega_{ij} \longmapsto w_{ij}.$$

The monomorphism $i(a)$ is a quasiisomorphism for general values of weights a, see [**SV3**]. More precise information is given in [**ESV**] and [**STV**]. Namely, let PV be the k-dimensional projective space compactifying V, and H_∞ the hyperplane at infinity. Let $C' = C \cup \{H_\infty\}$ be the configuration of hyperplanes in PV. Set

$$a(H_\infty) = -\sum_{H \in C} a(H). \tag{10.6.6}$$

Let L be any edge of C' of codimension r. Consider the projective localization PC_L of C at the edge L, see (2.6). PC_L is a configuration of hyperplanes in a complex $(r-1)$-dimensional projective space P^{r-1}. An edge L is called dense if

$$\chi(P^{r-1} \setminus \bigcup_{H \in PC_L} H) \neq 0, \tag{10.6.7}$$

where χ denotes the Euler characteristic.

This notion of density coincides with the notion of density in (2.6) if C is the complexification of a configuration of hyperplanes in a real affine space, see [BBR] and [GV].

10.6.8. Theorem [STV], cf. [ESV]. *The monomorphism (10.2.3) is a quasi-isomorphism if none of the quasiclassical weights of dense edges of C' is a positive integer.*

10.6.9. Example. For the configuration C in example (10.6.5), the monomorphism $i(a)$ is a quasiisomorphism if none of the numbers a_{12}, a_{13}, a_{23}, $\pm(a_{12} + a_{13} + a_{23})$ is a positive integer.

10.7. Hypergeometric Integrals

Let V, C, and $\mathcal{L}(a)$ be the same as in (10.6). Let $\mathcal{L}^*(a)$ be the line bundle dual to $\mathcal{L}(a)$ and $S^*(a)$ the sheaf of its horizontal sections.

Let $\Delta = (f : P \to Y, s \in \Gamma(P, f^*S^*(a)))$ be a p-cell with a coefficient in $S^*(a)$, see (2.2). Here P is an open p-dimensional convex polytope. Fix over P any branch of the multivalued function $f^*\ell_a$, where

$$\ell_a = \prod_{H \in C} \ell_H^{a(H)}. \tag{10.7.1}$$

Denote this branch by $b\ell_a$. Then $s = c \cdot b\ell_a$ for $c \in \mathbb{C}$, cf. (2.5).

Let $\omega \in \mathcal{A}^p(C, a)$ be a hypergeometric form. The integral,

$$\int_\Delta \omega = c \int_P b\ell_a f^*\omega,$$

is called a *hypergeometric integral*.

(10.7.2) Assume that $\{a(H)\}$ vary. Assume that $f : P \to Y$ and c are fixed, the closure of $f(P)$ lies in Y, the branch $b\ell_a$ analytically depends on a. Then for any $x \in \mathcal{A}^p(C)$, the integral,

$$\int_\Delta i(a)[x],$$

depends analytically on a.

Now let C be a configuration of hyperplanes in $\mathbb{R}^k$. Let $P \subset \mathbb{R}^k$ be a convex k-dimensional polytope such that the interior of P does not intersect $\bigcup_{H \in C} H$.

Say that P is in a general position with respect to C if (10.7.3) is true.

(10.7.3) Let Q be any closed facet of P and let L be any edge of C such that Q intersects L. Then Q intersects L transversally or $Q \subset \bigcup_{H \in C} H$.

For example of a polytope in general position (resp. not in general position), see Fig. 10.1a (resp. 10.1b).

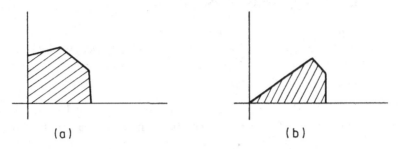

Fig. 10.1

For any $H \in \mathcal{C}$, fix an argument of $\ell_H|_P$. Hence a branch of $\ell_a|_P$ is fixed for arbitrary a. This branch defines a section $s_a \in \Gamma(P, \mathcal{S}^*(a)|_P)$. Set $\Delta_a = (P, s_a)$.

For any $H = (H_1, \ldots, H_k) \in \mathcal{A}^k(\mathcal{C})$, set

$$i[H] = d\log \ell_{H_1} \wedge \cdots \wedge d\log \ell_{H_k} \in \Omega^k(Y, \mathbb{C}). \tag{10.7.4}$$

The integral

$$I(P, H, a) = \int_{\Delta_a} i(a)[H] = \int_P \prod_{H \in \mathcal{C}} \ell_H^{a(H)} i[H] \tag{10.7.5}$$

is an example of a hypergeometric integral. We discuss its analytic properties as a function in a.

An edge L of $\mathcal{C}$ will be called trivial if there are affine coordinates $x_1, \ldots, x_p$, $y_1, \ldots, y_{k-p}$ in V for $1 < p < k$, such that any hyperplane of $\mathcal{C}$ containing L has an equation of the form $\ell(x_1, \ldots, x_p) = 0$ or $\ell(y_1, \ldots, y_{k-p}) = 0$, and the sets of hyperplanes of both types are nonempty.

An edge is trivial, if and only if it is dense in the sense of (10.6) or (2.6) see Crapo's theorem in [**STV**].

Let A be the complex affine space with coordinates $\{a_H | H \in \mathcal{C}\}$. The set of weights of $\mathcal{C}$ defines a point in $A : \{a_H = a(H) | H \in \mathcal{C}\}$.

For any edge L of $\mathcal{C}$, define a linear function on A by

$$a_L = \sum_{H \in \mathcal{C}, L \subset H} a_H. \tag{10.7.6}$$

Define two domains in A. Set $D_0 = D_0(\mathcal{C}) = \{a \in A | a_L(a) \notin \{0, -1, -2, \ldots\}$ for all nontrivial edges L of $\mathcal{C}\}$ and $D_{-1} = D_{-1}(\mathcal{C}) = \{a \in A | a_L(a) \notin \{-1, -2, -3, \ldots\}$ for all nontrivial edges L of $\mathcal{C}\}$.

10.7.7. Theorem. *For an arbitrary polytope P, the integral $I(P, H, a)$ is well-defined, if $\operatorname{Re} a(L) > 0$ for all edges L of $\mathcal{C}$. $I(P, H, a)$ is analytically continued to an analytic function in D_0. This function is a meromorphic function in A and has poles of at most first order on hyperplanes of $A - D_0$.*

The proof of the theorem is based on the following model example. Let $\varphi : \mathbb{R}^k \to \mathbb{R}$ be a smooth function with compact support. Consider the integral

$$J(a_1, \ldots, a_k) = \int_0^\infty \cdots \int_0^\infty t_1^{a_1} \ldots t_k^{a_k} \varphi(t_1 \ldots t_k) dt_1 \ldots dt_k. \qquad (10.7.8)$$

10.7.9. Lemma [GS], cf. **[AGV, (7.2.1)]**. *The integral J is well defined, if $\operatorname{Re} a_j > -1$ for all j, and is analytically continued to an analytic function in the domain $D = \{a \in \mathbb{C}^k | a_j \notin \{-1, -2, \ldots\}$ for all $j\}$. This function is a meromorphic function in $\mathbb{C}^k$. It has poles of at most first order at hyperplanes of $\mathbb{C}^k - D$.*

The theorem is reduced to the lemma by resolution of singularities of $\mathcal{C}$ and partition of unity, see (10.8.6).

Let $F = \{L^0 \supset L^1 \supset \cdots \supset L^k\} \in \operatorname{Flag}^k(\mathcal{C})$ be any k-flag of $\mathcal{C}$. Set

$$iS(F) = \sum a(H_1) \ldots a(H_k) d\log \ell_{H_1} \wedge \cdots \wedge d\log \ell_{H_k} \in \Omega^k(Y, \mathbb{C}), \qquad (10.7.10)$$

where the sum is taken over all $(H_1, \ldots, H_k)$ such that $L^1 \subset H_1, \ldots, L_k \subset H_k$.

Consider the integral

$$I(P, F, a) = \int_{\Delta_a} i(a) S[F] = \int_P \prod_{H \in \mathcal{C}} \ell_H^{a(H)} iS(F), \qquad (10.7.11)$$

see (10.6.4).

10.7.12. Theorem. *Let the polytope P be in general position with respect to $\mathcal{C}$. Then the integral $I(P, F, a)$ is analytically continued to an analytic function in D_{-1}.*

10.7.13. Example. The integral

$$\int_0^1 (t-2)^{a_1} d(t_1^{a_2})$$

is analytically continued to $\{(a_1, a_2) \in \mathbb{C}^2 | a_2 \notin \{-1, -2, \dots\}\}$.

The theorem is proved in (10.8).

At the end of this section, we formulate a generalization of Theorem (10.7.12).

Let $C' \subset C$ be a subconfiguration. Let $a, a' \in A$ be two sets of weights of C. We say that a and a' are equal with respect to C', if $a(H) = a'(H)$ for all $H \in C'$.

Any element $x \in \mathcal{A}^{\cdot}(C)$ may be considered as an element of the complex $(\mathcal{A}^{\cdot}(C), d(a))$ for arbitrary a.

Let $x \in \mathcal{A}^{\cdot}(C)$ be an element polynomially depending on variables $a(H)$, $H \in C$. Assume that there is another set of weights a' of the configuration C such that a' polynomially depends on $a(H)$, $H \in C$, the weights a and a' are equal with respect to $C' \subset C$, and x lies in the image of the contravariant form $S_{a'} : (F\ell^{\cdot}(C), d) \to (\mathcal{A}^{\cdot}(C), d(a'))$ for weights a'. We say that $i(a)[x] \in \mathcal{A}^{\cdot}(C, a)$ is a flag form with respect to a pair $C' \subset C$. Such a flag form $i(a)[S_{a'}(F)]$ corresponding to a flag $F \in \mathrm{Flag}^{\cdot}(C)$ will be denoted by $\omega(F, C' \subset C, a, a')$.

Assume that $F \in \mathrm{Flag}^k(C)$ and $\Delta_a = (P, s_a)$ is the cell defined above. Consider the integral

$$I(a) = \int_{\Delta_a} \omega(F, C' \subset C, a, a') \qquad (10.7.14)$$

as a function of a. We will discuss its analytic properties.

Introduce a domain $D(C' \subset C) \subset A$ as the set of all $a \in A$ such that

(i) $a_L(a) \notin \{0, -1, -2, \dots\}$ for any nontrivial edge L of C lying in $\bigcup_{H \in C-C'} H$,

(ii) $a_L(a) \notin \{-1, -2, -3, \dots\}$ for any nontrivial edge L of C not lying in $\bigcup_{H \in C-C'} H$.

10.7.15. Theorem. *Let a polytope P be in general position with respect to C. Then the integral $I(a)$ is analytically continued to an analytic function in $D(C' \subset C)$.*

The proof coincides with the proof of Theorem (10.7.12) given below in (10.8).

10.8. Resolution of Singularities of a Configuration of Hyperplanes

Let C be a configuration of hyperplanes in a real or complex affine k-dimensional space V, $Y = V - \bigcup_{H \in \mathcal{C}} H$. A resolution of singularities of $\mathcal{C}$ is a nonsingular algebraic manifold W and a proper algebraic map

$$\pi : W \longrightarrow V \qquad (10.8.1)$$

such that

(10.8.2) $\pi^{-1}(V - Y)$ is a divisor with normal crossings,

(10.8.3) π is an analytic isomorphism over the complement to the union of all edges of $\mathcal{C}$ of codimension ≥ 2.

There is a canonical resolution of singularities of a configuration of hyperplanes. Namely, let $\mathcal{L}_p$ be the union of all nontrivial edges L of dimension p, here $p = 0, \ldots, k-2$. Let $W_0 = V$, and let

$$\pi_p : W_p \longrightarrow W_{p-1} \quad \text{for } p = 1, \ldots, k-1$$

be the blow up along the proper transform of $\mathcal{L}_{p-1}$ under $\pi_1 \circ \pi_2 \circ \cdots \circ \pi_{p-1}$. Then $W = W_{k-1}$ and $\pi = \pi_1 \circ \cdots \circ \pi_{k-1}$ is the canonical resolution, cf. [**ESV**]. Below, we consider the canonical resolution.

Let $D = \pi^{-1}(\bigcup_{H \in \mathcal{C}} H)$. Let $D = \cup D_j$ be the decomposition into irreducible components. Let $\mathcal{C}' = \{D_j\}$. The collection $\mathcal{C}'$ is canonically identified with the set $\mathcal{C} \cup \mathcal{L}_0 \cup \mathcal{L}_1 \cup \cdots \cup \mathcal{L}_{k-2}$.

Let $\mathcal{C}$ be weighted. Consider the function ℓ_a, defined in (10.7.1), and its lifting $\pi^* \ell_a$ to W. We describe singularities of $\pi^* \ell_a$ on $D \subset W$. Namely, let $x^0 \in D$, and let $x_1, \ldots, x_k$ be local coordinates in a neighborhood of x^0 such that D in this neighborhood is a union of coordinate hyperplanes:

$$\{x_1 = 0\} \cup \{x_2 = 0\} \cup \cdots \cup \{x_\ell = 0\}$$

for some ℓ, $0 \leq \ell \leq k$. The divisor in $\mathcal{C}'$ determined by equation $\{x_j = 0\}$ for $j \in \{1, \ldots, \ell\}$ corresponds to some edge of $\mathcal{C}$. Denote this edge by L_j.

10.8.4. Lemma. *In a neighborhood of x^0, the function $\pi^* \ell_a$ has the form*

$$\pi^* \ell_a(x_1, \ldots, x_k) = \varphi(x_1, \ldots, x_k) \prod_{j=1}^{\ell} x_j^{a(L_j)}.$$

Here $a(L_j)$ is the quasiclassical weight of L_j and

$$\varphi(x_1, \ldots, x_k) = \prod_\alpha \varphi_\alpha(x_1, \ldots, x_k)^{b_a},$$

where $\{b_\alpha\}$ are some complex numbers, $\{\varphi_\alpha\}$ are some holomorphic functions, and each φ_α has no zeros in a neighborhood of x^0.

The proof is by an easy direct verification.

For $H \in \mathcal{C}$, consider the differential form $i(H) = d\ell_H/\ell_H$. Let x^0 and $x_1, \ldots, x_k$ be as above. Assume that for some m, $0 \leq m \leq \ell$, each of the edges $L_1, \ldots, L_m$ lies in H, and each of the edges $L_{m+1}, \ldots, L_\ell$ does not lie in H.

10.8.5. Lemma. *In a neighborhood of x^0, the differential form $\pi^* i(H)$ may be written as*

$$\pi^* i(H) = \omega_0 + \omega_1 dx_1/x_1 + \cdots + \omega_m dx_m/x_m,$$

where $\omega_1, \ldots, \omega_m$ are some analytic functions and ω_0 is some analytic differential one-form.

(10.8.6) As a corollary of (10.3.9) and (10.4.5), we get Theorem (10.3.7).

Proof of Theorem (10.7.12). For any edge L of $\mathcal{C}$, denote by H_L the hyperplane in A given by $a_L = 0$. To prove Theorem (10.7.12), it suffices to prove (10.8.7).

(10.8.7) For an arbitrary edge L of $\mathcal{C}$, $I(P, F, a)$ is an analytic function in a neighborhood of a generic point of H_L.

We prove (10.8.7) by induction on k. For $k = 1$, the statement (10.8.7) is obvious, see example (10.3.13). Assume that the induction hypotheses are satisfied and prove (10.8.7).

Let L be a vertex of $\mathcal{C}$. In this case, the statements (10.7.9), (10.8.4), and (10.8.5) show that it suffices to prove (10.8.7) for a central configuration $\mathcal{C}$. (This is an essential remark.)

For a central configuration $\mathcal{C}$, we have

$$\prod_{H \in \mathcal{C}} \ell_H^{a(H)} iS(F) = d\ell_1 \wedge \ldots d\ell_k, \qquad (10.8.8)$$

where

$$\ell_j = \prod_{\substack{H \in \mathcal{C} \\ H \supset L^j, L^{j-1} \not\subset H}} \ell_H^{a(H)}. \qquad (10.8.9)$$

Hence,

$$I(P, F, a) = \int_P d\ell_1 \wedge \cdots \wedge d\ell_k = \pm \int_{\partial P} \ell_k d\ell_1 \wedge \cdots \wedge d\ell_{k-1} \qquad (10.8.10)$$

for positive values of a.

Let $\partial P = \bigcup_i P_i$, where $\{P_i\}$ are $(k-1)$-dimensional facets of P. Let V_i be the affine $(k-1)$-dimensional subspace of V generated by P_i. Let C_i be the weighted configuration in V_i cut by C. The integral $\int_{P_i} \ell_k d\ell_1 \wedge \cdots \wedge d\ell_{k-1}$ is zero, if $P_i \subset \bigcup_{H \in C} H$, and is an integral of a suitable flag form over a polytope in general position with respect to C_i, if $P_i \not\subset \bigcup_{H \in C} H$. Applying (10.8.7) to this integral, we get (10.8.7) for the $I(P, F, a)$. Hence, $I(P, F, a)$ has no pole at $a_L = 0$ for dim $L = 0$.

If dim $L > 0$, then the proof of (10.8.7) is similar. We give the proof for the case dim $L = k - 1$, that is, L is a hyperplane of C. Let $D_L \subset W$ be the proper transform of L under π. Lift the integral $I(P, F, a)$ to the integral over the proper transform of P under π. Use a partition of unity. Then in a neighborhood of D_L, the integral may be represented as a linear combination of integrals of the form

$$\int_0^\infty \cdots \int_0^\infty \varphi(x_1, \ldots, x_k, a) a_L x_1^{a_L - 1} x_2^{b_2(a)} \ldots x_k^{b_k(a)} dx_1 \ldots dx_k \qquad (10.8.11)$$

and

$$\int_0^\infty \cdots \int_0^\infty \varphi(x_1, \ldots, x_k, a) x_1^{a_L} x_2^{b_2(a)} \ldots x_k^{b_k(a)} dx_1 \ldots dx_k \qquad (10.8.12)$$

Here φ is a smooth function in x with compact support analytically depending on a; $b_1, \ldots, b_k$ are suitable linear functions on A different from a_L. It is easy to see that the integrals are analytic functions at general points of the hyperplane $a_L = 0$. The second integral is analytic on $a_L = 0$ for trivial reasons. The first integral is an integral of a flag form $dx_1^{a_L}$ in variable x_1 and we use the induction hypothesis to prove analyticity of the first integral on $a_L = 0$. The general case $0 <$ dim $L < k - 1$ is similar.

10.9. Integration Pairings and Symmetric Frames

Let $C_\mathbb{R}$ be a configuration of hyperplanes in a real affine space $V_\mathbb{R}$. Fix affine equations for hyperplanes of $C_\mathbb{R}$. For each facet F of $C_\mathbb{R}$, fix a point $^F\omega \in F$. The set $\{^F\omega | F \subset C_\mathbb{R}\}$ will be called a frame of the $C_\mathbb{R}$.

Let V (resp. C) be the complexification of $V_\mathbb{R}$ (resp. $C_\mathbb{R}$), cf. (2.2). A frame of $C_\mathbb{R}$ will be called a frame of C.

Let $\mathcal{C}_\mathbb{R}$ be weighted and cooriented. Let $\mathcal{X}.(\mathcal{C}_\mathbb{R})$, $\mathcal{Q}.(\mathcal{C}_\mathbb{R})$ be the abstract complexes of $\mathcal{C}_\mathbb{R}$ defined in (2.9). We will omit the subscript '$\mathbb{R}$'.

Let
$$\omega : \mathcal{X}.(\mathcal{C}) \longrightarrow C.(X, \mathcal{S}^*(a))$$
$$\omega : \mathcal{Q}.(\mathcal{C}) \longrightarrow C.(Q'/Q'', \mathcal{S}^*(a)) \tag{10.9.1}$$

be the realizations of the abstract complexes associated with the chosen frame and with the chosen affine equations, see (2.9.6). Here a is the set of weights of $\mathcal{C}$.

By (2.9.7), the realization (10.9.1) combined with the natural inclusions $C.(X, \mathcal{S}^*(a)) \longrightarrow C.(V - \bigcup_{H \in \mathcal{C}} H, \mathcal{S}^*(a))$ and $C.(Q'/Q'', \mathcal{S}^*(a)) \longrightarrow C_!.(V - \bigcup_{H \in \mathcal{C}} H, \mathcal{S}^*(a))$ induces isomorphism (2.9.7.3) of the corresponding homology groups.

Monomorphism (10.6.2) and the integration of hypergeometric differential forms induce the integration pairing

$$I : \mathcal{A}^p(\mathcal{C}) \otimes \mathcal{X}_p(\mathcal{C}) \longrightarrow \mathbb{C},$$
$$y \otimes x \longmapsto \int_{\omega(x)} i(a)(y) \tag{10.9.2}$$

for all p.

This pairing is a pairing of complexes:

$$I(dy, x) = I(y, dx) \tag{10.9.3}$$

for all $y \in \mathcal{A}^{p-1}$, $x \in \mathcal{X}_p$.

(10.9.4) The pairing depends on the choice of affine equations for hyperplanes. If an equation $\ell_H = 0$ for a hyperplane $H \in \mathcal{C}$ is replaced by $\lambda \ell_H = 0$ where $\lambda \in \mathbb{R}$, then the pairing is multiplied by $|\lambda|^{a(H)}$, see the choice of sections in (2.5).

Fix affine equations for hyperplanes of $\mathcal{C}$. The pairing depends on the choice of the frame $\{{}^F\omega\}$.

(10.9.5) The induced pairing

$$I : H^p(\mathcal{A}^{\cdot}(\mathcal{C}), d) \otimes H_p(\mathcal{X}.(\mathcal{C}), d) \longrightarrow \mathbb{C}$$

is independent of the frame for all p.

(10.9.6) If weights a of C satisfy the hypothesis of Theorem (10.6.8), then the pairing (10.9.5) is nondegenerate.

The integration of hypergeometric forms over bounded components of the complement of $C_{\mathbb{R}}$ in $V_{\mathbb{R}}$ induces a pairing

$$H^{\cdot}(\mathcal{A}^{\cdot}(C), d) \otimes H_{\cdot}(Q_{\cdot}, d) \longrightarrow \mathbb{C}, \qquad (10.9.7)$$

see (2.3.6). This pairing is well-defined if weights a of C define a point in the domain D_0, see the definition in (10.7) and Theorem 10.7.7.

The integration of the flag hypergeometric forms over bounded components induces a pairing

$$H^{\cdot}(F\ell^{\cdot}(C), d) \otimes H_{\cdot}(Q_{\cdot}, d) \longrightarrow \mathbb{C},$$
$$y \otimes x \longmapsto \int_{\omega(x)} i(a)(S(y)), \qquad (10.9.8)$$

see (10.6.4). This pairing is well defined if $a \in D_{-1}$, see (10.7.2).

A new phenomenon appears if the frame of C is symmetric in the following sense.

Let F^i be any facet of $C_{\mathbb{R}}$ of codimension i. Consider the set $\omega(F^i) = \{{}^F\omega | F \geq F^i\}$. Let span (F^i) be the smallest affine subspace of $V_{\mathbb{R}}$ containing $\omega(F^i)$.

(10.9.9) The frame will be called *symmetric* if dim span $\omega(F^i) = i$ for any facet F^i of $C_{\mathbb{R}}$.

10.9.10. Example. Let $C_{n,k}(z)$ be a discriminantal configuration defined in (3.1) and $\{T_j^i\}$ its equipment introduced in (3.3.1). Let $\{{}^F\omega\}$ be the frame of $C_{n,k}$ defined by the equipment $\{T_j^i\}$ in (3.3) and (3.5). Then this frame is symmetric, see Lemmas (3.5.1) and (3.6.1). The following lemma is very important.

10.9.11. Lemma. *Assume that* $\operatorname{Re} a(H) > 0$ *for all* $H \in C$. *Let* $S : \mathcal{X}_{\cdot}(C) \to Q_{\cdot}(C)$ *be the homomorphism of the projection on the real part defined in (2.9). Then, for any p and $x \in \mathcal{X}_p$, $y \in \mathcal{A}^p$, we have*

$$\int_{\omega(x)} i(a)(y) = \int_{\omega(S(x))} i(a)(y).$$

Proof. Consider an arbitrary p-cell $E(F^0 \geq F^p)$ of the realization of $\mathcal{X}_p(\mathcal{C})$ and the deformation $R_t : E(F^0 \geq F^p) \to V$ for $t \in [0,1]$ defined in (2.2.6). It follows from the symmetry of the frame that the set $\bigcup_{t\in[0,1]} R_t E(F^0 \geq F^p)$ lies in a p-dimensional complex affine subspace of V. Now the lemma follows from the Stokes theorem.

10.9.12. Corollary. *Let S be the homomorphism of the projection on the real part, $(\mathcal{X}/\ker S)$. the quotient complex. Assume that the frame of C is symmetric. Then the integration pairing*

$$I : (\mathcal{X}/\ker S). \otimes \mathcal{A}^{\cdot} \longrightarrow \mathbb{C}$$

it well defined if $a \in D_0$.

10.9.13. Corollary. *Let S be the homomorphism of the projection on the real part, $S : F\ell^{\cdot} \to \mathcal{A}^{\cdot}$ the contravariant form defined in (10.3). Let $(\mathcal{X}/\ker S)$. and $(F\ell/\ker S)^{\cdot}$ be the corresponding quotient complexes. Assume that a frame of C is symmetric. Then the integration pairing*

$$I : (\mathcal{X}/\ker S). \otimes (F\ell/\ker S)^{\cdot} \longrightarrow \mathbb{C}$$

is well defined if $a \in D_{-1}$.

Let

$$I : H.((\mathcal{X}/\ker S).) \otimes H^{\cdot}((F\ell/\ker S)^{\cdot}) \longrightarrow \mathbb{C} \qquad (10.9.14)$$

be the induced pairing.

It is plausible that the pairing (10.9.14) is nondegenerate if none of the quasiclassical weights of dense edges of the compactification of $\mathcal{C}$ is a positive integer, cf. the hypothesis of (10.6.8).

We will prove some special cases of this conjecture in (12.2.8) and (12.5). It would be interesting to understand the topological meaning of the groups $H.((\mathcal{X}/\ker S).)$ and $H^{\cdot}((F\ell/\ker S)^{\cdot})$.

10.10. Hypergeometric Differential Equations

Let V^{n+k} and B^n be complex affine spaces of dimension $n+k$ and n, respectively. Let $pr : V \to B$, $\mathcal{C}$, and $\mathcal{C}'$ satisfy (10.4.a) and (10.4.b).

Set $U(\mathcal{C}) = \bigcup_{H \in \mathcal{C}} H$, $Y(\mathcal{C}) = V - U(\mathcal{C})$, and $Y(\mathcal{C}') = B - U(\mathcal{C}')$.

As in (10.4), for any $z \in B$, set $P(z) = pr^{-1}(z)$, $U(z) = U(\mathcal{C}) \cap P(z)$, and $Y(z) = Y(\mathcal{C}) \cap P(z)$. Set $V' = pr^{-1}(Y(\mathcal{C}'))$, $U'(\mathcal{C}) = U(\mathcal{C}) \cap V'$. The restriction of pr to V' or to $Y(\mathcal{C})$ will be denoted by pr_1 and pr_2, respectively.

The maps
$$pr_1 : (V', U'(\mathcal{C})) \longrightarrow Y(\mathcal{C}'),$$
$$pr_2 : Y(\mathcal{C}) \longrightarrow Y(\mathcal{C}') \qquad (10.10.1)$$

are locally trivial bundles with fiber $(P(z), U(z))$ and $Y(z)$, respectively.

Assume that $\mathcal{C}$ is weighted and a is the set of its weights. Let $\mathcal{L}(a)$ and $\mathcal{L}^*(a)$ be the line bundles with the integrable connection defined over $Y(\mathcal{C})$ in (2.5). Let $\mathcal{S}(a)$ and $\mathcal{S}^*(a)$ be their sheaves of horizontal sections. With the locally trivial bundles (10.10.1), there are associated four holomorphic vector bundles over $Y(\mathcal{C}')$:
$$R.pr_1, R.pr_2, R^{\cdot}pr_1, R^{\cdot}pr_2 \qquad (10.10.2)$$
with fibers $H.(P(z), U(z), \mathcal{S}^*(a,z))$, $H.(Y(z), \mathcal{S}^*(a,z))$, $H^{\cdot}(P(z), U(z), \mathcal{S}(a,z))$, $H^{\cdot}(Y(z), \mathcal{S}(a,z))$, respectively. Here $\mathcal{S}(a,z)$ (resp. $\mathcal{S}^*(a,z)$) is the restriction of $\mathcal{S}(a)$ (resp., $\mathcal{S}^*(a)$) to $Y(z)$.

The bundles $R.pr_i$ and $R^{\cdot}pr_i$ are dual for $i = 1, 2$ and have natural dual Gauss–Manin connection. Denote this Gauss–Manin connection by $\nabla(a)$.

There are natural morphisms
$$S : R.pr_1 \longrightarrow R.pr_2$$
$$S^* : R^{\cdot}pr_2 \longrightarrow R^{\cdot}pr_1 \qquad (10.10.3)$$

of bundles with connections induced by the natural homomorphisms of the corresponding (co)homologies.

Now let us consider the complexes of the vector bundles $(\mathcal{A}_h^{\cdot}(\mathcal{C}), d_h(a))$ and $(F\ell_h^{\cdot}(\mathcal{C}), d_h(a))$ defined in (10.5). These are bundles over $Y(\mathcal{C}')$ with the hypergeometric connection $\nabla(a)$. Let $H^{\cdot}(\mathcal{A}_h^{\cdot}(\mathcal{C}), d_h(a))$ and $H^{\cdot}(F\ell_h^{\cdot}(\mathcal{C}), d_h(a))$ be the bundles of cohomologies of these complexes,
$$S_h(a) : H^{\cdot}(F\ell_h^{\cdot}(\mathcal{C}), d_h) \longrightarrow H^{\cdot}(\mathcal{A}_h^{\cdot}(\mathcal{C}), d_h) \qquad (10.10.4)$$

the morphism induced by the horizontal contravariant form.

10.10.5. Theorem. *Monomorphism (10.6.3) induces a morphism*
$$i(a) : H^{\cdot}(\mathcal{A}_h^{\cdot}(\mathcal{C}), d_h(a)) \longrightarrow R^{\cdot}pr_2$$

of holomorphic vector bundles with connections.

Proof. For any $x \in \mathcal{A}_h^{\cdot}(\mathcal{C})$, $i(a)(x) \in \Omega^{\cdot}(\mathcal{L}(a))$, and $d(a)i(a)(x) = i(a)(d_v(a)x + d_h(a)x)$. The restriction of $i(a)(d_v(a)x)$ to any fiber $P(z)$ is zero by (10.5.9). This proves that $i(a)$ is a morphism of vector bundles. The fact that $i(a)$ is a morphism of vector bundles with connections follows from the definition of the operator Ω given in (10.5.10). In fact, the definition of Ω was chosen to get this morphism of connections.

10.10.6. Corollary of (10.4.20) and (10.6.8). *The morphism $i(a)$ of (10.10.5) is an isomorphism if none of the quasiclassical weights of dense edges of the compactification of $\mathcal{C}(z)$ is a positive integer.*

Note that this condition does not depend on $z \in Y(\mathcal{C}')$.

Let $\boldsymbol{H}^{\cdot}(\boldsymbol{\mathcal{A}}_h^{\cdot}(\boldsymbol{\mathcal{C}}), d_h(a))^*$ (resp., $\boldsymbol{H}^{\cdot}(\boldsymbol{F\ell}_h^{\cdot}(\boldsymbol{\mathcal{C}}), d_h(a))^*$) be the bundle with the integrable connection dual to $\boldsymbol{H}^{\cdot}(\boldsymbol{\mathcal{A}}_h^{\cdot}(\boldsymbol{\mathcal{C}}), d_h(a))$ (resp., $\boldsymbol{H}^{\cdot}(\boldsymbol{F\ell}_h^{\cdot}(\boldsymbol{\mathcal{C}}), d_h)$).

The morphism $i(a)$ of (10.10.5) induces a morphism

$$i(a)^* : R.pr_2 \longrightarrow \boldsymbol{H}^{\cdot}(\boldsymbol{\mathcal{A}}_h^{\cdot}(\boldsymbol{\mathcal{C}}), d_h(a))^*. \qquad (10.10.7)$$

We reformulate the fact that (10.10.7) is a morphism of bundles with connections in terms of the following differential equation.

Let $\gamma(z)$ be a horizontal section of $R.pr_2$, that is, $\gamma(z) \in H.(Y(z), \mathcal{S}^*(a))$ and $\gamma(z)$ depends horizontally on z. Let $[x] \in H^{\cdot}(\mathcal{A}_h^{\cdot}(\mathcal{C}), d_h(a))$ and $x \in \mathcal{A}_h^{\cdot}(\mathcal{C})$, its representative. Let $i(a)(x) \in \Omega^{\cdot}(\mathcal{L}(a))$ be the corresponding hypergeometric differential form. Then the integral

$$\text{Int}(x, \gamma(z)) = \int_{\gamma(z)} i(a)(x) \qquad (10.10.8)$$

is a well-defined multivalued holomorphic function on $Y(\mathcal{C}')$.

Fix $z \in Y(\mathcal{C}')$ and a branch of $\gamma(z)$ at z. Then (10.10.8) determines a linear function $i^*(a)(\gamma(z))$ on $H^{\cdot}(\mathcal{A}_h^{\cdot}(\mathcal{C}), d_h(a))$. This linear function analytically depends on z and satisfies the following hypergeometric differential equation:

$$d_z \text{Int}(x, \gamma(z)) = \sum_{H \in \mathcal{C}'} \text{Int}(\Omega(H)x, \gamma(z)) d\ell_H / \ell_H. \qquad (10.10.9)$$

Here $d_z : \mathcal{O}(Y(\mathcal{C}')) \to \Omega^1(Y(\mathcal{C}'))$ is the standard differential and $\Omega(H)$ is defined in (10.5.4).

10.10.10. Example. Consider $pr : \mathbb{C}^{n+1} \to \mathbb{C}^n$, $(t, z_1, \ldots, z_n) \mapsto (z_1, \ldots, z_n)$. Let $\mathcal{C} = \mathcal{C}_{n+1}$ be the configuration of all diagonal hyperplanes in $\mathbb{C}^{n+1}$ and

$\mathcal{C}' = \mathcal{C}_n$ the configuration of all diagonal hyperplanes in $\mathbb{C}^n$. Define the weights of $\mathcal{C}$ as follows. Let $\mathfrak{h} \subset s\ell_2$ be the Cartan subalgebra, $\alpha \in \mathfrak{h}^*$ the simple root. Fix $\Lambda_1, \ldots, \Lambda_n \in \mathfrak{h}^*$ and a number $\kappa \in \mathbb{C}^*$. Denote by H^i (resp., by H^{ij}) the hyperplane $t = z_j$ (resp. $z_i = z_j$) of $\mathcal{C}$. Set

$$a(H^i) = -(\alpha, \Lambda_i)/\kappa$$
$$a(H^{ij}) = (\Lambda_i, \Lambda_j)/\kappa.$$

The horizontal complex $(\mathcal{A}_h^{\cdot}(\mathcal{C}), d_h(a))$ has the form

(i) $\quad \mathbb{C} \longrightarrow \bigoplus_{i=1}^n \mathbb{C} H^i, \quad 1 \longmapsto \sum a(H^i) H^i.$

Set

(ii) $\quad \ell_a(t, z) = \prod_{i=1}^n (t - z_i)^{a(H^i)} \prod_{1 \leq i < j \leq n} (z_i - z_j)^{a(H^{ij})}$

and $\eta_i = \ell_a(t, z) d(t - z_i)/(t - z_i)$ for $i = 1, \ldots, n$.

Let $\gamma(z) \in H_1(\mathbb{C} \setminus \{z_1, \ldots, z_n\}, \mathcal{S}^*(a, z))$ be a horizontal family. Set

$$I_i = \mathrm{Int}\,(H^i, \gamma(z)) = \int_{\gamma(z)} \eta_i.$$

Then

(iii) $\quad a(H^1) I_1 + \cdots + a(H^n) I_n = 0.$

Set $I(z) = (I_1, \ldots, I_n)$. Then

(iv) $\quad dI(z) = \frac{1}{\kappa} \sum_{i<j} \Omega_{ij} I(z) d(z_i - z_j)/(z_i - z_j).$

Here $\Omega_{ij} \in \mathrm{Mat}_n(\mathbb{C})$ and the matrix Ω_{ij} has zero elements with the following exceptions:

$$(\Omega_{ij})_{pp} = (\Lambda_i, \Lambda_j) \quad \text{if } p \notin \{i, j\},$$
$$(\Omega_{ij})_{ii} = (\Lambda_j, \Lambda_i - \alpha), \quad (\Omega_{ij})_{jj} = (\Lambda_i, \Lambda_j - \alpha)$$
$$(\Omega_{ij})_{ij} = (\Lambda_j, \alpha), \quad (\Omega_{ij})_{ji} = (\Lambda_i, \alpha).$$

The differential equation (iv) is the hypergeometric differential equation (10.10.9) for these examples.

(10.10.11) Assume that $(pr : V \to B, \mathcal{C}, \mathcal{C}')$ have a real structure. Namely, they are the complexifications of $(pr_\mathbb{R} : V_\mathbb{R} \to B_\mathbb{R}, \mathcal{C}_\mathbb{R}, \mathcal{C}'_\mathbb{R})$, where $pr_\mathbb{R} : V_\mathbb{R} \to B_\mathbb{R}$

is an epimorphism of real spaces, $V = V_{\mathbb{R}} \otimes \sqrt{-1}V_{\mathbb{R}}$, $B = B_{\mathbb{R}} \oplus \sqrt{-1}B_{\mathbb{R}}$, and all hyperplanes of C and C' may be determined by affine equations with real coefficients.

As an example, see example (10.10.10).

Let A be the complex affine space with coordinates $\{a_H\}$, where H runs through the set of all horizontal hyperplanes of C. The set of weight of C defines a point in $A : \{a_H = a(H)\}$.

For any horizontal edge L of C, define a linear function on A by

$$a_L = \sum_{H \in C, L \subset H} a_H. \tag{10.10.12}$$

Define two domains in A. Set $D_0 = \{a \in A | a_L(a) \notin \{0, -1, -2, \ldots\}$ for all nontrivial horizontal edges L of $C\}$ and $D_{-1} = \{a \in A | a_L(a) \notin \{-1, -2, -3, \ldots\}$ for all nontrivial horizontal edges L of $C\}$.

10.10.13. Theorem. *Assume (10.10.11). Assume that* Re $a(L) > 0$ *for all horizontal edges L of C. The integration of hypergeometric forms well defines a natural morphism,*

$$i(a)^* : R.pr_1 \longrightarrow H^{\cdot}(A_h^{\cdot}(C), d_h(a))^*,$$

of bundles with connections. This morphism analytically depends on a and may be analytically continued to a holomorphic function in D_0. This function is a meromorphic function in A and has poles of at most first order on the hyperplanes of $A - D_0$.

Proof. A fiber of $R.pr_1$ is described in (2.3.6). Its dimension does not depend on a. Now the theorem follows from (10.7.7).

10.10.14. Corollary of (10.10.5). *The composition of morphisms of (10.10.4) and (10.10.5) induces a morphism*

$$iS(a) : H^{\cdot}(F\ell_h^{\cdot}(C), d_h) \longrightarrow R^{\cdot}pr_2$$

of bundles with connections.

Let

$$(iS(a))^* : R.pr_2 \longrightarrow H^{\cdot}(F\ell_h^{\cdot}(C), d_h)^* \tag{10.10.15}$$

be the dual morphism.

The fact that (10.10.15) is a morphism of bundles with connections may be formulated in terms of the following differential equation.

Let $\gamma(z)$ be a horizontal section of $R.pr_2$. Let $[x] \in H^{\cdot}(F\ell_h^{\cdot}(\mathcal{C}), d_h)$ and $x \in F\ell_h^{\cdot}(\mathcal{C})$ its representative. Let $i(a)S(a)(x) \in \Omega^{\cdot}(\mathcal{L}(a))$ be the corresponding flag hypergeometric differential form. Then the integral

$$\text{Int}\,(x, \gamma(z)) = \int_{\gamma(z)} i(a)S(a)(x) \tag{10.10.16}$$

is a well-defined multivalued holomorphic function on $Y(\mathcal{C}')$.

Fix $z \in Y(\mathcal{C}')$ and a branch of $\gamma(z)$ at z. Then (10.10.16) determines a linear function $iS^*(a)(\gamma(z))$ on $H^{\cdot}(F\ell_h^{\cdot}(\mathcal{C}), d_h)$. This function analytically depends on z and satisfies the following differential equation:

$$d_z \text{Int}\,(x, \gamma(z)) = \sum_{H \in \mathcal{C}'} \text{Int}\,(\Omega(H)x, \gamma(z)) d\ell_H / \ell_H, \tag{10.10.17}$$

where $\Omega(H)$ is defined in (10.5.20).

10.10.18. Example. Consider $pr : \mathbb{C}^{n+1} \to \mathbb{C}^n$, $\mathcal{C}$, $\mathcal{C}'$ as in (10.10.10). Let $F(H^i)$ be the flag of H^i. The horizontal complex $(F\ell_h^{\cdot}(\mathcal{C}), d_h)$ has the form

(i) $\mathbb{C} \longrightarrow \bigoplus_{i=1}^n \mathbb{C} F(H^i), \quad 1 \longmapsto \sum F(H^i).$

Set $\omega_i = a(H^i)\eta_i$ for $1 = 1, \ldots, n$, see (10.10.10). Let $\gamma(z) \in H_1(\mathbb{C} \setminus \{z_1, \ldots, z_n\}, S^*(a, z))$ be a horizontal family. Set $I_i = \int_{\gamma(z)} \omega_i$. Then

(ii) $I_1 + \cdots + I_n = 0.$

Set $I(z) = (I_1, \ldots, I_n)$. Then,

(iii) $dI(z) = \dfrac{1}{\kappa} \sum_{i<j} \Omega_{ij} I(z) d(z_i - z_j)/(z_i - z_j),$

where the matrix Ω_{ij} has zero elements with the following exceptions:

$$(\Omega_{ij})_{pp} = (\Lambda_i, \Lambda_j), \qquad \text{if } p \notin \{i,j\}$$
$$(\Omega_{ij})_{ii} = (\Lambda_j, \Lambda_i - \alpha), \quad (\Omega_{ij})_{jj} = (\Lambda_i, \Lambda_j - \alpha)$$
$$(\Omega_{ij})_{ij} = (\Lambda_i, \alpha), \qquad (\Omega_{ij})_{ji} = (\Lambda_j, \alpha).$$

The differential equation (iii) is the hypergeometric differential equation (10.10.17) for the given example.

10.10.19. Theorem. *Assume (10.10.11). Assume that* Re $a(L) > 0$ *for* *all horizontal edges L of C. The integration of flag hypergeometric forms well defines a natural morphism*

(i) $\qquad i^*(a) : R.pr_1 \to (H^{\cdot}(F\ell_h^{\cdot}(\mathcal{C}), d_h))^*$

of bundles with connections. Morphism (i) can be analytically continued to a holomorphic function in D_{-1}.

This theorem follows from (10.7.12).

11

Kac–Moody Lie Algebras without Serre's Relations and Their Doubles

Let $z = (z_1, \ldots, z_n) \in \mathbb{C}^n$, $z_1, \ldots, z_n$ pairwise distinct. Let $\mathcal{C}_{n,k}(z)$ be the discriminantal configuration of hyperplanes in $\mathbb{C}^k$ defined in (3.1). In Sec. 12, we recall the description in [SV3] of the complexes $(\mathcal{A}^\cdot(\mathcal{C}_{n,k}), d(a))$, $(\mathcal{F}\ell^\cdot(\mathcal{C}_{n,k}), d)$ and the homomorphism $S(a) : (\mathcal{F}\ell^\cdot(\mathcal{C}_{n,k}), d) \to (\mathcal{A}^\cdot(\mathcal{C}_{n,k}), d(a))$ in terms of the representation theory of Kac–Moody algebras. In this section, we recall the necessary information about Kac–Moody algebras without Serre's relations, see [K], [SV3].

11.1. Kac–Moody Lie Algebras without Serre's Relations [K], [SV3]

Let us fix the following data (cf. (4.1)):

(a) a finite-dimensional complex vector space $\mathfrak{h}$;

(b) a nondegenerate symmetric bilinear form (,) on $\mathfrak{h}$;

(c) linearly independent covectors ("simple roots") $\alpha_1, \ldots, \alpha_r \in \mathfrak{h}^*$.

As in (4.1), we denote by $b : \mathfrak{h} \to \mathfrak{h}^*$ the isomorphism induced by (,). We transfer the form (,) to $\mathfrak{h}^*$ using b. Set $b_{ij} = (\alpha_i, \alpha_j)$, $h_i = b^{-1}(\alpha_i) \in \mathfrak{h}$.

Denote by $\mathfrak{g}$ the Lie algebra generated by e_i, f_i for $i = 1, \ldots, r$ and $\mathfrak{h}$

subject to the relations

$$[h, e_i] = \langle h, \alpha_i \rangle e_i, \quad [h, f_i] = -\langle h, \alpha_i \rangle f_i,$$
$$[e_i, f_j] = \delta_{ij} h_i, \qquad (11.1.1)$$
$$[h, h'] = 0,$$

for all $i, j = 1, \ldots, r$ and $h, h' \in \mathfrak{h}$.

We denote by $\mathfrak{n}_-$ (resp. by $\mathfrak{n}_+$) the subalgebra of $\mathfrak{g}$ generated by f_i (resp. e_i) for $i = 1, \ldots, r$. We have

$$\mathfrak{g} = \mathfrak{n}_- \oplus \mathfrak{h} \oplus \mathfrak{n}_+.$$

Set $\mathfrak{b}_\pm = \mathfrak{n}_\pm \oplus \mathfrak{h}$. These are subalgebras of $\mathfrak{g}$.

For $\Lambda \in \mathfrak{h}^*$, denote by $M(\Lambda)$ the Verma module over $\mathfrak{g}$ generated by a vector v subject to the relations $\mathfrak{n}_+ v = 0$ and $hv = \langle \Lambda, h \rangle v$ for all $h \in \mathfrak{h}$.

For $\lambda = (k_1, \ldots, k_r) \in \mathbb{N}^r$, set

$$(\mathfrak{n}_\pm)_\lambda = \left\{ x \in \mathfrak{n}_\pm \,\Big|\, [h, x] = \Big\langle \pm \sum k_i \alpha_i, h \Big\rangle x \quad \text{for all } h \in \mathfrak{h} \right\}$$

$$M(\Lambda)_\lambda = \left\{ x \in M(\Lambda) \,\Big|\, hx = \Big\langle \Lambda - \sum k_i \alpha_i, h \Big\rangle x \quad \text{for all } h \in \mathfrak{h} \right\}.$$

We have $\mathfrak{n}_\pm = \oplus_\lambda (\mathfrak{n}_\pm)_\lambda$, $M(\Lambda) = \oplus_\lambda M(\Lambda)_\lambda$.

Let $\tau: \mathfrak{g} \to \mathfrak{g}$ be the Lie algebra automorphism such that $\tau(e_i) = -f_i$, $\tau(f_i) = -e_i$, $\tau(h) = -h$ for $h \in \mathfrak{h}$.

Set $M(\Lambda)^* = \oplus_\lambda M(\Lambda)^*_\lambda$. Define a structure of a $\mathfrak{g}$-module on $M(\Lambda)^*$ by the rule $\langle g\varphi, x \rangle = \langle \varphi, -\tau(g)x \rangle$ for $\varphi \in M^*$, $g \in \mathfrak{g}$, $x \in M$.

There is a unique bilinear form $K(\ ,\ )$ on $\mathfrak{g}$ such that

(11.1.2) K coincides with $(\ ,\ )$ on $\mathfrak{h}$; K is zero on $\mathfrak{n}_-$ and on $\mathfrak{n}_+$; $\mathfrak{h}$ and $\mathfrak{n}_- \oplus \mathfrak{n}_+$ are orthogonal;

(11.1.3) $K(f_i, e_j) = K(e_j, f_i) = \delta_{ij}$ for $i, j = 1, \ldots, r$;

(11.1.4) K is $\mathfrak{g}$-invariant, that is

$$K([x, y], z) = K(x, [y, z])$$

for all $x, y, z \in \mathfrak{g}$.

K is symmetric, and $K(\tau x, \tau y) = K(x, y)$.

Define a bilinear form S on $\mathfrak{g}$ by the rule

$$S(x, y) = -K(\tau x, y). \qquad (11.1.5)$$

Remark. In [**SV3**], the definition was different: $S(x,y) = K(\tau x, y)$.

The form S is symmetric, $S([x,y],z) = S(x,[\tau y, z])$, $S(\tau x, \tau y) = S(x,y)$. The subspaces $\mathfrak{n}_-$, $\mathfrak{h}$, $\mathfrak{n}_+$ are pairwise orthogonal with respect to S. The restriction of S to $\mathfrak{n}_-$ is uniquely determined by the properties:

$$\begin{aligned} S(f_i, f_j) &= \delta_{ij}, \\ S([f_i, x], y) &= S(x, [e_i, y]). \end{aligned} \tag{11.1.6}$$

The subspaces $(\mathfrak{n}_-)_\lambda$ for $\lambda \in \mathbb{N}^r$ are pairwise orthogonal with respect to S.

Similarly, the restriction of S to $\mathfrak{n}_+$ is uniquely determined by the properties:

$$\begin{aligned} S(e_i, e_j) &= \delta_{ij}, \\ S([f_i, x], y) &= S(x, [e_i, y]). \end{aligned}$$

Example. $S([f_1, f_2], [f_1, f_2]) = S([e_1, e_2], [e_1, e_2]) = -b_{12}$.

Let $M(\Lambda)$ be a Verma module with highest weight $\Lambda \in \mathfrak{h}^*$ and generating vector $v \in M(\Lambda)$. There is a unique bilinear form S on $M(\Lambda)$ such that

$$\begin{aligned} S(v,v) &= 1, \\ S(e_i x, y) &= S(x, f_i y), \\ S(f_i x, y) &= S(x, e_i y), \end{aligned} \tag{11.1.7}$$

for all $x, y \in M(\Lambda)$ and i.

For example, $S(f_1 v, f_1 v) = (\Lambda, \alpha_1)$.

S is symmetric. Subspaces $M(\Lambda)_\lambda$ are pairwise orthogonal with respect to S.

The form S induces a homomorphism of $\mathfrak{g}$-modules $S : M(\Lambda) \to M(\Lambda)^*$. The module $M(\Lambda)/\ker S$ is the irreducible $\mathfrak{g}$-module with highest weight Λ. More generally, define a symmetric form S on the space $\Lambda^p \mathfrak{n}_- \otimes M(\Lambda_1) \otimes \cdots \otimes M(\Lambda_n)$ for $\Lambda_i \in \mathfrak{h}^*$ by the rule

$$\begin{aligned} S(g_1 \wedge \cdots \wedge g_p &\otimes x_1 \otimes \cdots \otimes x_n, g'_1 \wedge \cdots \wedge g'_p \otimes x'_1 \otimes \cdots \otimes x'_n) \\ &= \det S(g_i, g'_j) \cdot \prod_{i=1}^n S(x_i, x'_i). \end{aligned} \tag{11.1.8}$$

This form induces a map

$$S: \Lambda^p \mathfrak{n}_- \otimes M(\Lambda_1) \otimes \cdots \otimes M(\Lambda_n) \to (\Lambda^p \mathfrak{n}_- \otimes M(\Lambda_1) \otimes \cdots \otimes M(\Lambda_n))^*. \quad (11.1.9)$$

The form S is called the contravariant form. The linear map (11.1.9) analytically depends on (b_{ij}) for a fixed r and on $\Lambda_1, \ldots, \Lambda_n \in \mathfrak{h}^*$. This map is nondegenerate for general values of (b_{ij}), $\Lambda_1, \ldots, \Lambda_n$, see Theorem 3.7 in [SV3].

11.1.10. Remark. Suppose that $b_{ii} \neq 0$ for all i. Set $a_{ij} = 2b_{ij}/b_{ii}$. The quotient $\bar{\mathfrak{g}} = \mathfrak{g}/\ker S$ is the Kac-Moody Lie algebra associated with the matrix (a_{ij}). Here S is the form defined in (11.1.5). Suppose that $a_{ij} \in \mathbb{Z}$, $a_{ij} \leq 0$ for all $i \neq j$. Then $\ker S$ is generated by the Serre elements $(\operatorname{ad} f_i)^{n_{ij}}(f_i)$ and $(\operatorname{ad} e_i)^{n_{ij}}(e_j)$ for all $i \neq j$, here $n_{ij} = -a_{ij} + 1$.

11.2. Complexes [SV], cf. (5.1)

For a Lie algebra $\mathfrak{g}$ and a $\mathfrak{g}$-module M, denote by $C.(\mathfrak{g}, M)$ the standard chain complex of $\mathfrak{g}$ with coefficients in M. $C_p(\mathfrak{g}, M) = \Lambda^p \mathfrak{g} \otimes M$ and

$$\begin{aligned} d: g_p \wedge \cdots \wedge g_1 \otimes x &= \sum_{i=1}^{p}(-1)^{i-1} g_p \wedge \cdots \wedge \widehat{g_i} \wedge \cdots \wedge g_1 \otimes g_i x \\ &+ \sum_{1 \leq i < j \leq p}(-1)^{i+j} g_p \wedge \cdots \wedge \widehat{g_j} \wedge \cdots \wedge \widehat{g_i} \wedge \cdots \wedge g_1 \wedge [g_j, g_i] \otimes x. \end{aligned} \quad (11.2.1)$$

Let $\Lambda_1, \ldots, \Lambda_n \in \mathfrak{h}^*$. Set $M = M(\Lambda_1) \otimes \cdots \otimes M(\Lambda_n)$. Consider the complex $C.(\mathfrak{n}_-, M)$. We have the weight decomposition

$$C.(\mathfrak{n}_-, M) = \bigoplus_{\lambda \in \mathbb{N}^r} C.(\mathfrak{n}_-, M)_\lambda. \quad (11.2.2)$$

11.2.3. Example. $f_2 \wedge f_1 \otimes v \in \Lambda^2 \mathfrak{n}_- \otimes M(\Lambda_1)$. $d: f_2 \wedge f_1 \otimes v = f_2 \otimes f_1 v - f_1 \otimes f_2 v - [f_2, f_1] \otimes v$.

11.2.4. Remark. The operator d does not depend on the parameters (b_{ij}), $\Lambda_1, \ldots, \Lambda_n$ of this complex.

Following [SV3], define the maps

$$\begin{aligned} \Lambda^2 \mathfrak{n}_-^* &\longrightarrow \mathfrak{n}^*, \\ \mathfrak{n}_-^* \otimes M^* &\longrightarrow M^* \end{aligned} \quad (11.2.5)$$

by the condition of commutativity of the diagrams

$$\begin{array}{ccc} \Lambda^2 \mathfrak{n}_- & \xrightarrow{[,]} & \mathfrak{n}_- \\ \downarrow S & & \downarrow S \\ \Lambda^2 \mathfrak{n}_-^* & \longrightarrow & \mathfrak{n}_-^* \end{array} \qquad (11.2.6)$$

$$\begin{array}{ccc} \mathfrak{n}_- \otimes M & \longrightarrow & M \\ \downarrow S & & \downarrow S \\ \mathfrak{n}_-^* \otimes M^* & \longrightarrow & M^* . \end{array} \qquad (11.2.7)$$

Here S is the contravariant form.

In [**SV3**], it is shown that S is an isomorphism for general values of parameters (b_{ij}), $\Lambda_1, \ldots, \Lambda_n$. Therefore, the commutativity condition defines (11.2.5) for general values of the parameters. In [**SV3**], it is shown that the maps (11.2.5) may be defined for arbitrary values of the parameters by continuity. A direct definition of maps (11.2.5) is given in (11.3), see [**SV3**].

The maps (11.2.5) determine a Lie algebra structure on $\mathfrak{n}_-^*$ and a $\mathfrak{n}_-^*$-module structure on M^*. Let $C.(\mathfrak{n}_-^*, M^*)$ be the corresponding standard chain complex:

$$C.(\mathfrak{n}_-^*, M^*) = \bigoplus_{\lambda \in \mathbb{N}^r} C.(\mathfrak{n}_-^*, M^*)_\lambda . \qquad (11.2.8)$$

The contravariant form induces a graded homomorphism of complexes,

$$S : C.(\mathfrak{n}_-, M) \longrightarrow C.(\mathfrak{n}_-^*, M^*) . \qquad (11.2.9)$$

11.2.10. Remark. Set $\bar{\mathfrak{g}} = \mathfrak{g}/\ker S$. It is the Kac–Moody algebra associated with the matrix (a_{ij}) defined in (11.1.10). Set $\bar{\mathfrak{n}}_- = \mathfrak{n}_-/\ker S \subset \bar{\mathfrak{g}}$. Set $L(\Lambda_i) = M(\Lambda_i)/\ker S$. These are irreducible $\bar{\mathfrak{g}}$-modules. Set $L = L(\Lambda_1) \otimes \cdots \otimes L(\Lambda_n)$. The quotient complex $C.(\mathfrak{n}_-, M)/\ker S$ is isomorphic to $C.(\bar{\mathfrak{n}}_-, L)$.

11.2.11. Example. $f_2^* \wedge f_1^* \otimes v^* \in \Lambda^2 \mathfrak{n}_-^* \otimes M(\Lambda)^*$.

$$d : f_2^* \wedge f_1^* \otimes v^* \mapsto (\Lambda, \alpha_1) f_2^* \otimes (f_1 v)^* - (\Lambda, \alpha_2) f_1^* \otimes (f_2 v)^* + b_{12}[f_2, f_1]^* \otimes v^*,$$

$[f_1, f_2]^* \otimes v^* \in \mathfrak{n}_-^* \otimes M(\Lambda)^*$. $d : [f_1, f_2]^* \otimes v^* \mapsto (\Lambda, \alpha_2)(f_1 f_2 v)^* - (\Lambda, \alpha_1)(f_2 f_1 v)^*$. Here $\{f_I^*\}$ (resp. $\{(f_I v)^*\}$) is the basis in $\mathfrak{n}_-^*$ (resp. in $M(\Lambda)^*$) dual to the monomial basis $\{f_I\}$ in $\mathfrak{n}_-$ (resp. $\{f_I v\}$ in $M(\Lambda)$).

11.3. The Double [D1], [SV3]

A Lie bialgebra is a vector space $\mathfrak{g}$ with a Lie algebra structure and a Lie coalgebra structure. These structures are compatible in the following sense: the cocommutator mapping $\nu : \mathfrak{g} \to \mathfrak{g} \wedge \mathfrak{g}$ must be a one-cocycle:

$$x\nu(y) - y\nu(x) = \nu([x,y]), \qquad (11.3.1)$$

for all $x, y \in \mathfrak{g}$. Here the action of $\mathfrak{g}$ on $\mathfrak{g} \wedge \mathfrak{g}$ is the adjoint one: $a(b \wedge c) = [a,b] \wedge c + b \wedge [a,c]$.

Let $\mathfrak{g}$ be a bialgebra. The double of $\mathfrak{g}$ is the Lie algebra equal to $\mathfrak{g} \oplus \mathfrak{g}^*$ as a space, with the bracket on $\mathfrak{g}$ and $\mathfrak{g}^*$ defined by the Lie algebra structures on $\mathfrak{g}$ and $\mathfrak{g}^*$ and, for $x \in \mathfrak{g}$ and $\ell \in \mathfrak{g}^*$,

$$[\ell, x] = \bar{\ell} + \bar{x},$$

where $\bar{x} \in \mathfrak{g}$ and $\bar{\ell} \in \mathfrak{g}^*$ are defined by the rules

$$\begin{aligned}\bar{\ell}(y) &= \ell([x,y]), \\ m(\bar{x}) &= [m, \ell](x),\end{aligned} \qquad (11.3.2)$$

for all $y \in \mathfrak{g}$ and $m \in \mathfrak{g}^*$. The double is denoted by $\mathcal{D}(\mathfrak{g})$.

The image r of the canonical element of $\mathfrak{g} \otimes \mathfrak{g}^*$ under the imbedding $\mathfrak{g} \otimes \mathfrak{g}^* \hookrightarrow \mathcal{D}(\mathfrak{g}) \otimes \mathcal{D}(\mathfrak{g})$ satisfies the classical Yang–Baxter equation:

$$[r^{12}, r^{13}] + [r^{12}, r^{23}] + [r^{13}, r^{23}] = 0, \qquad (11.3.3)$$

where $r^{ij} \in \mathcal{D}(\mathfrak{g})^{\otimes 3}$ is r positioned in the ith and jth factors, see [D1].

Now let $\mathfrak{g}$ be the Lie algebra associated with data (11.1.a)–(11.1.c) and defined by (11.1.1). $\mathfrak{g}$ is a Lie bialgebra with respect to the following cobracket. There exists a unique map $\nu : \mathfrak{g} \to \mathfrak{g} \wedge \mathfrak{g}$ such that ν satisfies (11.3.1) and

$$\begin{aligned}\nu(h) &= 0, \\ \nu(f_i) &= \frac{1}{2} f_i \wedge h_i, \\ \nu(e_i) &= \frac{1}{2} e_i \wedge h_i,\end{aligned} \qquad (11.3.4)$$

for all $h \in \mathfrak{h}$ and $i = 1,\ldots,r$, see [D1, Example 3.2] and [SV3].

11.3.5. Remarks. $\mathfrak{b}_+$ and $\mathfrak{b}_-$ are subbialgebras.

The map $\nu : \mathfrak{g} \to \mathfrak{g} \wedge \mathfrak{g}$ has the following property:
$$\nu\tau + \tau\nu = 0,$$
where τ is defined in (11.1). Thus, if $\rho : \mathfrak{b}_-^* \to \mathrm{End}(V)$ is a representation of the Lie algebra $(\nu|_{\mathfrak{b}_-})^* : \Lambda^2 \mathfrak{b}_-^* \to \mathfrak{b}_-^*$, then $-\rho \circ \tau : \mathfrak{b}_+^* \to \mathrm{End}(V)$ is a representation of the Lie algebra $(\nu|_{\mathfrak{b}_+})^* : \Lambda^2 \mathfrak{b}_+^* \to \mathfrak{b}_+^*$.

11.3.6. Example.
$$\nu([f_1, f_2]) = -\frac{1}{2}(h_1 + h_2) \wedge [f_1, f_2] - b_{12} f_1 \wedge f_2$$
$$\nu([e_1, e_2]) = -\frac{1}{2}(h_1 + h_2) \wedge [e_1, e_2] + b_{12} e_1 \wedge e_2.$$

Set
$$\mu : \mathbf{N}^r \longrightarrow \mathfrak{h}, \quad (\ell_1, \ldots, \ell_r) \longmapsto \sum_{i=1}^{r} \ell_i h_i.$$

For every $x \in (\mathfrak{n}_-)_\lambda$ and $y \in (\mathfrak{n}_+)_\lambda$, we have
$$\begin{aligned}\nu(x) &= -\frac{1}{2}\mu(\lambda) \wedge x + \nu(x)_-, \\ \nu(y) &= -\frac{1}{2}\mu(\lambda) \wedge y + \nu(y)_+,\end{aligned} \qquad (11.3.7)$$
where $\nu(x)_- \in \mathfrak{n}_- \wedge \mathfrak{n}_-$ and $\nu(y)_+ \in \mathfrak{n}_+ \wedge \mathfrak{n}_+$. Moreover,
$$\begin{aligned}\nu([f_i, x])_- &= f_i \cdot \nu(x)_- + f_i \wedge [h_i, x], \\ \nu([e_i, y])_+ &= e_i \cdot \nu(x)_+ + e_i \wedge [h_i, y],\end{aligned} \qquad (11.3.8)$$

The maps $\nu_- : \mathfrak{n}_- \to \mathfrak{n}_- \wedge \mathfrak{n}_-$ and $\nu_+ : \mathfrak{n}_+ \to \mathfrak{n}_+ \wedge \mathfrak{n}_+$ preserve the weight decomposition. Let $\nu_-^* : \mathfrak{n}_-^* \wedge \mathfrak{n}_-^* \to \mathfrak{n}_-^*$ and $\nu_+^* : \mathfrak{n}_+^* \wedge \mathfrak{n}_+^* \to \mathfrak{n}_+^*$ be the dual maps.

11.3.9. Lemma, cf. [SV3, Lemma 6.14.3]. *For every $x \in \mathfrak{n}_-$,*
$$S(\nu(x), a \wedge b) = \begin{cases} \frac{1}{2} S(x, [a, b]) & \text{if } a \in \mathfrak{n}_- \text{ and } b \in \mathfrak{h}, \\ S(x, [a, b]) & \text{if } a, b \in \mathfrak{n}_-. \end{cases}$$

For every $y \in \mathfrak{n}_+$,

$$S(\nu(y), a \wedge b) = \begin{cases} -\dfrac{1}{2}S(y, [a,b]) & \text{if } a \in \mathfrak{n}_+ \text{ and } b \in \mathfrak{h}, \\ -S(y, [a,b]) & \text{if } a, b \in \mathfrak{n}_+ \end{cases}$$

cf. (11.2.6).

The proof coincides with the proof of Lemma (6.14.3) in [**SV3**].

11.3.10. Corollary. *The following diagrams are commutative:*

$$\begin{array}{ccc} \Lambda^2 \mathfrak{n}_- & \xrightarrow{[\,,\,]} & \mathfrak{n}_- \\ \downarrow S & & \downarrow S \\ \Lambda^2 \mathfrak{n}_-^* & \xrightarrow{\nu_-^*} & \mathfrak{n}_-^* \end{array} \qquad \begin{array}{ccc} \Lambda^2 \mathfrak{n}_+ & \xrightarrow{[\,,\,]} & \mathfrak{n}_+ \\ \downarrow S & & \downarrow -S \\ \Lambda^2 \mathfrak{n}_+^* & \xrightarrow{\nu_+^*} & \mathfrak{n}_+^* \end{array}$$

S is an isomorphism for general values of parameters (b_{ij}), see [**SV3**, (3.7) and (6.6)]. Hence, $\nu_-^* : \Lambda^2 \mathfrak{n}_-^* \to \mathfrak{n}_-^*$ and $\nu_+^* : \Lambda^2 \mathfrak{n}_+^* \to \mathfrak{n}_+^*$ are Lie algebra structures on $\mathfrak{n}_-^*$ and $\mathfrak{n}_+^*$, respectively.

Comultiplication. Let $M = M(\Lambda_1) \otimes \cdots \otimes M(\Lambda_n)$, where $\Lambda_1, \ldots, \Lambda_n \in \mathfrak{h}^*$. Let $v = v_1 \otimes \cdots \otimes v_n \in M$ be the product of the generating vectors, $\Lambda = \Lambda_1 + \cdots + \Lambda_n$. Let $\mathfrak{b}_-$ act on $\mathfrak{b}_- \otimes M$ by the rule $a(b \otimes m) = [a, b] \otimes m + b \otimes am$.

For $1 \leq i \leq n$ and $a, b \in \mathfrak{g}, m = x_1 \otimes \cdots \otimes x_n \in M$, set $a^{(i)} m = x_1 \otimes \cdots \otimes x_{i-1} \otimes ax_i \otimes x_{i+1} \otimes \cdots \otimes x_n$ and $a^{(i)}(b \otimes m) = [a, b] \otimes m + b \otimes a^{(i)} m$.

There is a unique linear map $\nu_M : M \to \mathfrak{b}_- \otimes M$ such that

$$\nu_M(h \cdot x) = h \cdot \nu_M(x) \quad \text{for any } h \in \mathfrak{h} \text{ and } x \in M, \tag{11.3.11}$$

$$\nu_M(x) = \frac{1}{2}\left(b^{-1}(\Lambda) - \mu(\lambda)\right) \otimes x + \nu_M(x)_-. \tag{11.3.12}$$

Here, $x \in M_\lambda$, $\nu_M(x)_- \in \mathfrak{n}_- \otimes M$, and $b^{-1} : \mathfrak{h}^* \xrightarrow{\sim} \mathfrak{h}$ is defined in (11.1),

$$\nu_M(f_i^{(j)} x)_- = f_i \otimes h_i^{(j)} x + f_i^{(j)} \nu_M(x)_- \tag{11.3.13}$$

for $1 \leq i \leq r$ and $1 \leq j \leq n$.

Remark. This definition is different from the corresponding definition in [**SV3**, (6.15.1)]. The distinctions are explained by the distinctions in the definitions of the contravariant form S, see remark after (11.1.5).

11.3.14. Examples. $\nu(v) = \frac{1}{2} b^{-1}(\Lambda) \otimes v$,

$$\nu(f_i^{(k)} v) = \frac{1}{2}(b^{-1}(\Lambda) - h_i) \otimes f_i^{(k)} v + (\Lambda_k, \alpha_i) f_i \otimes v$$

$$\nu(f_j^{(k)} f_i^{(k)} v) = \frac{1}{2}(b^{-1}(\Lambda) - h_i - h_j) \otimes f_j^{(k)} f_i^{(k)} v + (\Lambda_k - \alpha_i, \alpha_j) f_j \otimes f_i^{(k)} v$$

$$+ (\Lambda_k, \alpha_i) f_i \otimes f_j^{(k)} v + (\Lambda_k, \alpha_i)[f_j, f_i] \otimes v.$$

We give two more explicit formulas for ν_M. First, for any $x \in M$,

$$\nu_M(x)_- = \sum_{i=1}^{r} f_i \otimes e_i x + \nu_M(x)_{--}, \qquad (11.3.15)$$

where $\nu_M(x)_{--} \in \mathfrak{n}_{--} \otimes M$ and $\mathfrak{n}_{--} = \oplus_{|\lambda|>1} (\mathfrak{n}_-)_\lambda$.

For an N-tuple of pairwise distinct integers $(i(1), \ldots, i(N))$, $1 \leq i(j) \leq r$ and $a(1), \ldots, a(N) \in \{1, \ldots, n\}$,

$$\nu_M \left(f_{i(N)}^{(a(N))} f_{i(N-1)}^{(a(N-1))} \cdots f_{i(1)}^{(a(1))} v \right)_-$$

$$= \sum_{p=1}^{N} \sum_{1 \leq j_1 < \cdots < j_p \leq N} \left[f_{i(j_p)}, [f_{i(j_{p-1})}, \ldots, [f_{i(j_2)}, f_{i(j_1)}] \cdots] \right] \qquad (11.3.16)$$

$$\otimes e_{i(j_1)} f_{i(N)}^{(a(N))} \cdots f_{i(j_p)}^{\widehat{(a(j_p))}} \cdots f_{i(j_2)}^{\widehat{(a(j_2))}} \cdots f_{i(1)}^{(a(1))} v.$$

Both formulas are proved by induction on $|\lambda|$, see examples (11.3.14).

11.3.17. Lemma, cf. [**SV3**, (6.15.2)]. *For any $x, y \in M$ and $a \in \mathfrak{b}_-$,*

$$S(\nu_M(x), a \otimes y) = \begin{cases} \dfrac{1}{2} S(x, ay) & \text{if } a \in \mathfrak{h} \\ S(x, ay) & \text{if } a \in \mathfrak{n}_-. \end{cases}$$

Here S is defined on $\mathfrak{b}_- \otimes M$ by the rule $S(a \otimes x, b \otimes y) = S(a, b) S(x, y)$, cf. (11.1.8).

Proof, cf. the proof of [**SV3**, (6.15.2)]. The first equality follows from (11.3.12).

First case. Let $a = f_p$ for $p \in \{1, \ldots, r\}$. We have

$$S(\nu_M(x), f_p \otimes y) = S\left(\sum f_i \otimes e_i x + \nu_M(x)_{--}, f_p \otimes y \right) = S(e_p x, y) = S(x, f_p y).$$

Second case. Let $x = f_{i(N)}^{(a(N))} \ldots f_{i(1)}^{(a(1))} v$ as in (11.3.16), $a \in \mathfrak{n}_-$. Prove (11.3.17) by induction on N. The formula is true for $N = 1$. We prove it for general N. Let $a \in (\mathfrak{n}_-)_\lambda$, and let $(\mathfrak{n}_-)_\lambda$ contain $F = \Big[f_{i(j_p)}, [f_{i(j_p-1)}, \ldots, [f_{i(j_2)}, f_{i(j_1)}] \ldots] \Big]$. Then

$$S(\nu_M(x), a \otimes y) = S(F, a) S\Big(e_{i(j_1)} f_{i(j_p)}^{(a(N))} \ldots f_{i(j_p)}^{(\widehat{a(j_p)})} \ldots f_{i(j_2)}^{(\widehat{a(j_2)})} \ldots f_{i(1)}^{(a(1))} v, y \Big)$$

$$= S\Big(\big[f_{i(j_p-1)}, [\ldots [f_{i(j_2)}, f_{i(j_1)}] \ldots] \big], [e_{i(j_p)}, a] \Big)$$

$$\times S\Big(e_{i(j_1)} f_{i(j_p)}^{(a(N))} \ldots f_{i(j_p)}^{(\widehat{a(j_p)})} \ldots f_{i(j_2)}^{(\widehat{a(j_2)})} \ldots f_{i(1)}^{(a(1))} v, y \Big)$$

$$= S\Big(f_{i(N)}^{(a(N))} \ldots f_{i(j_p)}^{(\widehat{a(j_p)})} \ldots f_{i(j_p-1)}^{(a(j_p-1))} \ldots f_{i(j_2)}^{(a(j_2))}$$
$$\ldots f_{i(1)}^{(a(1))} v, [e_{i(j_p)}, a] y \Big)$$

$$= S\Big(f_{i(N)}^{(a(N))} \ldots f_{i(j_p)}^{(a(j_p))} \ldots f_{i(1)}^{(a(1))} v, a y \Big).$$

The general case is treated similarly.

11.3.18. Corollary. *The following diagram is commutative*

$$\begin{array}{ccc} \mathfrak{n}_- \otimes M & \longrightarrow & M \\ \downarrow S & & \downarrow S \\ \mathfrak{n}_-^* \otimes M^* & \xrightarrow{\nu_{M-}^*} & M^* \end{array}$$

where $\mathfrak{n}_- \otimes M \to M$ *is the standard action.*

S is an isomorphism for general values of parameters (b_{ij}), $\Lambda_1, \ldots, \Lambda_n$, see [**SV3**, (3.7) and (6.6)]. Hence, $\nu_M^* : \mathfrak{b}_-^* \otimes M^* \to M^*$ is a $\mathfrak{b}_-^*$-module structure with respect to the Lie algebra structure $\nu^* : \Lambda^2 \mathfrak{b}_-^* \to \mathfrak{b}_-^*$. Moreover, $\nu_{M-}^* : \mathfrak{n}_-^* \otimes M^* \to M^*$ is an $\mathfrak{n}_-^*$-module structure with respect to the Lie algebra structure $\nu_-^* : \Lambda^2 \mathfrak{n}_-^* \to \mathfrak{n}_-^*$.

11.3.19. Remark. Assume that the parameters (b_{ij}) are such that $S : \mathfrak{n}_- \to \mathfrak{n}_-^*$ is nondegenerate. Let $\{a_I\}$ be any graded basis in $\mathfrak{n}_-$, $\{a_I^*\}$ the dual basis in $\mathfrak{n}_-^*$, $S_{IJ} := S(a_I, a_J)$, (S_{IJ}^{-1}) the matrix inverse to the matrix (S_{IJ}). Then

$$\nu_{M-}^* : a_I^* \otimes \varphi(\cdot) \longmapsto -\sum_J S_{IJ}^{-1} \varphi(\tau(a_J) \cdot)$$

for all $\varphi \in M^*$ and I, here $\tau : \mathfrak{g} \to \mathfrak{g}$ is defined in (11.1), cf. (5.1.6), (9.3.1).

11.3.20. Corollary. *The contravariant form S defined in (11.1.8) induces a homomorphism of complexes*

$$S : C.(\mathfrak{n}_-, M) \longrightarrow C.(\mathfrak{n}_-^*, M^*),$$

where $C.(\mathfrak{n}_-^*, M^*)$ is the standard chain complex of $\mathfrak{n}_-^*$ with coefficients in M^*.

Define an action of $\mathfrak{b}_-$ on M^* by

$$\mathfrak{b}_- \otimes M^* \xrightarrow{\tau \otimes id} \mathfrak{b}_+ \otimes M^* \longrightarrow M^*, \qquad (11.3.21)$$

where $\mathfrak{b}_+ \otimes M^* \to M^*$ is the action defined in (11.1). More explicitly, the action (11.3.21) is given by the rule $a \otimes \varphi(\cdot) \mapsto \varphi(-a\,\cdot)$.

The maps (11.3.21) and $\nu_M^* : \mathfrak{b}_-^* \otimes M^* \to M^*$ define an action of $\mathfrak{b}_- \oplus \mathfrak{b}_-^*$ on M^*.

11.3.22. Lemma, [**SV3**, (6.17.1)]. *The above rule defines on M^* a structure of a $D(\mathfrak{b}_-)$-module.*

For any $a \in \mathfrak{b}_-^*$, the action ν_M^* defines a map $\nu_M^*(a, \cdot) : M^* \to M^*$. Set $\rho(a) = -\left(\nu_M^*(a, \cdot)\right)^* : M \to M$. The map

$$\rho : \mathfrak{b}_-^* \longrightarrow \mathrm{End}\,(M) \qquad (11.3.23)$$

gives an action of $\mathfrak{b}_-^*$ on M. Define an action of $\mathfrak{b}_+^*$ on M by

$$\omega = -\rho \circ \tau : \mathfrak{b}_+^* \longrightarrow \mathrm{End}\,(M), \qquad (11.3.24)$$

see (11.3.5).

11.3.25. Examples. For any $h \in \mathfrak{h}$, $a \in \mathfrak{n}_+$, and $x \in M$, we have

$$\omega\bigl(S(\cdot, h)\bigr)x = -hx/2,$$
$$\omega\bigl(S(\cdot, a)\bigr)x = -ax.$$

Assume that the parameters (b_{ij}) are such that $S : \mathfrak{n}_+ \to \mathfrak{n}_+^*$ is nondegenerate. Let $\{a_I\}$ be any graded basis in $\mathfrak{n}_+$, let $\{a_I^*\}$ be the dual basis in $\mathfrak{n}_+^*$, $S_{IJ} := S(a_I, a_J)$, (S_{IJ}^{-1}) the matrix dual to the matrix (S_{IJ}). Then

$$\omega : a_I^* \otimes x \mapsto -\sum_J S_{IJ}^{-1} a_J x, \qquad (11.3.26)$$

cf. (5.1.6), (9.3.1).

Define an action of $\mathfrak{b}_+$ on M by the rule:

$$a \otimes x \longmapsto \tau(a)x \quad \text{for } a \in \mathfrak{b}_+, x \in M. \tag{11.3.27}$$

This action and the map ω define an action of $\mathfrak{b}_+ \oplus \mathfrak{b}_+^*$ on M.

11.3.28. Lemma. *The above rule defines on M a structure of a $D(\mathfrak{b}_+)$-module, cf. (9.3.3).*

Proof. It is sufficient to prove the lemma for general values of the parameters (b_{ij}). In this case, $\mathfrak{b}_+^*$ is generated by linear functions of the form $S(\cdot, c)$, where $c \in \mathfrak{b}_+$. The Lie algebra structure on $\mathfrak{b}_+^*$ is given by the rule: $[S(\cdot, a), S(\cdot, b)]$ is $S(\cdot, [b, a])$ if $a, b \in \mathfrak{n}_+$, $S(\cdot, [b, a])/2$ if $a \in \mathfrak{n}_+$ and $b \in \mathfrak{h}$, 0 if $a, b \in \mathfrak{h}$. The Lie algebra structure on the double is given by (11.3.2). Now the lemma is proved by direct verification, see [**SV3**, (6.17.1)].

11.3.29. Remark. Let $\mathfrak{g}$ be any Lie algebra of the form (11.1) and let V be a $\mathfrak{g}$-module. Define a $\mathfrak{g}$-module structure on V^* by the rule $g\varphi(\cdot) = \varphi(-g\cdot)$ for all $g \in \mathfrak{g}$ and $\varphi \in V^*$. With respect to this module structure, M becomes a $D(\mathfrak{b}_-)$-module and M^* become a $D(\mathfrak{b}_+)$-module. Namely, $D(\mathfrak{b}_-) = \mathfrak{b}_- \oplus \mathfrak{b}_-^*$ and $\mathfrak{b}_-$ acts on M by the standard action. The elements of $\mathfrak{b}_-^*$ having the form $S(a, \cdot)$ act by the rule

(i)
$$S(a, \cdot) \otimes x = \tau(a)x \quad \text{if } a \in \mathfrak{n}_-$$
$$S(a, \cdot) \otimes x = \tau(a)x/2 \quad \text{if } a \in \mathfrak{h}.$$

$D(\mathfrak{b}_+) = \mathfrak{b}_+ \oplus \mathfrak{b}_+^*$ and $\mathfrak{b}_+$ acts on M^* by the rule $a\varphi(\cdot) = \varphi(-\tau(a)\cdot)$ for all $a \in \mathfrak{b}_+$ and $\varphi(\cdot) \in M^*$, that is, as defined in (11.1). The elements of $\mathfrak{b}_+^*$ having the form $S(a, \cdot)$ act by the rule

(ii)
$$S(a, \cdot) \otimes \varphi(\cdot) = \varphi(a\cdot) \quad \text{if } a \in \mathfrak{n}_+$$
$$S(a, \cdot) \otimes \varphi(\cdot) = \varphi(a\cdot)/2 \quad \text{if } a \in \mathfrak{h}.$$

For a vector space V, denote by $\Omega(V) \in V \otimes V^*$ the tautological element. For $\lambda \in \mathbb{N}^r$, set $\Omega_{\lambda,\pm}^- := \Omega((\mathfrak{n}_\pm)_\lambda) \in (\mathfrak{n}_\pm)_\lambda \otimes (\mathfrak{n}_\pm)_\lambda^*$, $\Omega_{\lambda,\pm}^+ := \Omega((\mathfrak{n}_\pm)_\lambda^*) \in (\mathfrak{n}_\pm)_\lambda^* \otimes (\mathfrak{n}_\pm)_\lambda$, $\Omega^0 = (\Omega(\mathfrak{h}) + \Omega(\mathfrak{h}^*)) \in \mathfrak{h} \otimes \mathfrak{h}^* + \mathfrak{h}^* \otimes \mathfrak{h}$. Set

$$\Omega_\pm = \sum_\lambda \Omega_{\lambda,\pm}^- + \Omega^0 + \sum_\lambda \Omega_{\lambda,\pm}^+ \in \mathcal{D}(\mathfrak{b}_\pm) \otimes \mathcal{D}(\mathfrak{b}_\pm). \tag{11.3.30}$$

11.3.31. Lemma. *For any $x \in D(\mathfrak{b}_+)$ and $y \in D(\mathfrak{b}_-)$, we have $[\Omega_+, x \otimes 1 + 1 \otimes x] = 0$ and $[\Omega_-, y \otimes 1 + 1 \otimes y] = 0$.*

The proof is by direct verification.

For any $\Lambda, \Lambda' \in \mathfrak{h}$, Ω_+ (resp. Ω_-) acts on $M(\Lambda) \otimes M(\Lambda')$ (resp. $M(\Lambda)^* \otimes M(\Lambda')^*$) by means of $\mathfrak{b}_+ \oplus \mathfrak{b}_+^*$-action (resp. $\mathfrak{b}_- \otimes \mathfrak{b}_-^*$-action) introduced above.

11.3.32. Lemma. *The following diagram is commutative:*

$$\begin{array}{ccc} M(\Lambda) \otimes M(\Lambda') & \xrightarrow{\Omega_+} & M(\Lambda) \otimes M(\Lambda') \\ \downarrow S & & \downarrow S \\ M(\Lambda)^* \otimes M(\Lambda')^* & \xrightarrow{-\Omega_-} & M(\Lambda)^* \otimes M(\Lambda')^* . \end{array}$$

The proof follows from (11.3.26) and remark (11.3.19).

11.3.33. Example. Let $\mathfrak{g} = s\ell_2$ be generated by e, f, h. Then $D(\mathfrak{b}_-)$ is generated by h, f, h^*, f^* subject to the relations $[f, h] = 2f$, $[f^*, h^*] = f^*/2$, $[f^*, h] = -2f^*$, $[f^*, f] = 2h^* - h/2$, $[h^*, f] = f/2$. Representation (11.3.22) of $D(\mathfrak{b}_-)$ in M^* is given by the formulas: $f \mapsto -e$, $h \mapsto -h$, $h^* \mapsto h/4$, $f^* \mapsto f$, where operators h, e, f on M^* are described in (11.1). The operator $\Omega_- = f \otimes f^* + f^* \otimes f + h \otimes h^* + h \otimes h^*$ acts on $M^*(\Lambda) \otimes M(\Lambda')^*$ as $-(e \otimes f + f \otimes e + h \otimes h/2)$.

$D(\mathfrak{b}_+)$ is generated by h, e, h^*, e^* subject to the relations $[e, h] = -2e$, $[e^*, h^*] = e^*/2$, $[e^*, h] = 2e^*$, $[e^*, e] = -2h^* - h/2$, $[h^*, e] = e/2$. Representation (11.3.28) of $D(\mathfrak{b}_+)$ in M is given by the formulas: $e \mapsto -f$, $h \mapsto -h$, $h^* \mapsto -h/4$, $e^* \mapsto -e$. The operator $\Omega_+ = e \otimes e^* + e^* \otimes e + h \otimes h^* + h^* \otimes h$ acts on $M(\Lambda) \otimes M(\Lambda')$ as $f \otimes e + e \otimes f + h \otimes h/2$.

The operator Ω_- acts on $M(\Lambda) \otimes M(\Lambda')$ by means of the $D(\mathfrak{b}_-)$-action introduced in (11.3.29). Analogously, Ω_+ acts on $M(\Lambda)^* \otimes M(\Lambda')^*$ by means of the $D(\mathfrak{b}_+)$-action introduced in (11.3.29).

11.3.34. Lemma. *$\Omega_+ + \Omega_-$ acts as zero on $M(\Lambda) \otimes M(\Lambda')$ and on $M(\Lambda)^* \otimes M(\Lambda')^*$, cf. (11.3.33).*

Proof. It suffices to check the lemma for $M(\Lambda) \otimes M(\Lambda')$. If the contravariant form is nondegenerate, then the lemma follows from (11.3.26), (11.3.27), and (11.3.29.i). The operator $\Omega_+ + \Omega_-$ depends analytically on parameters of the contravariant form. Hence this operator acts as zero for all values of parameters.

Now let us consider the Kac–Moody algebra $\bar{\mathfrak{g}} = \mathfrak{g}/\ker S$. From (11.3.9), it follows that $\ker S \subset \mathfrak{g}$ is an ideal. The form S induces a nondegenerate form on $\bar{\mathfrak{g}}$. Set $K(x,y) = -S(\tau(x),y)$ for all $x,y \in \bar{\mathfrak{g}}$. K is nondegenerate on $\bar{\mathfrak{g}}$ and has properties (11.1.2)–(11.1.4).

For $\Lambda \in \mathfrak{h}^*$, set $L(\Lambda) = M(\Lambda)/\ker(S: M(\Lambda) \to M(\Lambda)^*)$. From (11.3.7) it follows that $\bar{\mathfrak{g}}$ naturally acts on $L(\Lambda)$. $L(\Lambda)$ is the irreducible $\bar{\mathfrak{g}}$-module with highest weight Λ.

From (11.3.31), it follows that Ω_+ naturally acts on $L(\Lambda) \otimes L(\Lambda')$, where $\Lambda, \Lambda' \in \mathfrak{h}^*$. We describe this action. The Killing from K induces a nondegenerate pairing between root spaces $\bar{\mathfrak{g}}_\alpha$ and $\bar{\mathfrak{g}}_{-\alpha}$ [K,Th. 9.11]. Let $\Omega_\alpha \in \bar{\mathfrak{g}}_\alpha \otimes \bar{\mathfrak{g}}_{-\alpha}$ be the corresponding element generated by dual bases. Set

$$\bar{\Omega} = \sum_\alpha \Omega_\alpha. \qquad (11.3.35)$$

11.3.36. Example. For $\mathfrak{g} = s\ell_2$, we have $\bar{\mathfrak{g}} = \mathfrak{g}$, $\bar{\Omega} = f \otimes e + e \otimes f + h \otimes h/2$.

11.3.37. Lemma. *The following diagram is commutative*

$$\begin{array}{ccc} M(\Lambda) \otimes M(\Lambda') & \xrightarrow{\Omega_+} & M(\Lambda) \otimes M(\Lambda') \\ \downarrow & & \downarrow \\ L(\Lambda) \otimes L(\Lambda') & \xrightarrow{\bar{\Omega}} & L(\Lambda) \otimes L(\Lambda'). \end{array}$$

Proof. Ω_+ is defined in (11.3.29). Check that Ω^0 acts as $\Omega_{\alpha=0} \in \mathfrak{h} \otimes \mathfrak{h}$, see (11.3.30). In fact, let $\{h^i\}$ be a basis in $\mathfrak{h}$, $\{h^{i*}\}$ the dual basis in $\mathfrak{h}^*$, $S_{ij} = S(h^i, h^j)$, (S_{ij}^{-1}) the matrix inverse to (S_{ij}). Then $\Omega^0 = \sum h^i \otimes h^{i*} + \sum h^{i*} \otimes h^i = \sum S_{ij}^{-1}\bigl(h^i \otimes S(h^j, \cdot) + S(h^j, \cdot) \otimes h^i\bigr)$ acts on $S\bigl(L(\Lambda) \otimes L(\Lambda')\bigr)$ as $\sum S_{ij}^{-1}\bigl((-h^i) \otimes (-h^j/2) + (-h^j/2) \otimes (-h^i)\bigr) = \sum S_{ij}^{-1} h^i \otimes h^j = \Omega_{\alpha=0}$, cf. (11.3.25) and (11.3.27).

Choose a graded basis in $\mathfrak{n}_+$, $\{x_\alpha, y_\beta | \alpha \in A, \beta \in B\}$ such that $\{y_\beta | \beta \in B\}$ form a basis of $\ker S$ on $\mathfrak{n}_+$. Let $\{x_\alpha^*, y_\beta^*\}$ be the dual basis in $\mathfrak{n}_+^*$.

For any β, the element $y_\beta \otimes y_\beta^*$ acts as zero in $L(\Lambda) \otimes L(\Lambda')$, because $y_\beta M(\Lambda) \subset \ker S|_{M(\Lambda)}$.

$\{x_\alpha\}$ induces a basis in $\bar{\mathfrak{n}}_+ \subset \bar{\mathfrak{g}}$, and $\{-\tau(x_\alpha)\}$ induces a basis in $\bar{\mathfrak{n}}_- \subset \bar{\mathfrak{g}}$. $\{K(-\tau(x_\ell), x_m)\} = \{S(x_\ell, x_m)\}$ is a nondegenerate matrix. Let $(K_{\ell m}^{-1}) = (S_{\ell m}^{-1})$ be the matrix inverse to $\{S(x_\ell, x_m)\}$.

The element $\sum_\lambda \Omega^-_{\lambda,+}$, see (11.3.29), acts on $S(L(\Lambda) \otimes L(\Lambda'))$ as $\sum_\ell x_\ell \otimes x_\ell^* = \sum S_{\ell m}^{-1} x_\ell \otimes S(x_m, \cdot) = \sum S_{\ell m}^{-1} \tau(x_\ell) \otimes (-x_m) = \sum K_{\ell m}^{-1}(-\tau(x_\ell)) \otimes x_\ell = \sum_{\alpha>0} \Omega_\alpha$. Analogously, $\sum_\lambda \Omega^+_{\lambda,+}$ acts on $L(\Lambda) \otimes L(\Lambda')$ as $\sum_{\alpha<0} \Omega_\alpha$. The lemma is proved.

11.4. Homology in Degree Zero

Consider the complex $C.(\mathfrak{n}_-^*, M^*)_\lambda$ defined in (11.2) and (11.3). Then

$$H_0(C.(\mathfrak{n}_-^*, M^*)_\lambda) = M_\lambda^* / \operatorname{Im}\left((\mathfrak{n}_-^* \otimes M^*)_\lambda \longrightarrow M_\lambda^*\right).$$

The dual space has the form

$$H_0(C.(\mathfrak{n}_-^*, M^*)_\lambda)^* = \ker\left(\nu_{M,-} : M \longrightarrow (\mathfrak{n}_- \otimes M)_\lambda\right). \tag{11.4.1}$$

The Lie algebra $D(\mathfrak{b}_-)$ acts on M as explained in (11.3.29), $D(\mathfrak{b}_-) = \mathfrak{b}_- \otimes \mathfrak{b}_-^*$ as vector spaces.

11.4.2. Lemma

$$H_0(C.(\mathfrak{n}_-^*, M^*)_\lambda)^* = \{m \in M_\lambda \,|\, \mathfrak{n}_-^* m = 0\}.$$

The lemma follows easily from (11.3.29) and Remark (11.3.19).

11.4.3. Remarks

(i) If $S : \mathfrak{n}_- \to \mathfrak{n}_-^*$ is nondegenerate, then

$$H_0(C.(\mathfrak{n}_-^*, M^*)_\lambda)^* = \{m \in M_\lambda \,|\, e_1 m = \cdots = e_r m = 0\}.$$

(ii) The natural projection

$$\pi : M \longrightarrow M/\ker S = L$$

sends $\{m \in M_\lambda \,|\, \mathfrak{n}_-^* m = 0\}$ to $\{m \in L_\lambda \,|\, e_1 m = \cdots = e_r m = 0\}$. Here L is the tensor product of the corresponding irreducible representations, see (11.2.10).

An interesting problem is to describe the image $\pi\{m \in M_\lambda \,|\, \mathfrak{n}_-^* m = 0\} \subset \{m \in L_\lambda \,|\, e_1 m = \cdots = e_r m = 0\}$. For example, for $\mathfrak{g} = s\ell_2$, the image coincides with $\{m \in L_\lambda \,|\, e_1 m = 0\}$.

(iii) Lemma (11.4.2) is an analog of Remark (9.3.8.3).

11.5. Knizhnik–Zamolodchikov Differential Equation

As in (11.3), let $M = M(\Lambda_1) \otimes \cdots \otimes M(\Lambda_n)$. Denote by $\Omega_{+,ij}$ the operator on M acting as Ω_+ on $M(\Lambda_i) \otimes M(\Lambda_j)$ and as the identity on the other factors. These operators respect the weight decomposition.

Let $\Omega_{+,ij}$ act on spaces $\Lambda^p\mathfrak{n}_- \otimes M$ as the identity on $\Lambda^p\mathfrak{n}_-$ and as above on M.

11.5.1. Lemma. *The above action respects the differential in $C.(\mathfrak{n}_-, M)$.*

The lemma is a corollary of (11.3.30), cf. (11.3.27).

Analogously, denote by $\Omega_{+,ij}$ the operator in M^* acting as Ω_+ on $M(\Lambda_i)^* \otimes M(\Lambda_j)^*$ and as the identity on the other factors. $\Omega_{+,ij}$ acts as above on $\Lambda^p\mathfrak{n}_-^* \otimes M^*$ and respects the differential in $C.(\mathfrak{n}_-^*, M^*)$.

Let U_n be the union of all diagonal hyperplanes in $\mathbb{C}^n$, see (8.1.1). Denote by $\mathbf{M}$, $\mathbf{M}_\lambda$, $\mathbf{C.(\mathfrak{n}, M)}$ etc. the trivial bundles over $\mathbb{C}^n \setminus U_n$ with fibers M, M_λ, $C.(\mathfrak{n}, M)$ etc., resp.

The Knizhnik–Zamolodchikov connection ∇_{KZ} on $\mathbf{C.(\mathfrak{n}_-, M)}$ with parameter $\kappa \in \mathbb{C}^*$ is defined by

$$\nabla_{\text{KZ}}\left(\frac{\partial}{\partial z_i}\right) = \frac{\partial}{\partial z_i} - \frac{1}{\kappa}\sum_{j\neq i}\frac{\Omega_{+,ij}}{z_i - z_j} \quad \text{for } i = 1,\ldots,n. \qquad (11.5.2)$$

The Knizhnik–Zamolodchikov connection ∇_{KZ} on $\mathbf{C.(\mathfrak{n}_-^*, M^*)}$ with parameter $\kappa \in \mathbb{C}^*$ is defined by

$$\nabla_{\text{KZ}}\left(\frac{\partial}{\partial z_i}\right) = \frac{\partial}{\partial z_i} - \frac{1}{\kappa}\sum_{j\neq i}\frac{\Omega_{+,ij}}{z_i - z_j} \quad \text{for } i = 1,\ldots,n. \qquad (11.5.3)$$

(11.5.4) $C.(\mathfrak{n}_-^*, M^*)_\lambda$ is dual to $C.(\mathfrak{n}_-, M)_\lambda$ as a vector space. The KZ connection with parameter κ on $C.(\mathfrak{n}_-^*, M^*)_\lambda$ is dual to the KZ connection with parameter $-\kappa$ on $C.(\mathfrak{n}_-, M)_\lambda$ with respect to this duality.

(11.5.5) The contravariant form S induces a homomorphism,

$$S : \mathbf{C.(\mathfrak{n}_-, M)} \longrightarrow \mathbf{C(\mathfrak{n}_-^*, M^*)},$$

of bundles with the KZ connection with the parameter κ, see (11.3.32) and (11.3.34).

11.5.6. Lemma. *The KZ connection is an integrable connection.*

Proof. Integrability conditions are:

$$[\Omega_{ij}, \Omega_{ik} + \Omega_{jk}] = [\Omega_{ij} + \Omega_{ik}, \Omega_{jk}] = 0 \quad \text{for } i < j < k,$$
$$[\Omega_{ij}, \Omega_{k\ell}] = 0 \quad \text{for distinct } i, j, k, \ell,$$

see [K1,K2]. The integrability conditions are satisfied since we have (11.3.31).

The KZ connection respects the weight decomposition and the differentials in complexes $C.(\mathfrak{n}_-, M)_\lambda$ and $C.(\mathfrak{n}_-^*, M^*)_\lambda$, where $\lambda \in \mathbb{N}^r$. Hence, it induces an integrable connection on homology bundles which is denoted by the same symbol ∇_{KZ}.

Let $L = L(\Lambda_1) \otimes \cdots \otimes L(\Lambda_n)$, where $L(\Lambda_j) = M(\Lambda_j)/\ker S$. There is a natural epimorphism $M \to L$. The KZ connection respects the kernel of this epimorphism and induces an integrable connection on L.

(11.5.7) The KZ connection on L with parameter κ induced by the KZ connection (11.5.2) on M is given by

$$\nabla_{KZ}\left(\frac{\partial}{\partial z_i}\right) = \frac{\partial}{\partial z_i} - \frac{1}{\kappa}\sum_{j \neq i} \frac{\bar{\Omega}_{ij}}{z_i - z_j} \quad \text{for } i = 1, \ldots, n,$$

where $\bar{\Omega}_{ij}$ is the operator on L acting as $\bar{\Omega}$ on $L(\Lambda_i) \otimes L(\Lambda_j)$ and as the identity on the other factors, see (11.3.31).

This connection was discovered by Knizhnik and Zamolodchikov in [KZ].

The contravariant form induces an isomorphism, $S: L \to L^*$, of $\bar{\mathfrak{g}}$-modules, where the $\bar{\mathfrak{g}}$-module structure on L^* is defined as in (11.1). We have $S(M) = S(L) = L^* \subset M^*$. The KZ connection on M^* respects the image of S, see (11.3.31). Hence, the KZ connection on M^* induces an integrable connection on L^*.

(11.5.8) The KZ connection on L^* with parameter $\kappa \in \mathbb{C}^*$ induced by the KZ connection (11.5.3) on M^* is given by

$$\nabla_{KZ}\left(\frac{\partial}{\partial z_i}\right) = \frac{\partial}{\partial z_i} - \frac{1}{\kappa}\sum_{j \neq i} \frac{\bar{\Omega}_{ij}}{z_i - z_j} \quad \text{for } i = 1, \ldots, n.$$

The contravariant form induces an isomorphism

$$S : L \longrightarrow L^*, \tag{11.5.9}$$

of bundles with the KZ connection.

In the next section, we shall recall from [SV3] how to construct horizontal sections of the KZ connection in terms of hypergeometric integrals.

12

Hypergeometric Integrals of a Discriminantal Configuration

In this section, we recall the description in [SV3] of the flag complex and the complex of hypergeometric differential forms of a discriminantal configuration in terms of Kac-Moody Lie algebras without Serre's relations.

12.1. Complexes of a Discriminantal Configuration

Fix data (11.1.a)-(11.1.c). Let $\mathfrak{g}$ be the Lie algebra associated with these data. Fix $\Lambda_1, \ldots, \Lambda_n \in \mathfrak{h}^*$, $\lambda = (\ell_1, \ldots, \ell_r) \in \mathbb{N}^r$, $\kappa \in \mathbb{C}^*$.

Set $\ell = \ell_1 + \cdots + \ell_r$. Consider the space $\mathbb{C}^\ell$ with coordinates $t_1(1), \ldots, t_{\ell_1}(1)$, $t_1(2), \ldots, t_{\ell_2}(2), \ldots, t_1(r), \ldots, t_{\ell_r}(r)$. Consider a discriminantal configuration $\mathcal{C}_{n,\ell}(z)$ in $\mathbb{C}^\ell$, where $z = (z_1, \ldots, z_n)$ are pairwise distinct complex numbers.

Let $\mathcal{C}_{n,\ell}(z)$ be weighted. Denote the weight of a hyperplane $t_a(i) = t_b(j)$ by $a(t_a(i), t_b(j))$, the weight of a hyperplane $t_a(i) = z_j$ by $a(t_a(i), z_j)$. Define the weights by

$$a(t_a(i), t_b(j)) = b_{ij}/\kappa = (\alpha_i, \alpha_j)/\kappa, \qquad (12.1.1)$$
$$a(t_a(i), z_j) = -(\Lambda_j, \alpha_i)/\kappa,$$

where $\alpha_1, \ldots, \alpha_r$ are simple roots of $\mathfrak{g}$, cf. (8.13.4).

The group $\Sigma' = \Sigma_{\ell_1} \times \cdots \times \Sigma_{\ell_r}$ acts on $\mathbb{C}^\ell$ permuting coordinates $\{t_a(i)\}$ preserving the index i. The action of Σ' respects the weights of $\mathcal{C}_{n,\ell}(z)$.

The group Σ' naturally acts on the flag complex $(F\ell^\cdot, d)$ of $\mathcal{C}_{n,\ell}$ permuting edges and flags. Denote by

$$(F\ell\,\dot{},d)^{-} \subset (F\ell\,\dot{},d) \tag{12.1.2}$$

the subcomplex of all x such that $\sigma x = (-1)^{|\sigma|} x$ for all $\sigma \in \Sigma'$.

The group Σ' naturally acts on the complex $(\mathcal{A}\,\dot{},d(a))^{-}$ of $\mathcal{C}_{n,\ell}$ permuting hyperplanes. Denote by

$$(\mathcal{A}\,\dot{},d(a))^{-} \subset (\mathcal{A}\,\dot{},d(a))^{-} \tag{12.1.3}$$

the subcomplex of all x such that $\sigma x = (-1)^{|\sigma|} x$ for all $\sigma \in \Sigma'$.

Let $F\ell^{p-}$, $\mathcal{A}^{p-}$ for $p = 0, \ldots, \ell$ be the vector spaces constituting $(F\ell\,\dot{},d)^{-}$, $(\mathcal{A}\,\dot{},d(a))^{-}$, resp. The space $F\ell^{p-}$ is dual to $\mathcal{A}^{p-}$ for all p, see (10.1.10).

The restriction of the quasiclassical contravariant form of $\mathcal{C}_{n,\ell}$ to $(F\ell\,\dot{},d)^{-}$ induces a homomorphism of complexes,

$$S : (F\ell\,\dot{},d)^{-} \longrightarrow (\mathcal{A}\,\dot{},d(a))^{-}. \tag{12.1.4}$$

In [**SV3**], isomorphisms of complexes $C.(\mathfrak{n}_{-}, M)_\lambda$, $C.(\mathfrak{n}_{-}^{*}, M^{*})_\lambda$ and $(F\ell\,\dot{}, d)^{-}$, $(\mathcal{A}\,\dot{},d(a))^{-}$, resp., were constructed. These isomorphisms identify the contravariant form S defined in (11.1.9) with the quasiclassical contravariant form of $\mathcal{C}_{n,\ell}$.

More precisely, in [**SV3**, Theorems (5.13), (6.6), (6.13)], isomorphisms,

$$\begin{aligned}
\psi_p &: C_p(\mathfrak{n}_{-}, M)_\lambda \longrightarrow F\ell^{\ell-p-}, \\
\eta_p &= (\psi_p^{*})^{-1} : C_p(\mathfrak{n}_{-}^{*}, M^{*})_\lambda \longrightarrow \mathcal{A}^{\ell-p-}
\end{aligned} \tag{12.1.5}$$

for all p were constructed. These isomorphisms have the following properties.

$$\begin{aligned}
d\psi &= \psi d, \\
-\kappa d(a)\eta &= \eta d, \\
\eta_p S &= (-\kappa)^{\ell-p} S \psi_p.
\end{aligned} \tag{12.1.6}$$

As a corollary, we get the following commutative diagram,

$$\begin{array}{ccc}
H_p C.(\mathfrak{n}_{-}, M)_\lambda & \xrightarrow{S} & H_p C.(\mathfrak{n}_{-}^{*}, M^{*})_\lambda \\
\psi_p \downarrow & & \downarrow \eta_p \\
H^{\ell-p}(F\ell\,\dot{},d)^{-} & \xrightarrow{(-\kappa)^{\ell-p} S} & H^{\ell-p}(\mathcal{A}\,\dot{},d(a))^{-},
\end{array} \tag{12.1.7}$$

for all p. Here $H.$ (resp., $H^{\cdot}$) are homology (resp. cohomology) groups, and ψ and η are the induced isomorphisms.

12.1.8. Remark. Set $\bar{\mathfrak{g}} = \mathfrak{g}/\ker S$. Let $\bar{\mathfrak{n}}_- \subset \bar{\mathfrak{g}}$ be the nilpotent subalgebra generated by f_i's, $L(\Lambda_i) = M(\Lambda_i)/\ker S$, $L = L(\Lambda_1) \otimes \cdots \otimes L(\Lambda_n)$. Then η, restricted to the subcomplex $C.(\mathfrak{n}_-, L)_\lambda \simeq S(C.(\mathfrak{n}_-, M)_\lambda) \subset C.(\mathfrak{n}_-^*, M^*)_\lambda$, induces an isomorphism of this subcomplex and the subcomplex $S((F\ell^{-\cdot}, d)^-) \subset (\mathcal{A}^{\ell-\cdot}, d(a))^-$.

12.1.9. Example. $\bar{\mathfrak{g}} = \mathfrak{g} = s\ell_2$ is generated by e, f, h subject to the relations $[h,e] = 2e, [h,f] = -2f, [e,f] = h$. Fix $\lambda = \ell \in \mathbb{N}$ and $\Lambda_1, \ldots, \Lambda_n \in \mathfrak{h}^*$.

Consider a discriminantal configuration $\mathcal{C}_{n,\ell}(z)$ in $\mathbb{C}^\ell$ with weights

$$a(t_i, t_j) = (\alpha_1, \alpha_1)/\kappa = 2/\kappa, \quad \text{for } 1 \leq i < j \leq \ell,$$
$$a(t_i, z_j) = -(\alpha_1, \Lambda_j)/\kappa,$$

for $i = 1, \ldots, \ell$ and $j = 1, \ldots, n$. Let $\Sigma' = \Sigma_\ell$ be the group of all permutations of the coordinates $t_1 \ldots t_\ell$. For these data, we have isomorphisms (12.1.5)–(12.1.8).

We recall the construction of the isomorphisms (12.1.5) for the special case $\lambda = (1, \ldots, 1) \in \mathbb{N}^r$. The case of general $\lambda \in \mathbb{N}^r$ is treated by the standard symmetrization procedure, see [**SV3**, (5.11)], (4.7), (4.8), and Example (12.1.29) below.

For $\lambda = (1, \ldots, 1)$, the group Σ' is trivial and $\ell = r$. Denote coordinates in $\mathbb{C}^r$ by $t_1, \ldots, t_r$, the hyperplane $t_i = t_j$ by H_{ij}, the hyperplane $t_i = z_j$ by H_i^j.

Consider a discriminantal configuration $\mathcal{C}_{n,r}(z)$ in $\mathbb{C}^r$ with weights

$$\begin{aligned} a(t_i, t_j) &= (\alpha_i, \alpha_j)/\kappa, \\ a(t_i, z_j) &= -(\alpha_i, \Lambda_j)/\kappa, \end{aligned} \tag{12.1.10}$$

for all i and j.

For this configuration, we construct isomorphisms

$$\begin{aligned} \psi &= \psi_p : C_p(\mathfrak{n}_-, M)_\lambda \longrightarrow F\ell^{r-p}, \\ \eta &= \eta_p = (\psi_p^*)^{-1} : C_p(\mathfrak{n}_-^*, M^*)_\lambda \longrightarrow \mathcal{A}^{r-p} \end{aligned} \tag{12.1.11}$$

with properties

$$\begin{aligned} d\psi &= \psi d, \quad -\kappa d(a)\eta = \eta d, \\ \eta_p S &= (-\kappa)^{r-p} S \psi_p. \end{aligned} \tag{12.1.12}$$

It suffices to construct the isomorphism ψ since η is determined by ψ.

For any $I = (i_1, \ldots, i_N) \subset (1, \ldots, r)$, set

$$f_I = f_{i_N} \cdots f_{i_1},$$
$$[f_I] = [f_{i_N}, [f_{i_{N-1}}, [\cdots [f_{i_2}, f_{i_1}] \cdots]]] \in \mathfrak{n}_-. \quad (12.1.13)$$

For any p, the space $(\Lambda^p \mathfrak{n}_- \otimes M)_\lambda$ is generated by all elements

$$[f_{J_1}] \wedge [f_{J_2}] \wedge \cdots \wedge [f_{J_p}] \otimes f_{I_1} v_1 \otimes \cdots \otimes f_{I_n} v_n \quad (12.1.14)$$

such that $J_1, \ldots, J_p$ and $I_1, \ldots, I_n$ have no intersections and $J_1 \cup \cdots \cup J_p \cup I_1 \cup \cdots \cup I_n = \{1, \ldots r\}$. Here v_i is the generating vector for $M(\Lambda_i)$.

For a nonempty subset $I = \{i_1, \ldots, i_p\} \subset \{1, \ldots, r\}$, set

$$L_I = H_{i_1 i_2} \cap H_{i_2 i_3} \cap \cdots \cap H_{i_{p-1} i_p}. \quad (12.1.15)$$

L_I is an edge of $\mathcal{C}_{n,r}$. For $p = 1$, we set $L_I = \mathbb{C}^r$. For $i \in \{1, \ldots, n\}$, set

$$L_I^i = H_{i_1}^i \cap H_{i_2}^i \cap \cdots \cap H_{i_p}^i, \quad (12.1.16)$$

L_I^i is an edge of $\mathcal{C}_{n,r}$. Set $L_\emptyset^i = \mathbb{C}^r$.

For nonintersecting subsets $J_1, \ldots, J_p$ and $I_1, \ldots, I_n \subset \{1, \ldots, r\}$, set

$$L_{J_1, \ldots, J_p; I_1, \ldots, I_n} = \left[\bigcap_{k=1}^p L_{J_k} \right] \cap \left[\bigcap_{i=1}^n L_{I_i}^i \right]. \quad (12.1.17)$$

This is an edge of $\mathcal{C}_{n,r}$, and any edge has this form.

Multiplication of flags

Given two subsets $J \subset \{1, \ldots, r\}$ and $I \subset \{1, \ldots, n\}$, define a configuration $\mathcal{C}_{I,J}$ in $\mathbb{C}^r$ as the configuration consisting of all hyperplanes H_{j_1, j_2} with $j_1, j_2 \in J$ and H_j^i with $j \in J$ and $i \in I$.

Given subsets $J, J' \subset \{1, \ldots, r\}$ and $I, I' \subset \{1, \ldots, n\}$ such that $J \cap J' = \emptyset$, $I \cap I' = \emptyset$, define maps

$$\circ : \operatorname{Flag}^p(\mathcal{C}_{I,J}) \times \operatorname{Flag}^q(\mathcal{C}_{I',J'}) \longrightarrow \operatorname{Flag}^{p+q}(\mathcal{C}_{I \cup I', J \cup J'}) \quad (12.1.18)$$

as follows. For $F = F(H_1, \ldots, H_p) \in \operatorname{Flag}^p(\mathcal{C}_{I,J})$ and $F' = F(H_1', \ldots, H_q') \in \operatorname{Flag}^q(\mathcal{C}_{I',J'})$, set $F \circ F' = F(H_1, \ldots, H_p, H_1', \ldots, H_q')$, see notation in (10.1.8).

Map (12.1.18) well defines a map

$$F\ell^p(C_{I,J}) \otimes F\ell^q(C_{I',J'}) \longrightarrow F\ell^{p+q}(C_{I\cup I', J\cup J'}), \qquad (12.1.19)$$

see [**SV3**, (5.7.2)].

Define commutators $g \in \mathfrak{n}_-$ of length ℓ by induction on ℓ as follows. Commutators of length 1 are $f_1, \ldots, f_r$. A commutator of length ℓ is an expression of the form $g = [g_1, g_2]$, where g_i is a commutator of length ℓ_i, $\ell = \ell_1 + \ell_2$. We will denote the set of indices of all f_i's contained in g by $|g|$. So $\ell(g) = \#|g|$.

To any commutator g, we assign a flag $F\ell(g) \in F\ell^{\ell(g)-1}(C_{\emptyset, |g|})$ as follows. Set $F\ell(f_i) = \Box$, the flag of length 0. If $g = [g_2, g_1]$ and $F\ell(g_1)$ and $F\ell(g_2)$ are defined, then $F\ell(g_1) \circ F\ell(g_2)$ is a flag of length $\ell(g) - 2$. Let $F\ell(g)$ be the flag $F\ell(g_1) \circ F\ell(g_0)$ completed by the edge $L_{|g|}$ of dimension $\ell(g) - 1$.

12.1.20. Example. For $J = \{j_1, \ldots, j_\ell\}$, we have $F\ell([f_J]) = F(H_{j_1 j_2}, H_{j_2 j_3}, \ldots, H_{j_{\ell-1} j_\ell})$. $F\ell([[f_1, f_3], [f_2, f_1]]) = F(H_{12}, H_{34}, H_{13})$.

For $f_I = f_{i_\ell} \ldots f_{i_1}$ and i such that $1 \leq i \leq n$, set

$$F^i(f_I) = F(H_{i_1}^i, H_{i_2}^i, \ldots, H_{i_\ell}^i) \in F\ell^\ell(C_{\{i\}, I}). \qquad (12.1.21)$$

Let $z \in C_p(\mathfrak{n}_-, M)_\lambda$,

$$z = g_p \wedge g_{p-1} \wedge \cdots \wedge g_1 \otimes f_{I_1} v_1 \otimes \cdots \otimes f_{I_n} v_n, \qquad (12.1.22)$$

where all g_i are commutators, $\ell(g_i) = \ell_i$, $|g_i| = J_i$. Let $\{f_{i_1}, \ldots, f_{i_r}\}$ be the list of f_i's in z, read from the right to the left. Define a permutation $\sigma(z) \in \Sigma_r$ by $\sigma(z)j = i_j$. Set

$$\psi_p(z) = (-1)^{|\sigma(z)| + \sum_{i=1}^p (\ell_i - 1)} F^n(f_{I_n}) \circ F^{n-1}(f_{I_{n-1}}) \circ \qquad (12.1.23)$$
$$\cdots \circ F^1(f_{I_1}) \circ F(g_1) \circ \cdots \circ F(g_p).$$

12.1.24. Lemma. [**SV3**, (5.8.11)]. *Rule (12.1.23) well defines maps*

$$\psi_p : C_p(\mathfrak{n}_-, M)_\lambda \longrightarrow F\ell^{r-p}(C_{n,r})$$

for all p.

12.1.25. Example. For $n = 1$, we have $\psi(f_r \wedge \cdots \wedge f_{k+1} \otimes f_k \ldots f_1 v) = F(H_1^1, \ldots, H_k^1)$.

12.1.26. Remark. The coefficient in (12.1.23) is uniquely determined by the formula $\psi_r(f_r \wedge f_{r-1} \wedge \cdots \wedge f_1 \otimes v_1 \otimes \cdots \otimes v_n) = \Box$ and the condition

$\psi_p(z) = \text{const}\,(z) \cdot F^n(f_{I_n}) \circ F^{n-1}(f_{I_{n-1}}) \circ \cdots \circ F^1(f_{I_1}) \circ F(g_1) \circ \cdots \circ F(g_p),$
see [**SV3**, (5.8.15)].

12.1.27. Example. Let $\lambda = (1, 0, \ldots, 0) \in \mathbf{N}^r$. Set

$$f^{(0)} = f_1 \otimes v_1 \otimes \cdots \otimes v_n, \quad f^{(i)} = v_1 \otimes \cdots \otimes f_1 v_i \otimes \cdots \otimes v_n \quad \text{for } 1 \leq i \leq n.$$

Then $f^{(0)}$ forms a basis in $C_1(\mathbf{n}_-, M)_\lambda$; $f^{(1)}, \ldots, f^{(n)}$ form a basis in $C_0(\mathbf{n}_-, M)_\lambda$. Denote by $(\ )^*$ the elements of the dual bases. The differentials are defined by $df^{(0)} = f^{(1)} + \cdots + f^{(n)}$, $df^{(0)^*} = (\Lambda_1, \alpha_1)f^{(1)^*} + \cdots + (\Lambda_n, \alpha_1)f^{(n)^*}$. The contravariant form is given by $f^{(0)} \mapsto f^{(0)^*}$, $f^{(i)} = (\Lambda_i, \alpha_1)f^{(i)^*}$ for all i, see (11.3.14). The isomorphisms ψ and η are given by $f^{(0)} \mapsto \square$, $f^{(0)^*} \mapsto \square^*$, $f^{(i)} = F(H_1^i)$, $f^{(i)^*} \mapsto H_1^i$.

$\mathbb{C}\square$ and $\mathbb{C}\square^*$ may be identified with $\mathbb{C}$.

12.1.28. Example. Let $n = 1$ and $\lambda = (1, 1, 0, \ldots, 0) \in \mathbf{N}^r$. Then $f_2 \wedge f_1 \otimes v$; $[f_2, f_1] \otimes v$, $f_2 \otimes f_1 v$, $f_1 \otimes f_2 v$; $f_2 f_1 v$, $f_1 f_2 v$ form bases in $C_2(\mathbf{n}_-, M)_\lambda$, $C_1(\mathbf{n}_-, M)_\lambda$, $C_0(\mathbf{n}_-, M)_\lambda$, respectively. The differentials are defined by $d(f_2 \wedge f_1 \otimes v) = -[f_2, f_1] \otimes v + f_2 \otimes f_1 v - f_1 \otimes f_2 v$, $d([f_2, f_1] \otimes v) = [f_2, f_1]v$, $d(f_2 \otimes f_1 v) = f_2 f_1 v$, $d(f_1 \otimes f_2 v) = f_1 f_2 v$; $d((f_2 \wedge f_1 \otimes v)^*) = (\alpha_1, \alpha_2)([f_2, f_1] \otimes v)^* + (\Lambda, \alpha_1)(f_2 \otimes f_1 v)^* - (\Lambda, \alpha_2)(f_1 \otimes f_2 v)^*$, $d(([f_2, f_1] \otimes v)^*) = (\Lambda, \alpha_2)(f_1 f_2 v)^* - (\Lambda, \alpha_1)(f_2 f_1 v)^*$, $d((f_2 \otimes f_1 v)^*) = (\Lambda - \alpha_1, \alpha_2)(f_2 f_1 v)^*$, $d((f_1 \otimes f_2 v)^*) = (\Lambda - \alpha_2, \alpha_1)(f_1 f_2 v)^*$, see (11.2.3) and (11.2.11). The contravariant form is defined by $f_2 \wedge f_1 \otimes v \mapsto (f_1 \wedge f_1 \otimes v)^*$, $[f_2, f_1] \otimes v \mapsto -(\alpha_1, \alpha_2)([f_2, f_1] \otimes v)^*$, $f_2 \otimes f_1 v \mapsto (\Lambda, \alpha_1)(f_2 \otimes f_1 v)^*$, $f_1 \otimes f_2 v \mapsto (\Lambda, \alpha_2)(f_1 \otimes f_2 v)^*$, $f_2 f_1 v \mapsto (\Lambda - \alpha_1, \alpha_2)(\Lambda, \alpha_1)(f_2 f_1 v)^* + (\Lambda, \alpha_1)(\Lambda, \alpha_2)(f_1 f_2 v)^*$, $f_1 f_2 v \mapsto (\Lambda, \alpha_1)(\Lambda, \alpha_2)(f_2 f_1 v)^* + (\Lambda - \alpha_2, \alpha_1)(\Lambda, \alpha_2)(f_1 f_2 v)^*$. The isomorphisms ψ and φ are given by $f_2 \wedge f_1 \otimes v \mapsto 1$, $(f_2 \wedge f_1 \otimes v)^* \mapsto 1$, $f_2 \otimes f_1 v \mapsto F(H_1^1)$, $f_1 \otimes f_2 v \mapsto -F(H_2^1)$, $[f_2, f_1] \otimes v \mapsto -F(H_{12})$, $(f_2 \otimes f_1 v)^* \mapsto H_1^1$, $(f_1 \otimes f_2 v)^* \mapsto -H_2^1$, $([f_2, f_1] \otimes v)^* \mapsto -H_{12}$, $f_2 f_1 v \mapsto F(H_1^1, H_2^1)$, $f_1 f_2 v \mapsto -F(H_2^1, H_1^1)$, $(f_2 f_1 v)^* \mapsto H_1^1 \wedge H_{12}$, $(f_1 f_2 v)^* \mapsto H_{12} \wedge H_2^1$.

12.1.29. Example. Let $n = 1$ and $\lambda = (2, 0, \ldots, 0) \in \mathbf{N}^r$. Then $f_1 \otimes f_1 v$ and $f_1^2 v$ form bases in $C_1(\mathbf{n}_-, M)_\lambda$ and $C_0(\mathbf{n}_-, M)_\lambda$, respectively. $d(f_1 \otimes f_1 v) = f_1^2 v$, $d((f_1 \otimes f_1 v)^*) = (\alpha_1, 2\Lambda - \alpha_1)(f_1^2 v)^*$, see (11.3.15). The contravariant form is defined by $f_1 \otimes f_1 v \mapsto (\Lambda, \alpha_1)(f_1 \otimes f_1 v)^*$, $f_1^2 v \mapsto (\alpha_1, 2\Lambda - \alpha_1)(\alpha_1, \Lambda)(f_1^2 v)^*$. The corresponding configuration is $\mathcal{C}_{1,2}(z)$, the configuration of lines $H_1^1 : t_1(1) = z_1$, $H_2^1 : t_2(1) = z_1$, $H_{12} : t_1(1) = t_2(1)$ with weights $-(\alpha_1, \Lambda)/\kappa$ and $-(\alpha_1, \Lambda)/\kappa$, $(\alpha_1, \alpha_1)/\kappa$, resp. The isomorphisms ψ and φ of (12.1.5) are given by $f_1 \otimes f_1 v \mapsto F(H_1^1) - F(H_2^1)$, $(f_1 \otimes f_1 v)^* \mapsto H_1^1 - H_2^1$, $f_1^2 v \mapsto F(H_1^1, H_2^1) -$

$F(H_2^1, H_1^1)$, $(f_1^2 v)^* \mapsto H_1^1 \wedge H_{12} + H_{12} \wedge H_2^1$, cf. (12.1.28).

12.2. Hypergeometric Pairings

Fix data (4.1.a)–(4.1.d). Fix $\Lambda_1, \ldots, \Lambda_n \in \mathfrak{h}^*$, $\lambda = (\ell_1, \ldots, \ell_r) \in \mathbb{N}^r$.

Let $\mathfrak{g}$ be the Lie algebra associated with these data in (11.1). Let $U_q \mathfrak{g}$ be the quantum group associated with these data in (4.1).

Consider the following morphisms of complexes:

$$S : C.(\mathfrak{n}_-, M)_\lambda \longrightarrow C.(\mathfrak{n}_-^*, M^*)_\lambda \qquad (12.2.1)$$

and

$$S : C.(^+U_q\mathfrak{n}_-; M(q); {}^+\Delta)_\lambda \longrightarrow C^*(^+U_q\mathfrak{n}_-; M(q); \mu)_\lambda . \qquad (12.2.2)$$

The first morphism is defined in (11.2.8). Here $M = M_1 \otimes \cdots \otimes M_n$, where M_j is the Verma module over $\mathfrak{g}$ with highest weight Λ_j, $j = 1, \ldots, n$. The second morphism is defined in (5.1.26). Here $M(q) = M_1(q) \otimes \cdots \otimes M_n(q)$, where $M_j(q)$ is the Verma module over $U_q\mathfrak{g}$ with highest weight Λ_j. As in (12.1), consider the configuration $\mathcal{C}_{n,\ell}(z)$ in $\mathbb{C}^\ell$ with weights defined by (12.1.1). Assume that $z = (z_1, \ldots, z_n) \in \mathbb{R}^n$ and $z_1 < z_2 < \cdots < z_n$. Fix an equipment $\{T_j^i\}$ for $\mathcal{C}_{n,\ell}(z)$.

Consider the isomorphisms

$$\begin{aligned} \psi : C.(\mathfrak{n}_-, M)_\lambda &\longrightarrow (F\ell^{\ell-\cdot}(\mathcal{C}_{n,\ell}), d)^-, \\ \eta : C.(\mathfrak{n}_-^*, M^*)_\lambda &\longrightarrow (\mathcal{A}^{\ell-\cdot}(\mathcal{C}_{n,\ell}), d)^-, \end{aligned} \qquad (12.2.3)$$

described in (12.1).

Consider the monomorphisms

$$\begin{aligned} \alpha : C.(^+U_q\mathfrak{n}_-; M(q); {}^+\Delta)_\lambda &\longrightarrow \mathcal{X}_{\ell-\cdot}(\mathcal{C}_{n,\ell})^-, \\ \alpha : C^*_\cdot(^+U_q\mathfrak{n}_-; M(q); \mu)_\lambda &\longrightarrow Q_{\ell-\cdot}(\mathcal{C}_{n,\ell})^-, \end{aligned} \qquad (12.2.4)$$

described in (5.11), see (5.11.1). These monomorphisms are quasiisomorphisms by (4.7.5), (5.4) and (5.5).

Consider the realization of the abstract complexes of $\mathcal{C}_{n,\ell}(z)$ associated with the equipment $\{T_j^i\}$,

$$\begin{aligned} \omega : \mathcal{X}.(\mathcal{C}_{n,\ell}(z))^- &\longrightarrow C.(X, \mathcal{S}^*(a))^-, \\ \omega : Q.(\mathcal{C}_{n,\ell}(z))^- &\longrightarrow C.(Q'/Q'', \mathcal{S}^*(a))^-. \end{aligned} \qquad (12.2.5)$$

This realization is described in (3.16.7) and (2.9.7), see also (5.11).

12.2.6. Theorem. *Integration pairing (10.9.2) and isomorphisms (12.2.3)–(12.2.5) induce a pairing of complexes*

(i) $I(z,\{T_j^i\}): C.(\mathfrak{n}_-^*, M^*)_\lambda \otimes C.(^+U_q\mathfrak{n}_-; M(q); {}^+\Delta)_\lambda \longrightarrow \mathbb{C},$

$$y \otimes x \longmapsto \int_{\omega\alpha(x)} \eta(y).$$

We have

(ii) $I(dy, x) = I(y, dx)$
for all $x \in C_{p-1}(^+U_q\mathfrak{n}_-; M(q); {}^+\Delta)_\lambda$, $y \in C_p(\mathfrak{n}_-^*, M^*)_\lambda$, and all p.

The theorem is a direct corollary of the formulated statements.

The pairing induces the maps

$$I(z,\{T_j^i\}): C_p(^+U_q\mathfrak{n}_-; M(q); {}^+\Delta)_\lambda \longrightarrow C_p^*(\mathfrak{n}_-^*, M^*)_\lambda. \qquad (12.2.7)$$

In other words, the maps

$$I(z,\{T_j^i\}): ((^+U_q\mathfrak{n}_-)^{\otimes p} \otimes M(q))_\lambda \longrightarrow (\Lambda^p \mathfrak{n}_- \otimes M)_\lambda \qquad (12.2.8)$$

for all p.

The pairing (12.2.6) induces a pairing

$$I(z): H.(C.(\mathfrak{n}_-^*, M^*)_\lambda) \otimes H.(C.(^+U_q\mathfrak{n}_-; M(q); {}^+\Delta)_\lambda) \longrightarrow \mathbb{C} \qquad (12.2.9)$$

and the corresponding maps

$$I_p(z): H_p(C.(^+U_q\mathfrak{n}_-; M(q); {}^+\Delta)_\lambda) \longrightarrow H_p(C.(\mathfrak{n}_-^*; M^*)_\lambda)^* \qquad (12.2.10)$$

for all p.

This pairing does not depend on the equipment $\{T_j^i\}$.

12.2.11. Example. For $\mathfrak{g} = s\ell_2$, we have

$$H_0(C.(^+U_q\mathfrak{n}_-; M(q); {}^+\Delta)_\lambda) = \{x \in M(q)_\lambda | ((1)_q^k/(k)_q!)e^k x = 0 \text{ for all } k > 0\},$$

cf. (5.1.7). In particular, if q is not a root of unity, then this space coincides with $\{x \in M(q)_\lambda | ex = 0\}$. We have

$$H_0(C.(\mathfrak{n}_-^*, M^*))^* = \{x \in M_\lambda | ex = 0\}.$$

Thus, (12.2.10) gives a map $I_0(z)$ of the first space to the second space.

More generally, for an arbitrary $\mathfrak{g}$, we have

(i) $\qquad H_0(C.(^+U_q\mathfrak{n}_-; M(q); {}^+\Delta)_\lambda) = \{x \in M(q)_\lambda | g_I x = 0 \text{ for all } I\}$,

where the operators g_I are defined in (5.1.6), (5.1.17) and (5.1.18), see (9.3.8.3). We have

(ii) $\qquad H_0(C.(\mathfrak{n}_-^*, M^*)_\lambda)^* = \{x \in M_\lambda | \nu_M(x)_- = 0\}$,

where the map $\nu_{M,-} : M \to \mathfrak{n}_- \otimes M$ is defined in (11.3.11)–(11.3.13).

The map $I_0(z)$ maps the space (i) to the space (ii).

(12.2.12) Let κ vary and let all other data be fixed. Then the weights of $C_{n,\ell}(z)$ become functions of κ, $a = a(\kappa)$. For almost all values of κ, the weights $a(\kappa)$ satisfy conditions of Theorem (10.6.8). Such κ will be called *generic* with respect to the data (4.1.a)–(4.1.d), $\Lambda_1, \ldots, \Lambda_n$, λ. The exceptional values form a subset Exc $\subset \mathbb{C}$. Exc has the property: the set $\{\kappa^{-1} | \kappa \in \text{Exc}\}$ is a discrete subset of $\mathbb{C}$ and is contained in a suitable arithmetic progression.

12.2.13. Theorem. *Pairing (12.2.9) is nondegenerate if κ is generic.*

The theorem is a direct corollary of the above formulated statements.

Let A be the complex affine space with coordinates $\{a_H | H \in \mathcal{C}_{n,\ell}\}$. Consider the domain $D_{-1}(\mathcal{C}_{n,\ell}) \subset A$ defined in (10.7).

Consider the quotient complex $C.(\mathfrak{n}_-, M)_\lambda / \ker S$, where S is given in (12.2.1). This complex is naturally identified with $C.(\bar{\mathfrak{n}}_-, L)_\lambda$, see (11.2.10). Here $\bar{\mathfrak{n}}$ is the nilpotent subalgebra of the Kac–Moody algebra $\bar{\mathfrak{g}} = \mathfrak{g}/\ker S$ and $L = L(\Lambda_1) \otimes \cdots \otimes L(\Lambda_n)$ is the tensor product of irreducible modules over $\bar{\mathfrak{g}}$ with highest weights $\Lambda_1, \ldots, \Lambda_n$, respectively.

Consider the quotient complex $C.(^+U_q\mathfrak{n}_-; M(q); {}^+\Delta)_\lambda / \ker S$, where S is given in (12.2.2). This complex is naturally identified with $C.(^+U_q\bar{\mathfrak{n}}_-; L(q); \delta)$, see (9.4.9). Here $U_q\bar{\mathfrak{n}}_-$ is the nilpotent subalgebra of the quantum group $U_q\bar{\mathfrak{g}} = U_q\mathfrak{g}/\ker S$ corresponding to the Kac–Moody algebra $\bar{\mathfrak{g}}$ and $L(q) = L(\Lambda_1, q) \otimes \cdots \otimes L(\Lambda_n, q)$ is the tensor product of irreducible modules over $U_q\bar{\mathfrak{g}}$ with highest weights $\Lambda_1, \ldots, \Lambda_n$, respectively, see (9.4). See also (9.5) for the description of $U_q\bar{\mathfrak{g}}$.

12.2.14. Theorem. *Assume that $a \in \mathfrak{D}_{-1}$ where a is the set of weights of*

$C_{n,\ell}(z)$. Then pairing (12.2.6) induces a well-defined pairing of complexes

$$J(z, \{T_j^i\}) : C.(\bar{\mathfrak{n}}_-; L)_\lambda \otimes C.(^+U_q\bar{\mathfrak{n}}_-; L(q); \delta)_\lambda \longrightarrow \mathbb{C},$$

$$[y] \otimes [x] \longrightarrow \int_{\omega\alpha(x)} \eta S(y),$$

where $[y] \in C.(\bar{\mathfrak{n}}_-; L)_\lambda$ and $y \in C.(\mathfrak{n}_-; M)_\lambda$ is its representative, $[x] \in C.(^+U_q\bar{\mathfrak{n}}_-; L(q); \delta)_\lambda$ and $x \in C.(^+U_q\mathfrak{n}_-; M(q), {}^+\Delta)_\lambda$ is its representative.

The theorem is a corollary of (10.9.10) and (10.9.13).

(12.2.15) Fix a Lie algebra $\mathfrak{g}$ as above and weights $\Lambda_1, \ldots, \Lambda_n$. Let y be a sum of monomials in $f_1, \ldots, f_r$, $y = \sum c_I f_I$, where $c_I = c_I(\kappa)$ are holomorphic functions of $\kappa \in \mathbb{C}^*$ for all I. Assume that $y \in \ker(S : U_q\mathfrak{n}_- \to (U_q\mathfrak{n}_-)^*)$ for all $q = \exp(2\pi i/\kappa)$.

12.2.16. Theorem. *Under assumptions (12.2.15), for any element* $x \in C.(^+U_q\mathfrak{n}_-; M(q); {}^+\Delta)_\lambda$ *of the form*

$$f_{i_1} \otimes \cdots \otimes f_{I_{s-1}} \otimes f_I y f_J \otimes f_{I_{s+1}} \otimes \cdots \otimes f_{I_k} \otimes f_{J_1} v_1 \otimes \cdots \otimes f_{J_n} v_n$$

or

$$f_{I_1} \otimes \cdots \otimes f_{I_k} \otimes f_{J_1} v_1 \otimes \cdots \otimes f_I y f_J v_s \otimes \cdots \otimes f_{J_n} v_n,$$

we have

$$I(z, \{T_j^i\}) : C.(\bar{\mathfrak{n}}_-; L)_\lambda \otimes x \longrightarrow 0$$

for all q. Here $C.(\bar{\mathfrak{n}}_-; L)_\lambda$ is the image of the contravariant form.

The Chevalley–Serre elements $R_{ij}(f)$ introduced in (4.1.11) for a quantum group corresponding to a generalized Cartan matrix serve as examples of the element y in (12.2.15).

Proof of the Theorem. Fix a flag. The integral of the corresponding flag form over the chain representing x is a holomorphic function of q. For generic values of q, this function is zero, by (10.9.13). Hence, this function is zero for all values of q.

(12.2.17) Let y be a sum of monomials in the Verma module $M(\Lambda_p, q)$ for some $p \in \{1, \ldots, n\}$, $y = \sum c_I f_I v_p$, where $c_I = c_I(\kappa)$ are holomorphic functions of $\kappa \in \mathbb{C}^*$ for all I. Assume that $y \in \ker(S : M(\Lambda_p, q) \to M(\Lambda_p, q)^*)$ for all

$q = \exp(2\pi i/\kappa)$.

12.2.18. Theorem. *Under assumptions (12.2.17), for any element $x \in C.(^+U_q\mathfrak{n}_-; M(q); {}^+\Delta)_\lambda$ of the form*

$$f_{I_1} \otimes \cdots \otimes f_{I_k} \otimes f_{J_1}v_1 \otimes \cdots \otimes f_I y \otimes \cdots \otimes f_{J_n}v_n,$$

we have

$$I(z, \{T_j^i\}) : C.(\tilde{\mathfrak{n}}_-; L)_\lambda \otimes x \longrightarrow 0$$

for all q.

The proof is similar to the proof of Theorem (12.2.16).

Assume that the number $k = 2(\Lambda_p, \alpha_j)/(\alpha_j, \alpha_j)$ belongs to N for some j. Then the element $f_j^{k+1}v_p$ is an example of the element y in (12.2.17).

(12.2.19) Assume that, for a given κ, there exists a subset $I \subset \{1, \ldots, r\}$ such that $(\alpha_i, \alpha_j)/\kappa > 0$ for all $i, j \subset I$. Assume that there exist $\mu = (m_1, \ldots, m_r) \in \mathbb{N}^r$ and an element $y \in (U_q\mathfrak{n}_-)_\mu$ such that

(i) $\qquad\qquad m_i = 0 \quad \text{if } i \notin I,$

(ii) $\qquad\qquad y \in \ker(S : U_q\mathfrak{n}_- \longrightarrow (U_q\mathfrak{n}_-)^*).$

12.2.20. Theorem. *Under assumptions (12.2.19), for any element $x \in C.(^+U_q\mathfrak{n}_-; M(q); {}^+\Delta)_\lambda$ of the form*

$$f_{I_1} \otimes \cdots \otimes f_I y f_J \otimes \cdots \otimes f_{I_k} \otimes f_{J_1}v_1 \otimes \cdots \otimes f_{J_n}v_n,$$

or

$$f_{I_1} \otimes \cdots \otimes f_{I_k} \otimes f_{J_1}v_1 \otimes \cdots \otimes f_I y f_J v_s \otimes \cdots \otimes f_{J_n}v_n,$$

we have

(i) $\qquad\qquad I(z, \{T_j^i\}) : C.(\mathfrak{n}_-; M^*)_\lambda \otimes x \longrightarrow 0.$

Assume that $(\alpha_j, \alpha_j)/\kappa$ is positive for some j, and $q^{(\alpha_j,\alpha_j)/2}$ is a primitive root of 1 of order t. Then the element f_j^t is an example of the element y in (12.2.19).

Proof of the Theorem. Assume that the numbers $-(\Lambda_m, \alpha_i)/\kappa$, $(\alpha_i, \alpha_j)/\kappa$ are positive for all i, j, m. Then (12.2.20.i) is true by (10.9.12). By analyticity, (12.2.20.i) is true for all values of $(\Lambda_m, \alpha_i)/\kappa$, $(\alpha_i, \alpha_j)/\kappa$.

(12.2.21) Assume that, for a given κ, there exists a number $p \in \{1, \ldots, n\}$ and a subset $I \subset \{1, \ldots, r\}$ such that

(i) $\qquad -(\Lambda_p, \alpha_i)/\kappa > 0 \quad$ for all $i \in I$,

(ii) $\qquad (\alpha_i, \alpha_j)/\kappa > 0 \quad$ for all $i, j \subset I$.

Assume that there exist $\mu = (m_1, \ldots, m_r) \in \mathbb{N}^r$ and $y \in M(\Lambda_p, q)_\mu$ such that

(iii) $\qquad m_i = 0 \quad \text{if } i \notin I$,

(iv) $\qquad y \in \ker(S : M(\Lambda_p, q) \longrightarrow M(\Lambda_p, q)^*)$.

12.2.22. Theorem. *Under assumptions (12.2.21), for any element $x \in C.(^+U_q\mathfrak{n}_-; M(q); {^+\Delta})_\lambda$ of the form*

$$f_{I_1} \otimes \cdots \otimes f_{I_k} \otimes f_{J_1} v_1 \otimes \cdots \otimes f_I y \otimes \cdots \otimes f_{J_n} v_n,$$

we have

$$I(z, \{T_j^i\}) : C.(\mathfrak{n}_-^*; M^*)_\lambda \otimes x \longrightarrow 0.$$

The proof is similar to the proof of Theorem (12.2.20).

Assume that $-(\Lambda_p, \alpha_i)/\kappa$ and $(\alpha_i, \alpha_i)/\kappa$ are positive, and the number $k = -2(\Lambda_p, \alpha_i)/(\alpha_i, \alpha_i)$ is a natural number. Assume that the number $q^{(\alpha_i, \alpha_i)/2}$ is a primitive root of 1 of order t. Assume that a natural number j is such that $k + j$ is divisible by t. Then the element $f_i^{j+1} v_p$ is an example of the element y in (12.2.21).

Now let us return to the pairing (12.2.14) and consider the induced pairing

$$J(z) : H.(C.(\bar{\mathfrak{n}}_-; L)_\lambda) \otimes H.(C.(^+U_q\bar{\mathfrak{n}}_-; L(q); \delta)_\lambda) \longrightarrow \mathbb{C} \qquad (12.2.23)$$

and the corresponding maps

$$J_p(z) : H_p(C.(U_q\bar{\mathfrak{n}}_-; L(q); \delta)_\lambda) \longrightarrow H_p(C.(\bar{\mathfrak{n}}_-; L)_\lambda)^* \qquad (12.2.24)$$

for all p.

12.2.25. Example.

$H_0(C.(^+U_q\bar{\mathfrak{n}}_-; L(q); \delta)_\lambda) = \text{Vac}\,(L(q))_\lambda := \{x \in L(q)_\lambda | e_1 x = \cdots = e_r x = 0\}$.

$H_0(C.(\bar{\mathfrak{n}}_-, L)_\lambda)^* = \{y \in L_\lambda^* | e_1 y = \cdots = e_r y = 0\}$.

12.2.26. Theorem. *Let* $\mathfrak{g} = s\ell_2$, α *the single positive root of* $s\ell_2$. *Assume that the numbers* $(\Lambda_1, \alpha), \ldots, (\Lambda_n, \alpha)$ *are non-negative integers and q is not a root of unity. Then the pairing*

(i) $\qquad J(z) : H_0(C.(\bar{\mathfrak{n}}_-; L)_\lambda) \otimes H_0(C.(^+U_q\bar{\mathfrak{n}}_-; L(q); \delta)_\lambda) \longrightarrow 0$

is nondegenerate.

12.2.27. Corollary. *The corresponding map*

$$J_0(z) : \{x \in L(q)_\lambda \mid ex = 0\} \longrightarrow \{x \in L_\lambda \mid ex = 0\}$$

is an isomorphism.

Conjecturally, the pairing (12.2.23) is nondegenerate if q is generic in the sense of (12.2.12), see (12.5).

If q is not generic, then the maps J_p may have nontrivial kernel and cokernel. It is a very interesting problem to describe degenerations of J_p. We will discuss this problem in Sec. 13.

Theorem (12.2.26) is proved in four steps:

First step. The natural map

$$S : H_0(C.(\bar{\mathfrak{n}}_-; L)_\lambda) \longrightarrow H_0(C.(\mathfrak{n}_-^*, M^*)_\lambda) \qquad (12.2.28)$$

is a monomorphism. This statement easily follows from the formula

$$\nu_{M-}^* : f^* \otimes \varphi^*(\cdot) \longmapsto \varphi^*(e \cdot),$$

see (11.3.15).

Second step. The natural map

$$\pi : H_0\big(C.(^+U_q\mathfrak{n}_-; M(q); ^+\Delta)_\lambda\big) \longrightarrow H_0\big(C.(^+U_q\bar{\mathfrak{n}}_-; L(q); \delta)_\lambda\big) \qquad (12.2.29)$$

is an epimorphism.

Third step.

$$\dim \{x \in L(q)_\lambda \mid ex = 0\} = \dim \{x \in L_\lambda \mid ex = 0\},$$

if q is not a root of unity, see [Lu1, R1].

Fourth step. The pairing

$$I : H_0\big(C.(\mathfrak{n}_-^*, M^*)_\lambda\big) \otimes H_0\big(C.(^+U_q\mathfrak{n}_-; M(q); {}^+\Delta)_\lambda\big) \longrightarrow \mathbb{C}$$

is nondegenerate by (12.2.13), and $J(y, \pi x) = I(Sy, x)$ for all x, y. Hence the pairing (12.2.26,i) is nondegenerate.

12.3. Hypergeometric Integrals and the Knizhnik–Zamolodchikov Connection

As in (12.1), fix data (11.1.a)–(11.1.c), the Lie algebra $\mathfrak{g}$ associated with these data. Fix $\Lambda_1, \ldots, \Lambda_n \in \mathfrak{h}$, $\lambda = (\ell_1, \ldots, \ell_r) \in \mathbb{N}^r$, $\kappa \in \mathbb{C}^*$. Set $\ell = \ell_1 + \cdots + \ell_n$.

Consider the space $\mathbb{C}^{n+\ell}$ with coordinates $t_1(1), \ldots, t_{\ell_1}(1), t_1(2), \ldots, t_{\ell_2}(2), \ldots, t_1(r), \ldots, t_{\ell_r}(r), t_{r+1}, \ldots, t_{r+n}$, see (8.13) and (8.14). Let $p_{n,\ell} : \mathbb{C}^{n+\ell} \to \mathbb{C}^n$, $(\{t_i(j)\}, t_{r+1}, \ldots, t_{r+n}) \mapsto \{t_{r+1}, \ldots, t_{r+n}\}$ be the projection.

Introduce the weights A of the configuration $\mathcal{C}_{n+\ell}$ in $\mathbb{C}^{n+\ell}$ of all diagonal hyperplanes:

$$\begin{aligned} a(t_i, t_j) &= (\Lambda_{i-r}, \Lambda_{j-r})/\kappa, \\ a(t_m(i), t_j) &= -(\alpha_i, \Lambda_{j-r})/\kappa, \\ a(t_m(i), t_s(j)) &= (\alpha_i, \alpha_j)/\kappa \end{aligned} \qquad (12.3.1)$$

for all i, j, m, s.

Let $\mathcal{S}(A)$ and $\mathcal{S}^*(A)$ be the local systems on $\mathbb{C}^{n+\ell} - \mathcal{C}_{n+\ell}$ associated with A, $\mathcal{L}(A)$ the line bundle on $\mathbb{C}^{n+\ell} - \mathcal{C}_{n+\ell}$ with the integrable connection $\nabla(A)$ associated with A, see (2.5).

The group $\Sigma' = \Sigma_{\ell_1} \times \cdots \times \Sigma_{\ell_r}$ naturally acts on $\mathbb{C}^{n+\ell} - \mathcal{C}_{n+\ell}$ permuting the coordinates $\{t_j(i)\}$ in such a way that the index i is preserved. The action of Σ' respects fibers of $p_{n,\ell}$, the local systems $\mathcal{S}(A)$, $\mathcal{S}^*(A)$, the integrable connection $\nabla(A)$.

The triple $(p_{n,\ell} : \mathbb{C}^{n+\ell} \to \mathbb{C}^n, \mathcal{C}_{n+\ell}, \mathcal{C}_n)$ satisfies the assumptions (10.4.a) and (10.4.b). For this triple, consider the horizontal complexes and the horizontal contravariant form introduced in (10.4),

$$S_h : \big(F\ell_h^{\cdot}(\mathcal{C}_{n+\ell}), d_h\big) \longrightarrow \big(\mathcal{A}_h^{\cdot}(\mathcal{C}_{n+\ell}), d_h(A)\big). \qquad (12.3.2)$$

The group Σ' naturally acts on these complexes and the contravariant form commutes with this action. Denote by

$$\begin{aligned}(F\ell_h^\cdot(\mathcal{C}_{n+\ell}), d_h)^- &\subset (F\ell_h^\cdot(\mathcal{C}_{n+\ell}), d_h),\\ (\mathcal{A}_h^\cdot(\mathcal{C}_{n+\ell}), d_h(A))^- &\subset (\mathcal{A}_h^\cdot(\mathcal{C}_{n+\ell}), d_h(A)),\end{aligned} \qquad (12.3.3)$$

the subcomplexes of skew-invariants of the Σ'-action defined as in (12.1.2) and (12.1.3).

The isomorphisms (10.4.20), (10.4.21), and (12.1.5) induce the isomorphisms

$$\begin{aligned}\psi_p &: C_p(\mathfrak{n}_-, M)_\lambda \longrightarrow F\ell_h^{\ell-p}(\mathcal{C}_{n+\ell})^-,\\ \eta_p &: C_p(\mathfrak{n}_-^*, M^*)_\lambda \longrightarrow \mathcal{A}^{\ell-p}(\mathcal{C}_{n+\ell})^-\end{aligned} \qquad (12.3.4)$$

for all p. Here the left spaces are defined in (11.2). The isomorphisms have the properties:

$$\begin{aligned} d_h\psi &= \psi d,\\ -\kappa d_h(a)\eta &= \eta d,\\ \eta_p S &= (-\kappa)^{\ell-p} S_h \psi_p.\end{aligned} \qquad (12.3.5)$$

As a corollary, we have the following commutative diagram

$$\begin{array}{ccc} H_p C.(\mathfrak{n}_-, M)_\lambda & \xrightarrow{S} & H_p C.(\mathfrak{n}_-^*, M^*)_\lambda \\ \psi \downarrow & & \downarrow \eta \\ H^{\ell-p}(F\ell_h^\cdot(\mathcal{C}_{n+\ell}), d_h)^-) & \xrightarrow{(-\kappa)^{\ell-p} S_h} & H^{\ell-p}((\mathcal{A}_h^\cdot(\mathcal{C}_{n+\ell}), d_h(a))^-) \end{array} \qquad (12.3.6)$$

for all p, see (12.1.7) and (12.1.8).

Let $\mathcal{A}_h^\cdot(\mathcal{C}_{n+\ell})^-$ (resp. $F\ell_h^\cdot(\mathcal{C}_{n+\ell})^-$) be the trivial bundle over $\mathbb{C}^n$ with fiber $\mathcal{A}^\cdot(\mathcal{C}_{n+\ell})^-$ (resp. $F\ell_h^\cdot(\mathcal{C}_{n+\ell})^-$). Consider the hypergeometric connection $\nabla(a)$ on these bundles defined in (10.5). This connection commutes with the horizontal contravariant form and the Σ'-action on fibers.

Let $C.(\mathfrak{n}_-, M)_\lambda$ (resp. $C.(\mathfrak{n}_-^*, M^*)_\lambda$) be the trivial bundle over $\mathbb{C}^n$ with fiber $C.(\mathfrak{n}_-, M)_\lambda$ (resp. $C.(\mathfrak{n}_-^*, M^*)_\lambda$). Consider the Knizhnik–Zamolodchikov connection ∇_{KZ} on these bundles defined in (11.5).

12.3.7. *Conjecture*, cf. [SV3]. Isomorphisms (12.3.4) induce isomorphisms of bundles,

$$\begin{aligned}\psi_p &: C_p(\mathfrak{n}_-, M)_\lambda \longrightarrow F\ell_h^{\ell-p}(\mathcal{C}_{n+\ell})^-,\\ \eta_p &: C_p(\mathfrak{n}_-^*, M^*)_\lambda \longrightarrow \mathcal{A}_h^{\ell-p}(\mathcal{C}_{n+\ell})^-,\end{aligned}$$

sending the Knizhnik–Zamolodchikov connection with parameter $-\kappa$ to the hypergeometric connection.

More precisely, let $x \in C.(\mathbf{n}_-, M)_\lambda$, $y \in C.(\mathbf{n}_-^*, M^*)_\lambda$. Let $\bar{x}$ and $\bar{y}$ be the constant sections of $C.(\mathbf{n}_-, M)_\lambda$ and $C.(\mathbf{n}_-^*, M^*)_\lambda$ generated by x and y, respectively. Then conjecturally we have

$$\nabla(a)\psi(\bar{x}) = \frac{1}{\kappa} \sum_{1 \leq i < j \leq n} \psi \Omega_{+,ij}(\bar{x}) d(z_i - z_j)/(z_i - z_j),$$

$$\nabla(a)\eta(\bar{y}) = \frac{1}{\kappa} \sum_{1 \leq i < j \leq n} \eta \Omega_{+,ij}(\bar{y}) d(z_i - z_j)/(z_i - z_j).$$

(12.3.8)

Here $z_1 = t_{r+1}, \ldots, z_n = t_{r+n}$ are coordinates in $\mathbb{C}^n$, $\Omega_{+,ij}$ are the operators defined in (11.5).

This conjecture has been proven in [**SV3**] for $p = 0$.

12.3.9. Theorem [**SV3**, Theorems 7.2.5′ and 7.2.5″]. *The isomorphism*

$$\eta_0 : C_0(\mathbf{n}_-^*, M^*)_\lambda \longrightarrow \mathcal{A}_h^\ell(\mathcal{C}_{n+\ell})^-$$

sends the KZ connection with parameter $-\kappa$ to the hypergeometric connection.

12.3.10. Corollary. *The isomorphism*

$$\psi_0 : C_0(\mathbf{n}_-, M)_\lambda \longrightarrow F\ell_h^\ell(\mathcal{C}_{n+\ell})^-$$

sends the KZ connection with parameter $-\kappa$ to the hypergeometric connection.

Proof of the Corollary. For generic values of weights, the contravariant forms are isomorphisms and, hence, (12.3.9) follows from (12.3.8). The KZ and hypergeometric connections depend analytically on weights, hence, (12.3.9) is true for arbitrary weights.

12.3.11. Example. Let $\mathfrak{g} = s\ell_2$ and $\lambda = 1$. The triple $(p_{n,1} : \mathbb{C}^{n+1} \to \mathbb{C}^n, \mathcal{C}_{n+1}, \mathcal{C}_n)$ corresponds to this case. The weights of $\mathcal{C}_{n+1}$ are defined by (12.3.1) and coincide with the weights defined in example (10.10.10).

The horizontal complex is described in (10.10.10.i). The hypergeometric connection on $\mathcal{A}_h^1(\mathcal{C}_{n+1})$ is given by

(i) $$\nabla(a)\bar{y} = \frac{1}{\kappa} \sum_{1 \leq i < j \leq n} \Omega_{+,ij} \bar{y} \, d(z_i - z_j)/(z_i - z_j),$$

where $z_1 = t_2, \ldots, z_n = t_{n+1}$ are coordinates in $\mathbb{C}^n$, $\bar{y}$ is a constant section of the trivial bundle $\mathcal{A}_h^1(\mathcal{C}_{n+1})$ and the operators Ω_{ij} are defined in (10.10.10).

$\mathcal{A}_h^1(\mathcal{C}_{n+1})$ is identified with $C_0(\mathfrak{n}_-^*, M^*)_{\lambda=1}$ in (12.1.27). The KZ connection with parameter $-\kappa$ is given by

(ii) $$\nabla_{\mathrm{KZ}}\bar{y} = \frac{1}{\kappa} \sum_{1 \leq i < j \leq n} \Omega_{+,ij}\bar{y}\, d(z_i - z_j)/(z_i - z_j),$$

where $\bar{y}$ is a constant section of $C_0(\mathfrak{n}_-^*, M^*)_{\lambda=1}$. For $\mathfrak{g} = sl_2$, we have $\Omega_+ = e \otimes f + f \otimes e + h \otimes h/2$, that is, Ω_+ is the Casimir operator, see example (11.3.33). It is easy to see that the isomorphism of (12.1.27) sends $\Omega_{+,ij}$ to Ω_{ij}.

12.3.12. Corollary. *There is a commutative diagram of morphisms of trivial bundles*

$$\begin{array}{ccc} H_0(C_\cdot(\mathfrak{n}_-, M)_\lambda) & \xrightarrow{S} & H_0(C_\cdot(\mathfrak{n}_-^*, M^*)_\lambda) \\ \psi \downarrow & & \downarrow \eta \\ H^\ell((F\ell_h^\cdot(\mathcal{C}_{n+\ell}), d_h)^-) & \xrightarrow{(-\kappa)^\ell S_h} & H^\ell((\mathcal{A}_h^\cdot(\mathcal{C}_{n+\ell}), d_h(a))^-). \end{array}$$

These are morphisms of bundles with connections: the upper bundles have the KZ connection with parameter $-\kappa$, the lower bundles have the hypergeometric connection $\nabla(a)$.

The triple $(p_{n,\ell} : \mathbb{C}^{n+\ell} \to \mathbb{C}^n, \mathcal{C}_{n+\ell}, \mathcal{C}_n)$ satisfies the assumptions of (10.10). As in (10.10), consider the following four vector bundles with the Gauss–Manin connection,

$$R_\ell pr_1, \ R_\ell pr_2, \ R^\ell pr_1, \ R^\ell pr_2.$$

These are bundles over $Y(\mathcal{C}_n) = \mathbb{C}^n \setminus \bigcup_{H \in \mathcal{C}_n} H$ with fibers $H_\ell(P(z), U(z), \mathcal{S}^*(a, z))$, $H_\ell(Y(z), \mathcal{S}^*(a, z))$, $H^\ell(P(z), U(z), \mathcal{S}(a, z))$, $H^\ell(Y(z), \mathcal{S}(a, z))$, see notation in (10.10).

The group Σ' naturally acts on fibers of these bundles. In each of these bundles, consider the subbundle of all vectors x such that $\sigma x = (-1)^{|\sigma|} x$ for all $\sigma \in \Sigma'$. These subbundles are denoted by

$$R_\ell pr_1^-, \ R_\ell pr_2^-, \ R^\ell pr_1^-, \ R^\ell pr_2^-,$$

respectively.

The isomorphisms η shown in (12.3.12) and morphism (10.10.7) induce a morphism of bundles with connections

$$I : R_\ell pr_2^- \longrightarrow H_0(C.(\mathbf{n}_-^*, M^*)_\lambda)^*. \tag{12.3.13}$$

Namely, the bundle $R_\ell pr_2$ has the Gauss–Manin connection. The bundle $H_0(C.(\mathbf{n}_-^*, M^*)_\lambda)^*$ has the connection dual to the KZ connection with parameter $-\kappa$. By (11.4.2) and (11.5.4), we get the following statement.

12.3.14. Corollary. *The isomorphism η and morphism (10.10.7) induce a morphism of bundles with connections*

$$I : R_\ell pr_2^- \longrightarrow \mathbf{Vac}\,(M)_\lambda,$$

where $\mathbf{Vac}\,(M)_\lambda$ is the trivial bundle over $\mathbb{C}^n$ with fiber

$$\mathrm{Vac}\,(M)_\lambda = \{m \in M \,|\, \mathbf{n}_-^* m = 0\}$$

and the KZ connection with parameter κ.

(12.3.15) By (10.4.20), (10.6.8), and (10.10.6), this morphism is an isomorphism for generic values of the parameter κ. Explicit conditions on κ are given in (10.6.8).

Statement (12.3.14) can be reformulated as follows. Let $\gamma(z) \in H_\ell(Y(z), \mathcal{S}^*(a,z))^-$ be a locally constant family of homology classes. Then the integral

$$\int_{\gamma(z)} i(a) \cdot \eta(x)$$

is a well-defined function on $x \in H_0(C.(n_-^*, M^*)_\lambda)$. Here η is shown in (12.3.12) and $i(a)$ is defined in (10.6.3). This function is an element $\varphi(z)$ of $\mathrm{Vac}\,(M)_\lambda$. It depends on $z \in \mathbb{C}^n$ and satisfies the KZ differential equation

$$d\varphi(z) = \frac{1}{\kappa} \sum_{i<j} \Omega_{+,\,ij} \varphi(z)\, d(z_i - z_j)/(z_i - z_j). \tag{12.3.16}$$

There is the natural projection $\mathrm{Vac}\,(M)_\lambda \to \mathrm{Vac}\,(L)_\lambda$, where $L = L(\Lambda_1) \otimes \cdots \otimes L(\Lambda_n)$ is the tensor product of irreducible representations with highest weights $\Lambda_1, \ldots, \Lambda_n$, resp., and $\mathrm{Vac}\,(L)_\lambda = \{x \in L \,|\, e_1 x = \cdots = e_r x = 0\}$. Let

Vac$(L)_\lambda$ be the trivial bundle over $\mathbb{C}^n$ with fiber $\text{Vac}(L)_\lambda$ and with the KZ connection defined in (11.5.7).

12.3.17. Corollary. *The natural projection* $\text{Vac}(M)_\lambda \to \text{Vac}(L)_\lambda$ *induces a morphism of bundles with connections*

$$J : R_\ell pr_2^- \longrightarrow \mathbf{Vac}\,(L)_\lambda\,,$$

where the left bundle has the Gauss-Manin connection and the right bundle has the KZ connection with parameter κ.

12.3.18. Remark. The important problem is to describe the image of this morphism. The image is a subbundle invariant under the KZ connection. Probably, for generic values of κ, J is an epimorphism. For special values of κ, the image may be a proper subbundle of $\text{Vac}(L)_\lambda$ and satisfies some algebraic equations apart from the KZ equation, see (12.4) and [**FSV**]. If $\mathfrak{g} = s\ell_2$, then J is an epimorphism for generic values of κ, see Remark (11.4.3, ii).

Consider $\bar{\mathfrak{g}} = \mathfrak{g}/\ker S$, $L = M/\ker S$, as in (11.2.10). Let $\bar{\mathfrak{n}}_- \subset \bar{\mathfrak{g}}$ be the corresponding nilpotent subalgebra. The contravariant form induces a natural map

$$H_0(C.(\bar{\mathfrak{n}}_-, L)_\lambda) \longrightarrow H_0(C.(\mathfrak{n}_-^*, M^*)_\lambda) \qquad (12.3.19)$$

where

$$H_0\big(C.(\bar{\mathfrak{n}}_-, L)_\lambda\big) = \big(L/(f_1 L + \cdots + f_r L)\big)_\lambda\,.$$

The contravariant form naturally identifies $H_0(C.(\bar{\mathfrak{n}}_-, L)_\lambda)^*$ with $\text{Vac}(L)_\lambda$ given above. Hence

12.3.20. Corollary. *Map (12.3.8), the isomorphism η in (12.3.12), and morphism (10.10.7) induce a morphism of bundles with connections*

$$J : R_\ell pr_2^- \longrightarrow \mathbf{Vac}\,(L)_\lambda$$

coinciding with (12.3.17).

Let us consider the real part of the base of our bundles: $\mathbb{R}Y(\mathcal{C}_n) := \mathbb{R}^n \cap Y(\mathcal{C}_n) \subset Y(\mathcal{C}_n) \subset \mathbb{C}^n$. The connected components of $\mathbb{R}Y(\mathcal{C}_n)$ are numerated by permutations $\sigma \in \Sigma_n$, the corresponding domain is

$$\mathfrak{D}_\sigma = \{(z_1, \ldots, z_n) \in \mathbb{R}^n | z_{\sigma(1)} < \cdots < z_{\sigma(n)}\}\,.$$

Fix such a domain. Restrict the bundle $p_{n,\ell} : \mathbb{C}^{n+\ell} \to \mathbb{C}^n$ to the domain $\mathfrak{D}_\sigma$. In Sec. 8.2, for any point $z^0 \in \mathfrak{D}_\sigma$, we have constructed chain complexes

$C.(X, S^*(A), z^0)$, $C.(Q'/Q'', S^*(A), z^0)$. The group Σ' naturally acts on these complexes. Denote by $^-$ the corresponding subcomplex generated by all skew-invariant elements of this action. The chain complexes compute the homology groups of the fiber. The first computes $H.(Y(z^0), S^*(a, z^0))$, the second computes $H.(P(z^0), U(z^0), S^*(a, z^0))$.

In Sec. 8.14, we have constructed the canonical quasi-isomorphisms

$$\varphi : C.(^+U_q\mathfrak{n}_-; M_\sigma(q); {}^+\Delta)_\lambda \longrightarrow C.(X, S^*(A), z^0)^- ,$$
$$\varphi : C_{\cdot}^*(^+U_q\mathfrak{n}_-; M_\sigma(q); \mu)_\lambda \longrightarrow C.(Q'/Q'', S^*(A), z^0)^- . \quad (12.3.21)$$

Here $M_\sigma(q) = M_{\sigma(1)}(q) \otimes \cdots \otimes M_{\sigma(n)}(q)$ is the tensor product of Verma modules over $U_q\mathfrak{g}$ with highest weights $\Lambda_1, \ldots, \Lambda_n$, respectively. These quasiisomorphisms are horizontal with respect to the Gauss–Manin connection, see (8.2.8).

In particular, we have a canonical isomorphism,

$$\varphi : H_0(C.(^+U_q\mathfrak{n}_-; M_\sigma(q); {}^+\Delta)_\lambda) \longrightarrow H_\ell(Y(z^0), S^*(a, z^0))^- , \quad (12.3.22)$$

where $H_0(C.(^+U_q\mathfrak{n}_-; M_\sigma(q); {}^+\Delta)_\lambda) = \text{Vac}\,(M_\sigma(q))_\lambda$ is described in (9.3.8.3).

Quasiisomorphism (12.3.20) and morphism (12.3.13) induce a map

$$I_0(z^0) : \text{Vac}\,(M_\sigma(q))_\lambda \longrightarrow \text{Vac}\,(M)_\lambda \quad (12.3.23)$$

for any $z^0 \in \mathfrak{D}_\sigma$, see (12.2.10).

12.3.24. Corollary. *For any* $x \in \text{Vac}\,(M_\sigma(q))_\lambda$, *the function*

$$I(\cdot, x) : z \longmapsto I_0(z)x$$

is a solution to the KZ equation (12.3.15).

For generic values of the parameter κ, all solutions may be represented in this form, see (12.2.13).

Thus, solutions of the KZ equation are parametrized by $\text{Vac}\,(M_\sigma(q))_\lambda$. The monodromy of the KZ equation is the action on $\text{Vac}\,(M_\sigma(q))_\lambda$ of the braid group given by the universal R-matrix, see the precise description in (8.14.6).

The fact that the monodromy representation of the KZ equation is equivalent to the R-matrix representation of the braid group constitutes the beautiful Kohno–Drinfeld theorem [**K1, K2, D3**]. Our identification of the KZ equation with the Gauss–Manin connection and of the monodromy of the Gauss–Manin connection with the R-matrix representation allow us to make explicit the equivalence of the Kohno–Drinfeld theorem. This equivalence is given by integration of the hypergeometric forms over cycles with twisted coefficients.

Note that $S : L \to L^*$ is an isomorphism of $\bar{\mathfrak{g}}$-modules and hence $H_0(C.(\bar{n}_-, L)_\lambda)^* \simeq \{y \in L_\lambda | e_1 y = \cdots = e_r y = 0\}$.

(12.3.25) Let A be the complex affine space with coordinates $\{a_H | H \in \mathcal{C}_{n,\ell}\}$. Consider the domain $D_{-1} \subset A$ defined in (10.7). Assume that the weights a of the configuration $\mathcal{C}_{n,\ell}$ in the fiber of the projection $p_{n,\ell} : \mathbb{C}^{n+\ell} \to \mathbb{C}^\ell$ belong to D_{-1}.

For any point $z^0 \in \mathfrak{D}_\sigma$, consider pairing (12.2.14) and the induced map

$$J_0(z^0) : \text{Vac}\,(L_\sigma(q))_\lambda \longrightarrow \text{Vac}\,(L)_\lambda, \qquad (12.3.26)$$

where $L_\sigma(q) = L(\Lambda_{\sigma(1)}, q) \otimes \cdots \otimes L(\Lambda_{\sigma(n)}, q)$, $\text{Vac}\,(L_\sigma(q))_\lambda = \{x \in L_\sigma(q)_\lambda | e_1 x = \cdots = e_r x = 0\}$, see (12.2.25).

12.3.27. Corollary. *For any $x \in \text{Vac}\,(L_\sigma(q))_\lambda$, the function*

$$J(\,\cdot\,, x) : z \longmapsto J_0(z) x$$

is a solution of the KZ equation (11.5.7).

For almost all κ, all solutions with values in $\text{Vac}\,(L)_\lambda$ may be represented in this form, see Theorem (12.2.26) and (12.5).

Thus, solutions to the KZ equation with values in $\text{Vac}\,(L)_\lambda$ are parametrized by $\text{Vac}\,(L_\sigma(q))_\lambda$.

12.4. Resonances, [FSV, TK]

Let $\bar{\mathfrak{g}}$ be a complex simple Lie algebra. Fix a system of Chevalley generators $e_1, \ldots, e_r, f_1, \ldots, f_r \in \bar{\mathfrak{g}}$. Let $\mathfrak{h} \subset \bar{\mathfrak{g}}$ be the standard Cartan subalgebra, and let $\alpha_1, \ldots, \alpha_r \in \mathfrak{h}^*$ be the simple roots. Fix the invariant scalar product on $\bar{\mathfrak{g}}$ such that $(\theta, \theta) = 2$, where θ is the highest root. Let $\Omega \in \bar{\mathfrak{g}} \otimes \bar{\mathfrak{g}}$ be the corresponding invariant form, the Casimir operator.

The above objects form data of type (11.1.a)–(11.1.c). Let $\mathfrak{g}$ be the Lie algebra defined by these data in (11.1). Thus, $\bar{\mathfrak{g}} = \mathfrak{g}/\ker S$, where S is the contravariant form.

For a weight $\Lambda \in \mathfrak{h}^*$, denote by $L(\Lambda)$ the irreducible $\bar{\mathfrak{g}}$-module with highest weight Λ.

Fix a positive integer k. Call a weight Λ *finite* if $L(\Lambda)$ is finite-dimensional and $(\Lambda, \theta) \leq k$.

Let $\Lambda_1, \ldots, \Lambda_n \in \mathfrak{h}^*$ be finite weights. Set $L = L(\Lambda_1) \otimes \cdots \otimes L(\Lambda_n)$. Fix $\lambda \in \mathbb{N}^r$, $\lambda = (\ell_1, \ldots, \ell_r)$. Set

$$\Lambda = \Lambda_1 + \cdots + \Lambda_n - \sum_{i=1}^{r} \ell_i \alpha_i \,.$$

Assume that Λ is finite. Set $p = k - (\Lambda, \theta) + 1$. Then p is a positive integer.

Fix a nonzero root vector $f_\theta \in \bar{\mathfrak{g}}_{-\theta}$. Fix complex numbers $z_1, \ldots, z_n$. Introduce an operator

$$z f_\theta = \sum_{j=1}^{n} z_j f_\theta^{(j)} : L \longrightarrow L, \qquad (12.4.1)$$

where $f_\theta^{(j)}$ is f_θ acting on the jth factor $L(\Lambda_j)$.

The maps

$$C_\cdot(\mathfrak{n}_-, M)_\lambda \xrightarrow{\psi} (\mathcal{F}\ell^{-\cdot}(\mathcal{C}_{n,\ell}(z)), d)^-$$
$$\xrightarrow{S} (\mathcal{A}^{\ell-\cdot}(\mathcal{C}_{n,\ell}(z)), d)^- \xrightarrow{i(a)} (\Omega^\cdot(\mathcal{L}(a), \mathcal{C}_{n,\ell}(z)), d(a)), \qquad (12.4.2)$$

described in (12.1), (10.3.8), and (10.6.3), induce a map

$$\varphi : C_\cdot(\bar{\mathfrak{n}}_-, L)_\lambda \longrightarrow (\Omega^\cdot(\mathcal{L}(a), \mathcal{C}_{n,\ell}(z)), d(a)), \qquad (12.4.3)$$

where $\bar{\mathfrak{n}}_- \subset \bar{\mathfrak{g}}$ is the corresponding nilpotent subalgebra. In particular, φ induces a map

$$\varphi : L_\lambda \longrightarrow \Omega^\ell(\mathcal{L}(a), \mathcal{C}_{n,\ell}(z)) \,. \qquad (12.4.4)$$

Here a is the set of the weights of the discriminantal configuration $\mathcal{C}_{n,\ell}(z_1, \ldots, z_n)$ described in (12.1.1). These weights depend on the parameter κ.

12.4.5. Theorem [FSV], [FSVI–II]. *Assume that $\kappa = k + g$, where k is fixed above, and g is the dual Coxeter number of $\bar{\mathfrak{g}}$. Then, for every $x \in L_\lambda$ having the form $x = (zf_\theta)^p y$, the differential form $\varphi(x)$ is exact, see its description below.*

12.4.6. Example [FSV]. Let $\mathfrak{g} = s\ell_2$, $\alpha = \alpha_1$, $(\alpha, \alpha) = 2$. Then $\kappa = k + 2$, $\theta = \alpha$, $f_\theta = f = f_1$. Assume that $\lambda = 1$, i.e., $\Lambda = \Lambda_1 + \cdots + \Lambda_n - \alpha$. Assume that $p = 1$. The equality $p = k - (\Lambda, \theta) + 1$ implies the resonance condition

(i) $$\sum_{i=1}^{n} (\alpha, -\Lambda_i)/\kappa = -1 \,.$$

Set $f^{(j)} = v_1 \otimes \cdots \otimes fv_j \otimes \cdots \otimes v_n \in L_\lambda$ for $j = 1, \ldots, n$. Set $x = zf_\theta(v_1 \otimes \cdots \otimes v_n) = z_1 f^{(1)} + \cdots + z_n f^{(n)}$. The differential form $\varphi(x)$ is a hypergeometric form for the configuration $C_{n,1}(z_1, \ldots, z_n)$. We have

(ii) $$\varphi(f^{(j)}) = \frac{(\alpha, -\Lambda_j)}{\kappa} \prod_{i=1}^n (t - z_i)^{(\alpha, -\Lambda_i)/\kappa} dt/(t - z_j)$$

and

(iii) $$\varphi(x) = d\left(t \prod_{i=1}^n (t - z_i)^{(\alpha, -\Lambda_i)/\kappa} \right).$$

Description of the differential form $\varphi(x)$.

Let $\theta = a_1 \alpha_1 + \cdots + a_r \alpha_r$, $\ell'_i = \ell_i - pa_i$. Set $\lambda' = (\ell'_1, \ldots, \ell'_r)$, $\ell' = \ell'_1 + \cdots + \ell'_r$, $a = a_1 + \cdots + a_r$.

The discriminantal configuration $C_{n,\ell}(z) \subset \mathbb{C}^\ell$ mentioned in (12.4.4) has the following properties: $\ell = \ell_1 + \cdots + \ell_r$, and the coordinates t in $\mathbb{C}^\ell$ correspond to the simple roots α_i belonging to $\ell_1 \alpha_1 + \cdots + \ell_r \alpha_r$, see (12.1).

We have
$$\ell_1 \alpha_1 + \cdots + \ell_r \alpha_r = \ell'_1 \alpha_1 + \cdots + \ell'_r \alpha_r + p(a_1 \alpha_1 + \cdots + a_r \alpha_r).$$

Divide the coordinates of $\mathbb{C}^\ell$ into $p + 1$ groups. The first group corresponds to the simple roots belonging to $\ell'_1 \alpha_1 + \cdots + \ell'_r \alpha_r$ and will be denoted by $u = (u_1, \ldots, u_{\ell'})$. The other p groups correspond to the simple roots belonging to p summands θ. We denote such a group by $v_*(j) = (v_1(j), \ldots, v_a(j))$ for $j = 1, \ldots, p$.

Consider the following multivalued function on $\mathbb{C}^\ell - C_{n,\ell}(z)$:

$$\ell = \prod (t_i(j) - t_c(d))^{(\alpha_j, \alpha_d)/\kappa} \prod (t_i(j) - z_c)^{-(\alpha_j, \Lambda_c)/\kappa}. \tag{12.4.7}$$

Here the second product is taken over all i, j, c, and the first is taken over all pairs (i, j), (c, d) such that $j < d$, or $j = d$ and $i < c$.

Using the previous notation, we may rewrite this function as

$$\ell = \ell_u \prod_{j=1}^p \ell_{v(j)} \ell_{v(j),u} \ell_{v(j),v(1)} \cdots \ell_{v(j),v(j-1)} \tag{12.4.8}$$

where

$$\ell_u = \prod (u_i - u_j)^{(\alpha(u_i),\alpha(u_j))/\kappa} \prod (u_i - z_m)^{-(\alpha(u_i),\Lambda_m)/\kappa},$$

$$\ell_{v(j)} = \prod (v_c(j) - v_d(j))^{(\alpha(v_c(j)),\alpha(v_d(j)))/\kappa} \prod (v_c(j) - z_m)^{-(\alpha(v_c(j)),\Lambda_m)/\kappa},$$

$$\ell_{v(j),u} = \prod (v_c(j) - u_d)^{(\alpha(v_c(j)),\alpha(u_d))/\kappa},$$

$$\ell_{v(j),v(d)} = \prod (v_i(j) - v_c(d))^{(\alpha(v_i(j)),\alpha(v_c(d)))/\kappa}.$$

Denote by $\sum$ the sum of the exponents of all linear factors in ℓ/ℓ_u. The condition $p = k - (\Lambda, \theta) + 1$ implies the resonance condition [**FSV**]:

$$\sum + p = 0. \qquad (12.4.9)$$

This is a generalization of (12.4.6.i).

Now let $x = (zf_\theta)^p y$. Then $y \in L_{\lambda'}$. By the construction of [**SV3**], see also (12.1), $\varphi(y)$ is a hypergeometric ℓ'-form in the space with coordinates u. Consider the form $\varphi(f_\theta v)$, where $v = v_1 \otimes \cdots \otimes v_n \in L$. By definition, it is a hypergeometric a-form in the space with coordinates $\bar{v} = (\bar{v}_1, \ldots, \bar{v}_a)$. This form has the following shape:

$$\varphi(f_\theta v) = \ell_{\bar{v}} d\bar{v} \operatorname{Sym} B(\bar{v}),$$

where $B(\bar{v})$ is a certain polynomial in

$$\frac{1}{\bar{v}_i - \bar{v}_j} \quad \text{and} \quad \frac{1}{\bar{v}_i - z_m},$$

and Sym denotes the symmetrization over the product of symmetric groups $\Sigma_{a_1} \times \cdots \times \Sigma_{a_r}$. It is easy to see that $B(\bar{v})$ has the form

$$B(\bar{v}) = A(\bar{v}) \sum_{i,m} \frac{-(\alpha(\bar{v}_i), \Lambda_m)}{\bar{v}_i - z_m}$$

where $A(\bar{v})$ is the polynomial only in $1/(\bar{v}_i - \bar{v}_j)$.

Define an $(a-1)$-form $\pi(\bar{v})$:

$$\pi(\bar{v}) = \sum_{b=1}^{a} (-1)^{b-1} \bar{v}_b d\bar{v}_1 \wedge \cdots \wedge \widehat{d\bar{v}_b} \wedge \ldots d\bar{v}_a.$$

Set

$$(12.4.10) \quad \omega = \mathrm{Sym}\,\{\varphi(y) \wedge d[\ell_{v(1)}\ell_{v(1),u}A(v(1))\pi(v(1))]$$
$$\wedge d[\ell_{v(2)}\ell_{v(2),u}\ell_{v(2),v(1)}A(v(2))\pi(v(2))]\ldots$$
$$\wedge d[\ell_{v(p)}\ell_{v(p),u}\ell_{v(p),v(1)}\cdots\ell_{v(p),v(p-1)}A(v(p))\pi(v(p))]\},$$

where Sym denotes the symmetrization over $\Sigma_{\ell_1} \times \cdots \times \Sigma_{\ell_r}$. It is clear that ω is exact.

12.4.11. Theorem [FSV]. *Under the above assumptions,*

$$\varphi((zf_\theta)^p y) = \mathrm{const}\,\omega + \left[\sum + p\right]\omega',$$

for some ω'.

See (12.4.6.iii) as an example.

Consider the morphism

$$J : R_\ell pr_2^- \longrightarrow \mathbf{Vac}\,(L)_\lambda$$

described in (12.3.6). This is a morphism of bundles over $\mathbb{C}^n - \mathcal{C}_n$.

12.4.12. Corollary of (12.4.5). *Under the assumptions of Theorem (12.4.5), the image of J in the fiber of $\mathbf{Vac}\,(L)_\lambda$ over a point z belongs to the subspace*

$$W(z) = \{x \in L_\lambda | e_1 x = \cdots = e_r x = 0,\, (ze_\theta)^p x = 0\},$$

where

$$ze_\theta = \sum_{j=1}^n z_j e_\theta^{(j)} : L \longrightarrow L,$$

and e_θ is a nonzero root vector in $\bar{\mathfrak{g}}_\theta$.

Thus, the solutions of the KZ equations given by hypergeometric integrals satisfy in addition the algebraic equation $(ze_\theta)^p x = 0$, if $\kappa = k + g$ and all weights are finite.

Consider the morphism

$$J_0(z) : \mathbf{Vac}\,(L_\sigma(q))_\lambda \longrightarrow \mathbf{Vac}\,(L)_\lambda \qquad (12.4.13)$$

described in (12.3.26). Here $\sigma \in \Sigma_n$ and z is a point of the domain $\mathfrak{D}_\sigma \subset \mathbb{R}^n$, see (12.3). This morphism is defined under the assumption that the weights of the configuration $C_{n,\ell}$ form a point of the domain D_{-1} defined in (10.7).

12.4.14. Corollary of (12.4.5). *Under the assumptions of Theorem (12.4.5), the image of $J_0(z)$ belongs to the subspace $W(z)$ defined in (12.4.12).*

12.5. Nondegeneracy of the Hypergeometric Pairing $J(z) : H_0(C.(\bar{\mathfrak{n}}_-;L)_\lambda) \otimes H_0(C.(^+U_q\bar{\mathfrak{n}}_-; L(q); \delta)_\lambda) \to \mathbb{C}$ for Almost All κ.
Asymptotics for $\kappa \to \infty$.

Let C be a weighted configuration of hyperplanes in $\mathbb{R}^k$. Let a be a set of weights of C.

Let $P \subset \mathbb{R}^k$ be an oriented convex κ-dimensional polytope such that the interior of P does not intersect C. Assume that P is in the general position with respect to C in the sense of (10.7.3).

Let $F = (L^0 \supset L^1 \supset \cdots \supset L^k)$ be a flag of C, where L^j is an edge of C of codimension j.

Say that P is adjacent to F if P has a sequence of facets $P^0 \supset P^1 \supset \cdots \supset P^k$, codim $P^j = j$, such that $P^j \subset L^j$ for all j.

Assume that P is adjacent to F. Then the pair (P, F) defines a new orientation of P. It is defined by any basis $v_1, \ldots, v_k$ in $\mathbb{R}^k$ such that v_j lies in L^j and is directed from P^{j+1} to P^j.

Define the index $\operatorname{ind}(P, F)$ as follows. If P is not adjacent to F, set $\operatorname{ind}(P, F) = 0$. If P is adjacent to F, set $\operatorname{ind}(P, F) = 1$ if the orientations of P and of the pair (P, F) coincide, and $\operatorname{ind}(P, F) = -1$ otherwise.

Let $\omega(F, a) \in F\ell^k(C, a)$ be the flag hypergeometric form corresponding to the flag F.

For any $H \in C$, fix a linear function ℓ_H on $\mathbb{R}^k$ whose kernel is H. For each H, fix an argument of $\ell_{H|P}$. Hence, a branch of the function

$$\ell_a = \prod_{H \in C} \ell_H^{a(H)}$$

is fixed over P for arbitrary weights a. This branch defines a section $s_a \in \Gamma(P, \mathcal{S}^*(a)|_P)$ of the local system $\mathcal{S}^*(a)$ associated with weights a, see (10.7).

Set $\Delta_a = (P, s_a)$. Consider the integral

$$I(P, F, a) = \int_{\Delta_a} \omega(F, a).$$

By (10.7.12), the integral is a holomorphic function of a in a neighborhood of the point $a = 0$.

12.5.1. Lemma.
$$I(P, F, a) = \text{ind}\,(P, F) + o(a).$$

The lemma easily follows by induction on k from the formula:
$$\lim_{a \to 0} \int_0^1 \varphi(t) d(t^a) = \varphi(0).$$

Now let us consider the hypergeometric pairing (12.2.14), in particular, its special case:
$$J[z, \{T_j^i\}, \kappa] : C_0(\bar{\mathfrak{n}}_-; L)_\lambda \otimes C_0(^+U_q\bar{\mathfrak{n}}_-; L(q); \delta)_\lambda \longrightarrow \mathbb{C}.$$

Here
$$C_0(\bar{\mathfrak{n}}_-; L) = L_\lambda = (L(\Lambda_1) \otimes \cdots \otimes L(\Lambda_n))_\lambda,$$
$$C_0(^+U_q\bar{\mathfrak{n}}_-; L(q); \delta)_\lambda = L(q)_\lambda = (L(\Lambda_1, q) \otimes \cdots \otimes L(\Lambda_n, q))_\lambda,$$
$$q = \exp(2\pi i / \kappa), \quad \lambda = (\ell_1, \ldots, \ell_r) \in \mathbb{N}^r.$$

We will describe $\lim J[z, \{T_j^i\}, \kappa]$ as κ tends to infinity.

Let v_j (resp. $v_j(q)$) be the highest vector in $L(\Lambda_j)$ (resp. $L(\Lambda_j, q)$). For any monomial $f_J = f_{J_1}v_1 \otimes \cdots \otimes f_{J_n}v_n \in L_\lambda$, set $f_J(q) = f_{J_1}v_1(q) \otimes \cdots \otimes f_{J_n}v_n(q) \in L(q)_\lambda$. Set $\ell = \ell_1 + \cdots + \ell_r$.

12.5.2. Lemma. *For any monomials $f_I, f_J \in L_\lambda$,*
$$\lim_{\kappa \to \infty} \left[\frac{\kappa}{2\pi\sqrt{-1}} \right]^\ell J[z, \{T_j^i\}, \kappa](f_I \otimes f_J(q)) = \ell_1! \ldots \ell_r! S(f_I, f_J)$$

where S is the contravariant form on L.

In particular, this limit is independent of $\{T_j^i\}$ and z.

The lemma is a direct corollary of (12.5.2) and (10.9.11).

12.5.3. Example. Let $n = 2$ and $\lambda = (1, 0, \ldots, 0)$. Let $f_I = v_1 \otimes f_1 v_2$. Then
$$J[z, \{T_j^i\}, \kappa](f_I \otimes f_I(q)) = \int_\gamma \omega,$$

where

$$\omega = (z_1 - z_2)^{(\Lambda_1,\Lambda_2)/\kappa}(t-z_1)^{-(\Lambda_1,\alpha_1)/\kappa}\,d(t-z_2)^{-(\Lambda_2,\alpha_1)/\kappa}$$

and γ is shown in Fig. 12.1.a.

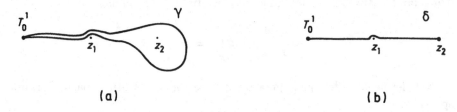

Fig. 12.1

We have

$$\int_\gamma \omega = [q^{(\Lambda_2,\alpha_1)/2} - q^{-(\Lambda_2,\alpha_1)/2}]\int_\delta \omega,$$

where the path δ is shown in Fig. 12.1.b. Thus

$$\lim_{\kappa\to\infty} \frac{\kappa}{2\pi\sqrt{-1}}\int_\delta \omega = (\Lambda_2,\alpha_1).$$

Set $b_{ij} = (\alpha_i,\alpha_j), a_{ij} = 2b_{ij}/b_{ii}$.

(12.5.4) Suppose that $a_{ij} \in \mathbb{Z}$ for all i and j, and $a_{ij} \leq 0$ for all $i \neq j$. Suppose that $2(\Lambda_i,\alpha_j)/(\alpha_j,\alpha_j) \in \mathbb{N}$ for $i = 1,\ldots,n$ and all j.

12.5.5. Theorem. *Under assumptions (12.5.4), the hypergeometric pairing*

$$J[z,\kappa] : H_0(C.(\bar{\mathfrak{n}}_-;L)_\lambda) \otimes H_0(C.(^+U_q\mathfrak{n}_-;L(q);\delta)_\lambda) \longrightarrow \mathbb{C}$$

is nondegenerate for almost all κ, see (12.2.25) for a description of these spaces.

Proof. By [Lu1],

$$\dim H_0(C.(\bar{\mathfrak{n}}_-;L)_\lambda) = \dim H_0(C.(^+U_q\bar{\mathfrak{n}}_-;L(q);\delta)_\lambda) \qquad (12.5.6)$$

if $(\alpha_i,\alpha_i)/\kappa$ is not a rational number for all i.

Hence equality (12.5.7) holds for all q in a neighborhood of $q = 1$.

For such q, the pairing $J[z,\kappa]$ holomorphically depends on z and κ. By (12.5.3), this pairing has the asymptotic expansion

$$J[z,\kappa] = \left[\frac{2\pi\sqrt{-1}}{\kappa}\right]^\ell (\ell_1!\ldots\ell_r!\, S + o(1/\kappa)),$$

where

$$S : \left(L/\sum f_i L\right)_\lambda \otimes \{x \in L_\lambda | e_1 x = \cdots = e_r x\} \longrightarrow \mathbb{C}$$

is the pairing induced by the contravariant form on L. The pairing S is nondegenerate. This proves the theorem.

Remark. We have proved that the pairing is nondegenerate for almost all $1/\kappa$, and the exceptions form a discrete subset of the complex line. It would be interesting to find explicitly exceptional values of $1/\kappa$, cf. [**V2, V3**].

13

Resonances at Infinity

13.1. Projective Transformations of the Complex Line and Discriminantal Configurations

Set
$$f = \prod_{1 \leq i < j \leq \ell} (t_i - t_j)^{a_{i,j}} \prod_{\substack{1 \leq i \leq \ell \\ 1 \leq m \leq n}} (t_i - z_m)^{a_{i,\ell+m}}. \qquad (13.1.1)$$

This is a holomorphic multivalued function on $\mathbb{C}^\ell - C_{n,\ell}(z)$, where $z_1, \ldots, z_n$ are pairwise distinct complex numbers and $a_{i,j}$'s are parameters.

Fix $p \in \{1, \ldots, n\}$. Define a rational map
$$\pi = \pi_p : \mathbb{C}^\ell \longrightarrow \mathbb{C}^\ell,$$
$$(u_1, \ldots, u_\ell) \longmapsto (t_1, \ldots, t_\ell) = (u_1^{-1} + z_p, \ldots, u_\ell^{-1} + z_p). \qquad (13.1.2)$$

Then
$$\pi^* f = \text{const} \prod_{1 \leq i < j \leq \ell} (u_i - u_j)^{a_{i,j}} \prod_{1 \leq i \leq n} u^{b_i}$$
$$\times \prod_{\substack{m=1 \\ m \neq p}}^{n} \prod_{i=1}^{\ell} \left[u_i - \frac{1}{z_m - z_p} \right]^{a_{i,\ell+m}}, \qquad (13.1.3)$$

where
$$b_i = -\sum_{\substack{j=1 \\ j \neq i}}^{\ell} a_{i,j} - \sum_{m=1}^{n} a_{i,\ell+m} \qquad (13.1.4)$$

for all i.

The map π_p defines a biholomorphism
$$\pi_p : \mathbb{C}^{\ell} - C_{n,\ell}(z^p) \longrightarrow \mathbb{C}^{\ell} - C_{n,\ell}(z), \qquad (13.1.5)$$
where
$$z^p = \left(0, (z_1 - z_p)^{-1}, \ldots, (z_n - z_p)^{-1}\right).$$

(13.1.6) This biholomorphism transforms the hypergeometric forms of $C_{n,\ell}(z)$ into the hypergeometric forms of $C_{n,\ell}(z^p)$:

$$d\ln(t_i - t_j) \longmapsto d\ln(u_j - u_i) - d\ln u_i - d\ln u_j,$$
$$d\ln(t_i - z_m) \longmapsto d\ln\left(u_i - (z_m - z_p)^{-1}\right) - d\ln u_i,$$
$$d\ln(t_i - z_p) \longmapsto d\ln u_i.$$

According to the results of Secs. 2–5, the homology groups of $\mathbb{C}^{\ell} - C_{n,\ell}(z)$ with coefficients in the local system defined by f are identified with the homology groups of a suitable quantum group with coefficients in a suitable module. The same is true for the homology groups of $\mathbb{C}^{\ell} - C_{n,\ell}(z^p)$ with coefficients in the local system defined by $\pi^* f$. Moreover, the quantum group corresponding to $C_{n,\ell}(z^p)$ is the same as the quantum group corresponding to $C_{n,\ell}(z)$. The homology groups of $\mathbb{C}^{\ell} - C_{n,\ell}(z)$ and $\mathbb{C}^{\ell} - C_{n,\ell}(z^p)$ are canonically isomorphic. Therefore, the homology groups of the corresponding quantum group with coefficients in the corresponding modules are canonically isomorphic.

We will describe this isomorphism in Subsecs. 13.3–13.4. This isomorphism is important for studying the hypergeometric pairing of Sec. 12.

13.2. Complementary Weight

Let $U_q \mathfrak{g}$ be a quantum group of type (4.1.a)–(4.1.d). Let $\Lambda_1, \ldots, \Lambda_n \in \mathfrak{h}^*$ and $\lambda = (\ell_1, \ldots, \ell_r) \in \mathbf{N}^r$.

(13.2.1) A vector $\Lambda_\infty \in \mathfrak{h}^*$ is called the *weight complementary to* $\Lambda_1, \ldots, \Lambda_n$ *at level* λ, if

$$(\Lambda_\infty + \Lambda_1 + \cdots + \Lambda_n - \ell_1 \alpha_1 - \cdots - \ell_r \alpha_r, \alpha_j) = -(\alpha_j, \alpha_j)$$

for all $j = 1, \ldots, r$.

Obviously, a complementary weight does exist.

13.2.2. Example. Let $\mathfrak{h}$ and $\alpha_1, \ldots, \alpha_r$ be the Cartan subalgebra and the simple positive roots of a simple Lie algebra. Then

$$\Lambda_\infty = -\rho - \sum \Lambda_m + \sum k_j \alpha_j,$$

where ρ is the sum of all positive roots of the Lie algebra.

Fix a complementary weight Λ_∞.

Consider a discriminantal configuration $\mathcal{C}_{n,\ell}(z)$, $z = (z_1 < \cdots < z_n) \subset \mathbb{R}$, with the weights a given by (12.1.1). Let $\mathcal{S}^*(a)$ be the local system defined on $\mathbb{C}^\ell - \mathcal{C}_{n,\ell}(z)$ by a.

Consider the discriminantal configuration $\mathcal{C}_{n,\ell}(z^p)$,

$$z^p = (z_1^p < z_2^p < \cdots < z_n^p):$$
$$= \left((z_{p-1} - z_p)^{-1} < (z_{p-2} - z_p)^{-1} < \cdots < (z_1 - z_p)^{-1} \right.$$
$$\left. < 0 < (z_n - z_p)^{-1} < \cdots < (z_{p+1} - z_p)^{-1} \right),$$

with the weights a^p:

$$a(t_i(j), t_u(v)) = (\alpha_j, \alpha_v)/\kappa,$$

$$a(t_i(j), z_u^p) = \begin{cases} -(\alpha_j, \Lambda_{p-u})/\kappa & \text{for } u < p, \\ -(\alpha_j, \Lambda_{n+p+1-u})/\kappa & \text{for } u > p, \\ -(\alpha_j, \Lambda_\infty)/\kappa & \text{for } u = p. \end{cases} \qquad (13.2.3)$$

Let $\mathcal{S}^*(a^p)$ be the local system defined on $\mathbb{C}^\ell - \mathcal{C}_{n,\ell}(z^p)$ by a^p.

13.2.4. Lemma. *The biholomorphism π_p defined in (13.1.5) induces a canonical isomorphism*

$$\mathcal{S}^*(a^p) \simeq \mathcal{S}^*(a).$$

Proof. See (13.1.3) and (13.1.4).

Let $M_1, \ldots, M_n, M_\infty$ be the $U_q\mathfrak{g}$-Verma modules with highest weights $\Lambda_1, \ldots, \Lambda_n, \Lambda_\infty$, respectively. Set

$$\begin{aligned} M &= M_1 \otimes \cdots \otimes M_n, \\ M^p &= M_{p-1} \otimes \cdots \otimes M_1 \otimes M_\infty \otimes M_n \otimes \cdots \otimes M_{p+1}. \end{aligned} \qquad (13.2.5)$$

By (5.11.3), there are canonical isomorphisms

$$H.(\mathbb{C}^\ell - \mathcal{C}_{n,\ell}(z), \mathcal{S}^*(a))^- \simeq H_{\ell-.}(^+U_q\mathfrak{n}_-; M; {}^+\Delta)_\lambda,$$

$$H.(\mathbb{C}^\ell - \mathcal{C}_{n,\ell}(z^p), \mathcal{S}^*(a^p))^- \simeq H_{\ell-.}(^+U_q\mathfrak{n}_-; M^p; {}^+\Delta)_\lambda.$$

13.2.6. Corollary of (13.2.4). *The biholomorphism π_p induces a canonical isomorphism*

$$H.(^+U_q\mathfrak{n}_-; M; {}^+\Delta)_\lambda \simeq H.(^+U_q\mathfrak{n}_-; M^p; {}^+\Delta)_\lambda.$$

In fact, there is a canonical isomorphism of the corresponding complexes

$$C.(^+U_q\mathfrak{n}_-; M; {}^+\Delta)_\lambda \simeq C.(^+U_q\mathfrak{n}_-; M^p; {}^+\Delta)_\lambda.$$

This isomorphism will be called the *inversion isomorphism*. We will describe the inversion isomorphism for $n = 1$ and $p = 1$ in the next section. The general case is similar, see (13.4).

13.3. The Inversion Isomorphism for One Verma Module

Assume that $n = 1$ and $z = z_1 = 0$. Then $p = 1$ and $z^p = 0$. We compare the homology of the configuration $\mathcal{C}_{1,\ell}(z = 0)$ with the weights a given by (12.1.1) and the homology of the same configuration $\mathcal{C}_{1,\ell}(z^p = 0)$ with the weights given by (13.2.3) for $n = 1$, $p = 1$. Denote these configurations by $\mathcal{C}_{1,\ell}$ and $\mathcal{C}_{1,\ell}^\infty$, resp.

Let

$$T_0^1 < \cdots < T_0^\ell < 0 < T_1^1 < \cdots < T_1^\ell \qquad (13.3.1)$$

be an equipment of $\mathcal{C}_{1,\ell}$. Let $C.(X, \mathcal{S}^*(a))$ be the chain complex constructed for $\mathcal{C}_{1,\ell}$ and $\{T_j^i\}$ in Secs. 2 and 3. Let $C.(X, \mathcal{S}^*(a))^-$ denote its skew-invariant part with respect to the natural action of the group $\Sigma' = \Sigma_{\ell_1} \times \cdots \times \Sigma_{\ell_r}$, see (12.1).

Let

$$\varphi : C.(^+U_q\mathfrak{n}_-; M_1; {}^+\Delta)_\lambda \hookrightarrow C.(X, \mathcal{S}^*(a))^- \qquad (13.3.2)$$

be the realization described in (5.11). φ is a monomorphism inducing the isomorphism of homology groups.

Let

$$x = f_{I_1}| \ldots |f_{I_k}|f_{I_0}v_1 \qquad (13.3.3)$$

be a monomial in $C_k(^+U_q\mathfrak{n}_-; M_1; {}^+\Delta)_\lambda = \big((U_q\mathfrak{n}_-)^{\otimes k} \otimes M_1\big)_\lambda$. Here, v_1 is the generating vector of M_1 and $f_{I_s} = f_1(s)\ldots f_{u(s)}(s)$, for $s = 0,\ldots,k$. The image $\varphi(x)$ is an $(\ell - k)$ chain shown in Fig. 13.1. For the description of such a chain, see (5.11).

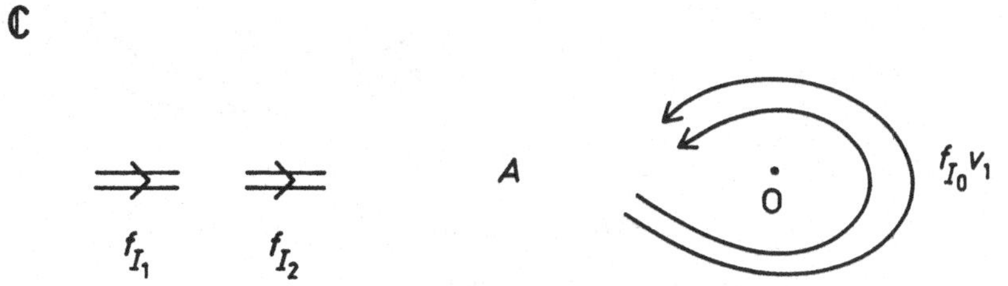

$$\varphi(f_{I_1}|\ldots|f_{I_k}|f_{I_0}v_1)$$

Fig. 13.1

Let

$$\bar{T}_0^1 < \cdots < \bar{T}_0^\ell < 0 < \bar{T}_1^1 < \cdots < \bar{T}_1^\ell \qquad (13.3.4)$$

be the equipment of $C_{1,\ell}^\infty$ such that $\bar{T}_j^i = \big(T_j^{\ell-i+1}\big)^{-1}$.

Let $C.(X^\infty, \mathcal{S}^*(a^1))$ be the chain complex constructed for $C_{1,\ell}^\infty$ and $\{\bar{T}_j^i\}$ in Secs. 2 and 3. Let $C.(X^\infty, \mathcal{S}^*(a^1))^-$ denote its skew-invariant part with respect to Σ'. Let

$$\bar{\varphi} : C.(^+U_q\mathfrak{n}_-; M_\infty; {}^+\Delta)_\lambda \longrightarrow C.(X^\infty, \mathcal{S}^*(a^1))^- \qquad (13.3.5)$$

be the realization described in (5.11).

Let $y = f_{J_1}|\ldots|f_{J_k}|f_{J_0}v_\infty$ be a monomial in $C_k(^+U_q\mathfrak{n}_-; M_\infty; {}^+\Delta)_\lambda$, where v_∞ is the generating vector of M_∞. The image $\bar{\varphi}(y)$ is an $(\ell - k)$ chain shown in Fig. 13.2.

The image $\pi^{-1}\bar{\varphi}(y)$ of the chain $\bar{\varphi}(y)$ under the biholomorphism $\pi^{-1} = (\pi_1)^{-1}$ is shown in Fig. 13.3.

It is easy to see that the chain complex $\pi^{-1}\bar{\varphi}(C.(^+U_q\mathfrak{n}_-; M_\infty; {}^+\Delta)_\lambda)$ may be continuously deformed to the chain complex $\varphi(C.(^+U_q\mathfrak{n}_-; M_1; {}^+\Delta)_\lambda)$ in such a way that any element of the first complex, see Fig. 13.3, is deformed into a union of elements of the second complex shown in Fig. 13.1.

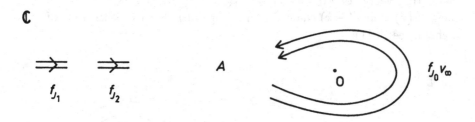

$$\bar\varphi(f_{J_1}|\ldots|f_{J_k}|f_{J_0}v_\infty)$$

Fig. 13.2

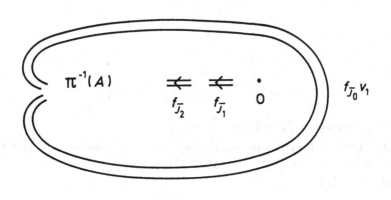

$$\pi^{-1}\bar\varphi(f_{J_1}|\ldots|f_{J_k}|f_{J_0}v_\infty)$$

Fig. 13.3

13.3.6. Example. Let $\lambda = (1,1,0,\ldots,0) \in \mathbf{N}^r$. Then the monomial $f_1|f_2v_\infty \in C_1({}^+U_q\mathfrak{n}_-; M_\infty; {}^+\triangle)_\lambda$ is shown in Fig. 13.4a. Its image under π and the decomposition of the image into a sum of monomials of $C_1({}^+U_q\mathfrak{n}_-; M_\infty; {}^+\triangle)_\lambda$ is shown in Fig. 13.4b.

Comparing the distinguished sections over chains of the complexes, we get the formula

$$(\pi^{-1})_* : f_1|f_2v_\infty \longmapsto -f_1|f_2v_1 + q^{(\alpha_1,\alpha_2)/4-(\Lambda_1,\alpha_2)/2}f_1f_2|v_1$$
$$- q^{-(\alpha_1,\alpha_2)/4+(\Lambda_1,\alpha_2)/2}f_2f_1|v_1.$$

The Inversion Isomorphism for n Verma Modules 301

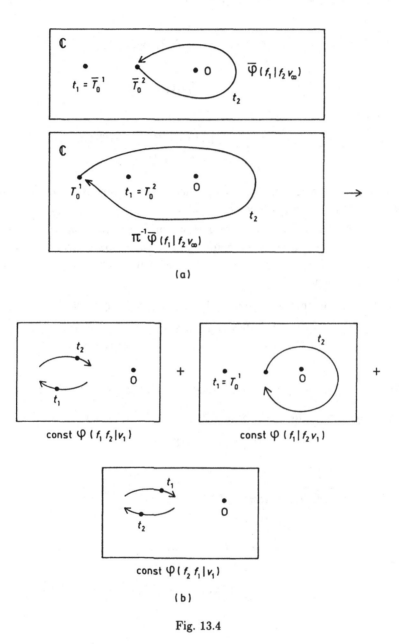

Fig. 13.4

The other elements of the complex $C.(^+U_q\mathfrak{n}_-; M_\infty; {}^+\Delta)_\lambda$ are transformed as follows:

$$f_i|f_j|v_\infty \longmapsto -f_i|f_j|v_1,$$
$$f_if_j|v_\infty \longmapsto -f_jf_i|v_1,$$
$$f_if_jv_\infty \longmapsto f_jf_iv_1,$$

for $(i,j) = (1,2), (2,1)$, and

$$f_2|f_1v_\infty \longmapsto -f_2|f_1v_1 + q^{(\alpha_1,\alpha_2)/4-(\Lambda_1,\alpha_1)/2}f_2f_1|v_1$$
$$- q^{-(\alpha_1,\alpha_2)/4+(\Lambda_1,\alpha_1)/2}f_1f_2|v_1.$$

For the general case, the matrix of π_*^{-1} has a triangular form. To formulate the statement, we need the following definition.

Let $x \in C_k(^+U_q\mathfrak{n}_-; M_1; {}^+\Delta)_\lambda$ be a monomial of form (13.3.3). Let $x^1 = f_{J_1}|\ldots|f_{J_k}|f_{J_0}v_1$ be another monomial of the same space. We say that x^1 is a *left monomial with respect to* x if the following conditions, (13.3.7) and (13.3.8), are satisfied.

(13.3.7) f_{J_0} may be obtained from f_{I_0} by deleting some symbols $\{f_{a_1},\ldots,f_{a_p}\}$.

(13.3.8) $f_{J_1}|\ldots|f_{J_k}$ may be obtained from $f_{I_1}|\ldots|f_{I_k}$ by adding the same symbols $\{f_{a_1},\ldots,f_{a_p}\}$ to the beginning or to the end of the monomials $f_{I_1},\ldots,f_{I_k}$.

Example. $f_3f_1f_2|f_1v_1$ is a left monomial with respect to $f_1|f_2f_1f_3v_1$.

13.3.9. Theorem. *There exists an isotopy*

$$h_t : \pi^{-1}(X^\infty) \longrightarrow \mathbb{C}^\ell - \mathcal{C}_{1,\ell}(0)$$

such that

(i) h_t *commutes with the action of* Σ'.

(ii) $h_0 = \mathrm{id}$ *and* $h_1\pi^{-1}(X^\infty) = X$.

The map $h_1\pi^{-1}$ sends any cell of X^∞ into a union of suitable cells of X. Therefore, $h_1\pi^{-1}$ induces a map of complexes:

$$(h_1\pi^{-1})_* : C.(X^\infty, \mathcal{S}^*(a^1))^- \longrightarrow C.(X, \mathcal{S}^*(a))^-.$$

(iii) *This map induces a map*

$$\psi : C.(^+U_q\mathfrak{n}_-; M_\infty; {}^+\Delta)_\lambda \longrightarrow C.(^+U_q\mathfrak{n}_-; M_1; {}^+\Delta)_\lambda$$

such that the following diagram is commutative:

$$
\begin{array}{ccc}
C.(^+U_q\mathfrak{n}_-;M_\infty;{}^+\Delta)_\lambda & \xrightarrow{\psi} & C.(^+U_q\mathfrak{n}_-;M_1;{}^+\Delta)_\lambda \\
\downarrow & & \downarrow \\
C.(X^\infty,\mathcal{S}^*(a^1))^- & \xrightarrow{(h_1\pi^{-1})_*} & C.(X,\mathcal{S}^*(a))^-.
\end{array}
$$

(iv) The map ψ is given by the formula:

$$f_{J_1}|\ldots|f_{J_k}|f_{J_0}v_\infty \longmapsto (-1)^{\ell+k(k+1)/2}f_{\overline{J_k}}|\ldots|f_{\overline{J_1}}|f_{\overline{J_0}}v_1 + \sum c_x x,$$

where $f_{\overline{J_p}}$ for $p \in \{0,\ldots,k\}$ is the monomial f_{J_p} written in the inverse order. The sum is taken over all monomials $x \in C_k(^+U_q\mathfrak{n}_-;M_1;{}^+\Delta)_\lambda$ that are left monomials with respect to $f_{\overline{J_k}}|\ldots|f_{\overline{J_1}}|f_{\overline{J_0}}v_1$; c_x's are suitable coefficients. Each c_x is a polynomial with integer coefficients in $q^{\pm b_{ij}/4}$, $q^{\pm(\Lambda_1,\alpha_j)/2}$.

As an example, see Example (13.3.6). Formula (iv) shows that ψ is an isomorphism.

13.3.10. Corollary.

$$f_{J_0}v_\infty \longmapsto (-1)^\ell f_{\overline{J_0}}v_1,$$
$$f_{J_1}|\ldots|f_{J_k}|v_\infty \longmapsto (-1)^{\ell+k(k+1)/2}f_{\overline{J_k}}|\ldots|f_{\overline{J_1}}|v_1$$

for all monomials $f_{J_0}v_\infty \in C_0(^+U_q\mathfrak{n}_-;M_\infty;{}^+\Delta)_\lambda$ and $f_{J_1}|\ldots|f_{J_k}|v_\infty \in C_k(^+U_q\mathfrak{n}_-;M_\infty;{}^+\Delta)_\lambda$.

The proof of the theorem is essentially the same as the proof of Theorem (8.3.4). It is based on the bundle properties of the discriminantal configuration described in Sec. 6.

It is easy to give an explicit formula for the coefficients c_x in the formula for ψ. The formula for c_x is similar to the formula for the action of the R-matrix, see (7.4.4).

13.4. The Inversion Isomorphism for n Verma Modules

In this section, we will consider the inversion isomorphism corresponding to the rational map π_1 defined in (13.1). We compare the homology groups of the configuration $C_{n,\ell}(z)$ with the weights a given by (12.1.1) and the homology groups of the configuration $C_{n,\ell}(z^1)$ with the weights a^1 given by (13.3.2). Let

$$T_0^1 < \cdots < T_0^\ell < z_1 < \cdots < z_n < T_n^1 < \cdots < T_n^\ell \qquad (13.4.1)$$

be an equipment of $C_{n,\ell}(z)$. Let $C.(X, \mathcal{S}^*(a))$ be the chain complex constructed for $C_{n,\ell}(z)$ and $\{T_j^i\}$ in Subsecs. 2 and 3. Let $C.(X, \mathcal{S}^*(a))^-$ be its Σ'-skew-invariant part. Let $\varphi : C.(^+U_q\mathfrak{n}_-; M; {^+}\Delta)_\lambda \to C.(X, \mathcal{S}^*(a))^-$ be the realization described in (5.11). Let

$$x = f_{I_1}|\ldots|f_{I_k}|f_{J_1}v_1|\ldots|f_{J_n}v_n \qquad (13.4.2)$$

be a monomial in $C_k(^+U_q\mathfrak{n}_-; M_1; {^+}\Delta)_\lambda$. The image $\varphi(x)$ is an $(\ell - k)$ chain shown in Fig. 13.5.

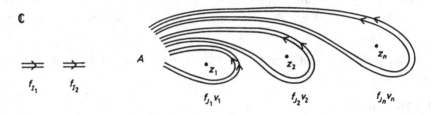

Fig. 13.5

Let

$$\bar{T}_0^1 < \cdots < \bar{T}_0^\ell < z_1^1 < \cdots < z_n^1 < \bar{T}_n^1 < \cdots < \bar{T}_n^\ell \qquad (13.4.3)$$

be the equipment of $C_{n,\ell}(z^1)$, where

$$\bar{T}_0^i = \left[T_0^{\ell-i+1} - z_1\right]^{-1} \quad \text{for all } i,$$
$$\bar{T}_j^i = \left[T_{n-j+1}^{\ell-i+1} - z_1\right]^{-1} \quad \text{for all } i \text{ and all } j > 0.$$

Recall that $z_1^1 = 0$ and $z_j^1 = (z_{n-j+2} - z_1)^{-1}$, see (13.2). The equipment $\{\bar{T}_j^i\}$ is the image of the equipment $\{T_j^i\}$ under π_1.

Let $C.(X^1, \mathcal{S}^*(a^1))$ be the chain complex constructed for $C_{n,\ell}(z^1)$ and $\{\bar{T}_j^i\}$ as in Secs. 2 and 3, $C.(X^1, \mathcal{S}^*(a^1))^-$ its Σ'-skew-invariant part. Let

$$\varphi^1 : C.(^+U_q\mathfrak{n}_-; M^1; {^+}\Delta)_\lambda \longrightarrow C.(X^1, \mathcal{S}^*(a^1))^-$$

be the realization, here $M^1 = M_\infty \otimes M_n \otimes \cdots \otimes M_2$, see (13.2). Let $y = f_{I_1}|\cdots|f_{I_k}|f_{J_\infty}v_\infty|f_{J_n}v_n|\cdots|f_{J_2}v_2$ be a monomial in $C_k(^+U_q\mathfrak{n}_-;M^1;{}^+\Delta)_\lambda$. The image $\varphi^1(y)$ is shown in Fig. 13.6.

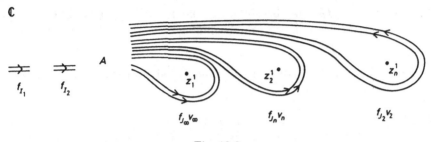

Fig. 13.6

The image $(\pi_1)^{-1}\varphi^1(y)$ of the chain $\varphi^1(y)$ under the biholomorphism $(\pi_1)^{-1}$ is shown in Fig. 13.7.

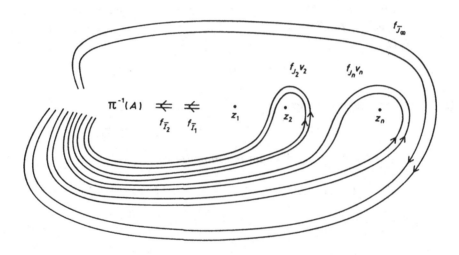

Fig. 13.7

To formulate an analog of Theorem (13.3.9), we introduce the corresponding definition.

Let $x' = f_{P_1}|\cdots|f_{P_k}|f_{Q_1}v_1|\cdots|f_{Q_n}v_n$ be a monomial in $C_k(^+U_q\mathfrak{n}_-;M;{}^+\Delta)_\lambda$. Let x be the monomial in (13.4.2). We say that x' is a left monomial with respect to x, if the following two conditions are satisfied:

(13.4.4) $f_{Q_1}v_1|\ldots|f_{Q_n}v_n$ may be obtained from $f_{J_1}v_1|\ldots|f_{J_n}v_n$ by deleting some symbols $\{f_{a_1},\ldots,f_{a_p}\}$.

(13.4.5) $f_{P_1}|\ldots|f_{P_k}$ may be obtained from $f_{I_1}|\ldots|f_{I_k}$ by adding the symbols $\{f_{a_1},\ldots,f_{a_p}\}$ to the beginning or to the end of the monomials $f_{I_1},\ldots,f_{I_k}$.

Example. $\hat{f}_1 f_4|f_3 f_2|v_1|v_2$ is a left monomial with respect to $f_1|f_2|f_3 v_1|f_4 v_2$.

13.4.6. Theorem. *There exists an isotopy*
$$h_t : (\pi_1)^{-1}(X^1) \longrightarrow \mathbb{C}^\ell - \mathcal{C}_{n,\ell}(z)$$
such that

(i) h_t *commutes with the action of* Σ'.

(ii) $h_0 = id$ *and* $h_1(\pi_1)^{-1}(X^1) = X$. *The map* $h_1(\pi_1)^{-1}$ *sends any cell of* X^1 *into a union of suitable cells of* X. *Therefore,* $h_1(\pi_1)^{-1}$ *induces a map of complexes:*
$$(h_1\pi^{-1})_* : C.(X^1, S^*(a^1))^- \longrightarrow C.(X, S^*(a))^-.$$

(iii) *This map induces a map*
$$\psi : C.(^+U_q\mathfrak{n}_-; M^1; {}^+\Delta)_\lambda \longrightarrow C.(^+U_q\mathfrak{n}_-; M; {}^+\Delta)_\lambda$$
such that the following diagram is commutative:

$$\begin{array}{ccc} C.(^+U_q\mathfrak{n}_-; M^1; {}^+\Delta)_\lambda & \xrightarrow{\psi} & C.(^+U_q\mathfrak{n}_-; M; {}^+\Delta)_\lambda \\ \varphi^1 \downarrow & & \downarrow \varphi \\ C.(X^1, S^*(a^1))^- & \xrightarrow{(h_1\pi^{-1})_*} & C.(X, S^*(a))^-. \end{array}$$

(iv) *The map* ψ *is given by the formula*
$$f_{I_1}|\ldots|f_{I_k}|f_{J_\infty}v_\infty|f_{J_n}v_n|\ldots|f_{J_2}v_2 \longmapsto cf_{\overline{I_k}}|\ldots|f_{\overline{I_1}}|y + \sum c_x x.$$
Here $y \in M_1 \otimes \cdots \otimes M_n$ is defined by
$$y = f_{\overline{J_\infty}}[(R^{n-1})(R^{n-2}R^{n-1})\ldots(R^k R^{k+1}\ldots R^{n-1})\ldots$$
$$\ldots (R^1 R^2 \ldots R^{n-1})f_{J_n}v_n|f_{J_{n-1}}v_{n-1}|\ldots|f_{J_2}v_2|v_1].$$

$f_{\overline{J}}$ is the monomial f_J written in the reverse order, R^p is the R-matrix operator introduced in (7.6.9), c is a nonzero constant, the sum in the formula for ψ is taken over all monomials x that are left monomials with respect to $f_{\overline{I_k}}|\ldots|f_{\overline{I_1}}|y$. The coefficient c is a product of suitable powers of $q^{\pm b_{ij}/4}$ and $q^{\pm(\Lambda_i,\alpha_j)/2}$, $q^{\pm(\Lambda_\infty,\alpha_j)/2}$ taken with a coefficient 1 or -1. Each coefficient c_x is a polynomial with integer coefficients in $q^{\pm b_{ij}/4}$ and $q^{\pm(\Lambda_i,\alpha_j)/2}$, $q^{\pm(\Lambda_\infty,\alpha_j)/2}$.

Formula (iv) shows that ψ is an isomorphism.

Sketch of the Proof. The monomial $u = f_{I_k}|\ldots|f_{I_1}|f_{J_n}v_n|f_{J_{n-1}}v_{n-1}|\ldots|f_{J_2}v_2|v_1$ is shown in Fig. 13.8.

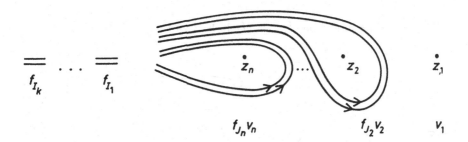

Fig. 13.8

The operator $R^1 R^2 \ldots R^{n-1}$ transforms this monomial into an element of $M_1 \otimes M_n \otimes M_{n-1} \otimes \cdots \otimes M_2$ shown in Fig. 13.9, see Theorem (8.3.4).

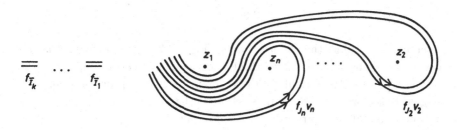

Fig. 13.9

The operator $R^{n-1}(R^{n-2}R^{n-1})\ldots(R^1R^2\ldots R^{n-1})$ transforms u into an element of $M_1 \otimes M_2 \otimes \cdots \otimes M_n$ shown in Fig. 13.10.

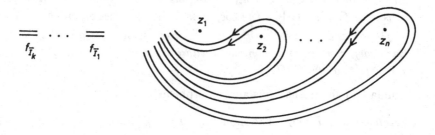

Fig. 13.10

Therefore, the monomial $f_{\overline{I_k}}|\ldots|f_{\overline{I_1}}|y$ is shown in Fig. 13.11.

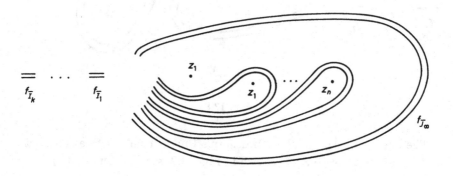

Fig. 13.11

One can show that the chain shown in Fig. 13.7 can be deformed into a union of the chain shown in Fig. 13.11 and the chains corresponding to the monomials listed in the formula of the theorem. A careful proof is completely analogous to the proof of Theorem (8.3.4) and is based on the bundle properties of a discriminantal configuration described in Sec. 6.

13.4.7. Corollary of (13.4.6). *Let $u = \sum c_I f_I$ be an element of $^+U_q\mathfrak{n}_-$. Let $\bar{u} = \sum c_I f_{\bar{I}} \in {}^+U_q\mathfrak{n}_-$ be the element obtained from u by reversing order in all monomials f_I composing u. Let x be an element of $C_0(^+U_q\mathfrak{n}_-; M^1; {}^+\Delta)_\lambda =$*

$(M_\infty \otimes M_n \otimes \cdots \otimes M_2)_\lambda$, which is a linear combination of the elements of the form

(i) $$a_\infty u v_\infty | a_n v_n | \ldots | a_2 v_2 .$$

Then $\psi(x)$ belongs to the image of the operator $M_1 \otimes \cdots \otimes M_n \xrightarrow{\bar{u}} M_1 \otimes \cdots \otimes M_n$ of the multiplication by $\bar{u}$.

13.5. Generic Sets

Under the assumptions of (13.2), let $\Lambda \in \mathfrak{h}^*$, $\lambda = (\ell_1, \ldots, \ell_r) \in \mathbf{N}^r$, and $\kappa \in \mathbb{C}^*$.

The pair Λ, κ will be called generic with respect to λ if for any $\mu = (m_1, \ldots, m_r) \in \mathbf{N}^r$, with $m_j \leq \ell_j$ for all j, the number

$$a(\mu) = -\sum_{i=1}^r m_i(\alpha_i, \Lambda)/\kappa + \sum_{i=1}^r m_i(m_i - 1)(\alpha_i, \alpha_i)/2\kappa + \sum_{i<j} m_i m_j (\alpha_i, \alpha_j)/\kappa$$

does not belong to the set $\{0, -1, -2, \ldots\}$.

Let $\Lambda_1, \ldots, \Lambda_n \in \mathfrak{h}^*$, $\lambda \in \mathbf{N}^r$, $\kappa \in \mathbb{C}^*$. Consider a discriminantal configuration $\mathcal{C}_{n,\ell}(z)$ with weights a given by (12.1.1). Given $p \in \{1, \ldots, n\}$, let L be any nontrivial edge of $\mathcal{C}_{n,\ell}(z)$ lying in the union of hyperplanes $x_i(j) = z_p$ for all i, j.

13.5.1. Lemma. *If Λ_p, κ is a generic pair with respect to λ, then the weight of L does not belong to the set $\{0, -1, -2, \ldots\}$.*

Proof. It is easy to see that any nontrivial edge L lying in the union of these hyperplanes has the following form. Fix a vector $\mu = (m_1, \ldots, m_r) \in \mathbf{N}^r$ with $m_j \leq \ell_j$ for all j. For any j, fix a subset $I_j \subset \{1, \ldots, \ell_j\}$ such that $|I_j| = m_j$. Then $L = \{x \in \mathbb{C}^\ell | x_i(j) = z_p$ for $j = 1, \ldots, r$ and $i \in I_j\}$. The weight of such L is the number $a(\mu)$.

Now we will formulate the main theorem of this section.

(13.5.2) We say that $\Lambda_1, \ldots, \Lambda_n \in \mathfrak{h}^*$, $\lambda \in \mathbf{N}^r$, and $\kappa \in \mathbb{C}^*$ form a generic set if the following two conditions are fulfilled:

(i) Consider a configuration $C_{n,\ell}(z)$ with weights defined by (12.1.1). Then this set of weights defines a point in D_{-1}, where the domain $D_{-1}(C_{n,\ell})$ is defined in (10.7).

(ii) The weight Λ_∞, complementary to $\Lambda_1, \ldots, \Lambda_n$, λ, and the number κ form a generic pair with respect to λ.

For any $p = 1, \ldots, n$, we have the canonical isomorphism $H.(\mathbb{C}^\ell - C_{n,\ell}(z), \mathcal{S}^*(a))^- \cong H.(\mathbb{C}^\ell - C_{n,\ell}(z^p), \mathcal{S}^*(a^p))^-$ induced by the map π_p, see notations in (13.2). Let

$$S : H.(\mathbb{C}^\ell - C_{n,\ell}(z^p), \mathcal{S}^*(a^p))^- \longrightarrow H.(\mathbb{C}^\ell, C_{n,\ell}(z^p), \mathcal{S}^*(a^p))^-$$

be the natural homomorphism. Let

$$S_p : H.(\mathbb{C}^\ell - C_{n,\ell}(z), \mathcal{S}^*(a))^- \longrightarrow H.(\mathbb{C}^\ell, C_{n,\ell}(z^p), \mathcal{S}^*(a^p))^-$$

be the composition of these homomorphisms.

13.5.3. Theorem. *Let $\Lambda_1, \ldots, \Lambda_n$, λ, and κ form a generic set. Let $x \in H_\ell(\mathbb{C}^\ell - C_{n,\ell}(z), \mathcal{S}^*(x))^-$ belong to $\ker S_p$ for some p. Then for any flag hypergeometric form $\omega \in F\ell^\ell(C_{n,\ell}(z), a)$ of the configuration $C_{n,\ell}(z)$ with weights a, we have*

$$\int_x \omega = 0.$$

The theorem is proved in (13.6).

Theorem (13.5.3) produces elements of the kernel of the hypergeometric pairing. In Sec. 14, we give other statements producing elements of the kernel, see (14.4).

13.6. Transformations of Flag Forms

For any p, define a subconfiguration C^p in $C_{n,\ell}(z^p)$ as follows. We have $z^p = (z_1^p, \ldots, z_n^p)$, where $z_p^p = 0$, see (13.2). Hence $C_{n,\ell}(z^p)$ contains the set of all coordinate hyperplanes $x_i(j) = 0$ for all i, j. Let C^p be the configuration of all noncoordinate hyperplanes of $C_{n,\ell}(z^p)$.

Let $\omega \in \mathcal{A}^\cdot(C_{n,\ell}(z), a)$ be a hypergeometric form of the configuration $C_{n,\ell}(z)$ with weights a. Then the form $\pi_p^* \omega$ is a hypergeometric form of the configuration $C_{n,\ell}(z^p)$ with weights a^p.

13.6.1. Theorem. *For any flag form $\omega \in F\ell^\cdot(C_{n,\ell}(z), a)$, the form $\pi_p^* \omega$ is a* linear combination of flag hypergeometric forms of the pair $\mathcal{C}^p \subset \mathcal{C}_{n,\ell}(z^p)$ with weights a^p, see the definition in (10.7).

We will prove the theorem for an example. The general case is similar.

Let $n = 2$, $p = 2$, $\lambda = (1,1,0,\ldots,0) \in \mathbf{N}^r$. Let $\omega = a_1^1 d\ln(x_1 - z_1) \wedge (a_2^1 d\ln(x_2 - z_1) + a_{12} d\ln(x_2 - x_1))$ be a flag form. Here a_j^i (resp. a_{ij}) is the weight of a hyperplane $x_j = z_i$ (resp. $x_i = x_j$). ω is the flag form of the flag $\{x_1 = z_1\} \supset \{x_1 = x_2 = z_1\}$. Then

$$\begin{aligned}
\pi_2^* w &= a_1^1(d\ln(1 + (z_2 - z_1)u_1) - d\ln u_1) \wedge (a_2^1 d\ln(1 + (z_2 - z_1)u_2) \\
&\quad + a_{12} d\ln(u_1 - u_2) - (a_2^1 + a_{12}) d\ln u_2) \\
&= a_1^1 d\ln(1 + (z_2 - z_1)u_1) \wedge (a_2^1 d\ln(1 + (z_2 - z_1)u_2) + a_{12} d\ln(u_1 - u_2)) \\
&\quad - a_1^1 d\ln u_1 \wedge (-(a_2^1 + a_{12}) d\ln u_2 + a_{12} d\ln(u_1 - u_2)) \\
&\quad - a_1^1 d\ln u_1 \wedge a_2^1 d\ln(1 + (z_2 - z_1)u_2) \\
&\quad - a_1^1 d\ln(1 + (z_2 - z_1)u_1) \wedge (a_2^1 + a_{12}) d\ln u_2.
\end{aligned}$$

Each of the four summands on the right-hand side is a flag form of the pair $\mathcal{C}^2 \subset \mathcal{C}_{2,2}(z^2)$.

Note that the second, third, and fourth summands are not the flag hypergeometric forms of $\mathcal{C}_{2,2}(z^2)$ with weights a^2, since the coefficients of $d\ln u_1$ and $d\ln u_2$ in these summands are not the weights of the hyperplanes $u_1 = 0$ and $u_2 = 0$.

13.6.2. Proof of Theorem (13.5.3). Consider the complexes X and Q'/Q'' constructed for $\mathcal{C}_{n,\ell}(z^p)$ in Secs. 2 and 3. The homology class x can be represented by a chain $[x] \in C_\ell(X, \mathcal{S}^*(a^p))$. By assumptions, $[x] \in \ker S$, where

$$S : C_\ell(X, \mathcal{S}^*(a^p)) \longrightarrow C_\ell(Q'/Q'', \mathcal{S}^*(a^p))$$

is the homomorphism of the projection on the real part. By Theorem (13.6.1), $\pi_p^*\omega$ is a linear combination of flag forms of the pair $\mathcal{C}^p \subset \mathcal{C}_{n,\ell}(z^p)$. By Theorem (10.7.15),

$$\int_{[x]} \pi_p^*\omega = \int_{S[x]} \pi_p^*\omega = 0.$$

13.6.3. Remark. Theorem (13.5.3) has a generalization to homology classes of dimension less than ℓ. Namely, let $x \in H_k(\mathbb{C}^\ell - \mathcal{C}_{n,\ell}(z), \mathcal{S}^*(a))$ for some k.

Assume that x has a representative $[x] \in C_k(X, \mathcal{S}^*(a^p))$ such that $S[x] = 0$. Then
$$\int_x \omega = 0$$
for any flag form $\omega \in F\ell^k(\mathcal{C}_{n,\ell}(z), a)$. The proof is the same.

13.7. The Kernel of the Hypergeometric Pairing for $s\ell_2$

Let $\mathfrak{g} = s\ell_2$, $\alpha = \alpha_1$, $f = f_1$, $e = e_1$, $(\alpha, \alpha) = 2$, cf. (12.1.9), (12.4.6), and (7.6.4). Set
$$[e^t]_q = \frac{(1)_q^t}{(t)_q!} e^t \quad \text{for } t \in \mathbb{N}.$$

For $\Lambda \in \mathfrak{h}^*$, let $M(\Lambda)$ (resp. $L(\Lambda)$) be the Verma module over $s\ell_2$ (resp., irreducible module) with highest weight Λ. Let $M(\Lambda, q)$ (resp., $L(\Lambda, q)$) be the Verma module over $U_q s\ell_2$ (resp., irreducible module) with highest weight Λ. For $\Lambda_1, \ldots, \Lambda_n \in \mathfrak{h}^*$ and $k \in \mathbb{C}$, set $M(q) = M(\Lambda_1, q) \otimes \cdots \otimes M(\Lambda_n, q)$, $L(q) = L(\Lambda_1, q) \otimes \cdots \otimes L(\Lambda_n, q)$, $\operatorname{Vac} M(q)_k = \{x \in M(q) | hx = kx, [e^t]_q x = 0$ for all $t\}$, $M = M(\Lambda_1) \otimes \cdots \otimes M(\Lambda_n)$, $L = L(\Lambda_1) \otimes \cdots \otimes L(\Lambda_n)$.

(13.7.1) Assume that κ is a natural number, and
$$\Lambda_1 + \cdots + \Lambda_n = \frac{m}{2}\alpha \quad \text{for } m \in \mathbb{N}.$$

(13.7.2) Assume that $k, \ell \in \mathbb{N}$ are numbers such that $k \leq \ell$, and $m - \ell + k + 1$ is divisible by κ.

13.7.3. Lemma. *Under assumptions (13.7.1) and (13.7.2),*
$$f^{\ell-k}(\operatorname{Vac} M(q)_{m-2k}) \subset \operatorname{Vac} M(q)_{m-2\ell}.$$

Proof. First assume that $n = 1$ and $k = 0$. Then the statement is that $f^{\ell+1} v_1$ is a singular vector, if $(\Lambda_1, \alpha) = m$, and $m - \ell + 1$ is divisible by κ. In fact,
$$[e^t]_q f^\ell v_1 = \frac{(\ell)_q (\ell-1)_q \cdots (\ell+1-t)_q}{(1)_q (2)_q \cdots (t)_q} (m-\ell+1)_q (m-\ell+2)_q$$
$$\cdots (m-\ell+t)_q f^{\ell+1-t} v_1,$$

and this is zero for all t by the assumptions.

Now, for arbitrary n, the lemma follows from the following formula. For $m \in \mathbb{C}$, $\ell, k \in \mathbb{N}$, and $x \in M(q)$, such that $hx = mx$, we have

$$[e^\ell]_q f^k x = \sum_{p=0}^{\ell} [k_q(k-1)_q \ldots (k-p+1)_q (m-k+\ell)_q (m-k+\ell-1)_q \quad (13.7.4)$$

$$\ldots (m-k+\ell-p+1)_q/(p)_q!] f^{k-p} [e^{\ell-p}]_q x.$$

This formula is easily proved by induction.

For any $\ell \in \mathbb{N}$, consider a discriminantal configuration $\mathcal{C}_{n,\ell}(z)$ with $z = (z_1, \ldots, z_n)$, $z_1 < \cdots < z_n$, and weights a given by

$$a(x_i, x_j) = 2/\kappa,$$
$$a(x_i, z_s) = -m_s/\kappa, \quad (13.7.5)$$

for $i, j \in (1, \ldots, \ell)$ and $s \in \{1, \ldots, n\}$. The group Σ_ℓ acts on $\mathcal{C}_{n,\ell}(z)$. Denote the Σ_ℓ-skew-invariant subspace by the superscript '−'.

Let X and Q'/Q'' be the complexes constructed for $\mathcal{C}_{n,\ell}(z)$ in Secs. 2 and 3. Let

$$C.(^+U_q\mathfrak{n}_-; M(q); {}^+\Delta)_\lambda \hookrightarrow C.(X, \mathcal{S}^*(a))^-,$$
$$C.^*(^+U_q\mathfrak{n}_-; M(q); \mu)_\lambda \hookrightarrow C.(Q'/Q'', \mathcal{S}^*(a))^-, \quad (13.7.6)$$

be the quasiisomorphisms constructed in (5.11), here $\lambda = \ell \in \mathbb{N}$. Let

$$C.(\mathfrak{n}_-^*; M^*)_\lambda \simeq \mathcal{A}^\cdot(\mathcal{C}_{n,\ell}(z), a)^-,$$
$$C.(\mathfrak{n}_-; L)_\lambda \simeq F\ell^\cdot(\mathcal{C}_{n,\ell}(z), a)^-, \quad (13.7.7)$$

be the isomorphisms constructed in (12.1), here $\mathcal{A}^\cdot$ and $F\ell^\cdot$ are the complexes of hypergeometric forms and flag hypergeometric forms. Let

$$I = I(z, \{T_j^i\}) : C.(\mathfrak{n}_-^*, M^*)_\lambda \otimes C.(^+U_q\mathfrak{n}_-; M(q); {}^+\Delta)_\lambda \longrightarrow \mathbb{C}$$

be the hypergeometric pairing, see (12.2).

13.7.8. Theorem. *Assume (13.7.1) and (13.7.2). Assume that $m - 2\ell + 2 > 0$. Then, for any $x \in f^{\ell-k}(\text{Vac } M(q)_{m-2k+2})$, we have*

$$I : C_0(\mathfrak{n}_-^*; M^*)_\lambda \otimes x \longrightarrow 0.$$

Proof. Denote by Λ_∞ the weight, complementary to $\Lambda_1, \ldots, \Lambda_n$ at level ℓ. Then $\Lambda_\infty = -\Lambda_1 - \cdots - \Lambda_n + (\ell-1)\alpha$, and $-(\Lambda_\infty, \alpha) > 0$, by assumptions. Let v_∞ be the generating vector of $M(\Lambda_\infty, q)$, then

$$f^{\ell-k} v_\infty \in \operatorname{Vac} M(\Lambda_\infty, q)_{-m+2k-2}.$$

For the configuration $C_{n,\ell}(z)$, consider the configuration $C_{n,\ell}(z^1)$ with the weights a^1 defined in (13.2). Consider the hypergeometric pairing for this configuration:

$$I^1 : \big(M(\Lambda_\infty) \otimes M(\Lambda_n) \otimes \cdots \otimes M(\Lambda_2)\big)^*_\lambda \otimes \big(M(\Lambda_\infty, q) \otimes M(\Lambda_n, q) \\ \otimes \cdots \otimes M(\Lambda_2, q)\big)_\lambda \longrightarrow \mathbb{C}. \qquad (13.7.9)$$

Let $y \in \big(M(\Lambda_\infty, q) \otimes \ldots M(\Lambda_2, q)\big)_\lambda$ be any element of the form

$$f^p v_\infty \otimes f^{t_n} v_n \otimes \cdots \otimes f^{t_2} v_2, \qquad (13.7.10)$$

where $p \geq \ell - k$, then we have

$$I^1 : \big(M(\Lambda_\infty) \otimes \cdots \otimes M(\Lambda_2)\big)_\lambda \otimes y \longrightarrow 0 \qquad (13.7.11)$$

by (12.2.22).

Let $\psi : \big(M(\Lambda_\infty, q) \otimes M(\Lambda_n, q) \otimes \cdots \otimes M(\Lambda_2, q)\big)_\lambda \to \big(M(\Lambda_1, q) \otimes \cdots \otimes M(\Lambda_n, q)\big)_\lambda$ be the isomorphism described in Theorem (13.4.6). Let $x \in f^{\ell-k}(\operatorname{Vac} M(q)_{m-2k+2})$. By (13.4.6), $\psi^{-1}(x)$ is a sum of monomials of the form (13.7.10). Now (13.7.11) implies the theorem.

(13.7.12) Fix a positive integer κ. Let $\Lambda_1, \ldots, \Lambda_n \in \mathfrak{h}^*$ be weights such that, for any i, the number $m_i = (\Lambda_i, \alpha)$ is a non-negative integer and $m_i \leq \kappa - 2$.

This means that $\Lambda_1, \ldots, \Lambda_n$ are finite weights with respect to the number $k = \kappa - 2$ in the sense of (12.4).

Set $m = m_1 + \cdots + m_n$ as before.

(13.7.13) Fix $\ell \in \mathbb{N}$ such that $\kappa - 2 \geq m - 2\ell \geq 0$. Let

$$I(z, \{T^i_j\}) : L_\lambda \otimes M(q)_\lambda \longrightarrow \mathbb{C} \qquad (13.7.14)$$

be the hypergeometric pairing for $C_{n,\ell}(z)$. Here $\lambda = \ell$, and L_λ is embedded in M^*_λ by the contravariant form.

Set $p = \kappa - 1 - m + 2\ell$,

$$U(z)_\lambda = \left(L/\left(\operatorname{Im} f + \operatorname{Im}(zf)^p\right)\right)_\lambda, \qquad (13.7.15)$$

here $\operatorname{Im} f$ (resp., $\operatorname{Im}(zf)^p$) is the image of the multiplication by f (resp., of the operator $(zf)^p$ defined in (12.4)).

By (12.2.6) and (12.4.5), the hypergeometric pairing induces a pairing

$$I(z) : U(z)_\lambda \otimes \operatorname{Vac} M(q)_{m-2\ell} \longrightarrow \mathbb{C}. \qquad (13.7.16)$$

By (13.7.3), we have $f^p(\operatorname{Vac} M(q)_{m+2p-2\ell}) \subset \operatorname{Vac} M(q)_{m-2\ell}$. Set

$$\begin{aligned} F(M(q))_{m-2\ell} = {} & \operatorname{Vac} M(q)_{m-2\ell}/(f^p(\operatorname{Vac} M(q)_{m+2p-2\ell}) \\ & + \operatorname{Vac} M(q)_{m-2\ell} \cap \ker S)\,, \end{aligned} \qquad (13.7.17)$$

where $S : M(q) \to M(q)^*$ is the contravariant form, and $\lambda = \ell$ as before.

13.7.18. Lemma. *Pairing (13.7.16) induces a pairing*

$$I(z) : U(z)_\lambda \otimes F_{m-2\ell} \longrightarrow \mathbb{C}.$$

The lemma follows from (12.2.18) and (13.7.8).

The contravariant form induces an isomorphism

$$S : U(z)_\lambda^* \simeq W(z)_\lambda\,,$$

where $W_\lambda(z) = \{x \in L_\lambda | ex = 0, (ze)^p x = 0\}$, see (12.4). Hence the pairing (13.7.18) induces a map,

$$I(z) : F_{m-2\ell} \longrightarrow W(z)_\lambda\,,$$

where the first space depends on q and the second depends on z, as explained above. For a fixed $x \in F_{m-2\ell}$, the map $z \mapsto I(z)x$ is a solution of the KZ equation with values in $W(z)_\lambda \subset \operatorname{Vac} L_\lambda$, see (12.3.25) and (12.3.26), and the map $I(z)$ sends the R-matrix representation on $F_{m-2\ell}$ into the monodromy representation of the KZ equation on $W(z)_\lambda$.

13.7.19. Theorem. *The pairing (13.7.18) is nondegenerate.*

The theorem is proved in Sec. 14, see (14.5).

At the end of this section, we describe $F_{m-2\ell}$.

13.7.20. Lemma. *For $\Lambda \in \mathfrak{h}^*$ and $t \in \mathbf{N}$ such that $t \geq \kappa$ and $(\Lambda, \alpha)/2 \in \mathbf{Z}$, the operator $[e^t]_q : M(\Lambda, q) \to M(\Lambda, q)$ is identical zero.*

The lemma is proved by direct verification.

13.7.21. Corollary. *The kernel of the contravariant form on $M(\Lambda, q)$ is invariant under the action of operators $[e^t]_q$ for $t \in \mathbf{N}$, and operators $[e^t]_q$ for $t \in \mathbf{N}$ are well defined on the irreducible module $L(\Lambda, q)$.*

Under assumptions (13.7.12) and (13.7.13), set

$$\operatorname{Vac} L(q)_t = \{ x \in L(q) \mid hx = tx, ex = [e^\kappa]_q x = 0 \}. \qquad (13.7.22)$$

13.7.23. Lemma. *The contravariant form induces an epimorphism*

$$S : \operatorname{Vac} M(q)_t \longrightarrow \operatorname{Vac} L(q)_t$$

for all t.

Proof. For all j, the space $K_j = \ker(S : M(\Lambda_j, q) \to M(\Lambda_j, q))$ is generated by vectors $f^k v_j, f^{k+1} v_j, \ldots$, where k is the smallest natural number with the property $e f^k v_j = 0$. Set $J_j = \operatorname{span}\{v_j, fv_j, \ldots, f^{k-1} v_j\} \subset M(\Lambda_j, q)$. Then $M(\Lambda_j, q) = J_j \oplus K_j$. These decompositions induce a decomposition

$$M(q) = \ker S \oplus (J_1 \otimes \cdots \otimes J_n).$$

The contravariant form sends isomorphically $J_1 \otimes \cdots \otimes J_n$ onto $L(q)$. Now let $[x] \in \operatorname{Vac} L(q)_t$, $x \in J_1 \otimes \cdots \otimes J_n$, and $Sx = [x]$. Then $ex \in J_1 \otimes \cdots \otimes J_n$, $[e^\kappa]_q x \in J_1 \otimes \cdots \otimes J_n$, $Sex = 0$, $S[e^\kappa]_q x = 0$. Hence $x \in \operatorname{Vac} M(q)_t$, and the lemma is proved.

13.7.24. Corollary.

$$F_{m-2\ell} = \operatorname{Vac} L(q)_{m-2\ell} / f^p (\operatorname{Vac} L(q)_{m+2p-2\ell}),$$

where $\operatorname{Vac}(\ )$ is defined in (13.7.22).

13.7.25. Lemma [RT]. *Let $m_1, m_2 \in \mathbf{N}$ and $0 \leq m_1, m_2 \leq \kappa - 2$. Then*

$$\dim \operatorname{Vac} \left(L\!\left(\frac{m_1}{2}\alpha, q\right) \otimes L\!\left(\frac{m_2}{2}\alpha, q\right) \right)_t$$

is 1, if $m_1 + m_2 = t \mod 2$, and $|m_1 - m_2| \leq t \leq m_1 + m_2$, and is 0, otherwise.

13.7.26. Lemma. Let $m_1, m_2, t \in \mathbb{N}$ and $0 \leq m_1, m_2, t \leq \kappa - 2$. Then

$$\dim F\left(L\left(\frac{m_1}{2}\alpha, q\right) \otimes L\left(\frac{m_2}{2}\alpha, q\right)\right)_t$$

is 1, if

(i) $m_1 + m_2 = t \mod 2$,

(ii) $|m_1 - m_2| \leq t \leq \min(m_1 + m_2, 2\kappa - m_1 - m_2 - 4)$, and is 0, otherwise.
If $\dim F = 1$, then there is a nonzero homomorphism

$$C^t_{m_1, m_2} : L\left(\frac{t}{2}\alpha, q\right) \to L\left(\frac{m_1}{2}\alpha, q\right) \otimes L\left(\frac{m_2}{2}\alpha, q\right).$$

The lemma is a direct corollary of Sec. 8.4 in [RT].

Let $\kappa, \ell \in \mathbb{N}$ and $\Lambda_1, \ldots, \Lambda_n \in \mathfrak{h}^*$ be as in (13.7.12) and (13.7.13). Let $n > 2$. Set $s = m - 2\ell$. Define the path space $P^s_{m_1,\ldots,m_n}$ as the space of complex linear combinations of sequences $(p_1, \ldots, p_{n-2})$ of integers in $\cdot\{0, 1, \ldots, \kappa - 2\}$ such that all triples $(m_{n-1}, m_n; p_{n-2})$, $(m_1, p_1; s)$, $(m_i, p_i; p_{i-1})$, for $i = 2, \ldots, n-2$, satisfy conditions (i) and (ii) above.

13.7.27. Theorem. The homomorphism

$$P^s_{m_1,\ldots,m_n} \longrightarrow \operatorname{Vac} L(q)_s,$$

$$(p_1, \ldots, p_{n-2}) \longmapsto (1 \otimes \cdots \otimes 1 \otimes C^{p_{n-2}}_{m_{n-1}, m_n}) \ldots (1 \otimes C^{p_1}_{m_2, p_2}) C^s_{m_1, p_1} v_s,$$

where $v_s \in L(\frac{s}{2}\alpha, q)$ is the generating singular vector, composed with the canonical projection $\operatorname{Vac} L(q)_s \to F(L(q))_s$, gives an isomorphism

$$P^s_{m_1,\ldots,m_n} \longrightarrow F(L(q))_s.$$

The theorem is proved in (13.8.26).

13.8. Remarks on the Representation Theory of the Quantum Double of the Subalgebra $U_q\mathfrak{b}_- \subset U_q s\ell_2$

Throughout this section, we assume that κ is a fixed natural number and $q = \exp(2\pi i/\kappa)$.

For any $a \in \mathbb{C}$, set $L(a) := L(\frac{a}{2}\alpha, q)$.

Denote by $\mathcal{A}$ the algebra generated by f, h, $[e^t]_q$ for $t \in \mathbb{N}$ with relations:

$$\begin{aligned}
[e, f] &= q^{h/2} - q^{-h/2}, \\
[[e^t]_q, f] &= [e^{t-1}]_q [q^{(t-1)/2} q^{h/2} - q^{-(t-1)/2} q^{-h/2}], \\
[h, f] &= -2f, \\
[h, e] &= 2e, \\
[h, [e^t]_q] &= 2t[e^t]_q, \\
[e^1]_q &= e, \\
[e^a]_q [e^b]_q &= \begin{bmatrix} a+b \\ a \end{bmatrix}_q [e^{a+b}]_q,
\end{aligned} \qquad (13.8.1)$$

for $t, a, b \in \mathbb{N}$. As $\kappa \in \mathbb{N}$, this algebra is generated by f, h, e, $[e^\kappa]_q$.

$\mathcal{A}$ is a Hopf algebra with the comultiplication

$$\begin{aligned}
\Delta h &= h \otimes 1 + 1 \otimes h, \\
\Delta f &= f \otimes q^{h/2} + q^{-h/2} \otimes f, \\
\Delta [e^t]_q &= \sum_{a=0}^{t} [e^a]_q q^{(-t+a)h/4} \otimes [e^{t-a}]_q q^{ah/4}.
\end{aligned} \qquad (13.8.2)$$

There is a natural homomorphism $U_q s\ell_2 \to \mathcal{A}$ sending f, h, e to f, h, e, resp.

Let $U_q \mathfrak{b}_-$ be the subalgebra of $U_q s\ell_2$ generated by f, h. Let $\mathfrak{D}(U_q \mathfrak{b}_-)$ be its quantum double. By Theorem (9.3.3), there is a natural epimorphism $\mathfrak{D}(U_q \mathfrak{b}_-) \to \mathcal{A}$.

Let V be a representation of $\mathcal{A}$. A vector $x \in V$ will be called $\mathcal{A}$-*singular* (resp., *singular*) if $[e^t]_q x = 0$ for all $t \in \mathbb{N}$ (resp., $ex = 0$).

As $[e^t]_q$ and e commute, $[e^t]_q$ acts on the space of singular vectors.

(13.8.3) By (13.7.21), the algebra $\mathcal{A}$ (and hence the quantum double $\mathfrak{D}(U_q \mathfrak{b}_-)$) acts on the tensor products of irreducible modules $\{L(a)\}$ with integral weights $a \in \mathbb{Z}$.

In this section, we consider a tensor product of the form

$$L = L(a_1) \otimes \cdots \otimes L(a_n), \qquad (13.8.4)$$

where $a_j \in \mathbb{N}$, $0 \leq a_j \leq \kappa - 2$ for all j, and study its $\mathcal{A}$-singular vectors.

Tensor product (13.8.4) as an $U_q s\ell_2$-representation was studied in [**RT**]. We formulate necessary results from [**RT**].

It is convenient to use a graphical representation for the structure of $U_q s\ell_2$-modules. We represent vectors from the module by vertices ordered vertically according to the values of the weights. Arrows pointing down show the action of f. Arrows pointing up show the action of e. The absence of arrows coming out of a vertex means that the corresponding vector is annihilated by one of the generators e or f, see [**RT**].

For any $a \in \mathbb{Z}$, $a \neq \kappa - 1 \bmod \kappa$, we define below a 2κ-dimensional $U_q s\ell_2$-module $X(a)$. The structure of this module is given in Fig. 13.12.

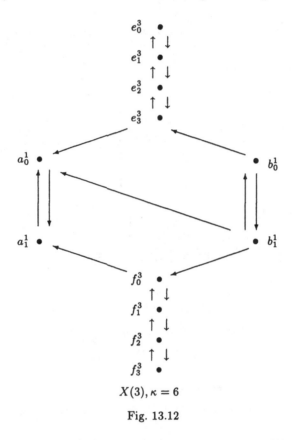

$X(3), \kappa = 6$

Fig. 13.12

Let $a = j \bmod \kappa$ and $j \in \{0, 1, \ldots, \kappa - 2\}$. Let e_n^j for $n = 0, \ldots, j$ and a_n^j, b_n^j for $n = 0, 1, \ldots, \kappa - 2 - j$ and f_n^j for $n = 0, \ldots, j$ be the elements of a weight basis of $X(a)$, see Fig. 13.12. The action of the generators f, h, e of $U_q s\ell_2$ has the following form in this basis:

$$\left.\begin{aligned} he_n^j &= (a - 2n)e_n^j, \\ ee_n^j &= \frac{(n)_q}{(1)_q}(a - n + 1)_q e_{n-1}^j, \\ fe_n^j &= e_{n+1}^j \end{aligned}\right\} n = 0, \ldots, j$$

$$\left.\begin{aligned} ha_n^j &= (a - 2j - 2n - 2)a_n^j, \\ ea_n^j &= \frac{(n+j+1)_q}{(1)_q}(a - j - n)_q a_{n-1}^j, \\ fa_n^j &= a_{n+1}^j \end{aligned}\right\} n = 0, \ldots, \kappa - 2 - j$$

$$\left.\begin{aligned} hb_n^j &= (a - 2j - 2n - 2)b_n^j, \\ eb_n^j &= \frac{(n+j+1)_q}{(1)_q}(a - j - n)_q b_{n-1}^j + a_{n-1}^j, \\ fb_n^j &= b_{n+1}^j \end{aligned}\right\} n = 0, \ldots, \kappa - 2 - j \quad (13.8.5)$$

$$\left.\begin{aligned} hf_n^j &= (a - 2\kappa - 2n)f_n^j, \\ ef_n^j &= \frac{(n)_q}{(1)_q}(a - n + 1)_q f_{n-1}^j + \delta_{n,0} a_{\kappa-2-j}^j, \\ ff_n^j &= f_{n+1}^j \end{aligned}\right\} n = 0, \ldots, j.$$

Here $e_{j+1}^j = a_0^j$, $a_{-1}^j = e_j^j$, $b_{\kappa-1-j}^j = f_0^j$, and $a_{\kappa-1-j}^j = 0$.

Define the action of $[e^\kappa]_q$ on $X(a)$ by the formula

$$\begin{aligned}[e^\kappa]_q f_n^j &= c e_n^j, \\ [e^\kappa]_q a_\ell^j &= [e^\kappa]_q b_m^j = [e^\kappa]_q e_n^j = 0 \end{aligned} \quad (13.8.6)$$

for all ℓ, m, n, where c is the coefficient of the proportionality in the formula

$$[e^{\kappa-1}]_q \left[q^{-1/2} q^{h/2} - q^{1/2} q^{-h/2}\right] f_j^j = c a_0^j.$$

Note that c is not zero, see Fig. 13.12.

13.8.7. Lemma. *Formulas (13.8.5) and (13.8.6) define an $\mathcal{A}$-module structure on $X(a)$. The action of $[e^\kappa]_q$ on $X(a)$ is uniquely determined by (13.8.5).*

The lemma follows from the equality

$$[[e^\kappa]_q, f] = [e^{\kappa-1}]_q [q^{-1/2} q^{h/2} - q^{1/2} q^{-h/2}].$$

13.8.8. Lemma [RT]. *Let $0 \leq a_1, a_2 \leq \kappa - 2$. Then*

1. *If $a_1 + a_2 \leq \kappa - 2$,*

$$L(a_1) \otimes L(a_2) = \bigoplus_{\substack{|a_1-a_2| \leq a \leq a_1+a_2 \\ a = a_1+a_2 \bmod 2}} L(a)$$

as $U_q s\ell_2$-modules.

2. *If $a_1 + a_2 > \kappa - 2$,*

$$L(a_1) \otimes L(a_2) = \left[\bigoplus_{\substack{|a_1-a_2| \leq a \leq 2\kappa-4-a_1-a_2 \\ a = a_1+a_2 \bmod 2}} L(a) \right] \oplus \left[\bigoplus_{\substack{\kappa \leq a \leq a_1+a_2 \\ a = a_1+a_2 \bmod 2}} X(a) \right]$$

$$\oplus \begin{cases} \emptyset & \text{if } a_1 + a_2 = \kappa \bmod 2 \\ L(\kappa - 1) & \text{if } a_1 + a_2 = \kappa - 1 \bmod 2 \end{cases}$$

as $U_q s\ell_2$-modules.

The lemma has an obvious reformulation for $a_1, a_2 \in \mathbb{Z}$, $a_1 \neq \kappa - 1 \bmod \kappa$, $a_2 \neq \kappa - 1 \bmod \kappa$.

The space $L(a_1) \otimes L(a_2)$ has the $\mathcal{A}$-module structure, indicated in (13.8.3).

13.8.9. Lemma. *Each of the summands on the right-hand side of the decompositions of Lemma (13.8.8) is $\mathcal{A}$-invariant.*

Proof. $[e^\kappa]_q$ acts as zero on the left-hand side of (13.8.8.1). This proves the $\mathcal{A}$-invariance of the summands of (13.8.8.1).

Let $L(a)$ be an irreducible $U_q s\ell_2$-module on the right-hand side of (13.8.8.2). Obviously, $[e^\kappa]_q$ acts as zero on $L(a)$, because it increases eigenvalues of h by 2κ. Hence $L(a)$ is $\mathcal{A}$-invariant.

Let Y be a summand on the right-hand side of (13.8.8.2). Let Z be the sum of all summands different from Y. $L(a_1) \otimes L(a_2) = Y \oplus Z$. Let $pr_Y : L(a_1) \otimes L(a_2) \to Y$ be the projection along the second summand.

For any two different summands X and Y, set

$$f_{X,Y} := pr_Y \cdot [e^\kappa]_q|_X : X \longrightarrow Y.$$

The map $f_{X,Y}$ is a morphism of $U_q s\ell_2$-modules. The summand X is preserved by $\mathcal{A}$ iff $f_{X,Y} \equiv 0$ for all Y.

Let $X(a)$ be a summand on the right-hand side of (13.8.8.2) and $\{e_n^j, a_n^j, b_n^j, f_n^j\}$ its basis. One can generate $X(a)$ acting on b_0^j by e and f. $[e^\kappa]_q b_0^j = 0$ because $[e^\kappa]_q$ increases eigenvalues of h by 2κ. Hence $f_{X(a),Y} = 0$ for all Y, and $X(a)$ is $\mathcal{A}$-invariant.

13.8.10. Lemma [RT]. Let $a_1 \in \{0, \ldots, \kappa - 2\}$ and $a_2 \in \mathbb{Z}$, $a_2 = -1 \bmod \kappa$. Then

$$L(a_1) \otimes L(a_2) = \bigoplus_{\substack{a_2+1 \leq a \leq a_1+a_2 \\ a = a_1+a_2 \bmod 2}} X(a) \oplus \begin{cases} \emptyset, & \text{if } a_1 = 1 \bmod 2 \\ L(a_2), & \text{if } a_1 = 0 \bmod 2 \end{cases}$$

as $U_q s\ell_2$-modules.

13.8.11. Lemma [RT]. Let $a_1, j_2 \in \{0, 1, \ldots, \kappa - 2\}$. Let $a_2 \in \mathbb{Z}$ and $a_2 = j_2 \bmod \kappa$.

1. If $a_1 + j_2 \leq \kappa - 2$, $a_1 \leq j_2$, then,

$$L(a_1) \otimes X(a_2) = \bigoplus_{\substack{a_2-a_1 \leq a \leq a_1+a_2 \\ a = a_1+a_2 \bmod 2}} X(a)$$

as $U_q s\ell_2$-modules.

2. If $a_1 + j_2 \leq \kappa - 2$, $a_1 > j_2$, then

$$L(a_1) \otimes X(a_2) = \bigoplus_{\substack{a_1+a_2-2j_2 \leq a \leq a_1+a_2 \\ a = a_1+a_2 \bmod 2}} X(a) \oplus \bigoplus_{\substack{a_2-j_2 \leq a \leq a_1+a_2-2j_2-2 \\ a = a_1+a_2 \bmod 2}} \mathbb{C}^2 \otimes X(a)$$

$$\oplus \begin{cases} \emptyset, & \text{if } a_1 + j_2 = \kappa \bmod 2 \\ \mathbb{C}^2 \otimes L(a_2 - j_2 - 1), & \text{if } a_1 + j_2 = \kappa - 1 \bmod 2 \end{cases}$$

as $U_q s\ell_2$-modules.

3. If $a_1 + j_2 > \kappa - 2$, $a_1 \leq j_2$, then

$$L(a_1) \otimes X(a_2) = \bigoplus_{\substack{a_2 - a_1 \leq a \leq 2\kappa + a_2 - 2j_2 - a_1 - 4 \\ a = a_1 + a_2 \bmod 2}} X(a) \oplus$$

$$\bigoplus_{\substack{2\kappa + a_2 - 2j_2 - a_1 - 2 \leq a \leq a_1 + a_2 \\ a = a_1 + a_2 \bmod 2}} \bigl(X(a) \oplus X(a - 2\kappa)\bigr)$$

$$\oplus \begin{cases} \emptyset, & \text{if } a_1 + j_2 = \kappa \bmod 2 \\ L(a_2 - j_2 + \kappa - 1) \oplus L(a_2 - j_2 - \kappa - 1), & \text{if } a_1 + j_2 = \kappa - 1 \bmod 2 \end{cases}$$

as $U_q s\ell_2$-modules.

4. If $a_1 + j_2 > \kappa - 2$, $a_1 > j_2$, then

$$L(a_1) \otimes X(a_2) = \bigoplus_{\substack{a_1 + a_2 - 2j_2 \leq a \leq 2\kappa + a_2 - 2j_2 - a_1 - 4 \\ a = a_1 + a_2 \bmod 2}} X(a) \oplus$$

$$\bigoplus_{\substack{a_2 - j_2 \leq a \leq a_1 + a_2 - 2j_2 - 2 \\ a = a_1 + a_2 \bmod 2}} \mathbb{C}^2 \otimes X(a) \oplus$$

$$\bigoplus_{\substack{2\kappa + a_2 - 2j_2 - a_1 - 2 \leq a \leq a_1 + a_2 \\ a = a_1 + a_2 \bmod 2}} \bigl(X(a) \oplus X(a - 2\kappa)\bigr)$$

$$\oplus \begin{cases} \emptyset, & \text{if } a_1 + j_2 = \kappa \bmod 2 \\ L(a_2 - j_2 + \kappa - 1) \oplus L(a_2 - j_2 - \kappa - 1), & \text{if } a_1 + j_2 = \kappa - 1 \bmod 2 \end{cases}$$

$$\oplus \begin{cases} \emptyset, & \text{if } a_1 + j_2 = 0 \bmod 2 \\ \mathbb{C}^2 \otimes L(a_2 - j_2 - 1), & \text{if } a_1 + j_2 = 1 \bmod 2 . \end{cases}$$

Define on $L(a_1) \otimes L(a_2)$ an $\mathcal{A}$-module structure by the comultiplication.

13.8.12. Lemma. *Each of the summands in the right-hand side of the decomposition (13.8.10) is $\mathcal{A}$-invariant.*

The proof is analogous to the proof of (13.8.9).

Let $L(a_1)$ and $X(a_2)$ be the $U_q s\ell_2$-modules of Lemma (13.8.11). Consider the $\mathcal{A}$-module structure on $L(a_1)$ and $X(a_2)$ described in (13.8.3) and (13.8.7). Induce an $\mathcal{A}$-module structure on $L(a_1) \otimes X(a_2)$ by the comultiplication.

13.8.13. Lemma. *All the summands of the decompositions of (13.8.11.1) and (13.8.11.2) are $\mathcal{A}$-invariant. All the summands of the decompositions of (13.8.11.3) and (13.8.11.4) different from the summands of the form $X(a-2\kappa)$ and $L(a_2 - j_2 - \kappa - 1)$ are $\mathcal{A}$-invariant.*

The proof is analogous to the proof of (13.8.9).

Let $a_1, j_2 \in \{0, \ldots, \kappa - 2\}$. Let $a_2 \in \mathbb{Z}$ be such that $a_2 = j_2 \bmod \kappa$. We discuss properties of singular vectors in $L(a_1) \otimes X(a_2)$.

Let $\{f^n v\}$ for $n = 0, \ldots, a_1$ be a basis of $L(a_1)$, where v is the generating vector. Introduce a new basis

$$E_n = e^{a_1 - n} f^{a_1} v \quad \text{for } n = 0, \ldots, a_1. \tag{13.8.14}$$

Then
$$eE_{n+1} = E_n,$$
$$hE_n = (a_1 - 2n)E_n. \tag{13.8.15}$$

Let $\{e_n^{j_2}, a_n^{j_2}, b_n^{j_2}, f_n^{j_2}\}$ be the basis of $X(a_2)$ described in (13.8.5). Introduce a new basis:

$$A_n = e^{\kappa - 1 - n} b_{\kappa - 2 - j_2}^{j_2}, \quad B_n = e^{\kappa - 1 - n} f_{j_2}^{j_2}, \tag{13.8.16}$$

for $n = 0, \ldots, \kappa - 1$. Then
$$eA_{n+1} = A_n, \quad eB_{n+1} = B_n$$
$$hA_n = (a - 2n)A_n, \tag{13.8.17}$$
$$hB_n = (a - 2j_2 - 2n - 2)B_n.$$

Consider the action of $\mathcal{A}$ on $X(a_2)$ defined by (13.8.7). Then

$$[e^\kappa]_q A_n = 0 \quad \text{for all } n,$$
$$[e^\kappa]_q B_n = 0 \quad \text{for } n \le \kappa - j_2 - 2, \tag{13.8.18}$$
$$[e^\kappa]_q B_n = d A_{j_2 + n + 1 - \kappa} \quad \text{for } n > \kappa - j_2 - 2,$$

where d is a suitable nonzero number.

Denote by X_1 (resp., X_2) the subspace of $X(a_2)$ generated by $\{A_n\}$ (resp., $\{B_n\}$). Then
$$X(a_2) = X_1 \oplus X_2.$$

13.8.19. Lemma.

$$\mathrm{Vac}\,\big(L(a_1)\otimes X(a_2)\big) = \mathrm{Vac}\,\big(L(a_1)\otimes X_1\big) \oplus \mathrm{Vac}\,\big(L(a_2)\otimes X_2\big)$$

where $\mathrm{Vac}\,(M) := \{x\in M\,|\,ex=0\}$.

The lemma follows from (13.8.15) and (13.8.17).

Set $\mathrm{Vac}\,(M)_a = \{x\in M\,|\,ex=0,\,hx=ax\}$ for any $a\in\mathbb{C}$.

13.8.20. Lemma.

$$\mathrm{Vac}\,\big(L(a_1)\otimes X_1\big) = \bigoplus_{\substack{a_1+a_2-2j_1\le a\le a_1+a_2 \\ a=a_1+a_2\bmod 2}} \mathrm{Vac}\,\big(L(a_1)\otimes X_1\big)_a,$$

$$\mathrm{Vac}\,\big(L(a_2)\otimes X_2\big) = \bigoplus_{\substack{a_1+a_2-2j_1-2j_2-2\le a\le a_1+a_2-2j_2-2 \\ a=a_1+a_2\bmod 2}} \mathrm{Vac}\,\big(L(a_2)\otimes X_2\big)_a.$$

All spaces on the right-hand side of these decompositions are one-dimensional.

The lemma easily follows from (13.8.15) and (13.8.17). A vector generating a nontrivial space $\mathrm{Vac}\,(L(a_1)\otimes X_1)_a$ has the form:

$$x = \sum_{p=0}^{t} c_p E_{t-p}\otimes A_p, \qquad (13.8.21)$$

where $t=(a_1+a_2-a)/2$, and $\{c_p\}$ are suitable nonzero numbers. A vector generating a nontrivial space $\mathrm{Vac}\,(L(a_1)\otimes X_2)_a$ has the form

$$y = \sum_{p=0}^{t} c_p E_{t-p}\otimes B_p, \qquad (13.8.22)$$

where $t=(a_1+a_2-2j_2-2-a)/2$, and $\{c_p\}$ are suitable nonzero numbers.

Consider the $\mathcal{A}$-module structure on $L(a_1)$ and $X(a_2)$ described in (13.8.3) and (13.8.7). Consider the $\mathcal{A}$-module structure on $L(a_1)\otimes X(a_2)$ induced by the comultiplication.

13.8.23. Lemma. *For all a,*

$$[e^\kappa]_q\big(\mathrm{Vac}\,(L(a_1)\otimes X_2)_{a-2\kappa}\big) \subset \mathrm{Vac}\,\big(L(a_1)\otimes X_1\big)_a,$$

$$[e^\kappa]_q\big(\mathrm{Vac}\,(L(a_1)\otimes X_1)_a\big) = 0.$$

Moreover, if the spaces $\mathrm{Vac}\,(L(a_1) \otimes X_2)_{a-2\kappa}$ and $\mathrm{Vac}\,(L(a_1) \otimes X_1)_a$ are both one-dimensional, then the map

$$[e^\kappa]_q : \mathrm{Vac}\,(L(a_1) \otimes X_2)_{a-2\kappa} \longrightarrow \mathrm{Vac}\,(L(a_1) \otimes X_1)_a$$

is nonzero.

Proof. The first statement is valid because $[e^\kappa]_q$ increases eigenvalues of h by 2κ.

The spaces $\mathrm{Vac}\,(L(a_1) \otimes X_2)_{a-2\kappa}$ and $\mathrm{Vac}\,(L(a_1) \otimes X_1)_a$ are both one-dimensional if
$$\begin{gathered} a_1 + j_2 \geq \kappa - 1, \\ 2\kappa + a_2 - 2j_2 - a_1 - 2 \leq a \leq a_1 + a_2, \\ a = a_1 + a_2 \bmod 2. \end{gathered} \tag{13.8.24}$$

Let a satisfy (13.8.24). Let y generate $\mathrm{Vac}\,(L(a_1) \otimes X_2)_{a-2\kappa}$. Then y has the form (13.8.22), where $t \geq \kappa - j_2 - 1$. By (13.8.18), we have

$$[e^\kappa]_q y = d \sum_{p=0}^{t} c_p q^{-\kappa h/4} E_{t-p} \otimes A_{j_2+p+1-\kappa} \neq 0,$$

where d is the constant in (13.8.18). The lemma is proved.

Let $a_1, \ldots, a_m \in \{0, \ldots, \kappa - 2\}$. By (13.8.8), (13.8.10), and (13.8.11), we have

$$L(a_1) \otimes \cdots \otimes L(a_m) = \left[\bigoplus_{a \in A_1} L(a) \right] \oplus \left[\bigoplus_{a \in A_2} X(a) \right] \tag{13.8.25}$$

as $U_q s\ell_2$-modules, where A_1 and A_2 are some sets of integers.

Let Y be one of the summands on the right-hand side of (13.8.25). The summand Y will be called *admissible* if, for any singular vector $y \in Y$, we have $[e^\kappa]_q y \neq 0$. Here the $\mathcal{A}$-module structure on (13.8.25) is induced by the comultiplication.

Let Z be the sum of the summands in (13.8.25) different from Y. $L(a_1) \otimes \cdots \otimes L(a_m) = Y \oplus Z$.

Let $a_0 \in \{0, \ldots, \kappa - 2\}$. Consider the tensor product $L(a_0) \otimes \cdots \otimes L(a_m) = (L(a_0) \otimes Y) \oplus (L(a_0) \otimes Z)$. By (13.8.8), (13.8.10), and (13.8.11), we have

$$L(a_0) \otimes Y = \left[\bigoplus_{a \in A_3} L(a) \right] \oplus \left[\bigoplus_{a \in A_4} X(a) \right] \tag{13.8.26}$$

as $U_q s\ell_2$-modules, where A_3 and A_4 are suitable sets of integers.

13.8.27. Lemma.

1. *If Y is admissible, then, for every $b_3 \in A_3$ and $b_4 \in A_4$, the summands $L(b_3)$ and $X(b_4)$ of $L(a_0) \otimes \cdots \otimes L(a_m)$ are admissible.*

2. *If Y is an $\mathcal{A}$-submodule of $L(a_1) \otimes \cdots \otimes L(a_m)$, then each summand in (13.8.26) is either an $\mathcal{A}$-submodule of $L(a_0) \otimes \cdots \otimes L(a_m)$ or an admissible summand of $L(a_0) \otimes \cdots \otimes L(a_m)$.*

Proof of (13.8.27.1). Let $Y(b) = X(b)$. Let $\{A_n, B_n\}$ be the basis in $X(b)$ introduced in (13.8.16). The space of singular vectors of $X(b)$ is generated by A_0 and B_0. By assumptions, $[e^\kappa]_q A_0 \neq 0$ and $[e^\kappa]_q B_0 \neq 0$.

Let $\{E_n\}$ be the basis in $L(a_0)$ introduced in (13.8.14). Singular vectors in $L(a_0) \otimes X(b)$ have the form shown in (13.8.21) and (13.8.22). The action of $[e^\kappa]_q$ on $L(a_0) \otimes X(b)$ is given by (13.8.2):

$$\Delta [e^\kappa]_q = [e^\kappa]_q \otimes q^{h/4} + \cdots + q^{-\kappa h/4} \otimes [e^\kappa]_q .$$

Only the last term of this sum may act nontrivially on the elements of the form shown in (13.8.21) and (13.8.22). Moreover, the result of this action is nontrivial. This proves (13.8.27.1) for $Y = X(b)$. The proof for $Y = L(b)$ is similar.

Proof of (13.8.27.2). $Y \subset L(a_1) \otimes \cdots \otimes L(a_m)$ is an $\mathcal{A}$-submodule. The $\mathcal{A}$-module structure on Y coincides with the structure described in (13.8.3), if $Y = L(b)$ for some b, and coincides with the structure described in (13.8.7), if $Y = X(b)$ for some b. $L(a_0) \otimes Y$ is an $\mathcal{A}$-submodule of $L(a_0) \otimes \cdots \otimes L(a_m)$. The $\mathcal{A}$-module structure on $L(a_0) \otimes Y$ coincides with the structure induced by the comultiplication from the structures on $L(a_0)$ and Y.

Let $Y = L(b)$ for some b. Then all summands in (13.8.26) are $\mathcal{A}$-submodules by (13.8.9) and (13.8.12).

Let $Y = X(b)$ for some b. Let $j \in \{0, \ldots, \kappa - 2\}$ be such that $j = b \bmod \kappa$. If $a_0 + j \leq \kappa - 2$, then all summands in (13.8.26) are $\mathcal{A}$-submodules by (13.8.13).

If $a_0 + j > \kappa - 2$, then the decomposition of $L(a_0) \otimes X(b)$ as an $U_q s\ell_2$-module is given in (13.8.11.3) and (13.8.11.4). All summands of these decompositions are $\mathcal{A}$-submodules with the following exceptions. The exceptions are formed by the second summands in the pairs of the form $X(a) \oplus X(a - 2\kappa)$ and

$L(b-j+\kappa-1) \oplus L(b-j-\kappa-1)$, see (13.8.11). These second summands are admissible by (13.8.23).

13.8.28. *Proof of* (13.7.27). Let $a_1, \ldots, a_n \in \{0, \ldots, \kappa-2\}$. Consider the tensor product
$$L = L(a_1) \otimes \cdots \otimes L(a_n).$$
By (13.8.8), (13.8.10), (13.8.11), and (13.8.27), we have
$$\begin{aligned} L = \bigoplus_{a \in A_1} L(a) \oplus \Big[\bigoplus_{a \in A_2} L(a) \Big] \oplus \Big[\bigoplus_{a \in A_3} X(a) \Big] \\ \oplus \Big[\bigoplus_{a \in A_4} L(a) \Big] \oplus \Big[\bigoplus_{a \in A_5} X(a) \Big] \end{aligned} \quad (13.8.29)$$
as $U_q s\ell_2$-modules. Here $A_1, \ldots, A_5$ are some sets of integers with the following properties. $A_1 \subset \{0, \ldots, \kappa-2\}$. For any $a \in A_2$, we have $a = -1 \mod \kappa$. Each summand of the first, second, and third sums is an $\mathcal{A}$-submodule. Each summand of the fourth and fifth sums is admissible.

It is easy to see that, if $a \in A_1 \cup A_2$, then $a \geq 0$, and, if $a \in A_3$, then $a \geq \kappa$.

Let $s = \{0, \ldots, \kappa-2\}$. Then
$$\{x \in L \,|\, hx = sx, [e^t]_q x = 0 \text{ for all } t \in \mathbf{N}\}$$
$$= \Big\{ x \in \Big[\bigoplus_{a \in A_1} L(a) \Big] \oplus \Big[\bigoplus_{a \in A_3} X(a) \Big] \,\Big|\, hx = sx, [e^t]_q x = 0 \text{ for all } t \in \mathbf{N} \Big\}.$$

Set $p = \kappa - 1 - s$. Then
$$f^p \{x \in L \,|\, hx = (\kappa-1+p)x, [e^t]_q x = 0 \text{ for all } t \in \mathbf{N}\}$$
$$= \Big\{ x \in \bigoplus_{a \in A_3} X(a) \,\Big|\, hx = sx, [e^t]_q x = 0 \text{ for all } t \in \mathbf{N} \Big\}.$$

Hence
$$F(L)_s \simeq \Big\{ x \in \bigoplus_{a \in A_1} L(a) \,\Big|\, hx = sx, [e^t]_q x = 0 \text{ for all } t \in \mathbf{N} \Big\},$$
and this proves Theorem (13.7.27).

14

Degenerations of Discriminantal Configurations

In this section, we study asymptotics of the hypergeometric pairing for a discriminantal configuration $\mathcal{C}(z_1, \ldots, z_n)$ when several of $\{z_j\}$ tend to the same limit.

14.1. Composition of Singular Vectors

Assume that data (4.1.a)–(4.1.d) are given. Let $\mathfrak{g}$ be the Lie algebra defined by data (4.1.a)–(4.1.c) in (11.1). Let $U_q\mathfrak{g}$ be the quantum group defined by data (4.1.a)–(4.1.d) in (4.1).

For $\Lambda \in \mathfrak{h}^*$, let $M(\Lambda, q)$ (resp. $M(\Lambda)$) be the Verma module over $U_q\mathfrak{g}$ (resp. $\mathfrak{g}$) with highest weight Λ.

For $\Lambda^1, \ldots, \Lambda^p, \Lambda_2, \ldots, \Lambda_n \in \mathfrak{h}^*$ and $\lambda, \mu \in \mathbf{N}^r$, $\lambda = (\ell_1, \ldots, \ell_r)$, $\mu = (m_1, \ldots, m_r)$, set

$$\Lambda_1 = \Lambda_1(\lambda) = \Lambda^1 + \cdots + \Lambda^p - \ell_1 \alpha_1 - \cdots - \ell_r \alpha_r,$$

$$M_I(q) = M(\Lambda^1, q) \otimes \cdots \otimes M(\Lambda^p, q),$$

$$M_{II}(q) = M_{II}(\lambda, q) = M(\Lambda_1, q) \otimes \cdots \otimes M(\Lambda_n, q),$$

$$M(q) = M(\Lambda^1 q) \otimes \cdots \otimes M(\Lambda^p, q) \otimes M(\Lambda_2, q) \otimes \cdots \otimes M(\Lambda_n, q).$$

(14.1.1)

Define a linear map

$$\circ : M_I(q)_\lambda \otimes M_{II}(q)_\mu \longrightarrow M(q)_{\lambda+\mu},$$

(14.1.2)

where $M_I(q)_\lambda$ is the weight component of $M_I(q)$ corresponding to weight λ, and so on. Namely, for arbitrary monomials

$$F = f_{I_1} v^1 \otimes \cdots \otimes f_{I_p} v^p,$$
$$G = f_{J_1} v_1 \otimes \cdots \otimes f_{J_n} v_n, \qquad (14.1.3)$$

set

$$F \circ G = f_{J_1}(f_{I_1} v^1 \otimes \cdots \otimes f_{I_p} v^p) \otimes f_{J_2} v_2 \otimes \cdots \otimes f_{J_n} v_n, \qquad (14.1.4)$$

where v^i (resp. v_i) is the generating vector of $M(\Lambda^i, q)$ (resp. $M(\Lambda_i, q)$), and f_I, f_J are monomials in $U_q \mathfrak{n}_-$.

Consider the complexes $C.(^+U_q\mathfrak{n}_-; M_I(q); {^+\Delta})_\lambda$, $C.(^+U_q\mathfrak{n}_-; M_{II}(q); {^+\Delta})_\mu$, $C.(^+U_q\mathfrak{n}_-; M(q); {^+\Delta})_{\lambda+\mu}$. The spaces $M_I(q)_\lambda$, $M_{II}(q)_\mu$, $M(q)_{\lambda+\mu}$ are the terms in degree zero of these complexes. Let d_I, d_{II}, d be the differentials of these complexes, resp.

14.1.5. Lemma. For $x \in M_I(q)_\lambda$ and $y \in M_{II}(q)_\mu$, assume that $d_I x = 0$ and $d_{II} y = 0$. Then

$$d(x \circ y) = 0.$$

Proof. $d_I x = 0$ and $d_{II} y = 0$ mean that x and y lie in the kernel of all operators g_I introduced in (5.1). $x \circ y$ lies in the kernel of all g_I's by (5.1.8) and the remark after (5.1.17).

It is easy to see that there are analogously defined linear maps

$$\circ : C_a(^+U_q\mathfrak{n}_-; M_I(q); {^+\Delta})_\lambda \otimes C_b(^+U_q\mathfrak{n}_-; M_{II}(q); {^+\Delta})_\mu \longrightarrow \qquad (14.1.6)$$
$$C_{a+b}(^+U_q\mathfrak{n}_-; M(q); {^+\Delta})_{\lambda+\mu}$$

for all a, b such that

$$d(x \circ y) = (d_I x) \circ y + (-1)^a x \circ (d_{II} y)$$

for $x \in C_a$, $y \in C_b$.

14.1.7. Example. For $p = n = a = b = 1$, we have

$$(f_1 \otimes f_2 v^1) \circ (f_3 \otimes f_4 v_1) = f_3 \otimes f_1 \otimes f_4 f_2 v^1 + C_1 f_3 \otimes f_4 f_1 \otimes f_2 v^1$$
$$+ C_2 f_3 \otimes f_1 f_4 \otimes f_2 v^1,$$

where C_1 and C_2 are suitable numbers, see Fig. 14.1, cf. (5.11).

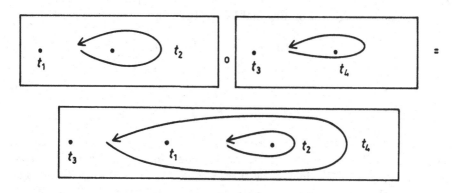

Fig. 14.1

Set $\ell = |\lambda| = \ell_1 + \cdots + \ell_r$, $m = |\mu| = m_1 + \cdots + m_r$.

Consider the space $\mathbb{C}^{\ell+p}$ with coordinates $z^1, \ldots, z^p, t_1(1), \ldots, t_{\ell_1}(1), t_1(2)$, $\ldots, t_{\ell_2}(2), \ldots, t_1(r), \ldots, t_{\ell_r}(r)$. Define weights a_I of diagonal hyperplanes in $\mathbb{C}^{\ell+p}$ by

$$a(t_a(i), t_b(j)) = (\alpha_i, \alpha_j)/\kappa,$$
$$a(t_a(i), z^j) = -(\Lambda^j, \alpha_i)/\kappa, \quad (14.1.8)$$
$$a(z^i, z^j) = (\Lambda^i, \Lambda^j)/\kappa,$$

for all a, b, i, j. Let $\mathcal{S}_I := \mathcal{S}^*(a_I)$ be the local system on the complement to diagonal hyperplanes in $\mathbb{C}^{\ell+p}$ defined by these weights, see (2.5.5) and (2.5.6). The group $\Sigma_I = \Sigma_{\ell_1} \times \cdots \times \Sigma_{\ell_r}$ acts on $\mathbb{C}^{\ell+p}$ permuting $t_i(j)$'s. Let $\pi_I : \mathbb{C}^{\ell+p} \to \mathbb{C}^p$ be the projection on $\{z^j\}$. Set $z_I = (z^1, \ldots, z^p)$.

Consider the space $\mathbb{C}^{n+m}$ with coordinates $z_1, \ldots, z_n, u_1(1), \ldots, u_{m_1}(1), u_1(2), \ldots, u_{m_2}(2), \ldots, u_1(r), \ldots, u_{m_r}(r)$. Define weights a_{II} of diagonal hyperplanes in $\mathbb{C}^{m+n}$ by

$$a(u_a(i), u_b(j)) = (\alpha_i, \alpha_j)/\kappa,$$
$$a(u_a(i), z_j) = -(\Lambda_j, \alpha_i)/\kappa, \quad (14.1.9)$$
$$a(z_i, z_j) = (\Lambda_i, \Lambda_j)/\kappa,$$

for all a, b, i, j. Let $\mathcal{S}_{II} := \mathcal{S}^*(a_{II})$ be the local system on the complement to diagonal hyperplanes in $\mathbb{C}^{m+n}$ defined by these weights, see (2.5.5) and (2.5.6). The group $\Sigma_{II} = \Sigma_{m_1} \times \cdots \times \Sigma_{m_r}$ acts on $\mathbb{C}^{m+n}$ permuting $u_i(j)$'s. Let $\pi_{II} : \mathbb{C}^{m+n} \to \mathbb{C}^n$ be the projection on $\{z_j\}$. Set $z_{II} = \{z_1, \ldots, z_n\}$.

Consider the space $\mathbb{C}^N$, $N = \ell + m + p + n - 1$, with coordinates $z^1, \ldots, z^p$, $z_2, \ldots, z_n$, $t_1(1), \ldots, t_{\ell_1}(1)$, $u_1(1), \ldots, u_{m_1}(1), \ldots, t_1(r), \ldots, t_{\ell_r}(r)$, $u_1(r), \ldots, u_{m_r}(r)$. Define weights a of diagonal hyperplanes in $\mathbb{C}^N$ by (14.1.8), (14.1.9), and

$$a(t_a(i), u_b(j)) = (\alpha_i, \alpha_j)/\kappa,$$
$$a(t_a(i), z_j) = -(\Lambda_j, \alpha_i)/\kappa,$$
$$a(u_a(i), z^j) = (\Lambda^j, \alpha_i)/\kappa, \qquad (14.1.10)$$
$$a(z_i, z^j) = (\Lambda_i, \Lambda^j)/\kappa,$$
$$a(t_a(i), t_b(j)) = a(u_a(i), u_b(j)) = (\alpha_i, \alpha_j)/\kappa,$$

for all a, b, i, j. Let $\mathcal{S} := \mathcal{S}^*(a)$ be the local system on the complement to diagonal hyperplanes in $\mathbb{C}^N$ defined by these weights, see (2.5.5) and (2.5.6). The group $\Sigma = \Sigma_{\ell_1 + m_1} \times \cdots \times \Sigma_{\ell_r + m_r}$ naturally acts on $\mathbb{C}^N$. Let $\pi : \mathbb{C}^N \to \mathbb{C}^{p+n-1}$ be the projection on $z^1, \ldots, z^p, z_2, \ldots, z_n$. Set $z = (z^1, \ldots, z^p, z_2, \ldots, z_n)$.

Fix $z \in \mathbb{R}^{p+n-1}$ such that $z^1 < \cdots < z^p < z_2 < \cdots < z_n$. Thus $z_I \in \mathbb{R}^p$. Choose $z_1 \in \mathbb{R}$ such that $z_1 < z_2$. Then $z_{II} \in \mathbb{R}^n$ is defined and $z_1 < \cdots < z_n$. We call such z, z_I and z_{II} compatible, see Fig. 14.2.

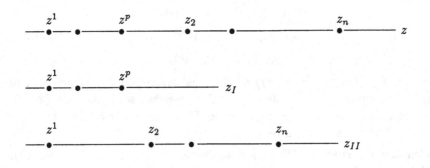

Fig. 14.2

Consider the fiber $\pi_I^{-1}(z_I) = \mathbb{C}^\ell$ with coordinates $\{t_i(j)\}$. The diagonal hyperplanes cut the discriminantal configuration $\mathcal{C}_{p,\ell}(z_I)$ in the fiber. Let X_I be the complex constructed for $\mathcal{C}_{p,\ell}(z_I)$ in Secs. 2 and 3. The complex lies in $\pi_I^{-1}(z_I) - \mathcal{C}_{p,\ell}(z_I)$ and computes homology groups of this complement. Let

$$\varphi_I = \varphi_I[z_I] : C.(^+U_q\mathfrak{n}_-; M_I(q); {}^+\Delta)_\lambda \longrightarrow C_{\ell - .}(X_I, \mathcal{S}_I)^- \qquad (14.1.11)$$

be the quasiisomorphism constructed in (8.2). Here "$-$" denotes the Σ_I-skew-invariant part.

Analogously consider the fiber $\pi_{II}^{-1}(z_{II}) = \mathbb{C}^m$ with coordinates $\{u_i(j)\}$. The diagonal hyperplanes cut the discriminantal configuration $\mathcal{C}_{n,m}(z_{II})$ in the fiber. Let $X_{II} \subset \pi_{II}^{-1}(z_{II})$ be the complex constructed for $\mathcal{C}_{n,m}(z_{II})$ in Secs. 2 and 3. Let

$$\varphi_{II} = \varphi_{II}[z_{II}] : C.(^+U_q\mathfrak{n}_-; M_{II}(q); {}^+\Delta)_\mu \longrightarrow C_{m-\cdot}(X_{II}, \mathcal{S}_{II})^- \quad (14.1.12)$$

be the quasiisomorphism constructed in (8.2). Here "−" denotes the Σ_I-skew-invariant part.

Consider the fiber $\pi^{-1}(z) = \mathbb{C}^{\ell+m}$ with coordinates $\{t_i(j), u_a(b)\}$. The diagonal hyperplanes cut the discriminantal configuration $\mathcal{C}_{p+n-1,\ell+m}(z)$ in the fiber. Let $X \subset \pi^{-1}(z) - \mathcal{C}_{p+n-1,\ell+m}(z)$ be the complex constructed for $\mathcal{C}_{p+n-1,\ell+m}(z)$ in Secs. 2 and 3. Let

$$\varphi = \varphi[z] : C.(^+U_q\mathfrak{n}_-; M(q); {}^+\Delta)_{\lambda+\mu} \longrightarrow C_{\ell+m-\cdot}(X, \mathcal{S})^- \quad (14.1.13)$$

be the quasiisomorphism constructed in (8.2). Here "−" denotes the Σ_I-skew-invariant part.

Let
$$C.(X, \mathcal{S})^- \hookrightarrow C.(\pi^{-1}(z) - \mathcal{C}_{p+n-1,\ell+m}(z), \mathcal{S})^- \quad (14.1.14)$$

be the canonical inclusion. Maps (14.1.6), (14.1.13), and (14.1.14) induce the maps

$$H_a\big(C.(^+U_q\mathfrak{n}_-; M_I(q); {}^+\Delta)_\lambda\big) \otimes H_b\big(C.(^+U_q\mathfrak{n}_-; M_{II}(q); {}^+\Delta)_\mu\big) \longrightarrow \\ H_{\ell+m-a-b}\big(\pi^{-1}(z) - \mathcal{C}_{p+n-1,\ell+m}(z), \mathcal{S}\big)^- \quad (14.1.15)$$

for all a, b. Here "−" denotes the Σ_I-skew-invariant part.

Below we will give another construction of these maps.

(14.1.16) Assume that $z^1, \ldots, z^p$ are very close to z_1. Assume that X_I is very close to the point $\{t_i(j) = z_1 \text{ for all } i, j\}$ of the fiber $\pi_I^{-1}(z_I)$.

Consider $X_I \times X_{II}$ as a complex in $\pi^{-1}(z) = \mathbb{C}^{\ell+m}$ with coordinates $\{t_i(j)\}$ and $\{u_i(j)\}$. The above assumptions imply that $X_I \times X_{II}$ lies in $\pi^{-1}(z) - \mathcal{C}_{p+n-1,\ell+m}(z)$.

For any cell Δ_1 of X_I, denote by s_1 the section of $\mathcal{S}_I$ over Δ_1 distinguished in (2.5) and (8.2). Analogously, for any cell Δ_2 of X_{II}, denote by s_2 the section of $\mathcal{S}_{II}$ over Δ_2 distinguished in (2.5) and (8.2). We distinguish a section s of $\mathcal{S}$ over $\Delta_1 \times \Delta_2$ as follows.

The sections s_1, s_2 and s have the form

$$s_1 = \text{const}_1 \prod br(z^i - z^j)^{(\Lambda^i, \Lambda^j)/\kappa} \times$$

$$\prod br\big(t_i(j) - z^k\big)^{-(\alpha_j, \Lambda^k)/\kappa} \prod br\big(t_i(j) - t_a(b)\big)^{(\alpha_j, \alpha_b)/\kappa},$$

$$s_2 = \text{const}_2 \prod_{1 \leq i < j \leq n} br(z_i - z_j)^{(\Lambda_i, \Lambda_j)/\kappa} \times$$

$$\prod_{\substack{k \geq 1 \\ i,j}} br\big(u_i(j) - z_k\big)^{-(\alpha_j, \Lambda_k)/\kappa} \prod br\big(u_i(j) - u_a(b)\big)^{(\alpha_j, \alpha_b)/\kappa},$$

$$s = \text{const} \prod br(z^i - z^j)^{(\Lambda^i, \Lambda^j)/\kappa} \prod_{\substack{j > 1 \\ i}} br(z^i - z_j)^{(\Lambda^i, \Lambda_j)/\kappa} \times \qquad (14.1.17)$$

$$\prod_{2 \leq i < j \leq n} br(z_i - z_j)^{(\Lambda_i, \Lambda_j)/\kappa} \prod br\big(t_i(j) - z^k\big)^{-(\alpha_j, \Lambda^k)/\kappa} \times$$

$$\prod br\big(t_i(j) - t_a(b)\big)^{(\alpha_j, \alpha_b)/\kappa} \prod_{\substack{k > 1 \\ i,j}} br\big(u_i(j) - z_k\big)^{-(\alpha_j, \Lambda_k)/\kappa} \times$$

$$\prod br\big(u_i(j) - u_a(b)\big)^{(\alpha_j, \alpha_b)/\kappa} \prod_{\substack{k > 1 \\ i,j}} br\big(t_i(j) - z_k\big)^{-(\alpha_j, \Lambda_k)/\kappa}$$

$$\prod br\big(t_i(j) - u_a(b)\big)^{(\alpha_j, \alpha_b)/\kappa} \prod br\big(u_i(j) - z^k\big)^{-(\alpha_j, \Lambda_k)/\kappa},$$

see (2.5.11). Here const_1, const_2, and const are complex numbers, and, for any function f, $br(f)$ denotes its suitably chosen univalued branch. The constants and the branches in the first and second formulas are defined. We define const and the branches in the third formula.

Set $\text{const} = \text{const}_1 \cdot \text{const}_2$. Choose the branches of $z^j - z^i$, $z_j - z_i$, $t_i(j) - z^k$, $t_i(j) - t_a(b)$, $u_i(j) - z_k$, $u_i(j) - u_a(b)$ the same as those in the first two formulas.

The functions $t_i(j) - z_k$ for $k > 1$ and $z_1 - z_k$ are approximately equal on $\Delta_1 \times \Delta_2$. Choose that argument of $t_i(j) - z_k$ which is close to the argument of $z_1 - z_k$ distinguished in the second formula of (14.1.17).

The functions $t_i(j) - u_a(b)$ and $z_1 - u_a(b)$ are approximately equal on $\Delta_1 \times \Delta_2$. Choose that argument of $t_i(j) - u_a(b)$ which is close to the argument of $z_1 - u_a(b)$ distinguished in the second formula of (14.1.17), and so on. The section s over $\Delta_1 \times \Delta_2$ is defined.

We have constructed a linear map

$$\nu : C_\cdot(X_I, \mathcal{S}_I) \otimes C_\cdot(X_{II}, \mathcal{S}_{II}) \longrightarrow C_\cdot\big(\pi^{-1}(z) - C_{p+n-1,\ell+m}(z), \mathcal{S}\big). \quad (14.1.18)$$

14.1.19. Lemma. *This linear map is a homomorphism of complexes.*

Let

$$\begin{aligned}\text{Asym} : C_\cdot(\mathbb{C}^{\ell+m} - C_{p+n-1,\ell+m}(z), \mathcal{S}) \\ \longrightarrow C_\cdot\big(\pi^{-1}(z) - C_{p+n-1,\ell+m}(z), \mathcal{S}\big)^{-}\end{aligned} \quad (14.1.20)$$

be the canonical projection on the Σ-skew-invariant part multiplied by the number $\prod_{i=1}^{r} \binom{\ell_i + m_i}{\ell_i}$. Maps (14.1.11), (14.1.12), (14.1.18), and (14.1.20) induce a monomorphism of complexes

$$\begin{aligned}C_\cdot({}^+U_q\mathfrak{n}_-; M_I(q); {}^+\Delta)_\lambda \otimes C_\cdot({}^+U_q\mathfrak{n}_-; M_{II}(q); {}^+\Delta)_\mu \longrightarrow \\ C_{\ell+m-\cdot-\cdot}\big(\pi^{-1}(z) - C_{p+n-1,\ell+m}(z), \mathcal{S}\big)^{-}.\end{aligned} \quad (14.1.21)$$

Hence, for all a and b, we have the homomorphism

$$\begin{aligned}H_a\big(C_\cdot({}^+U_q\mathfrak{n}_-; M_I(q); {}^+\Delta)\big)_\lambda \otimes H_b\big(C_\cdot({}^+U_q\mathfrak{n}_-; M_{II}(q); {}^+\Delta)_\mu\big) \longrightarrow \\ H_{\ell+m-a-b}\big(\pi^{-1}(z) - C_{p+n-1,\ell+m}(z), \mathcal{S}\big)^{-}.\end{aligned} \quad (14.1.22)$$

14.1.23. Lemma. *Under assumptions (14.1.16), maps (14.1.15) coincide with maps (14.1.22).*

The lemma follows from constructions of maps (14.1.11)–(14.1.13).

Set

$$\begin{aligned}M_I &= M(\Lambda^1) \otimes \cdots \otimes M(\Lambda^p), \\ M_{II} &= M_{II}(\lambda) = M(\Lambda_1) \otimes \cdots \otimes M(\Lambda_n), \\ M &= M(\Lambda^1) \otimes \cdots \otimes M(\Lambda^p) \otimes M(\Lambda_2) \otimes \cdots \otimes M(\Lambda_n).\end{aligned} \quad (14.1.24)$$

Recall that $M(\Lambda)$ is the Verma module over $\mathfrak{g}$ with highest weight Λ.

Consider the complexes $C_\cdot(\mathfrak{n}_-; M_I)_\lambda$, $C_\cdot(\mathfrak{n}_-; M_{II})_\mu$, and $C_\cdot(\mathfrak{n}_-; M)_{\lambda+\mu}$. The spaces $(M_I)_\lambda$, $(M_{II})_\mu$, and $M_{\lambda+\mu}$ are the terms in degree zero of these complexes, resp. There is a structure in these complexes analogous to that given in (14.1.6). We consider here only a piece of this structure.

Define a map
$$\circ : (M_I)_\lambda \otimes (M_{II})_\mu \longrightarrow M_{\lambda+\mu} \qquad (14.1.25)$$

by the formula (14.1.3), where v^i (resp. v_i) is the generating vector of $M(\Lambda^i)$ (resp. $M(\Lambda_i)$).

Let $F\ell^\cdot(\mathcal{C}_{p,\ell}(z_I), a_I)^-$ be the Σ_I-skew-invariant part of the complex of flag hypergeometric forms of the configuration $\mathcal{C}_{p,\ell}(z_I)$ with weights a_I. Denote by

$$\psi_I = \psi_I[z_I] : C.(\mathfrak{n}_-; M_I)_\lambda \longrightarrow F\ell^{\ell-\cdot}(\mathcal{C}_{p,\ell}(z_I), a_I)^- \qquad (14.1.26)$$

the canonical epimorphism constructed in (12.1), cf. (12.3).

Let $F\ell^\cdot(\mathcal{C}_{n,m}(z_{II}), a_{II})^-$ be the Σ-skew-invariant part of the complex of flag hypergeometric forms of $\mathcal{C}_{n,m}(z_{II})$ with weights a_{II}. Denote by

$$\psi_{II} = \psi_{II}[z_{II}] : C.(\mathfrak{n}_-; M_{II})_\mu \longrightarrow F\ell^{m-\cdot}(\mathcal{C}_{n,m}(z_{II}), a_{II})^- \qquad (14.1.27)$$

the canonical epimorphism constructed in (12.1), cf. (12.3). Finally, let

$$\psi = \psi[z] : C.(\mathfrak{n}_-; M)_{\lambda+\mu} \longrightarrow F\ell^{\ell+m-\cdot}(\mathcal{C}_{p+n-1,\ell+m}(z), a)^- \qquad (14.1.28)$$

be the canonical epimorphism constructed for $C.(\mathfrak{n}_-; M)_{\lambda+\mu}$ in (12.1) and (12.3).

Let $F \in (M_I)_\lambda$, $G \in (M_{II})_\mu$, and $F \circ G \in M_{\lambda+\mu}$ be the elements given by (14.1.3) and (14.1.4). We will compare three forms, $\psi_I(F) \in F\ell^\ell(\mathcal{C}_{p,\ell}(z_I), a_I)^-$, $\psi_{II}(G) \in F\ell^m(\mathcal{C}_{n,m}(z_{II}), a_{II})^-$, and $\psi(F \circ G) \in F\ell^{\ell+m}(\mathcal{C}_{p+n-1,\ell+m}(z), a)^-$.

14.1.29. Example. Let $F = f_1 v^1 \otimes v^2$, $G = f_2 v_1 \otimes v_2$. Then

$$F \circ G = f_2 f_1 v^1 \otimes v^2 \otimes v_2 + f_1 v^1 \otimes f_2 v^2 \otimes v_2,$$

$$\psi_I(F) = \left(-(\Lambda^1, \alpha_1)/\kappa\right) d(t_1 - z^1)/(t_1 - z^1),$$

$$\psi_{II}(G) = \left((-\Lambda^1 - \Lambda^2 + \alpha_1, \alpha_2)/\kappa\right) d(u_1 - z_1)/(u_1 - z_1),$$

$$\psi(F \circ G) = \left(-(\Lambda^1, \alpha_1)/\kappa\right) d(t_1 - z^1)/(t_1 - z^1)$$
$$\wedge \left((-\Lambda^1, \alpha_2)/\kappa\right) d(u_1 - z^1)/(u_1 - z^1)$$
$$+ \left((-\Lambda^2, \alpha_2)/\kappa\right) d(u_1 - z^2)/(u_1 - z^2)$$
$$+ \left((\alpha_1, \alpha_2)/\kappa\right) d(u_1 - t_1)/(u_1 - t_1)).$$

The general case is similar.

Namely, let $F \in (M_I)_\lambda$, $G \in (M_{II})_\mu$ be given by (14.1.3). Consider the differential $(\ell + m)$-form $w_1 = \psi_I(F) \wedge \psi_{II}(G)$. The m-form $\psi_{II}(G)$ is an exterior product of suitable sums of differential one-forms $\{\text{const } d\ell/\ell\}$, where ℓ is a linear function. Some of these one-forms have the shape

$$\begin{aligned}&((-\Lambda_1, \alpha_j)/\kappa)\, d(u_i(j) - z_1)/(u_i(j) - z_1) = ((-\Lambda^1 - \cdots - \Lambda^p \\ &+ \ell_1\alpha_1 + \cdots + \ell_r\alpha_r, \alpha_j)/\kappa)\, d(u_i(j) - z_1)/(u_i(j) - z_1).\end{aligned} \quad (14.1.30)$$

Replace each of these differential forms by the differential form

$$\begin{aligned}&\sum_{a=1}^p ((-\Lambda^a, \alpha_j)/\kappa)\, d(u_i(j) - z^a)/(u_i(j) - z^a) + \\ &\sum_{b=1}^r \sum_{a=1}^{\ell_b} ((\alpha_b, \alpha_j)/\kappa)\, d(u_i(j) - t_a(b))/(u_i(j) - t_a(b)).\end{aligned} \quad (14.1.31)$$

As a result, we get an $(\ell + m)$-form

$$w_2 = \psi_I(F) \wedge \widetilde{\psi_{II}(G)} \in F\ell^{\ell+m}(\mathcal{C}_{p+n-1,\ell+m}(z), a). \quad (14.1.32)$$

Let

$$\text{Asym} : F\ell^{\ell+m}(\mathcal{C}_{p+n-1,\ell+m}(z), a) \longrightarrow F\ell^{\ell+m}(\mathcal{C}_{p+n-1,\ell+m}(z), a)^-$$

be the canonical projection on the Σ-skew-invariant part multiplied by the number $\prod_{i=1}^r \binom{\ell_i + m_i}{\ell_i}$.

14.1.33. Lemma. $\psi(F \circ G) = \text{Asym}(w_2)$.

The proof easily follows from definitions.

14.2. Asymptotics of the Hypergeometric Pairing

Under assumptions of (14.1), assume that $z^1, \ldots, z^p$ depend on a parameter $\alpha \in \mathbb{R}_+$:
$$z^j(\alpha) = z_1 + w^j \alpha \quad \text{for } j = 1, \ldots, p,$$
where $w^1 < \cdots < w^p$ are some fixed real numbers. Thus $\{z^j(\alpha)\}$ tend to z_1 as α tends to zero. Set $w_I = (w^1, \ldots, w^p)$.

Fix an equipment $\{T_j^i(II)\}$ for $\mathcal{C}_{n,m}(z_{II})$. Fix an equipment $\{T_j^i(I, \alpha)\}$ for $\mathcal{C}_{p,\ell}(z_I(\alpha))$ such that each $T_j^i(I, \alpha)$ linearly depends on α and tends to z_1 as α

tends to zero. Then X_{II} and $X_I(\alpha)$ are defined, and conditions (14.1.16) are satisfied for small α. Let

$$I_I[z_I(\alpha)] : H_0\big(C.(\mathfrak{n}_-; M_I)_\lambda\big) \otimes H_0\big(C.({}^+U_q\mathfrak{n}_-; M_I(q); {}^+\Delta)_\lambda\big) \longrightarrow \mathbb{C},$$

$$I_{II}[z_{II}] : H_0\big(C.(\mathfrak{n}_-; M_{II})_\mu\big) \otimes H_0\big(C.({}^+U_q\mathfrak{n}_-; M_{II}(q); {}^+\Delta)_\mu\big) \longrightarrow \mathbb{C}, \quad (14.2.1)$$

$$I[z(\alpha)] : H_0\big(C.(\mathfrak{n}_-; M)_{\lambda+\mu}\big) \otimes H_0\big(C.({}^+U_q\mathfrak{n}_-; M(q); {}^+\Delta)_{\lambda+\mu}\big) \longrightarrow \mathbb{C}$$

be the hypergeometric pairing for $C_{p,\ell}(z_I(\alpha))$, $C_{n,m}(z_{II})$, and $C_{p+n-1,\ell+m}(z(\alpha))$, resp. We have

$$I_I[z_I(\alpha)](x_I \otimes y_I) = \int_{\varphi_I[z_I(\alpha)](y_I)} \psi_I[z_I(\alpha)](x_I),$$

$$I_{II}[z_{II}](x_{II} \otimes y_{II}) = \int_{\varphi_{II}[z_{II}](y_{II})} \psi_{II}[z_{II}](x_{II}), \quad (14.2.2)$$

$$I[z(\alpha)](x \otimes y) = \int_{\varphi[z(\alpha)](y)} \psi[z(\alpha)](x).$$

Set

$$d = \left[-\sum_{i=1}^p \Lambda^p, \sum_{j=1}^r \ell_j \alpha_j\right] / \kappa + \sum_{i=1}^r \ell_i(\ell_i - 1)(\alpha_i, \alpha_i)/2\kappa$$
$$+ \sum_{1 \leq i < j \leq r} \ell_i \ell_j (\alpha_i, \alpha_j)/\kappa. \quad (14.2.3)$$

14.2.4. Theorem. *For all x_I, x_{II}, y_I, y_{II}, we have*

$$I[z(\alpha)]\big((x_I \circ x_{II}) \otimes (y_I \circ y_{II})\big)$$

$$= \alpha^d \left[\prod_{i=1}^r \binom{\ell_i + m_i}{\ell_i} I_I[w_I](x_I \otimes y_I) \cdot I_{II}[z_{II}](x_{II} \otimes y_{II})\right.$$

$$\left. + \alpha F(z(\alpha), x_I, x_{II}, y_I, y_{II})\right],$$

where F is a holomorphic function for small α.

Proof. Consider the map

$$\nu : C.\big(X_I(z_I(\alpha)), S_I\big)^- \otimes C.\big(X_{II}(z_{II}(\alpha)), S_{II}\big)^-$$
$$\longrightarrow C.\big(\pi^{-1}(z(\alpha)) - C_{p+n-1,\ell+m}(z(\alpha)), S\big)$$

constructed in (14.1.18). Consider the hypergeometric $(\ell + m)$-form

$$\psi_I[z_I(\alpha)](x_I) \wedge \widetilde{\psi_{II}[z_{II}]}(x_{II})$$

constructed in (14.1.32).

14.2.5. Lemma. *For all x_I, x_{II}, y_I, y_{II}, we have*

$$\int_{\nu(\varphi_I[z_I(\alpha)](y_I) \otimes \varphi_{II}[z_{II}](y_{II}))} \psi_I[z_I(\alpha)](x_I) \wedge \widetilde{\psi_{II}[z_{II}]}(x_{II})$$

$$= \alpha^d \left[\int_{\varphi_I[w_I](y_I)} \psi_I[w_I](x_I) \cdot \int_{\varphi_{II}[z_{II}](y_{II})} \psi_{II}[z_{II}](x_{II}) \right.$$

$$\left. + \alpha F(z(\alpha), x_I, x_{II}, y_I, y_{II}) \right],$$

where F is a holomorphic function for small α.

Proof of the Lemma. First integrate over $\varphi_I[z_I(\alpha)](y_I)$, then over $\varphi_{II}[z_{II}](y_{II})$. The cycle $\varphi_{II}[z_{II}](y_{II})$ is fixed, the cycle $\varphi_I[z_I(\alpha)](y_I)$ homothetically tends to z_1 as α tends to zero. An arbitrary function of the form $(t_i(j) - z_a)^\gamma$, $a > 1$, restricted to $X_I(\alpha) \times X_{II}$ tends to $(z_1 - z_a)^\gamma$ as α tends to zero. An arbitrary function of the form $(t_i(j) - u_a(b))^\gamma$ restricted to $X_I(\alpha) \times X_{II}$ tends to $(z_1 - u_a(b))^\gamma$. An arbitrary function of the form $(t_i(j) - t_a(b))^\gamma$ or $(t_i(j) - z^a)^\gamma$ restricted to $X_I(\alpha) \times X_{II}$ gives a homogeneous function in α of degree γ. This proves the lemma.

By (14.1.23) and (14.1.33),

$$I[z_I(\alpha)]((x_I \circ x_{II})) \otimes (y_I \circ y_{II}) =$$

$$\int_{\text{Asym}\,\nu(\varphi_I[z_I(\alpha)](y_I) \otimes \varphi_{II}[z_{II}](y_{II}))} \text{Asym}\left(\psi_I[z_I(\alpha)](x_I) \wedge \widetilde{\psi_{II}[z_{II}]}(x_2)\right).$$

It is easy to see that exactly $\prod \binom{\ell_i + m_i}{\ell_i}$ terms of this antisymmetrization give a nontrivial impact of order α^d and each of these terms is given by (14.2.5).

Theorem (14.1.4) is proved.

14.2.6. Lemma. *For all $x \in M_{\lambda+\mu}$, $y_I \in M_I(q)_\lambda$, $y_{II} \in M_{II}(q)_\mu$, we have*

$$\int_{\varphi[z(\alpha)](y_I \circ y_{II})} \psi[z(\alpha)](x) = \alpha^d F(\alpha, x, y_I, y_{II}),$$

where F is a holomorphic function for small α.

The proof is analogous to the proof of Lemma (14.2.5).

Assume that $\lambda' = (\ell'_1, \ldots, \ell'_r)$, $\mu = (m'_1, \ldots, m'_r) \in \mathbb{N}^r$ and $\lambda' + \mu' = \lambda + \mu$. Assume that there exists j such that $\ell'_j < \ell_j$.

14.2.7. Lemma. *For any $y_I \in M_I(q)_\lambda$, $y_{II} \in M_{II}(q)_\mu$, $x_I \in (M_I)_{\lambda'}$, $x_{II} \in (M(\Lambda_2) \otimes \cdots \otimes M(\Lambda_n))_{\mu'}$, we have $x_I \otimes x_{II} \in M_{\lambda+\mu}$ and*

$$I[z(\alpha)]\big((x_I \otimes x_{II}) \otimes (y_I \circ y_{II})\big) = \alpha^{d+1} F(z(\alpha), x_I, x_{II}, y_I, y_{II}),$$

where F is a holomorphic function for small α.

The proof is analogous to the proof of Lemma (14.2.5).

14.2.8. Lemma. *For all $y_I \in M_I(q)_\lambda$, $y_{II} \in M_{II}(q)_\mu$, $x_I \in (M_I)_{\lambda'}$, $x_{II} \in (M_{II})_{\mu'}$, we have*

$$I[z(\alpha)]\big((x_I \circ x_{II}) \otimes (y_I \circ y_{II})\big) = \alpha^{d+1} F(z(\alpha), x_I, x_{II}, y_I, y_{II}),$$

where F is a holomorphic function for small α.

The lemma follows from (14.2.7).

Denote by $N(q)_\lambda$ the subspace

$$H_0\big(C.({}^+U_q\mathfrak{n}_-; M_I(q); {}^+\Delta)_\lambda\big) \circ H_0\big(C.({}^+U_q\mathfrak{n}_-; M_{II}(q); {}^+\Delta)_\mu\big)$$
$$\subset H_0\big(C.({}^+U_q\mathfrak{n}_-; M(q); {}^+\Delta)_{\lambda+\mu}\big).$$

Consider the subspace $(M_I)_\lambda \circ (M_{II})_\mu \subset M_{\lambda+\mu}$. Denote by N_λ the image of the canonical projection of this subspace to $H_0(C.(\mathfrak{n}_-; M)_{\lambda+\mu})$.

We have proved that the hypergeometric pairing $I[z(\alpha)]$ restricted to $H_0(C.(\mathfrak{n}_-; M)_{\lambda+\mu}) \otimes N(q)_\lambda$ has a decomposition into a power series

$$I[z(\alpha)] = \alpha^d (I_0 + \alpha I_1 + \ldots),$$

see (14.2.4) and (14.2.6). We have proved that I_0 restricted to $N_{\lambda'} \otimes N(q)_\lambda$ is zero, and I_0 restricted to $N_\lambda \otimes N(q)_\lambda$ is the tensor product of the hypergeometric pairing $I_I[w_I]$ and $I_{II}[z_{II}]$, see (14.2.4) and (14.2.8).

Dependence of the hypergeometric pairing $I[z(\alpha)]$ on $z(\alpha)$ is described by the KZ equation over $\mathbb{C}^{p+n-1} - \mathcal{C}_{p+n-1}$ with values in $\mathrm{Vac}\,(L(\Lambda^1) \otimes \cdots \otimes L(\Lambda^p) \otimes L(\Lambda_2) \otimes \cdots \otimes L(\Lambda_n))_{\lambda+\mu}$, see (12.3.16) and (12.3.19). Thus, statements (14.2.4),

(14.2.6), and (14.2.8) can be reformulated as statements about the KZ equation. Namely, these statements show that solutions parametrized by $\dot{N}(q)_\lambda$ have an asymptotic decomposition: $s(z(\alpha)) = \alpha^d(s_0(w_I, z_{II}) + \alpha s_1(w_I, z_{II}) + \ldots)$. The first term of asymptotics has a block form. The differential equation on the diagonal piece of the first term $s_0(w_I, z_{II})$ coincides with the tensor product of the KZ equations over $\mathbb{C}^p - \mathcal{C}_p$ and $\mathbb{C}^n - \mathcal{C}_n$.

14.3. Rank of the Hypergeometric Pairing

(14.3.1) Under assumptions of Secs. 14.1 and 14.2, define the rank $rk(\lambda; \Lambda^1, \ldots, \Lambda^p)$ of the hypergeometric pairing $I_I[z_I]$ in (14.2.1) as the maximal number t such that there exist $x^1, \ldots, x^t \in H_0(C.(\mathfrak{n}_-; M_I)_\lambda)$ and $y^1, \ldots, y^t \in H_0(C.({}^+U_q\mathfrak{n}_-; M_I(q); {}^+\Delta)_\lambda)$ with

$$\det(I_I[z_I](x^i \otimes y^j)) \neq 0.$$

For any $y \in H_0(C.(U_q\mathfrak{n}_-; M_I(q); {}^+\Delta)_\lambda$, the function $z_I \longmapsto I_I[z_I](\cdot \otimes y)$ is a solution to a suitable KZ equation, see (12.3.6) and (12.3.19). The rank of $I_I[z_I]$ is the dimension of the space of such solutions. Hence, the rank $rk(\lambda; \Lambda^1, \ldots, \Lambda^p)$ does not depend on z_I and is invariant under permutations of $\Lambda^1, \ldots, \Lambda^p$.

14.3.2. Theorem. *For any $\lambda \in \mathbf{N}^r$,*

$$rk(\lambda; \Lambda^1, \ldots, \Lambda^p, \Lambda_2, \ldots, \Lambda_n) \geq$$

$$\sum_{\lambda_1 + \lambda_2 = \lambda} rk(\lambda_1; \Lambda^1, \ldots, \Lambda^p) \cdot rk(\lambda_2; \Lambda_1(\lambda_1), \Lambda_2, \ldots, \Lambda_n),$$

where $\Lambda_1(\lambda_1)$ is given by (14.1.1).

Proof. For any λ_1 and λ_2, fix the corresponding $x_I^1(\lambda_1), \ldots, x_I^{t(\lambda_1)}(\lambda_1) \in H_0(C.(\mathfrak{n}_-; M_I)_{\lambda_1})$, $y_I^1(\lambda_1), \ldots, y_I^{t(\lambda_1)}(\lambda_1) \in H_0(C.({}^+U_q\mathfrak{n}_-; M_I(q); {}^+\Delta)_{\lambda_1})$, $x_{II}^1(\lambda_2), \ldots, x_{II}^{s(\lambda_2)}(\lambda_2) \in H_0(C.(\mathfrak{n}_-; M_{II}(\lambda_1))_{\lambda_2})$, $y_{II}^1(\lambda_2), \ldots, y_{II}^{s(\lambda_2)}(\lambda_2) \in H_0(C.({}^+U_q\mathfrak{n}_-; M_{II}(\lambda_1, q); {}^+\Delta)_{\lambda_2})$, with property (14.3.1). Here $t(\lambda_1) = rk(\lambda_1; \Lambda^1, \ldots, \Lambda^p)$, $s(\lambda_2) = rk(\lambda_2; \Lambda(\lambda_1), \Lambda_2, \ldots, \Lambda_n)$. $M_{II}(\lambda_1)$ and $M_{II}(\lambda_1, q)$ are defined in (14.1.24) and (14.1.1).

The matrix

$$\left\{ I[z(\alpha)]((x_I^i(\lambda_1) \circ x_{II}^j(\lambda_2)) \otimes (y_I^a(\lambda_1') \circ (y_{II}^b(\lambda_2'))) \right\} \qquad (14.3.3)$$

for $\lambda_1 + \lambda_2 = \lambda$, $\lambda_1' + \lambda_2' = \lambda$, $i \in \{1,\ldots,t(\lambda_1)\}$, $j = \{1,\ldots,s(\lambda_2)\}$, $a \in \{1,\ldots,t(\lambda_1')\}$, and $b \in \{1,\ldots,s(\lambda_2')\}$ has an asymptotic decomposition as α tends to zero, see (14.2.4) and (14.2.6). The first term of this decomposition has a block form, see (14.2.8). Therefore, the first term of the asymptotics of the determinant of this matrix is the product of the determinants of the diagonal blocks. By assumptions, the determinants of the diagonal blocks are not equal to zero. Hence the determinant of matrix (14.3.3) is not equal to zero. The theorem is proved.

14.4. Remarks on the Kernel of the Hypergeometric Pairing

Let $\lambda = (\ell_1,\ldots,\ell_r) \in \mathbf{N}^r$, $\Lambda \in \mathfrak{h}^*$, $\ell = \ell_1 + \cdots + \ell_r$. Define degree $d(\Lambda;\lambda)$ as the number

$$d(\Lambda;\lambda) = \left(-\Lambda, \sum_{j=1}^r \ell_j \alpha_j\right) / \kappa + \sum_{i=1}^r \ell_i(\ell_i - 1)(\alpha_i, \alpha_i)/2\kappa \qquad (14.4.1)$$
$$+ \sum_{1 \leq i < j \leq r} \ell_i \ell_j (\alpha_i, \alpha_j)/\kappa,$$

cf. (14.2.3).

Let $\mathcal{C}_{1,\ell}(z_1)$, $z_1 \in \mathbb{R}$, be a discriminantal configuration. Consider the hypergeometric pairing corresponding to $\mathcal{C}_{1,\ell}(z_1)$:

$$I[z_1] : H_0\bigl(C.(\mathfrak{n}_-^*; M(\Lambda)^*\bigr)_\lambda \otimes H_0\bigl(C.(^+U_q\mathfrak{n}_-; M(\Lambda,q); {}^+\Delta)_\lambda\bigr) \longrightarrow \mathbb{C},$$

see (12.2).

14.4.2. Lemma. $I[z_1] \equiv 0$ if $d(\Lambda,\lambda) \neq 0$.

Proof. Hypergeometric forms corresponding to elements of $M(\Lambda)_\lambda$ are homogeneous of degree $d(\Lambda,\lambda)$ with respect to homotheties with center at the vertex of $\mathcal{C}_{1,\ell}(z_1)$. An integral of the closed differential form over a cycle is invariant under homotheties of the cycle. Hence, the integral is zero if $d(\Lambda,\lambda) \neq 0$.

Now consider the procedure of composition of cycles discussed in (14.1) and (14.2). We use notations from (14.1) and (14.2).

Consider the special case of that procedure: $p = 1$, and n is arbitrary. Consider the hypergeometric pairing

$$I[z] : H_0\bigl(C.(\mathfrak{n}_-^*; M^*)_{\lambda+\mu}\bigr) \otimes H_0\bigl(C.(^+U_q\mathfrak{n}_-; M(q); {}^+\Delta)_{\lambda+\mu}\bigr) \longrightarrow \mathbb{C}, \quad (14.4.3)$$

corresponding to $C_{n,\ell+m}(z)$, where $z = (z^1, z_2, \ldots, z_n)$, cf. (14.2.1). Consider the restriction of that pairing to

$$H_0(C.(\mathfrak{n}_-^*; M^*)_{\lambda+\mu}) \otimes (H_0(C.(^+U_q\mathfrak{n}_-; M(\lambda^1, q); {}^+\Delta)_\lambda)$$
$$\circ H_0(C.(^+U_q\mathfrak{n}_-; M_{II}(q); {}^+\Delta)_\mu)).$$

14.4.4. Lemma. *The restriction is zero if* $d(\Lambda^1, \lambda) \notin \{0, -1, -2, \ldots\}$.

The proof easily follows from (14.4.2) and considerations of (14.2), see Lemma (14.2.6).

Now consider another special case of the composition procedure: $n = 1$, and p is arbitrary. Consider the hypergeometric pairing (14.4.3) corresponding to $C_{p,\ell+m}(z)$, $z = (z^1, \ldots, z^p)$. Consider the restriction of that pairing to

$$H_0(C.(\mathfrak{n}_-^*; M^*)_{\lambda+\mu}) \otimes (H_0(C.(^+U_q\mathfrak{n}_-; M_I(q); {}^+\Delta)_\lambda)$$
$$\circ H_0(C.(^+U_q\mathfrak{n}_-; M(\lambda_1, q); {}^+\Delta)_\mu)).$$

14.4.5. Lemma. *The restriction is zero if* $d(\Lambda_1, \mu) \notin \{0, 1, 2, \ldots\}$.

The proof easily follows from (14.4.2) and considerations of (14.2), see Lemma (14.2.6).

The lemmas of this section produce elements of the kernel of the hypergeometric pairing and may be used to estimate the dimension of the kernel from above.

14.5. The Selberg Formula

$$\ell! \int_\Delta \prod_{j=1}^\ell t_j^\alpha (1-t_j)^\beta \prod_{1 \leq i < j \leq \ell} (t_j - t_i)^{2\gamma} dt_1 \wedge \cdots \wedge dt_\ell$$
$$= \prod_{k=0}^{\ell-1} (\Gamma(\alpha + \beta + (2\ell - 2 - k)\gamma + 1)^{-1} \Gamma(\gamma + 1)^{-1} \Gamma((k+1)\gamma + 1) \quad (14.5.1)$$
$$\Gamma(\alpha + k\gamma + 1)\Gamma(\beta + k\gamma + 1)),$$

where $\Delta = \{t \in \mathbb{R}^\ell | 0 < t_1 < \cdots < t_\ell \leq 1\}$, see [**Ao, As, DF, Se, V3**].

The Selberg formula implies the following formula

$$(-\kappa)^{-\ell}\ell!m_2(m_2-1)\ldots(m_2-\ell+1)\cdot$$

$$\int_\Delta \prod_{j=1}^{\ell} t_j^{-m_1/\kappa}(1-t_j)^{-m_2/\kappa-1} \prod_{1\leq i<j\leq \ell}(t_j-t_i)^{2/\kappa}dt_1\ldots dt_\ell \qquad (14.5.2)$$

$$=\prod_{k=0}^{\ell-1}(\Gamma(-(m_1+m_2-(2\ell-2-k))\kappa)^{-1}\Gamma(1/\kappa+1)^{-1}$$

$$\cdot\,\Gamma((k+1)/\kappa+1)\Gamma(-(m_1-k)/\kappa+1)\Gamma(-(m_2-k)/\kappa+1)).$$

14.5.3. Lemma. *Let κ be a natural number. Let m_1, m_2, $\ell \in \mathbb{N}$ be numbers such that $0 \leq m_1$, m_2, $m_1 + m_2 - 2\ell \leq \kappa - 2$ and $\ell \leq \min(m_1, m_2)$. Then the left-hand side of (14.5.2) is not zero if and only if $m - 2\ell < 2\kappa - 2 - m_1 - m_2$.*

Proof. The right-hand side of (14.1.2) is zero iff $m_1 + m_2 - 2\ell + 2 \leq \kappa \leq m_1 + m_2 - \ell + 1$ and this implies the lemma.

14.6. The Hypergeometric Pairing for $\mathfrak{g} = s\ell_2$ and Two Modules

Let $\mathfrak{g} = s\ell_2$. Let e, f, h be its generators, α its positive root, and $(\alpha, \alpha) = 2$. Fix a natural κ. Denote by the same symbols e, f, h generators of $U_q s\ell_2$, $q = \exp(2\pi i/\kappa)$.

Let $\Lambda_1, \Lambda_2 \in \mathfrak{h}^*$ and $\lambda \in \mathbb{N}$. Set $M = M(\Lambda_1) \otimes M(\Lambda_2)$, $M(q) = (\Lambda_1, q) \otimes M(\Lambda_2, q)$, $\ell = \lambda$, $\Lambda_1 = \frac{m_1}{2}\alpha$, $\Lambda_2 = \frac{m_2}{2}\alpha$. Assume that $m_1, m_2, m_1 + m_2 - 2\ell \in \{0, 1, \ldots, \kappa - 2\}$.

Consider a discriminantal configuration $\mathcal{C}_{2,\ell}(z)$ in $\mathbb{C}^\ell$, where $z = (z_1, z_2)$ and $z_1 < z_2$. Define weights a of $\mathcal{C}_{2,\ell}(z)$ by

$$\begin{aligned} a(t_i, t_j) &= 2/\kappa, \\ a(z_1, t_j) &= -m_1/\kappa, \\ a(z_2, t_j) &= -m_2/\kappa. \end{aligned} \qquad (14.6.1)$$

Let $\mathcal{S}^*(a)$ be the corresponding local system over the complement of the configuration, see (2.5). Let $C.(Q'/Q'', \mathcal{S}^*(a))$ be the complex constructed for $\mathcal{C}_{2,\ell}(z)$ in Secs. 2 and 3, $C.(Q'/Q'', \mathcal{S}^*(a))^-$ its part skew-invariant with respect to the permutation group Σ_ℓ.

Consider the hypergeometric pairing

$$I[z] : H_0(C.(\mathfrak{n}_-; M)_\lambda) \otimes H_0(C.(^+U_q\mathfrak{n}_-; M(q); ^+\Delta)_\lambda) \longrightarrow \mathbb{C}$$

corresponding to this configuration.

14.6.2. Theorem. $I[z]$ *is nondegenerate iff*

$$|m_1 - m_2| \leq m_1 + m_2 - 2\ell \leq 2\kappa - m_1 - m_2 - 4.$$

Proof. Let $L = L(\Lambda_1) \otimes L(\Lambda_2)$ be the tensor product of the corresponding irreducible $s\ell_2$-modules, $H_0(C.(\mathfrak{n}_-; M)_\lambda) \to H_0(C.(\mathfrak{n}_-; L)_\lambda)$ the canonical epimorphism. The kernel of this epimorphism lies in the kernel of the hypergeometric pairing. The dimension of $H_0(C.(\mathfrak{n}_-; L)_\lambda)$ is one if $m_1 + m_2 - 2\ell \geq |m_1 - m_2|$ and is zero otherwise.

Let $L(q) = L(\Lambda_1, q) \otimes L_2(\Lambda_2, q)$ be the tensor product of the corresponding irreducible $s\ell_2$-modules. The kernel of the canonical projection

$$H_0(C.(^+U_q\mathfrak{n}_-; M(q); ^+\Delta)_\lambda) \longrightarrow H_0(C.(^+U_q\mathfrak{n}_-; L(q); ^+\Delta)_\lambda)$$

lies in the kernel of $I(z)$ by (12.2.18). The dimension of $H_0(C.(^+U_q\mathfrak{n}_-; L(q); ^+\Delta)_\lambda)$ is one if $m_1 + m_2 - 2\ell \geq |m_1 - m_2|$, and is zero otherwise.

Assume that $m_1 + m_2 - 2\ell \geq |m_1 - m_2|$. The monomial $v_1 \otimes f^\ell v_2 \in L(\Lambda_1) \otimes L(\Lambda_2)$ generates a nontrivial element x of $H_0(C.(\mathfrak{n}_-; L)_\lambda)$. A nontrivial element y of $H_0(C.(^+U_q\mathfrak{n}_-; L(q); ^+\Delta)_\lambda)$ is realized by a pair $(P, s) \in C.(Q'/Q'', \mathcal{S}^*(a))^-$, where P is the cube $\{t \in \mathbb{R}^\ell | z_1 \leq t_i \leq z_2\}$, and s is a nontrivial Σ-invariant section of $\mathcal{S}^*(a)$ over P. Thus, $I[z](x \otimes y) = \text{const} \cdot S$, where S is the Selberg integral (14.5.2) and const $\neq 0$. This integral is not zero iff $m_1 + m_2 - 2\ell \leq 2\kappa - m_1 - m_2 - 4$. The theorem is proved.

Let $\Lambda_1, \ldots, \Lambda_n \in \mathfrak{h}^*$ and $\Lambda_j = \frac{m_j}{2}\alpha$, $j = 1, \ldots, n$. Set $m = m_1 + \cdots + m_n$. Assume that $m_1, \ldots, m_n \in \{0, \ldots, \kappa - 2\}$. Let $\lambda = \ell \in \mathbb{N}$ be a number such that $m - 2\ell \in \{0, \ldots, \kappa - 2\}$. Let $P^{m-2\ell}_{m_1, \ldots, m_n}$ be the path space defined in (13.7).

14.6.3. Corollary of Theorems 14.3.2 and 14.6.2.

$$rk(\lambda; \Lambda_1, \ldots, \Lambda_n) \geq \dim P^{m-2\ell}_{m_1, \ldots, m_n}.$$

14.6.4. Theorem.

$$rk(\lambda; \Lambda_1, \ldots, \Lambda_n) = \dim P^{m-2\ell}_{m_1, \ldots, m_n}.$$

Proof. We use notations of (13.7). The following diagram is commutative,

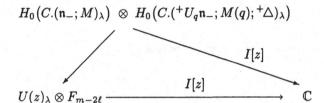

see (13.7.18). Here $I[z]$ are the hypergeometric pairings, and the left map is the canonical epimorphism.

We have $\dim F_{m-2\ell} = \dim P^{m-2\ell}_{m_1,\ldots,m_n}$, by Theorem (13.7.27). This equality and (14.6.3) imply the theorem.

14.6.5. Proof of Theorem (13.7.9). $\dim U(z)_\lambda = \dim P^{m-2\ell}_{m_1,\ldots,m_n}$, by Theorem 1 in [**FSV**], [**FSVI–II**] and Theorem 3.3 in [**TK**]. Hence $\dim U(z)_\lambda = \dim F_{m-2\ell} = rk(\lambda; \Lambda_1, \ldots, \Lambda_n)$, and the horizontal pairing in the above diagram is nondegenerate.

15

Remarks on Homology Groups of a Configuration with Coefficients in Local Systems More General than Complex One-Dimensional

In previous sections, we have been studying homology groups of the complement to the union of hyperplanes with coefficients in a one-dimensional complex local system of coefficients. Such a local system is defined by a product of powers of linear functions defining the hyperplanes of the configuration: $\ell = \prod \ell_H^{a(H)}$ for $H \in \mathcal{C}$. Such a local system has the distinguished family of local sections which are univalued branches of the function ℓ. These distinguished local sections form a module over a suitable ring of Laurent polynomials. All our constructions are formulated in terms of these distinguished sections. Therefore, our statements about homology groups with coefficients in a one-dimensional complex local system may be reformulated as statements about homology groups with coefficients in a local system of free modules of rank 1 over a suitable ring of Laurent polynomials.

15.1. Complexified Real Configuration

Let $\mathcal{C}$ be a configuration in $\mathbb{C}^k$ being the complexification of a real configuration of hyperplanes in $\mathbb{R}^k$. Let $Y \subset \mathbb{C}^k$ be the complement to the union of hyperplanes of $\mathcal{C}$. Let $\mathcal{A} = \mathbb{Z}[b(H), b(H)^{-1}]_{H \in \mathcal{C}}$ be the ring of Laurent polynomials. Let M be an $\mathcal{A}$-module.

We define the local system $\mathcal{S}(M)$ over Y as the local system with fiber M and the monodromy around $H \in \mathcal{C}$ equal to multiplication by $b(H)^4$.

Remark. The symbol $b(H)$ is the symbol replacing the number $\exp(\pi i a(H)/2)$ in constructions of Sec. 2.

More precisely, let X, Q, Q', Q'' be the complexes constructed for $\mathcal{C}$ in Sec. 2. Fix a trivialization of $\mathcal{S}(M)$ over cells of X and Q'/Q''. We define transition functions of $\mathcal{S}(M)$ for neighboring trivializations as follows.

(15.1.1) The transition function between the trivializations over cells $D_1, D_2 \in Q'/Q''$, such that $D_1 \subset \partial D_2$, is the identity operator.

Let $E(F^0 \geq F^j)$ be a cell of X, see Sec. 2. The boundary of $E(F^0 \geq F^j)$ is the union of cells $E(\pi_{F^j, F^{j-1}}(F^0) \geq F^{j-1})$ numerated by all facets F^{j-1} such that $F^{j-1} > F^j$, see (2.4.4) and (2.5.15). Introduce the twisting number $B(F^0, F^{j-1}; F^j)$ by

$$B(F^0, F^{j-1}; F^j) = \prod_{\substack{H \in \mathcal{C} \\ F^j \subset H, F^{j-1} \not\subset H}} b(H)^{\pm 1}, \qquad (15.1.2)$$

where $b(H)$ is taken, if H separates F^0 and F^{j-1}, and $b(H)^{-1}$ is taken otherwise, see (2.5.14).

(15.1.3) The transition function from the trivialization over $E(F^0 \geq F^j)$ to the trivialization over $E(\pi_{F^j, F^{j-1}}(F^0) \geq F^{j-1})$ is the operator of multiplication by the monomial $B(F^0, F^{j-1}; F^j) \in \mathcal{A}$.

(15.1.4) The complex $C.(X, \mathcal{S}(M))$ computes $H.(Y, \mathcal{S}(M))$, the complex $C.(Q'/Q'', \mathcal{S}(M))$ computes $H.(\mathbb{C}^k, \cup H, \mathcal{S}(M))$, see (2.3).

In (2.3), the homomorphism of the projection on the real part was defined as

$$S : C.(X, \mathcal{S}(M)) \longrightarrow C.(Q'/Q'', \mathcal{S}(M)).$$

The formula for S is given by (2.7.2). Namely, any cell $E(F^0 > F^j)$ is continuously deformed into the union of cells of the form $D(G^0 > F^j)$ for all $G^0 > F^j$. This deformation induces a transition function from the trivialization over $E(F^0 > F^j)$ to the trivialization over $D(G^0 > F^j)$.

(15.1.5) This transition function is the operator of multiplication by monomial $B'_{F^j}(G^0, F^0)$. This monomial is defined preceding Theorem (2.7.2), see also formula (2.6.5). In this formula, $\exp(\pm a(H)\pi i/2)$ should be replaced by

$b(H)^{\pm 1}$.

Denote by $\mathcal{S}(\mathcal{C})$ the local system $\mathcal{S}(M)$ for $M = \mathcal{A}$. $\mathcal{S}(\mathcal{C})$ has rank 1. Let $\mathcal{C}$ be cooriented, and let Or be the corresponding compatible orientation of cells of X and Q'/Q''.

Take an oriented cell $E(F^0 > F^j)$ with coefficient $1 \in \mathcal{A}$. This gives an element of $C.(X, \mathcal{S}(\mathcal{C}))$. Denote it by $E(F^0 > F^j, \mathcal{A}, Or)$. Such elements form a basis. The boundary operator on such cells is given by (2.5.15).

Take an oriented cell $D(F^0 > F^j)$ with coefficient $1 \in \mathcal{A}$. Denote this element of $C.(Q'/Q'', \mathcal{S}(\mathcal{C}))$ by $D(F^0 > F^j, \mathcal{A}, Or)$. These elements form a basis in the complex. The boundary operator is given by (2.5.13).

The homomorphism of the projection on the real part is given by (2.7.2).

Let a finite group G act on $\mathbb{R}^k$ by affine transformations. Naturally extend this action to an action of G on $\mathbb{C}^k$. Assume that G preserves the configuration $\mathcal{C}$. Let $\mathcal{A} = \mathbb{Z}[b(H), b(H)^{-1}]_{H \in \mathcal{C}}$ be as above. Set $b(H) = b(H')$, if H and H' belong to the same orbit of the action of G on $\mathcal{C}$. Denote the resulting ring of Laurent polynomials by $\mathcal{A}^G$.

Let M be an $\mathcal{A}^G$-module. Define the local system $\mathcal{S}(M)$ over Y by the construction described above. It has fiber M and monodromy $b(H)^4$ around $H \in \mathcal{C}$.

Define the action of G on $\mathcal{S}(M)$ as the action of G on Y extended by the identity operator on fibers of $\mathcal{S}(M)$. Denote by $\mathcal{S}(\mathcal{C}^G)$ the local system $\mathcal{S}(M)$ for $M = \mathcal{A}^G$. Define the complexes $C.(X, \mathcal{S}(\mathcal{C}^G))$ and $C.(Q'/Q'', \mathcal{S}(\mathcal{C}^G))$ as above. The action of G on these complexes is given by (2.8.5). The action commutes with the homomorphism of the projection on the real part.

15.2. Universal Quantum Group

In accordance with the algebraic tradition to replace numbers by letters, we replace all numbers $q^{(\alpha_i, \alpha_j)/4}$, $q^{(\alpha_i, \Lambda_m)/4}$, $q^{(\Lambda_\ell, \Lambda_m)/4}$ appearing in our considerations of quantum groups by symbols a_{ij}, λ_{im}, and $\mu_{\ell m}$, respectively. This annihilates the dependence of a quantum group on its Cartan matrix and the dependence of a Verma module on its highest weight. Therefore there is a unique quantum group of a given rank and this quantum group has a unique Verma module.

Fix $r \in \mathbb{N}$. Let $R_r = \mathbb{Q}(t_{ij}, \lambda_{im}, \mu_{\ell m})$ be the ring of rational functions in variables t_{ij}, λ_{im}, and $\mu_{\ell m}$, where $i, j \in \{1, \ldots, r\}$ and $\ell, m \in \mathbb{N}$. We set $t_{ij} = t_{ji}$ and $\mu_{\ell m} = \mu_{m\ell}$ for all i, j, ℓ, m.

The *universal quantum group* $\mathcal{V}_r$ of rank r is the R_r-algebra with the unit generated by f_i, e_i, $K_i^\pm$ for $i = 1, \ldots, r$.

15.2.1. Remarks.

(i) $K_i^\pm$ is $q^{\pm h_i/4}$ in our previous definitions.

(ii) It is convenient sometimes to take as generators the elements $x_i^+ = e_i$, $x_i^- = (t_{ii}^2 - t_{ii}^{-2})f_i$, $K_\pm^\pm$ for $i = 1, \ldots, r$.

These generators are subject to the relations:

$$\begin{aligned} K_i^\pm e_j K_i^\mp &= t_{ij}^{\pm 1}, \\ K_i^\pm f_j K_i^\mp &= t_{ij}^{\mp 1}, \\ [e_i, f_j] &= \delta_{ij}\left((K_i^+)^2 - (K_i^-)^2\right), \\ K_i^\pm K_j^\pm &= K_j^\pm K_i^\pm, \\ K_i^\mp K_j^\pm &= K_j^\pm K_i^\mp, \\ K_i^+ K_i^- &= 1 \end{aligned} \qquad (15.2.2)$$

for all i and j.

Define the comultiplication $\Delta : \mathcal{V}_r \to \mathcal{V}_r \otimes \mathcal{V}_r$ by the rule

$$\begin{aligned} \Delta(K_i^\pm) &= K_i^\pm \otimes K_i^\pm, \\ \Delta(f_i) &= f_i \otimes K_i^+ + K_i^- \otimes f_i, \\ \Delta(e_i) &= e_i \otimes K_i^+ + K_i^- \otimes e_i. \end{aligned} \qquad (15.2.3)$$

$\mathcal{V}_r$ is a Hopf algebra.

Define the counit $\epsilon : \mathcal{V}_r \to R_r$ and the antipode $A : \mathcal{V}_r \to \mathcal{V}_r$ by

$$\begin{aligned} \epsilon(f_i) &= \epsilon(e_i) = 0, \quad \epsilon(K_i^\pm) = 1, \\ A(K_i^\pm) &= K_i^\mp, \\ A(e_i) &= -t_{ii} e_i, \\ A(f_i) &= -t_{ii}^{-1} f_i. \end{aligned} \qquad (15.2.4)$$

Denote by $\mathcal{N}_+$ (resp. $^+\mathcal{N}_+$, $\mathcal{N}_-$, and $^+\mathcal{N}_-$) the subalgebra generated by 1, $e_1, \ldots, e_r$ (resp. by $e_1, \ldots, e_r$, by 1, $f_1, \ldots, f_r$, and by $f_1, \ldots, f_r$).

(15.2.5) For $m \in \mathbb{N}$, we define the Verma module $\mathcal{M}_m$ over $\mathcal{V}_r$ as the module generated by one vector v_m, subject to the relations:

$$\begin{aligned} K_i^\pm v_m &= \lambda_{im}^{\pm 1} v_m, \\ {}^+\mathcal{N}_+ v_m &= 0. \end{aligned}$$

We impose that $\mathcal{M}_m$ is free over $\mathcal{N}_-$.

For $\lambda = (\ell_1, \ldots, \ell_r) \in \mathbf{N}^r$, set

$$(\mathcal{N}_-)_\lambda = \left\{ x \in \mathcal{N}_- \,|\, K_j^+ x K_j^- = \prod_{i=1}^r t_{ij}^{-\ell_i} \text{ for all } j \right\},$$

$$(\mathcal{N}_+)_\lambda = \left\{ x \in \mathcal{N}_+ \,|\, K_j^+ x K_j^- = \prod_{i=1}^r t_{ij}^{\ell_i} \text{ for all } j \right\},$$

$$(\mathcal{M}_m)_\lambda = \left\{ x \in \mathcal{M}_m \,|\, K_j^+ x = \lambda_{jm} \prod_{i=1}^r t_{ij}^{-\ell_i} x \text{ for all } j \right\}.$$

We have the corresponding weight decompositions. The module $\mathcal{M}_m^*$ contragradient to $\mathcal{M}_m$ is defined as in (4.1). All constructions, notions, and statements of Secs. 4, 5, 7 and 9 have their natural reformulation for the quantum group $\mathcal{V}_r$ and its Verma modules and contragradient to Verma modules. In particular, there are contravariant forms

$$\begin{aligned} S : \mathcal{N}_\pm &\longrightarrow (\mathcal{N}_\pm)^*, \\ S : \mathcal{M}_n &\longrightarrow (\mathcal{M}_n)^*, \end{aligned} \qquad (15.2.6)$$

defined as in (4.2), R-matrix operators

$$\begin{aligned} R : \mathcal{M}_\ell \otimes \mathcal{M}_m &\longrightarrow \mathcal{M}_\ell \otimes \mathcal{M}_m, \\ R : \mathcal{M}_\ell^* \otimes \mathcal{M}_m^* &\longrightarrow \mathcal{M}_\ell^* \otimes \mathcal{M}_m^*, \end{aligned} \qquad (15.2.7)$$

defined as in (7.6), complexes $C.(^+\mathcal{N}_-; \mathcal{M}; {}^+\Delta)_\lambda$ and $C^{\cdot *}(^+\mathcal{N}_-; \mathcal{M}; \mu)_\lambda$, defined as in (5.1).

The definition of $\mathcal{V}_r$ only involves variables t_{ij}, see (15.2.2)–(15.2.4). Therefore, $\mathcal{V}_r$ can be defined as an R-algebra, where R is any ring containing the ring $\mathcal{A}_r = \mathbb{Z}[t_{ij}, t_{ij}^{-1}]$, here $i, j \in [1, \ldots, r]$ and $t_{ij} = t_{ji}$. The definition of the Verma module $\mathcal{M}_m$ involves variables λ_{im}. Therefore, the complexes $C.(^+\mathcal{N}_-; \mathcal{M}; {}^+\Delta)_\lambda$ and $C^{\cdot *}(^+\mathcal{N}_-; \mathcal{M}; \mu)_\lambda$ for $M = M_1 \otimes \cdots \otimes M_n$ can be defined if the quantum group $\mathcal{V}_r$ is defined as an R-algebra over any ring R containing the ring $\mathcal{A}_{r,n} = \mathbb{Z}[t_{ij}, t_{ij}^{-1}, \lambda_{im}, \lambda_{im}^{-1}]$, here $i, j \in \{1, \ldots, r\}$, $m \in \{1, \ldots, n\}$ and $t_{ij} = t_{ji}$. The action of the R-matrix operators R^p on these complexes involves variables $\mu_{\ell m}$. Therefore, the action of the R-matrix operators R^p on these complexes is defined if the quantum group $\mathcal{V}_r$ is defined as an R-algebra, where R is any ring containing the ring $\mathcal{A}_{r,n,n} = \mathbb{Z}[t_{ij}, t_{ij}^{-1}, \lambda_{im}, \lambda_{im}^{-1}, \mu_{\ell m}, \mu_{\ell m}^{-1}]$, here $i, j \in \{1, \ldots, r\}$, $\ell, m \in \{1, \ldots, n\}$ and $t_{ij} = t_{ji}$, $\mu_{\ell m} = \mu_{m\ell}$.

For the quantum group $\mathcal{V}_n(R)$ defined as an R-algebra, the corresponding complexes will be denoted by

$$C.(^+\mathcal{N}_-; \mathcal{M}; {}^+\Delta; R)_\lambda \quad \text{and} \quad C^*(^+\mathcal{N}_-; \mathcal{M}; \mu; R)_\lambda. \tag{15.2.8}$$

15.3. Discriminantal Configuration

Fix $r, n \in \mathbb{N}$ and $\lambda = (\ell_1, \ldots, \ell_r) \in \mathbb{N}^r$. Set $\ell = \ell_1 + \cdots + \ell_r$, $\Sigma' = \Sigma_{\ell_1} \times \cdots \times \Sigma_{\ell_r}$.

Let $\mathcal{C}_{n,\ell}(z)$ be a discriminantal configuration in $\mathbb{C}^\ell$ such that $z = (z_1, \ldots, z_n) \in \mathbb{R}^n$ and $z_1 < \cdots < z_n$. Let $\{t_i(j)\}$ be coordinates in $\mathbb{C}^\ell$, where $j = 1, \ldots, r$ and $i = 1, \ldots, \ell_j$. Σ' naturally acts on $\mathbb{C}^\ell$ preserving $\mathcal{C}_{n,\ell}(z)$. Let $\mathcal{S} = \mathcal{S}(\mathcal{C}_{n,\ell}^{\Sigma'})$ be the local system over $\mathbb{C}^\ell - \mathcal{C}_{n,\ell}(z)$ of $\mathcal{A}^{\Sigma'}$-rings defined in (15.1). We have

$$\mathcal{A}^{\Sigma'} = \mathcal{A}_{r,n} = \mathbb{Z}[t_{ij}, t_{ij}^{-1}, \lambda_{im}, \lambda_{im}^{-1}],$$

where $\mathcal{A}_{r,n}$ is defined in (15.2). The monodromy of $\mathcal{S}$ around hyperplanes $t_i(j) = t_a(b)$ and $t_i(j) = z_m$ is equal to multiplication by t_{jb}^4 and λ_{jm}^4, respectively.

The complexes $C.(X, \mathcal{S})$ and $(Q'/Q'', \mathcal{S})$ compute homology groups $H.(\mathbb{C}^\ell - \mathcal{C}_{n,\ell}(z), \mathcal{S})$ and $H.(\mathbb{C}^\ell, \mathcal{C}_{n,\ell}(z), \mathcal{S})$, resp., see (15.1).

The group Σ' naturally acts on the local system $\mathcal{S}$ and the above complexes. Denote by the upper script "$-$" the skew-invariant part of these complexes.

15.3.1. Theorem. *The constructions of Secs. 5.2 and 5.3 define monomorphisms*

$$\varphi : C.(^+\mathcal{N}_-; \mathcal{M}; {}^+\Delta; \mathcal{A}_{r,n})_\lambda \longrightarrow C_{\ell-.}(X, \mathcal{S})^-,$$
$$\varphi : C^*(^+\mathcal{N}_-; \mathcal{M}; \mu; \mathcal{A}_{r,n})_\lambda \longrightarrow C_{\ell-.}(Q'/Q'', \mathcal{S})^-,$$

where $\mathcal{M} = \mathcal{M}_1 \otimes \cdots \otimes \mathcal{M}_n$. These monomorphisms are quasiisomorphisms and the following diagram is commutative

$$\begin{array}{ccc} C.(^+\mathcal{N}_-; \mathcal{M}; {}^+\Delta; \mathcal{A}_{r,n})_\lambda & \longrightarrow & C_{\ell-.}(X, \mathcal{S})^- \\ S \downarrow & & \downarrow S \\ C^*(^+\mathcal{N}_-; \mathcal{M}; \mu; \mathcal{A}_{r,n})_\lambda & \longrightarrow & C_{\ell-.}(Q'/Q'', \mathcal{S})^-, \end{array}$$

where the left S is the contravariant form and the right S is the homomorphism of the projection on the real part.

The proof is given in (5.4)–(5.9).

Now let us consider the projection $p : \mathbb{C}^{n+\ell} \to \mathbb{C}^n$ and the configuration $\mathcal{C}_{n+\ell}$ of all diagonal hyperplanes in $\mathcal{C}^{n+\ell}$, see (12.3). $\mathbb{C}^{n+\ell}$ has coordinates $t_i(j), t_{r+1},\ldots,t_{r+n}$. The projection sends $\{t_i(j); t_{r+m}\}$ to $\{t_{r+m}\}$. Σ' acts on $\mathbb{C}^{n+\ell}$ preserving $\mathcal{C}_{n+\ell}$ and fibers of the projection.

Let $\mathcal{S} = \mathcal{S}(\mathcal{C}_{n,\ell}^{\Sigma'})$ be the local system over $\mathbb{C}^{n+\ell} - \mathcal{C}_{n,\ell}$ of $\mathcal{A}^{\Sigma'}$-rings defined in (15.1). We have
$$\mathcal{A}^{\Sigma'} = \mathcal{A}_{r,n,n} = \mathbb{Z}[t_{ij}^{\pm 1}, \lambda_{im}^{\pm 1}, \mu_{\ell m}^{\pm 1}],$$
where $\mathcal{A}_{r,n,n}$ is defined in (15.2). The monodromy of $\mathcal{S}$ around hyperplanes $t_i(j) = t_a(b)$, $t_i(j) = t_{r+m}$, and $t_{r+\ell} = t_{r+m}$ is equal to multiplication by t_{jb}^4, λ_{jm}^4, and $\mu_{\ell m}^4$, respectively.

We compute the skew-invariant part of the homology of a fiber of the projection with coefficients in $\mathcal{S}$.

Let $z^0 = (z_1^0,\ldots,z_n^0) \in \mathbb{R}^n \subset \mathbb{C}^n$, and $z_{\delta(1)}^0 < z_{\delta(2)}^0 < \cdots < z_{\delta(n)}^0$ for some permutation $\delta \in \Sigma_n$. Let $P_{n,\ell}(z^0)$ be the fiber over z^0, and $U_{n,\ell}(z^0) \subset P_{n,\ell}(z^0)$ be the union of hyperplanes of $\mathcal{C}_{n,\ell}(z^0)$, see notation in (8.1). Let $C_\cdot(P_{n,\ell}(z^0) - U_{n,\ell}(z^0), \mathcal{S})$ and $C_{!\cdot}(P_{n,\ell}(z^0) - U_{n,\ell}(z^0), \mathcal{S})$ be the chain complexes calculating $H_\cdot(P_{n,\ell}(z^0) - U_{n,\ell}(z^0), \mathcal{S})$ and $H_\cdot(P_{n,\ell}(z^0), U_{n,\ell}(z^0), \mathcal{S})$, respectively, see (8.1) and (2.3).

15.3.2. Theorem. *The constructions of (8.2) define monomorphisms*
$$\varphi : C_\cdot({}^+\mathcal{N}_-; \mathcal{M}_\delta; {}^+\Delta; \mathcal{A}_{r,n,n})_\lambda \longrightarrow C_{\ell-\cdot}(P_{n,\ell}(z^0) - U_{n,\ell}(z^0), \mathcal{S})^-,$$
$$\varphi : C_\cdot^*({}^+\mathcal{N}_-; \mathcal{M}_\delta; \mu; \mathcal{A}_{r,n,n})_\lambda \longrightarrow C_{!\ell-\cdot}(P_{n,\ell}(z^0) - U_{n,\ell}(z^0), \mathcal{S})^-,$$
where $\mathcal{M}_\delta = \mathcal{M}_{\delta(1)} \otimes \cdots \otimes \mathcal{M}_{\delta(n)}$. These monomorphisms are quasiisomorphisms. The contravariant form and the homomorphism of the projection on the real part commute with these monomorphisms.

The homology groups of neighboring fibers are canonically isomorphic. We describe monodromy of this isomorphism.

Let $z^u \in \mathbb{R}^n$ be the point obtained from z^0 by the transposition of its $\delta(u)$th and $\delta(u+1)$th coordinates for $u = 1,\ldots,n-1$. Fix a curve $\gamma^u : [0,1] \to \mathbb{C}^n \backslash U_n$ connecting z^0 and z^u as in (8.3.1).

Let $X(z^0)$ be the union of cells of the complex $\varphi(C_\cdot({}^+\mathcal{N}_-; \mathcal{M}_\delta; {}^+\Delta; \mathcal{A}_{r,n,n})_\lambda)$. Let $Q(z^0)$ be the union of cells of the complex $\varphi(C_\cdot^*({}^+\mathcal{N}_-; \mathcal{M}_\delta; \mu; \mathcal{A}_{r,n,n})_\lambda)$.

Set $\mathcal{M}_{\delta\mu} = \mathcal{M}_{\delta(1)} \otimes \cdots \otimes \mathcal{M}_{\delta(u-1)} \otimes \mathcal{M}_{\delta(u+1)} \otimes \mathcal{M}_{\delta(u)} \otimes \mathcal{M}_{\delta(u+2)} \otimes \cdots \otimes \mathcal{M}_{\delta(n)}$.

15.3.3. Theorem. *There exist continuous homotopies*
$$T_t^u : X(z^0) \longrightarrow P_{n,\ell}(\gamma^u(t)) - U_{n,\ell}(\gamma^u(t)),$$

$$T^u_{!t} : (Q(z^0) \cap U_{n,\ell}(z^0), Q(z^0)) \longrightarrow (U_{n,\ell}(\gamma^u(t)), P_{n,\ell}(\gamma^u(t))),$$

for $t \in [0,1]$ such that

(i) T^u_1 sends any cell of $\varphi(C.(^+\mathcal{N}_-; \mathcal{M}_\delta; {}^+\Delta; \mathcal{A}_{r,n,n})_\lambda)$ to a union of suitable cells of $\varphi(C.(^+\mathcal{N}_-; \mathcal{M}_{\delta^u}; {}^+\Delta; \mathcal{A}_{r,n,n})_\lambda)$.

(ii) $T^u_{!1}$ sends any cell of $\varphi(C.^*(^+\mathcal{N}_-; \mathcal{M}_\delta; \mu; \mathcal{A}_{r,n,n})_\lambda)$ to a union of suitable cells of $\varphi(C.^*(^+\mathcal{N}_-; \mathcal{M}_{\delta^u}; \mu; \mathcal{A}_{r,n,n})_\lambda)$. Hence the construction of the Gauss-Manin connection applied to T^u_t, $T^u_{!t}$ gives homomorphisms of complexes

$$T^u = T^u_1 : \varphi(C.(^+\mathcal{N}_-; \mathcal{M}_\delta; {}^+\Delta; \mathcal{A}_{r,n,n})_\lambda)$$
$$\longrightarrow \varphi(C.(^+\mathcal{N}_-; \mathcal{M}_{\delta^u}; {}^+\Delta; \mathcal{A}_{r,n,n})_\lambda),$$
$$T^u = T^u_{!1} : \varphi(C.^*(^+\mathcal{N}_-; \mathcal{M}_\delta; \mu; \mathcal{A}_{r,n,n})_\lambda)$$
$$\longrightarrow \varphi(C.^*(^+\mathcal{N}_-; \mathcal{M}_{\delta^u}; \mu; \mathcal{A}_{r,n,n})_\lambda).$$

For these homomorphisms,

$$T^u \varphi = \varphi R^u,$$

where the R-matrix operators R^u are described in (14.2).

The theorem is proved in Sec. 8.

Now let us consider the following two local systems of complexes over $\mathbb{C}^n - U_n$, where U_n is the union of all hyperplanes of $\mathcal{C}_n$. These local systems will be denoted by $\mathcal{R}$ and $\mathcal{R}^*$. The fiber of $\mathcal{R}$ (resp. $\mathcal{R}^*$) over a point $z^0 \in \mathbb{R}^n \subset \mathbb{C}^n$, such that $z^0_{\delta(1)} < \cdots < z^0_{\delta(n)}$ for some $\delta \in \Sigma_n$, is the complex $C.(^+\mathcal{N}_-; \mathcal{M}_\delta; {}^+\Delta; \mathcal{A}_{r,n,n})_\lambda$ (resp. $C.^*(^+\mathcal{N}_-; \mathcal{M}_\delta; \mu; \mathcal{A}_{r,n,n})_\lambda$). The identification of the fibers of these local systems over points z^0 and z^u along the curve γ^u is given by the R-matrix operator R^u as in Theorem (15.3.3), here $u = 1, \ldots, n-1$.

The contravariant form induces a homomorphism $S : \mathcal{R} \to \mathcal{R}^*$ of these local systems.

$\mathcal{C}_n$ is a discriminantal configuration: $\mathcal{C}_n = \mathcal{C}_{n,0}$. Fix an equipment $T^1 < \cdots < T^n$ of $\mathcal{C}_n$. Let X be the cellular complex constructed for $\mathcal{C}_n$ and $\{T^i\}$ in Secs. 2 and 3. X is a homotopy retract of $\mathbb{C}^n - U_n$. Therefore, the complexes $C.(X, \mathcal{R})$ and $C.(X, \mathcal{R}^*)$ calculate homology groups of $\mathbb{C}^n - U_n$ with coefficients in $\mathcal{R}$ and $\mathcal{R}^*$. We describe elements and differentials of the first complex. The description for the second complex is similar.

Discriminantal Configuration 355

We will describe $C.(X, \mathcal{R})$ for $n = 2$ and $n = 3$. The general case is similar.

15.3.4. Example.

Let $n = 2$. Then $\mathbb{C}^2 - U_2 = \{(t_1, t_2) \in \mathbb{C}^2 | t_1 \neq t_2\}$. The complex X consists of 4 oriented cells $E(t_1|t_2)$, $E(t_2|t_1)$, $E(t_1t_2)$, $E(t_2t_1)$, see Fig. 15.1 and Sec. 3. The fiber of $\mathcal{R}$ over the point $E(t_1|t_2) = (T^1, T^2)$ is the complex $C.(1,2) := C.(^+\mathcal{N}_-; \mathcal{M}_1 \otimes \mathcal{M}_2; {}^+\Delta; \mathcal{A}_{r,2,2})_\lambda$. The fiber of $\mathcal{R}$ over the point $E(t_2|t_1) = (T^2, T^1)$ is the complex $C.(2,1) := C.(^+\mathcal{N}_-; \mathcal{M}_2 \otimes \mathcal{M}_1; {}^+\Delta; \mathcal{A}_{r,2,2})_\lambda$.

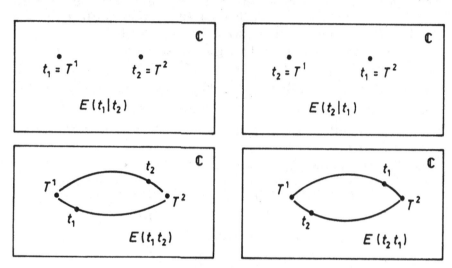

Fig. 15.1

The identification of these fibers along $E(t_1t_2)$ is $R^1 : C.(1,2) \to C.(2,1)$. The identification along $E(t_2t_1)$ is $R^1 : C.(2,1) \to C.(1,2)$.

Therefore, $C.(X, \mathcal{R})$ has the form:

$C_0 : E(t_1|t_2) \otimes x + E(t_2|t_1) \otimes y$ for $x \in C.(1,2)$ and $y \in C.(2,1)$.

$C_1 : E(t_1|t_2) \otimes (x_1 \oplus y_1) + E(t_2|t_1) \otimes (x_2 \oplus y_2)$ for $x_i \in C.(1,2)$, $y_i \in C.(2,1)$ such that $y_1 = R^1 x_1$ and $x_2 = R^1 y_2$.

The differential has the form

$$E(t_1t_2) \otimes (x_1 \oplus y_1) \longmapsto E(t_1|t_2) \otimes x_1 + E(t_2|t_1) \otimes y_1,$$
$$E(t_2t_1) \otimes (x_2 \oplus y_2) \longmapsto E(t_1|t_2) \otimes x_2 + E(t_2|t_1) \otimes y_2.$$

In particular,

$$H_0(\mathbb{C}^2 - U_2, \mathcal{R}) \simeq \{\cdots \to (^+\mathcal{N}_-)^{\otimes k} \otimes \text{cok} \to (^+\mathcal{N}_-)^{\otimes (k+1)} \otimes \text{cok} \to \ldots\},$$

$$H_1(\mathbb{C}^2 - U_2, \mathcal{R}) \simeq \{\cdots \to (^+\mathcal{N}_-)^{\otimes k} \otimes \ker \to (^+\mathcal{N}_-)^{\otimes(k+1)} \otimes \ker \to \cdots\},$$

where $\mathrm{cok} := \mathcal{M}_1 \otimes \mathcal{M}_2 / \mathrm{Im}\,(I - (R^1)^2 : \mathcal{M}_1 \otimes \mathcal{M}_2 \to \mathcal{M}_1 \otimes \mathcal{M}_2)$ and $\ker := \ker(I - (R^1)^2 : \mathcal{M}_1 \otimes \mathcal{M}_2 \to \mathcal{M}_1 \otimes \mathcal{M}_2)$.

15.3.5. Example. Let $n = 3$.

Denote by $C.(i, j, k)$ the complex $C.(^+\mathcal{N}_-; \mathcal{M}_i \otimes \mathcal{M}_j \otimes \mathcal{M}_k; ^+\Delta; \mathcal{A}_{r,3,3})_\lambda$, where i, j, k is a permutation of 1, 2, 3. The complex X consists of the cells $E(t_i|t_j|t_k)$, $E(t_i t_j|t_k)$, $E(t_i|t_j t_k)$, $E(t_i t_j t_k)$. $E(t_i|t_j|t_k)$ is the point $t_i = T^1$, $t_j = T^2$, $t_k = T^3$. $E(t_i t_j|t_k)$ is the interval with vertices at $E(t_i|t_j|t_k)$ and $E(t_j|t_i|t_k)$, $E(t_i|t_j t_k)$ is the interval with vertices at $E(t_i|t_j|t_k)$ and $E(t_i|t_k|t_j)$. $E(t_i t_j t_k)$ is the two-dimensional polygon with six vertices $E(t_a|t_b|t_c)$. The fiber of $\mathcal{R}$ over $E(t_i|t_j|t_k)$ is $C.(i, j, k)$. $C.(X, \mathcal{R})$ has the following form.

$C_0 : E(t_i|t_j|t_k) \otimes x(i, j, k)$, where $(i, j, k) \in \Sigma_3$ and $x(i, j, k) \in C.(i, j, k)$.

$C_1 : E(t_i t_j|t_k) \otimes [x(i, j|k) \oplus x(j, i|k)] + E(t_i|t_j t_k) \otimes [x(i|j, k) \oplus x(i|k, j)]$,

where $x(ij|k)$, $x(i|jk) \in C.(i, j, k)$ for $(i, j, k) \in \Sigma_3$ and $x(j, i|k) = R^1 x(i, j|k)$, $x(i|k, j) = R^2 x(i|j, k)$.

The differential is given by the formulas:

$$E(t_i t_j|t_k) \otimes [x(i, j|k) \oplus x(j, i|k)] \longmapsto E(t_i|t_j|t_k)$$
$$\otimes x(i, j|k) + E(t_j|t_i|t_k) \otimes x(j, i|k),$$
$$E(t_i|t_j t_k) \otimes [x(i|j, k) \oplus x(i|k, j)] \longrightarrow -E(t_i|t_j|t_k)$$
$$\otimes x(i|j, k) - E(t_i|t_k|t_j) \otimes x(i|k, j).$$
$$C_2 = E(t_i t_j t_k) \otimes \bigoplus_{(a,b,c) \in \Sigma_3} x(a, b, c),$$

where $(i, j, k) \in \Sigma_3$ and $x(a, b, c) \in C.(a, b, c)$. The elements $\{x(a, b, c)\}$ are compatible in the following sense. Let (a, b) be one of the three pairs (i, j), (i, k), (j, k). Then $x(b, a, c) = R^1 x(a, b, c)$ and $x(c, b, a) = R^2 x(c, a, b)$.

The differential is given by the formulas:

$$E(t_i t_j t_k) \otimes \sum x(a, b, c) \longmapsto -E(t_i|t_j t_k) \otimes \big(x(i, j, k) \oplus x(i, k, j)\big)$$
$$- E(t_j t_k|t_i) \otimes \big(x(j, k, i) \oplus x(k, j, i)\big) - E(t_j|t_i t_k) \otimes \big(x(j, i, k) \oplus x(j, k, i)\big)$$
$$- E(t_i t_k|t_j) \otimes \big(x(i, k, j) \oplus x(k, i, j)\big) - E(t_k|t_i t_j) \otimes \big(x(k, i, j) \oplus x(k, j, i)\big)$$
$$- E(t_i t_j|t_k) \otimes \big(x(i, j, k) \oplus x(j, i, k)\big).$$

15.3.6. Remark. In Secs. 2 and 3, the cellular complex Q'/Q'' was constructed for $\mathcal{C}_n$ and $\{T^i\}$. $C.(Q'/Q'', \mathcal{R})$ calculates the homology groups $H.(\mathbb{C}^n,$

$U_n, \mathcal{R}$). The homomorphism of the projection on the real part $S : C.(X, \mathcal{R}) \to C.(Q'/Q'', \mathcal{R})$ is defined in (2.3). It would be interesting to find a topological meaning for the image of that homomorphism.

15.4. Remarks on Homology Groups of Braid Groups

It is well known that the fundamental group of the complement to the configuration $\mathcal{C}_n \subset \mathbb{C}^n$ is the pure braid group $\mathcal{P}_n$ on n strings, and this complement is a $K(\pi, 1)$-space [**TO**]. Similarly, the fundamental group of the complement to the configuration $\mathcal{C}_{n,\ell}(z_1, \ldots, z_n) \subset \mathbb{C}^\ell$ is a subgroup of the pure braid group $\mathcal{P}_{n+\ell}$, and this complement is a $K(\pi, 1)$-space. This subgroup $\mathcal{P}_{n,\ell} \subset \mathcal{P}_{n+\ell}$ is the subgroup of all pure braids such that its first n strings are pairwise unbraided, see Fig. 15.2.

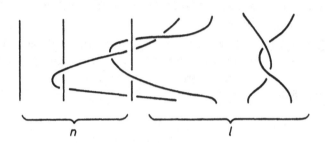

Fig. 15.2

Let M be a module over $\mathcal{P}_{n,\ell}$, then

$$H.(\mathcal{P}_{n,\ell}, M) \simeq H.\big(\mathbb{C}^\ell - \mathcal{C}_{n,\ell}(z)\big), \mathcal{S}(M)\big),$$

where $\mathcal{S}(M)$ is the local system of modules with fiber M [**FF**].

Therefore, the statements of Sec. 15.3 can be interpreted as statements on homology groups of $\mathcal{P}_{n,\ell}$ with coefficients in the corresponding modules. For example,

$$H.(\mathcal{P}_{n,\ell}, \mathcal{A}_{\ell,n}) \simeq H.\big(C.(^+\mathcal{N}_-; \mathcal{M}_1 \otimes \cdots \otimes \mathcal{M}_n; \Delta^+; \mathcal{A}_{r,n})_\lambda\big), \qquad (15.4.1)$$

where $^+\mathcal{N}_-$ is the nilpotent subalgebra of the quantum group $\mathcal{V}_\ell$, and $\mathcal{M}_1, \ldots, \mathcal{M}_n$ are its Verma modules, $\lambda = (1, \ldots, 1) \in \mathbf{N}^\ell$, see (15.3.1).

(15.4.2) Given $r \in \mathbf{N}$ and $\lambda \in \mathbf{N}^r$, let $^+\mathcal{N}_-$ be the nilpotent subalgebra of $\mathcal{V}_r$, and $\mathcal{M}_1, \ldots, \mathcal{M}_n$ its Verma modules. Then

$$H.\big(\mathcal{P}_n, C.(^+\mathcal{N}_-; \mathcal{M}_1 \otimes \cdots \otimes \mathcal{M}_n; {}^+\Delta; \mathcal{A}_{r,n,n})_\lambda\big) \simeq H.(C.(X, \mathcal{R})),$$

where $C.(X, \mathcal{R})$ is described in (15.3).

15.4.3. Remark. By (3.13.2), the complex $X \subset \mathbb{C}^\ell - C_{n,\ell}(z)$ is invariant under the action of the permutation group Σ_ℓ. Therefore this complex can be used to compute homology groups of the space $(\mathbb{C}^\ell - C_{n,\ell}(z))/\Sigma'$, where $\Sigma' = \Sigma_{\ell_1} \times \cdots \times \Sigma_{\ell_r}$, $\ell = \ell_1 + \cdots + \ell_r$. Let $S = S(C_{n,\ell}^{\Sigma'})$ be the Σ'-invariant local system over $\mathbb{C}^\ell - C_{n,\ell}(z)$ defined in (15.3). The Σ'-invariant part $C.(X, S)^+ \subset C.(X, S)$ computes the homology groups of the quotient with coefficients in the local system induced by S. The construction of Secs. 5.2 and 5.3 defines a quasiisomorphism

$$C.(^+\mathcal{N}_-; \mathcal{M}; {}^+\Delta; \mathcal{A}_{r,n})_\lambda^- \longrightarrow C.(X, S)^+,$$

where $^+\mathcal{N}_-$ is the nilpotent subalgebra of $\mathcal{V}_\ell$ and $\lambda = (1, \ldots, 1) \in \mathbf{N}^\ell$, the group Σ' acts permuting generators of $\mathcal{V}_\ell$ as in (4.8), and the supperscript $^-$ denotes the Σ'-skew-invariant subcomplex.

The space $(\mathbb{C}^\ell - C_{n,\ell}(z))/\Sigma'$ is a $K(\pi, 1)$-space. Its fundamental group is the subgroup $\mathcal{B}_{n,\ell_1,\ldots,\ell_r}$ of the braid group $\mathcal{B}_{n+\ell}$. This subgroup consists of all braids whose first n strings are unbraided and other ℓ strings are colored by r colors, see Fig. 15.3.

Fig. 15.3

15.4.4. Example. Consider the local system S over $\mathbb{C}^\ell - C_\ell$ with fiber $\mathbb{Z}[t, t^{-1}]$ and monodromy t^4 around each hyperplane $t_i = t_j$. The local system is Σ_ℓ-invariant and induces a local system on $(\mathbb{C}^\ell - C_\ell)/\Sigma_\ell$. Define a complex computing its homology.

For any $a > 1$, set

(i) $$\Delta F^a = \sum_{p=1}^{a-1} (\sqrt{-1})^{p(a-p)} \binom{a}{p}_q F^p \otimes F^{a-p},$$

where $q = \sqrt{-1}\,t^2$, and $(\ )_q$ is the q-binomial coefficient. Set $\Delta F = 0$. It is easy to see that the coefficients in (i) belong to $\mathbb{Z}[t, t^{-1}]$.

For any $k = 0, \ldots, \ell - 1$, let C_k be the free $\mathbb{Z}$-module with the basis $\{F^{a_1} \otimes \cdots \otimes F^{a_{\ell-k}} | a_1 + \cdots + a_{\ell-k} = \ell\}$. Define the map $d : C_k \to C_{k-1}$ by the formula

$$F^{a_1} \otimes \cdots \otimes F^{a_{\ell-k}} \longmapsto \sum_{j=1}^{\ell-k} (-1)^{j+1} F^{a_1} \otimes \cdots \otimes \Delta F^{a_j} \otimes \cdots \otimes F^{a_{\ell-k}}.$$

According to the previous remarks, the complex

$$0 \to C_{\ell-1} \to \cdots \to C_0 \to 0$$

computes the homology of $(\mathbb{C}^\ell - \mathcal{C}_\ell)/\Sigma_\ell$ (that is, the homology of the braid group $\mathcal{B}_\ell$) with coefficients in the local system induced by $\mathcal{S}$. In fact, this complex coincides with $C_\cdot({}^+\mathcal{N}_-(\mathcal{V}_\ell); {}^+\Delta; \mathbb{Z}[t, t^{-1}])^{\Sigma_\ell \text{ skew-inv}}_{(1,\ldots,1)}$, see (15.4.3).

15.4.5. Remark. In [Fu], a finite-dimensional complex of locally finite chains in $\mathbb{C}^\ell - \mathcal{C}_\ell$, computing cohomology of this space, was constructed. Our complex X is dual to the complex in [Fu].

15.5. Local Systems of Rank Greater than 1

Let $\mathcal{C}_{n,k}(z)$, $z_1 < \cdots < z_n$, be a discriminantal configuration in $\mathbb{C}^k$. Let $x_1, \ldots, x_k$ be coordinates in $\mathbb{C}^k$. Let X and Q'/Q'' be the complexes constructed for $\mathcal{C}_{n,k}(z)$ in Secs. 2 and 3.

Assume that a local system $\mathcal{S}$ over $Y = \mathbb{C}^k - \bigcup_{H \in \mathcal{C}_{n,\ell}(z)} H$ is given. Let V be its fiber.

The complexes $C_\cdot(X, \mathcal{S})$ and $C_\cdot(Q'/Q'', \mathcal{S})$ compute the homology groups $H_\cdot(Y, \mathcal{S})$ and $H_\cdot(\mathbb{C}^k, \cup H, \mathcal{S})$, resp. We will discuss similarities of these complexes and corresponding complexes of a quantum group. For a one-dimensional system $\mathcal{S}$ these similarities are discussed in Secs. 3–9 and in Subsecs. 15.1–15.4.

Let $\mathfrak{n} = \mathfrak{n}(k)$ be the free Lie algebra generated by $f_1, \ldots, f_k$, $U\mathfrak{n}$ its universal enveloping algebra. For any ℓ, let M_ℓ be the free left $U\mathfrak{n}$-module generated by one vector v_ℓ. Elements of the form $f_I v_\ell$, where f_I is a monomial in $U\mathfrak{n}$, form a basis in M_ℓ. Let $\overline{M}_\ell$ be the free two-sided module generated by one vector v_ℓ. Elements of the form $f_I v_\ell f_J$, where f_I and f_J are monomials in $U\mathfrak{n}$, form a basis in $\overline{M}_\ell$. Let $C_j = C_j(U\mathfrak{n}; \overline{M}_1, \ldots, \overline{M}_n; 2)$ be the direct sum of admissible tensor products, defined in (4.4.2.1). Let $C_j = \bigoplus_\lambda C_{j,\lambda}$ be the weight decomposition as in (4.1). Set $C_j^* = \oplus_\lambda C_{j,\lambda}^*$. Fix $\lambda = (1,\ldots,1) \in \mathbb{N}^k$. $(k-j)$-dimensional cells of X are numerated by monomials in $C_{j,\lambda}$. The cell corresponding to a monomial $f \in C_{j,\lambda}$ is denoted by $E(f)$ in Sec. 3.

$C_{j,\lambda}^*$ has the basis dual to the monomial basis in $C_{j,\lambda}$. $(k-j)$-dimensional cells of Q'/Q'' are numerated by elements of the basis dual to the monomial basis. The cell corresponding to a monomial $f^* \in C_{j,\lambda}^*$ is denoted by $D(f)$ in Sec. 3.

Fix the distinguished orientation of cells of X and Q'/Q'', defined in (3.10).

Fix a trivialization of $\mathcal{S}$ over cells of X and Q'/Q''. Then sections of $\mathcal{S}$ over cells are identified with elements of V. Therefore $C.(X,\mathcal{S})$ and $C.(Q'/Q'',\mathcal{S})$ are identified with complexes

$$\ldots \xrightarrow{d} C_{j,\lambda} \otimes V \xrightarrow{d} C_{j+1,\lambda} \otimes V \xrightarrow{d}$$

and

$$\ldots \xrightarrow{d} C_{j,\lambda}^* \otimes V \xrightarrow{d} C_{j+1,\lambda}^* \otimes V \xrightarrow{d},$$

respectively. We denote these complexes by $C.(U\mathfrak{n}, \overline{M}, 2, \mathcal{S})_\lambda$ and $C.(U\mathfrak{n}^*, \overline{M}^*, 2, \mathcal{S})_\lambda$.

The differentials in these complexes are given by the formulas

$$d : (f, v) \longmapsto \sum_{f \underset{\omega}{>} f'} \text{ind}\,(f \underset{\omega}{>} f')\,(f', T[f, f']v), \qquad (15.5.1)$$

$$d : (f^*, v) \longmapsto \sum_{f > f'} \text{ind}\,(f > f')\,(f'^*, T[f^*, f'^*]v), \qquad (15.5.2)$$

see (3.10.7), (3.12.1), and (3.12.2). Here we use the notations of (3.10), $v \in V$, $T[f, f'] : V \to V$ is the transition function of $\mathcal{S}$ from the trivialization over $E(f)$ to the trivialization over $E(f')$, $T[f^*, f'^*]$ is the transition function from $D(f)$ to $D(f')$.

The homomorphism of the projection on the real part

$$S : C.(U\mathfrak{n}; \overline{M}; 2; \mathcal{S}) \longrightarrow C.(U\mathfrak{n}^*; \overline{M}^*; 2; \mathcal{S})_\lambda$$

is given by the formula

$$S : (f, v) \longmapsto \sum_{f', f' \sim f} (f'^*, T[f, f'^*]v), \qquad (15.5.3)$$

see (3.12.5). Here we use the notations of (3.12.5), and $T[f, f'^*]$ is the transition function from $E(f)$ to $D(f')$ (the intersection of these cells is a point).

Complexes $C.(U\mathfrak{n}; \overline{M}; 2; \mathcal{S})_\lambda$ and $C.(U\mathfrak{n}^*; \overline{M}^*; 2; \mathcal{S})_\lambda$ have small quasiisomorphic subcomplexes.

Denote by $C_j(U\mathfrak{n}; M; \mathcal{S})_\lambda$ and $C_j(U\mathfrak{n}^*; M^*; \mathcal{S})_\lambda$ the spaces $(U\mathfrak{n}^{\otimes j} \otimes M_1 \otimes \cdots \otimes M_n)_\lambda \otimes V$ and $(U\mathfrak{n}^{\otimes j} \otimes M_1 \otimes \cdots \otimes M_n)^*_\lambda \otimes V$, resp., where $M_1 \otimes \cdots \otimes M_n$ are defined above.

For any j, define a map

$$\psi_j : C_j(U\mathfrak{n}; M; \mathcal{S})_\lambda \longrightarrow C_j(U\mathfrak{n}; \overline{M}; 2; \mathcal{S})_\lambda, \qquad (15.5.4)$$

cf. (5.2.4). First, for any monomial $F \in C_j(U\mathfrak{n}; M; \mathcal{S})_\lambda$, we define an oriented $(k-j)$-cell $\overline{E}(F)$ which is a union of cells of the form $E(f)$, where f is a monomial in $C_j(U\mathfrak{n}; \overline{M}; 2; \mathcal{S})_\lambda$. $\overline{E}(F)$ is shown in Fig. 5.7. More precisely, the monomial F may be considered as a monomial in $(U_q\mathfrak{n}_-^{\otimes j} \otimes M'_1 \otimes \cdots \otimes M'_n)_\lambda$. Here $U_q\mathfrak{n}_-$ is the nilpotent subalgebra of the quantum group $U_q\mathfrak{g}$ defined by an arbitrary matrix $\{(\alpha_i, \alpha_j)\}$, and $M'_1 \otimes \cdots \otimes M'_n$ are Verma modules with arbitrary highest weights. Let

$$\varphi : C.(^+U_q\mathfrak{n}_-; M'; ^+\Delta)_\lambda \longrightarrow C.(^+U_q\bar{\mathfrak{n}}'_-; \overline{M}'; 2; ^+\Delta)_\lambda$$

be the monomorphism constructed in (5.3). Let

$$p : C.(^+U_q\bar{\mathfrak{n}}'_-; \overline{M}'; 2; ^+\Lambda)_\lambda \longrightarrow C.(X, \mathcal{S}^*(a))$$

be the isomorphism constructed in Sec. 4, here $\mathcal{S}^*(a)$ is the local system constructed in Sec. 4. Then $p\varphi(F)$ is a linear combination of cells of the form $E(f)$, where f is a monomial in $C.(^+U_q\bar{\mathfrak{n}}'_-; \overline{M}'; 2; ^+\Delta)_\lambda$, with coefficients in $\mathcal{S}^*(a)$. The set of cells in this linear combination does not depend on generic values of parameters $\{(\alpha_i, \alpha_j), (\alpha_i, \Lambda_m)\}$. Forget about coefficients in $\mathcal{S}^*(a)$ and take the union of the cells of this linear combination. It is easy to see that this union is the image of a $(k-j)$-cell. We set $\overline{E}(F)$ to be this cell. The distinguished orientation of $\overline{E}(F)$ is defined as follows. F can be considered as a monomial in $C_j(U\mathfrak{n}; \overline{M}; 2; \mathcal{S})$. Therefore, the cell $E(F) \subset X$ is defined. It is easy to see that $E(F) \subset \overline{E}(F)$. The distinguished orientation of $E(F)$ induces a distinguished orientation of $\overline{E}(F)$.

Example. $\overline{E}(f_1|f_2v_1) = E(f_1|f_2v_1) - E(f_1|v_1f_2)$, see Fig. 15.4.

Fix a trivialization of $\mathcal{S}$ over cells of the form $\overline{E}(F)$. This induces a map

$$\eta : C_j(U\mathfrak{n}; M; \mathcal{S})_\lambda \longrightarrow C_{k-j}(X, \mathcal{S}),$$

and η induces map (15.5.4).

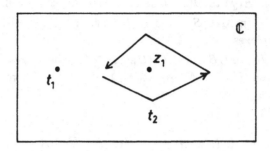

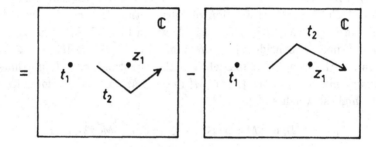

Fig. 15.4

15.5.5. Theorem. *The maps $\{\psi_j\}$ are monomorphisms. The image of $\{\psi_j\}$ is a subcomplex of $C.(U\mathfrak{n}; \overline{M}; 2; \mathcal{S})_\lambda$, hence $\{\psi_j\}$ induces a structure of a complex on $C.(U\mathfrak{n}; M; \mathcal{S})_\lambda$. The monomorphism $\psi : C.(U\mathfrak{n}; M; \mathcal{S})_\lambda \to C.(U\mathfrak{n}; \overline{M}; 2; \mathcal{S})_\lambda$ is a quasiisomorphism.*

The proof of the theorem repeats the proof of Theorem (5.5).

The differential in $C.(U\mathfrak{n}; M; \mathcal{S})_\lambda$ is analogous to the differential in $C.(U_q \mathfrak{n}_-; M_1 \otimes \cdots \otimes M_n; {}^+\Delta)_\lambda$ defined in (5.1).

15.5.6. Example. For $f_1 v_1 \in M_1 = C_0({}^+U_q\mathfrak{n}_-; M_1; {}^+\Delta)$, we have
$$d(f_1 v_1) = \left[q^{(\Lambda_1, \alpha_1)/2} - q^{-(\Lambda_1, \alpha_1)/2} \right] f_1 | v_1 .$$
For $(f_1 v_1, v) \in M_1 \otimes V = C_0(U\mathfrak{n}; M; \mathcal{S})$, we have
$$d(f_1 v_1, v) = (f_1 | v_1, T_1 v - T_2 v) .$$
Here T_1 and T_2 are two transition functions from the trivialization over $\overline{E}(f_1 v_1)$ to the trivialization over $\overline{E}(f_1 | v)$. ($\overline{E}(f_1 v_1)$ is an interval and the point $\overline{E}(f_1 | v_1)$ enters twice into its boundary.)

More generally, let $F \in C_j(^+U_q\mathfrak{n}_-; M_1 \otimes \cdots \otimes M_n; {^+\Delta})_\lambda$ be a monomial. F is a tensor product of $j+n$ factors: $F = F_1|\ldots|F_{j+n}$. We have

$$dF = -F_1|\ldots|F_j|^+\Delta(F_{j+1}|\ldots|F_{j+n}) + \sum_{p=1}^{j}(-1)^{j-p}F_1|\ldots|^+\Delta F_p|\ldots|F_{j+n},$$

where $^+\Delta$ is defined in (5.1). dF is a suitable linear combination of monomials in $C_{j+1}(^+U_q\mathfrak{n}_-; M_1 \otimes \cdots \otimes M_n; {^+\Delta})_\lambda$:

$$dF = \sum c_{F'} F'. \qquad (15.5.7)$$

The set of monomials F' in this linear combination does not depend on generic values of parameters $\{(\alpha_i, \alpha_j), (\alpha_i, \Lambda_p)\}$.

15.5.8. Lemma. *The differential in the complex $C.(U\mathfrak{n}; M; \mathcal{S})_\lambda$ is given by the formula*

$$d(F, v) = \sum (F', T[F, F']v),$$

where F' runs through the same set of monomials as in (15.5.7), and $T[F, F']$ is the transition function from $\overline{E}(F)$ to $\overline{E}(F')$.

For any j, define a map

$$\varphi_j : C_j(U\mathfrak{n}^*; M^*; \mathcal{S})_\lambda \longrightarrow C_j(U\mathfrak{n}^*; \overline{M}^*; 2; \mathcal{S})_\lambda \qquad (15.5.9)$$

analogously to (15.5.4), cf. (5.2.4). First, for any monomial $F^* \in C_j(U\mathfrak{n}^*; M^*; \mathcal{S})_\lambda$, define an oriented $(k-j)$-cell $\overline{D}(F)$ which is a union of cells of the form $D(F)$, where f is a monomial in $C_j(U\mathfrak{n}^*; \overline{M}^*; 2; \mathcal{S})_\lambda$. $\overline{D}(F)$ is shown in Fig. 5.17. Fix a trivialization of $\mathcal{S}$ over the cell $\overline{D}(F)$. This induces a map $C_j(U\mathfrak{n}^*; M^*; \mathcal{S})_\lambda \to C_{k-j}(Q'/Q'', \mathcal{S})$, and this map induces (15.5.9).

15.5.10. Theorem. *The maps $\{\varphi_j\}$ are monomorphisms. The image of $\{\varphi_j\}$ is a subcomplex of $C.(U\mathfrak{n}^*; \overline{M}^*; 2; \mathcal{S})_\lambda$, hence $\{\varphi_j\}$ induces a structure of a complex on $C.(U\mathfrak{n}^*; M^*; \mathcal{S})_\lambda$. The monomorphism $\varphi : C.(U\mathfrak{n}^*; M^*; \mathcal{S})_\lambda \longrightarrow C.(U\mathfrak{n}^*; \overline{M}^*; 2; \mathcal{S})_\lambda$ is a quasiisomorphism.*

The proof of the theorem repeats the proof of Theorem (5.4).

The differential in $C.(U\mathfrak{n}^*; M^*; \mathcal{S})_\lambda$ and the homomorphism of the projection on real part

$$S : C.(U\mathfrak{n}; M; \mathcal{S})_\lambda \longrightarrow C.(U\mathfrak{n}^*; M^*; \mathcal{S})_\lambda$$

are defined by the same formulas as the differential in $C^{\cdot *}(U_q \mathfrak{n}; M; \mu)_\lambda$ and the contravariant form

$$S : C_{\cdot}(^+U_q\mathfrak{n}; M; {}^+\Delta)_\lambda \longrightarrow C^{\cdot *}(^+U_q\mathfrak{n}; M; \mu)_\lambda \, .$$

In these formulas, numerical coefficients should be replaced by the corresponding transition functions.

15.5.11. Remarks.

(i) Local systems over the complement to $C_{n,k}$ are in one-to-one correspondence with $\mathcal{P}_{n,k}$-modules, where $\mathcal{P}_{n,k}$ is the fundamental group of the complement. An interesting example of such a module is $\mathbb{Z}[\mathcal{P}_{n,k}]$.

(ii) Compatible trivializations. Fix a base point $*$ in the complement to $C_{n,k}(z)$. For any cell in X and Q'/Q'', fix a path in the complement from $*$ to the cell. Fix a trivialization over the base point of the fiber of S. The distinguished paths and this trivialization induce trivializations of S over all cells of X and Q'/Q''. Such a family of trivializations will be called compatible. Consider any two cells of X and Q'/Q'' having a nonempty intersection outside $C_{n,k}$. The paths to these cells define an element γ of the fundamental group of the complement. For compatible trivializations, the transition function of S between these two cells is the action of the element γ.

(iii) The statements of this section show that the complexes $C_{\cdot}(U\mathfrak{n}; M; \mathcal{S})_\lambda$ and $C_{\cdot}(U\mathfrak{n}^*; M^*; \mathcal{S})_\lambda$ are an operator analog of the complexes $C_{\cdot}(^+U_q\mathfrak{n}_-; M; {}^+\Delta)_\lambda$ and $C^{\cdot *}(^+U_q\mathfrak{n}_-; M; \mu)_\lambda$, defined for an analog of a quantum group over a noncommutative algebra. Coefficients of such an analog of a quantum group form some extension of $\mathbb{Z}[\mathcal{P}_{n,k}]$. As Secs. 3–9 and 15.3 show, this analog of a quantum group turns into a quantum group for a local system $\rho : \mathcal{P}_{n,k} \to \mathrm{Aut}\,(V)$ such that, for any braid $\gamma \in \mathcal{P}_{n,k}$, $\rho(\gamma)$ only depends on the pairwise interaction of strings of γ.

Now consider a projection $p : \mathbb{C}^{n+k} \to \mathbb{C}^k$, the configuration $C_{n+\ell}$ of all diagonal hyperplanes in $\mathbb{C}^{n+k}$, and a local system $\mathcal{S}$ over the complement to the configuration $C_{n+\ell}$ in $\mathbb{C}^{n+k}$.

Theorems (15.5.5) and (15.5.10) give an algebraic description of homology groups of a fiber of p with coefficients in $\mathcal{S}$. There is an algebraic description of the monodromy representation of the Gauss–Manin connection of the pair $(p, \mathcal{S})$. Namely, there are analogs of Theorems (8.3.4), (8.14.4) and (8.14.5), the Gauss–Manin homotopies of cells are defined by suitable R-matrix operators.

These R-matrix operators are defined by formulas of Sec. 7. In these formulas, numerical coefficients of monomials should be replaced by the corresponding transition functions of the local system $\mathcal{S}$.

These statements indicate that the corresponding operator analog of a quantum group has quantum doubles of its nilpotent subalgebras. These quantum doubles act on Verma modules and their contragradient. The action of the canonical element of the quantum doubles on the tensor products of Verma modules or their contragradient coincides with elementary monodromy operators of the pair $(p, \mathcal{S})$, see Sec. 9.

References

[A] V. I. Arnol'd, *The cohomology ring of the coloured Braid group*, Mat. Zametki **5** (1969), 227–231.

[AGV1] V. I. Arnol'd, S. M. Gusein-Zade and A. N. Varchenko, *Singularities of Differential Maps*, Vol. I, Nauka, Moscow, 1982, Birkhäuser, 1985.

[AGV2] V. I. Arnol'd, S. M. Gusein-Zade and A. N. Varchenko, *Singularities of Differential Maps*, Vol. II, Nauka, Moscow, 1984, Birkhäuser, 1988.

[Ao1] K. Aomoto, *Jacobi polynomials associated with Selberg integrals*, SIAM J. Math Anal. **18** (1987), 545–549.

[Ao2] K. Aomoto, *On the complex Selberg integrals*, Quart. J. Math. Oxford **38** (1987), 385–399.

[As] R. S. Askey, *Some basic hypergeometric extensions of integrals of Selberg and Andrews*, SIAM J. Math. **11** (1980), 938–951.

[B] E. Brieskorn, *Sur les Groupes de Tresses (d'apres V. Arnold)*, Sem. Bourbaki, 24e Année, 1971/2, Lecture Notes in Math. **317**, Springer-Verlag, Berlin.

[Bj] A. Bjorner, *On the homology of geometric lattices*, Algebra Univ. **14** (1982), 107–128.

[BBR] M. Barnabei, A. Brini and G.-C. Rota, *Theory of Möbius functions*, Usp. Math. Nauk **41** no. 3 (1986), 113–157. English transl. in Russian Math. Surveys **41** (1986).

[BMP1] P. Bouwknegt, J. McCarthy and K. Pilch, *Free field approach to 2-dimensional conformal field theories*, Progr. Theor. Phys. Suppl. **102** (1990), 67–135.

[BMP2] P. Bouwknegt, J. McCarthy and K. Pilch, *Quantum group structure in the Fock space resolutions of $\widehat{sl}(n)$ representations*, Commun. Math. Phys. **131** (1990), 125–155.

[Ch1] I. Cherednik, *Monodromy representations for generalized Knizhnik-Zamolodchikov equation*, Publ. of RIMS **27**, no. 5 (1991), 711–726.

[Ch2] I. Cherednik, *Integral solutions of trigonometric Knizhnik-Zamolodchikov equations and Kac-Moody algebras*, Publ. of RIMS **27**, no. 5 (1991), 727–744.

[CF] P. Christe and R. Flume, *The four-point correlations of all primary operators of the $d = 2$ conformally invariant $SU(2)$ o-model with Wess-Zumino term*, Nucl. Phys. **B282** (1987), 466–494.

[DJMM] E. Date, M. Jimbo, A. Matsuo and T. Miwa, *Hypergeometric-type integrals and the $sl(2,\mathcal{C})$ Knizhnik-Zamolodchikov equation*, In: Yang–Baxter Equations, Conformal Invariance and Integrability in Statistical Mechanics and Field Theory, World Scientific.

[DeM] P. Deligne and G. D. Mostow, *Monodromy of hypergeometric functions and nonlattice integral monodromy*, Publ. Math IHES **63** (1986), 5–89.

[Do] V. Dotsenko, *Solving the $SU(2)$ conformal field theory with the Wakimoto free field representation*, Moscow (1990), 1–31.

[DF1] V. Dotsenko and V. Fateev, *Conformal algebra and multipoint correlation functions in 2D statistical models*, Nucl. Phys. **B240 [FS12]** (1984), 312–348.

[DF2] V. Dotsenko and V. Fateev, *Four-point correlation functions and the operator algebra in 2D conformal invariant theories with central charge $C \leq 1$*, Nucl. Phys. **B251** [**FS13**] (1985), 691–734.

[D1] V. Drinfeld, *Quantum groups*, Proc. ICM (Berkeley, 1986), vol. 1, Amer. Math. Soc. 1987, 798–820.

[D2] V. Drinfeld, *On almost cocommutative Hopf algebras*, (in Russian), Algebra and Analysis **1. no. 2** (1987), 30–46.

[D3] V. Drinfeld, *Quasi-Hopf algebras*, Algebra and Analysis **1** (1989), 30–46.

[DK] C. de Concini and V. Kac, *Representations of quantum groups at a root of 1*, in Colloque Dixmier 1989, Progress in Math. **92**, Birkhaüser, 1990, pp. 471–506.

[DKP] C. de Concini, V. Kac and C. Procesi, *Quantum adjoint action*, preprint, Scuola Normale Superiore Pisa, No. 95 (1991).

[ESV] H. Esnault, V. Schechtman and E. Viehweg, *Cohomology of local systems on the complement of hyperplanes*, Invent. Math. **109** (1992), 557–561; Erratum, ibid. **112** (1993), 447.

[FF] B. Feigin and D. Fuchs, *Cohomologies of Lie groups and Lie algebras*, Modern Problems in Mathematics, vol. **21** (in Russian), VINITI (1988) pp. 121–209.

[FSV] B. Feigin, V. V. Schechtman and A. N. Varchenko, *On algebraic equations satisfied by correlators in the WZW models*, Letters in Math. Phys. **20** (1990), 291–297.

[FSV1] B. Feigin, V. Schechtman and A. Varchenko, *On algebraic equations satisfied by hypergeometric correlators in WZW models*, I, Commun. Math. Phys. (1994).

[FSV2] B. Feigin, V. Schechtman and A. Varchenko, *On algebraic equations satisfied by hypergeometric correlators in WZW models*, II, preprint, 1994, 1–31.

[Fu] D. B. Fuchs, *Cohomologies of the braid group mod 2*, Functional Analysis and Its Appl. **4**, no. 2 (1970), 143–151.

[FW] G. Felder and C. Wieczerkowski, *Topological representations of the quantum group $U_q s\ell_2$*, Commun. Math. Phys. **138** (1991), 583–605.

[Ga] K. Gawedzki, *Geometry of Wess–Zumino–Witten models of conformal field theory*, Preprint IHES, 1990.

[GS] I. M. Gelfand and G. E. Shilov, *Generalized Functions and Actions on Them*, Vol. **1**, Moscow, 1959. English transl.: *Generalized Functions*, Academic Press, New Jersey, 1964.

[GV] I. M. Gelfand and A. N. Varchenko, *On Heaviside functions of a configuration of hyperplanes*, Functional Analysis and Its Applications **21**, no. 4 (1987), 1–18.

[Gi] A. Givental, *Twisted Picard–Lefschetz formulas*, Funct. Analysis and Its Appl. **22:1** (1988), 12–22.

[GiS] A. Givental and V. Schechtman, *Monodromy groups and Hecke algebras*, Usp. Mat. Nauk **42** (1987), 6–138. English transl. in Russian Math. Surveys **42** (1987).

[J] M. Jimbo, *A q-difference analog of $U\mathfrak{g}$ and the Yang–Baxter equation*, Lett. Math. Phys. **10** (1985), 63–69.

[K] V. G. Kac, *Infinite Dimensional Lie Algebras*, Cambridge University Press, Cambridge, 1985.

[Ka] M. Kapranov, *The permutoassociahedron, Maclane coherence theorem and asymptotic zones for the KZ equation*, J. Pure Appl. Algebra **85** (1993), 119–142.

[KZ] V. Knizhnik and A. Zamolodchikov, *Current algebras and Wess–Zumino models in two dimensions*, Nucl. Phys. **B247** (1984), 83–103.

[K1] T. Kohno, *Monodromy representations of braid groups and Yang–Baxter equations*, Ann. Inst. Fourier **37** (1987), 139–160.

[K2] T. Kohno, *Linear representations of braid groups and classical Yang–Baxter equations*, Contemp. Math. **78** (1988), 339–363.

[Ku] S. Kumar, *Representations of quantum groups at roots of unity*, preprint, 1991, 1–53.

[L1] R. Lawrence, *Homology Representations of Braid Groups*, PhD. Thesis, Oxford University Press, Oxford, 1987.

[L2] R. Lawrence, *A topological approach to the representations of the Iwahori–Hecke algebra*, Int. J. Modern Physics A **5**, no. 16 (1990), 3212–3219.

[L3] R. Lawrence, *The homological approach to higher representations*, Harvard University preprint, 1990, 1–14.

[L4] R. Lawrence, *A functorial approach to the one-variable Jones polynomial*, Harvard University preprint, 1990, 1–15.

[L5] R. Lawrence, *Homology representations of the Hecke algebra*, Commun. Math. Physics **135** (1990), 141–191.

[L6] R. Lawrence, *A note on two-row Hecke algebra representations*, Harvard University preprint, 1990, 1–6.

[L7] R. Lawrence, *Braid groups representations associated with $s\ell_m$*, Harvard University preprint, 1990, 1–14.

[Le] D. Lebedev, *Construction of quantum groups by Drinfeld's quantum double*, ITEP (1990), 1–24.

[LS] S. Levendorskii and Y. Soibelman, *Some applications of quantum Weyl groups I: The multiplicative formula for universal R-matrices for simple Lie algebras*, J. Geom. and Phys. **7**, no. 4 (1991).

[Lu1] G. Lusztig, *Quantum deformations of certain simple modules over enveloping algebras*, Adv. in Math. **70** (1988), 237–249.

[Lu2] G. Lusztig, *Quantum groups at a root of 1*, Geom. Dedicata **35** (1990), 89–144.

[Lu3] G. Lusztig, *Finite dimensional Hopf algebras arising from quantum groups*, J. Amer. Math. Soc. **3** (1990), 257–296.

[M1] A. Matsuo, *An application of Aomoto–Gelfand hypergeometric functions to the $SU(n)$ Knizhnik–Zamolodchikov equation*, Commun. Math. Phys. **134** (1991), 65.

[OS] P. Orlik and L. Solomon, *Combinatorics and topology of complements of hyperplanes*, Invent. Math. **56** (1980), 167–189.

[OT] P. Orlik and H. Terao, *Arrangements of hyperplanes*, Grundlehren der Math. Wiss. 300, Springer-Verlag, 1992.

[RT] N. Reshetikhin and V. Turaev, *Invariants of 3-manifolds via link polynomials and quantum groups*, Invent. Math. **103** (1991), 547–598.

[R1] M. Rosso, *Finite dimensional representations of the quantum analog of the enveloping algebra of a complex simple Lie algebra*, Comm. Math. Phys. **117** (1988), 581–593.

[R2] M. Rosso, *Analogue of P. B. W. theorem and the universal R-matrix for $U_h s\ell(N+1)$*, Commun. Math. Phys. **124** (1989), 307–318.

[R3] M. Rosso, *Analogues de la forme de Killing et du theoréme d'Harish–Chandra pour les groups quantiques*, Annales Scient. de l'Ecole Norm. Sup. **23** (1990), 445–467.

[R4] M. Rosso, *Certaines Formes Bilinéares sur les Groupes Quantiques et Une Conjecture de Schechtman et Varchenko*, preprint 1991.

[R5] M. Rosso, *Quantum groups at a root of 1 and tangle invariants*, Proc. of the Conf. on Topological and Geometrical Methods in Field Theory, Turku (Finland), May, 1991.

[S] M. Salvetti, *Topology of the complement of real hyperplanes in $\mathbb{C}^n$*, Invent. Math. **88** (1987), 603–618.

[Se] A. Selberg, *Bemerkninger om et multiplet integral*, Norsk Mat. Tidsskr. **26** (1944), 71–78.

[Sc] V. Schechtman, *Vanishing cycles and quantum groups*, Int. Math. Research Notices **3** (1992), 39–49.

[STV] V. Schechtman, H. Terao and A. Varchenko, *Local systems over complements of hyperplanes and the Kac–Kazhdan conditions for singular vectors*, preprint, 1994.

[SV1] V. V. Schechtman and A. N. Varchenko, *Integral representations of N-point conformal correlators in the WZW model*, preprint MPI/89-51, Bonn 1989.

[SV2] V. V. Schechtman and A. N. Varchenko, *Hypergeometric solutions of Knizhnik–Zamolodchikov equations*, Letters in Math. Phys. **20** (1990), 279–283.

[SV3] V. V. Schechtman and A. N. Varchenko, *Arrangements of hyperplanes and Lie algebras homology*, Invent. Math. **106** (1991), 139–194.

[SV4] V. V. Schechtman and A. N. Varchenko, *Quantum groups and homology of local systems*, Algebraic Geometry and Analytic Geometry, ICM-90, Satellite Conf. Proc., Tokyo, 1990, Springer-Verlag, 1991, 182–191.

[TK] A. Tsuchiya and Y. Kanie, *Vertex operators in conformal field theory on $\mathbb{P}^1$ and monodromy representations of braid groups*, Adv. Stud. Pure Math. **16** (1988), 297–372.

[TUY] A. Tsuchiya, K. Ueno and Y. Yamada, *Conformal field theory on universal family of stable curves with gauge symmetries*, Adv. Stud. Pure Math. **19** (1989), 459–566.

[V1] A. N. Varchenko, *Combinatorics and topology of a configuration of affine hyperplanes in a real space*, Functional Analysis and Its Applications **21** (1987), 9–19.

[V2] A. N. Varchenko, *The Euler beta-function, the Vandermonde determinant, Legendre's equation, and critical values of linear functions on a configuration of hyperplanes, I*, Math. USSR Izvestia **35** (1991), 543–571; *II*, Math. USSR Izvestia, **36** (1991), 155–168.

[V3] A. N. Varchenko, *Determinant formula for Selberg type integrals*, Func. Analysis and Its Appl. **4** (1991), 65–66.

[V4] A. N. Varchenko, *Bilinear form of a real configuration of hyperplanes*, Advances in Math. **97** (1993), 110–144.

[V5] A. N. Varchenko, *A geometric approach to representation theory of quantum groups*, in The Proc. of the XXth Int. Conf. on Differential Geometric Methods in Theoretical Physics, New York, 1991, vol. 1, World Scientific, pp. 152–175.

[WW] E. T. Whittaker and G. N. Watson, *A Course of Modern Analysis*, Cambridge Univ. Press, 1927.